Current Topics in Membranes, Volume 64

Leukocyte Adhesion

Current Topics in Membranes, Volume 64

Series Editors

Dale J. Benos
Department of Physiology and Biophysics
University of Alabama
Birmingham, Alabama

Sidney A. Simon
Department of Neurobiology
Duke University Medical Centre
Durham, North Carolina

Current Topics in Membranes, Volume 64

Leukocyte Adhesion

Edited by

Klaus Ley
Division of Inflammation Biology
La Jolla Institute for Allergy and Immunology
La Jolla, CA, USA

ELSEVIER

AMSTERDAM • BOSTON • HEIDELBERG • LONDON
NEW YORK • OXFORD • PARIS • SAN DIEGO
SAN FRANCISCO • SINGAPORE • SYDNEY • TOKYO
Academic Press is an imprint of Elsevier

Academic Press is an imprint of Elsevier
30 Corporate Drive, Suite 400, Burlington, MA 01803, USA
525 B Street, Suite 1900, San Diego, CA 92101-4495, USA

First edition 2009

ISBN: 978-0-12-374909-3
ISSN: 1063-5823

For information on all Academic Press publications
visit our website at elsevierdirect.com

Printed and bound in USA
08 09 10 11 12 10 9 8 7 6 5 4 3 2 1

Contents

CHAPTER 3 The Cytoskeleton and Deformability of White Blood Cells

Damir B. Khismatullin

SECTION 2 ADHESION MOLECULES

CHAPTER 4 Activation of Leukocyte Integrins

Eun Jeong Park and Motomu Shimaoka

Contributors

Numbers in parentheses indicate the pages on which the authors' contributions begin.

Ronen Alon (157) Department of Immunology, The Weizmann Institute of Science, Rehovot 76100, Israel

Olga Barreiro (277) Departamento de Biología Vascular e Inflamación, Centro Nacional de Investigaciones Cardiovasculares, 28029 Madrid, Spain; Servicio de Inmunología, Hospital Universitario de la Princesa, Universidad Autónoma de Madrid, 28006 Madrid, Spain

Christopher V. Carman (297) Division of Molecular and Vascular Medicine, Department of Medicine, Beth Israel Deaconess Medical Center, Center for Vascular Biology Research, Harvard Medical School, Boston, Massachusetts 02115

Damir B. Khismatullin (47) Department of Biomedical Engineering, Tulane University, New Orleans, Louisiana 70118, USA

Minsoo Kim (359) Department of Microbiology and Immunology, David H. Smith Center for Vaccine Biology and Immunology, University of Rochester, Rochester, New York 14642, USA

Craig T. Lefort (359) Department of Microbiology and Immunology, David H. Smith Center for Vaccine Biology and Immunology, University of Rochester, Rochester, New York 14642, USA

Klaus Ley (221) Division of Inflammation Biology, La Jolla Institute for Allergy and Immunology, La Jolla, California 92037, USA

Rodger P. McEver (195) Cardiovascular Biology Research Program, Oklahoma Medical Research Foundation, Oklahoma City, Oklahoma 73104, USA

William A. Muller (335) Department of Pathology, Northwestern University Feinberg School of Medicine, Chicago, Illinois 60611, USA

Eun Jeong Park (115) Immune Disease Institute, Program in Cellular and Molecular Medicine at Children's Hospital Boston, and Department of Anesthesia, Harvard Medical School, Boston, Massachusetts 02115, USA

Fredrick M. Pavalko (133) Department of Cellular and Integrative Physiology, Indiana University School of Medicine, Indianapolis, Indiana 46202, USA

Maria K. Pospieszalska (221) Division of Inflammation Biology, La Jolla Institute for Allergy and Immunology, La Jolla, California 92037, USA

Jin-Yu Shao (25) Department of Biomedical Engineering, Washington University, Saint Louis, Missouri 63011, USA

Motomu Shimaoka (115) Immune Disease Institute, Program in Cellular and Molecular Medicine at Children's Hospital Boston, and Department of Anesthesia, Harvard Medical School, Boston, Massachusetts 02115, USA

Richard E. Waugh (3) Department of Biomedical Engineering, University of Rochester, Rochester, New York, USA

Cheng Zhu (195) Coulter Department of Biomedical Engineering, Woodruff School of Mechanical Engineering, and Institute for Bioengineering and Biosciences, Georgia Institute of Technology, Atlanta, Georgia 30332, USA

Foreword

Klaus Ley* **and Dale Benos**[†]
*Division of Inflammation Biology, La Jolla Institute for Allergy and Immunology, La Jolla, California
[†]Department of Physiology, University of Alabama, Birmingham, Alabama

Studies of leukocyte adhesion have come a long way in the last 20 years, when leukocyte adhesion was still considered a nonspecific process. The discovery of leukocyte integrins (Harlan et al., 1985; Hemler et al., 1987; Springer, Thompson, Miller, Schmalstieg, & Anderson, 1984), their main endothelial ligands (Osborn et al., 1989; Rice & Bevilacqua, 1989; Rothlein, Dustin, Marlin, & Springer, 1986), the three selectins (Bevilacqua, Stengelin, Gimbrone, & Seed, 1989; Camerini, James, Stamenkovic, & Seed, 1989; Johnston, Cook, & McEver, 1989; Siegelman, van de Rijn, & Weissman, 1989; Tedder et al., 1989), and their main ligands PSGL-1 (Moore et al., 1992; Sako et al., 1995) and peripheral node addressins (Rosen, 1993) paved the way for a molecular understanding of leukocyte adhesion. This volume in the series *Current Topics in Membranes* presents a detailed account of our current understanding of the function of these molecules in leukocytes and endothelial cells. In keeping with the tradition of this series, the emphasis is on biophysical rather than biochemical or molecular biology aspects of this process.

In order for cells to discharge their functions, they must be able to sense and recognize their immediate surroundings. As the plasma membrane is the interface between the environment and the cell interior, its components, both proteins and lipids, are central to this recognition process. The leukocyte epitomizes membrane–substrate interactions, and therefore is a good model system with which to study these interactions and the proteins that mediate them. This volume's focus on the leukocyte coalesces novel experimental approaches and summarizes recent knowledge and controversies surrounding the phenomenon of adhesion that easily extends to other cell types. The contents of this volume provide new perspectives on one of the most fundamental properties of membranes, namely, recognition. Thus, it fits nicely into the *Current Topics in Membranes* series.

One of the most useful techniques in determining the spatial relation between different proteins, between subunits of heteromers, or between proteins and the lipid membrane is fluorescence resonance energy transfer.

Minsoo Kim and Craig Lefort provide a detailed account of this and related methods and their uses in leukocyte adhesion studies. The reader may find this chapter particularly useful to determine which experimental approach to take for a specific scientific question. In their chapter "Activation of Leukocyte Integrins," Eun Jeong Park and Motomu Shimaoka integrate data from crystallography, electron microscopy, mutagenesis, and epitope mapping studies to arrive at a model of how leukocyte integrins are activated. Their model is largely based on the $\alpha_L\beta_2$ integrin LFA-1, which is the best-studied leukocyte integrin in terms of its conformational changes. Ronen Alon then takes these integrin models and integrates them with chemokine receptors and signal transduction pathways to outline a mechanism by which chemokine binding to their ligands may activate leukocyte integrins. These membrane-cytoskeletal platforms for rapid chemokine signaling to integrins are of key importance for adhesion under flow. In most organs and tissues, selectins are indispensable to achieve leukocyte adhesion under flow. Cheng Zhu and Rodger McEver explore the biophysical regulation of selectin–ligand interactions under flow, with an emphasis on catch bonds. Catch bonds are characterized by a counterintuitive behavior in which the bonds become stronger when loaded with a force.

The main ligand for the endothelial E- and P-selectins is P-selectin Glycoprotein Ligand-1 (PSGL-1), also known as CD162. This molecule is expressed on the tips of leukocyte microvilli—thin structures that initiate the first contact with the endothelium. When pulling on PSGL-1, membrane tethers can be formed, which tend to reduce the force on the selectin–PSGL-1 bond by dissipating energy into pulling a tether away from the microvillus. Richard Waugh has studied this process in erythrocytes and leukocytes and provides a very lucid account of how pulling tethers works and why it matters for leukocyte adhesion. This chapter is complemented by Jin-Yu Shao's chapter on "Biomechanics of Leukocyte and Endothelial Cell Surface," which adds endothelial tethers to the mix and explains the relationship between microvilli and tethers.

Pulling tethers and extending microvilli only works because adhesion molecules are attached, in highly regulated and versatile ways, to the cytoskeleton. Fred Pavalko explores cytoskeletal interactions with leukocyte and endothelial cell adhesion molecules. The continuum between adhesion molecules and the viscoelastic cell body lends itself to modeling studies. Damir Khismatullin shows how the cytoskeleton and deformability of white blood cells can be integrated into a model of leukocyte adhesion. Maria Pospieszalska and Klaus Ley explore the different approaches to modeling leukocyte rolling, which constitutes one form of leukocyte adhesion. In recent years, the modeling efforts have yielded important predictions that were tested experimentally, but it is clear that much more needs to be done to fully understand the process of leukocyte rolling.

The next step in leukocyte–endothelial interactions is transendothelial migration. The current volume provides three accounts of this from different perspectives. Bill Muller explores how endothelial cells regulate transendothelial migration of leukocytes and discusses the molecules and mechanisms involved. His perspective comes from *in vitro* transmigration assays in which monocytes crawl through endothelial cell monolayers in the absence of flow. Olga Barreiro describes how endothelial adhesive platforms organize receptors to promote leukocyte extravasation. Her chapter integrates cell biology with biophysical data. In chapter 10, Chris Carman discusses how these structures form transmigratory cups and invadosome-like protrusions that ultimately allow the leukocyte to penetrate the endothelium and arrive at the site of inflammation or immune response.

Of course, a volume on leukocyte adhesion can never be complete. At least 38,820 articles on leukocyte adhesion have been published to date (PubMed, August 12, 2009), and it is impossible to cover all aspects. Nevertheless, we hope that the reader will find this volume a useful addition to the existing armamentarium of reviews on leukocyte adhesion. Finally, we would like to thank Gayathri Venkatasamy, the developmental editor for this volume, for putting it all together; Daisy Varbanova, the volume editor's assistant, for keeping track of the different versions of volume and all the figures; and the unnamed reviewers who spent their time to review all chapters.

References

Bevilacqua, M. P., Stengelin, S., Gimbrone , M. A., Jr., & Seed, B. (1989). Endothelial leukocyte adhesion molecule-1: An inducible receptor for neutrophils related to complement regulatory proteins and lectins. *Science, 243*, 1160–1165.

Camerini, D., James, S. P., Stamenkovic, I., & Seed, B. (1989). Leu-8/TQ1 is the human equivalent of the Mel-14 lymph node homing receptor. *Nature, 342*, 78–82.

Harlan, J. M., Killen, P. D., Senecal, F. M., Schwartz, B. R., Yee, E. K., Taylor, R. F., et al. (1985). The role of neutrophil membrane glycoprotein GP 150 in neutrophil adherence to endothelium *in vitro*. *Blood, 66*, 167–178.

Hemler, M. E., Huang, C., Takada, Y., Schwarz, L., Strominger, J. L., & Clabby, M. L. (1987). Characterization of the cell surface heterodimer VLA-4 and related peptides. *The Journal of Biological Chemistry, 262*, 11478–11485.

Johnston, G. I., Cook, R. G., & McEver, R. P. (1989). Cloning of GMP-140, a granule membrane protein of platelets and endothelium: Sequence similarity to proteins involved in cell adhesion and inflammation. *Cell, 56*, 1033–1044.

Moore, K. L., Stults, N. L., Diaz, S., Smith, D. F., Cummings, R. D., Varki, A., et al. (1992). Identification of a specific glycoprotein ligand for P-selectin (CD62) on myeloid cells. *The Journal of Cell Biology, 118*, 445–456.

Osborn, L., Hession, C., Tizard, R., Vassallo, C., Luhowskyj, S., Chi-Rosso, G., et al. (1989). Direct expression cloning of vascular cell adhesion molecule 1, a cytokine-induced endothelial protein that binds to lymphocytes. *Cell, 59*, 1203–1211.

Rice, G. E., & Bevilacqua, M. P. (1989). An inducible endothelial cell surface glycoprotein mediates melanoma adhesion. *Science, 246*, 1303–1306.

Rosen, S. D. (1993). L-selectin and its biological ligands. *Histochemistry, 100*, 185–191.

Rothlein, R., Dustin, M. L., Marlin, S. D., & Springer, T. A. (1986). A human intercellular adhesion molecule (ICAM-1) distinct from LFA-1. *Journal of Immunology, 137*, 1270–1274.

Sako, D., Comess, K. M., Barone, K. M., Camphausen, R. T., Cumming, D. A., & Shaw, G. D. (1995). A sulfated peptide segment at the amino terminus of PSGL-1 is critical for P-selectin binding. *Cell, 83*, 323–331.

Siegelman, M. H., van de Rijn, M., & Weissman, I. L. (1989). Mouse lymph node homing receptor cDNA clone encodes a glycoprotein revealing tandem interaction domains. *Science, 243*, 1165–1172.

Springer, T. A., Thompson, W. S., Miller, L. J., Schmalstieg, F. C., & Anderson, D. C. (1984). Inherited deficiency of the Mac-1, LFA-1, p150, 95 glycoprotein family and its molecular basis. *The Journal of Experimental Medicine, 160*, 1901–1918.

Tedder, T. F., Isaacs, C. M., Ernst, T. J., Demetri, G. D., Adler, A., & Disteche, C. M. (1989). Isolation and chromosomal localization of cDNAs encoding a novel human lymphocyte cell surface molecule, LAM-1: Homology with the mouse lymphocyte homing receptor and other human adhesion proteins. *The Journal of Experimental Medicine, 170*, 123–133.

Previous Volumes in Series

Current Topics in Membranes and Transport

Volume 23 Genes and Membranes: Transport Proteins and Receptors* (1985)
Edited by Edward A. Adelberg and Carolyn W. Slayman

Volume 24 Membrane Protein Biosynthesis and Turnover (1985)
Edited by Philip A. Knauf and John S. Cook

Volume 25 Regulation of Calcium Transport across Muscle Membranes (1985)
Edited by Adil E. Shamoo

Volume 26 Na^+–H^+ Exchange, Intracellular pH, and Cell Function* (1986)
Edited by Peter S. Aronson and Walter F. Boron

Volume 27 The Role of Membranes in Cell Growth and Differentiation (1986)
Edited by Lazaro J. Mandel and Dale J. Benos

Volume 28 Potassium Transport: Physiology and Pathophysiology* (1987)
Edited by Gerhard Giebisch

Volume 29 Membrane Structure and Function (1987)
Edited by Richard D. Klausner, Christoph Kempf, and Jos van Renswoude

Volume 30 Cell Volume Control: Fundamental and Comparative Aspects in Animal Cells (1987)
Edited by R. Gilles, Arnost Kleinzeller, and L. Bolis

Volume 31 Molecular Neurobiology: Endocrine Approaches (1987)
Edited by Jerome F. Strauss, III, and Donald W. Pfaff

**Part of the series from the Yale Department of Cellular and Molecular Physiology.*

Volume 32 Membrane Fusion in Fertilization, Cellular Transport, and Viral Infection (1988)
Edited by Nejat Düzgünes and Felix Bronner

Volume 33 Molecular Biology of Ionic Channels* (1988)
Edited by William S. Agnew, Toni Claudio, and Frederick J. Sigworth

Volume 34 Cellular and Molecular Biology of Sodium Transport* (1989)
Edited by Stanley G. Schultz

Volume 35 Mechanisms of Leukocyte Activation (1990)
Edited by Sergio Grinstein and Ori D. Rotstein

Volume 36 Protein–Membrane Interactions* (1990)
Edited by Toni Claudio

Volume 37 Channels and Noise in Epithelial Tissues (1990)
Edited by Sandy I. Helman and Willy Van Driessche

Current Topics in Membranes

Volume 38 Ordering the Membrane Cytoskeleton Trilayer* (1991)
Edited by Mark S. Mooseker and Jon S. Morrow

Volume 39 Developmental Biology of Membrane Transport Systems (1991)
Edited by Dale J. Benos

Volume 40 Cell Lipids (1994)
Edited by Dick Hoekstra

Volume 41 Cell Biology and Membrane Transport Processes* (1994)
Edited by Michael Caplan

Volume 42 Chloride Channels (1994)
Edited by William B. Guggino

Volume 43 Membrane Protein–Cytoskeleton Interactions (1996)
Edited by W. James Nelson

Volume 44 Lipid Polymorphism and Membrane Properties (1997)
Edited by Richard Epand

Volume 45 The Eye's Aqueous Humor: From Secretion to Glaucoma (1998)
Edited by Mortimer M. Civan

Volume 60 Computational Modelling of Membrane Bilayers
Edited by Scott E. Feller

Volume 61 Free Radical Effects on Membranes
Edited by Sadis Matalon

Volume 62 The Eye's Aqueous Humor
Edited by Mortimer M. Civan

Volume 63 Membrane Protein Crystallization
Edited by Larry DeLucas

SECTION 1

MEMBRANE COMPOSITION AND PROPERTIES

CHAPTER 1

Membrane Tethers

Richard E. Waugh
Department of Biomedical Engineering, University of Rochester, Rochester, New York, USA

I. OVERVIEW

Membrane tethers are thin cylinders of bilayer membrane pulled from the surfaces of cells or membrane vesicles under mechanical force. These structures have been well characterized in pure lipid systems and in red blood cells, where the process of tether formation can take place under well-defined conditions. From those studies detailed mechanical descriptions of the process have been validated and fundamental physical properties of the membrane bilayer and the strength of its association with the underlying cytoskeleton have been characterized. The dynamics of tether formation

Current Topics in Membranes, Volume 64 1063-5823/09 $35.00

DOI: 10.1016/S1063-5823(09)64001-1

can be characterized in terms of an effective viscosity coefficient that relates the dynamic force contribution to the rate of tether formation. More detailed mechanical analysis reveals that the effective viscosity is expected to scale as the square of the tether radius, which in turn decreases with increasing force. This leads to lower apparent viscosities at higher rates of tether formation and the appearance of "shear thinning," even though the drag coefficient that characterizes the rate-dependent force generated from interactions between the bilayer and the cytoskeleton remains constant. Tether formation from leukocytes has been found to be an important modulating factor in the rolling adhesive interactions between leukocytes and endothelium, where they act to prolong bond lifetime and maintain cell–cell contact. It is anticipated that properly accounting for the physical characteristics of tether formation will help lead to reliable predictions of cell behavior as they roll along and are detached from adhesive substrates.

II. INTRODUCTION

Membrane tethers are thin strands of bilayer membrane formed under mechanical force from the surface of a cell or bilayer vesicle. They were first observed during experiments designed to determine the elasticity of red blood cell membrane using a flow channel (Hochmuth, Mohandas, & Blackshear, 1973). Red blood cells were attached to a glass slide under conditions that promoted single-point attachments, and the cells were deformed under fluid forces due to the controlled flow rate in the chamber. When a threshold force was exceeded, cells began to creep along the glass forming thin membrane strands between the cell body and the attachment site on the glass. Early electron micrographs of tethers placed their diameter at approximately 100 nm, but their length could easily exceed several tens of micrometers (Fig. 1). The tethers exhibited elastic behavior. When fluid forces were reduced abruptly, the length of the tether rapidly decreased, and then the length rebounded when the force was restored. Tether formation was also found to be at least partially reversible. Reduction of the force below a threshold value resulted in a gradual decrease in tether length, as the tether was restored to the cell surface. These elastic and quasielastic behaviors led to the conclusion that these tethers must contain a membrane-associated cytoskeleton because it did not seem possible that a fluid bilayer could exhibit such behavior. As will be seen in the following sections, this early conclusion proved to be incorrect, and in disproving it, investigators learned a great deal about bilayer membrane mechanical behavior, and the stability of the bilayer–skeletal interface.

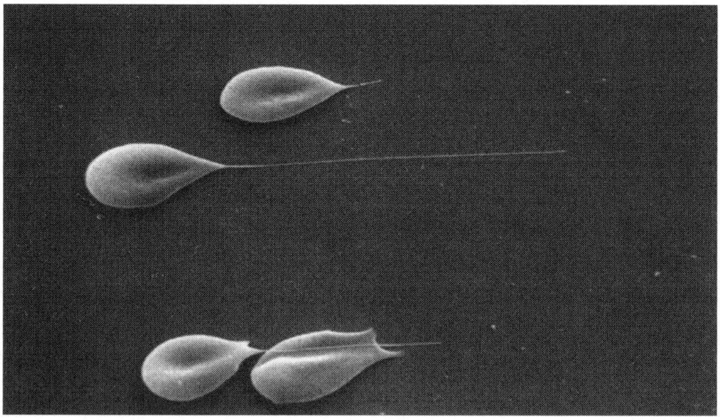

FIGURE 1 Scanning electron micrograph of red blood cells subjected to fluid shear stress in a flow chamber. Note the very long tether pulled from the second cell from the top. The length of the tether is approximately 30 μm. Cells were subjected to shear stress then fixed under flow with glutaraldehyde, critical point dried and coated with gold.

It was recognized from their earliest discovery that membrane tethers provided a unique opportunity to learn about the physical properties of cell membranes, and much of the literature reflects researchers' efforts to understand and characterize their physical properties and the mechanisms that accounted for their interesting behavior. It was only relatively recently that investigators became aware that tether formation could have physiological importance in the context of leukocyte adhesion to the vascular endothelium.[1] The seminal observation that led to this realization was that tethers do in fact form from the surfaces of leukocytes as they are detached from the vessel wall during rolling interactions (Schmidtke & Diamond, 2000). It was subsequently recognized that tether formation could lead to a reduction in the rate of loading on molecular bonds between cells and substrates, thus prolonging bond lifetime and lengthening the duration of adhesion between the cell and the endothelium (Park et al., 2002).

In the following sections, we review the physical characteristics of tethers, focusing first on what has been learned from model systems and red blood cells, where the conditions of tether formation can be controlled precisely.

[1]An unfortunate confusion of nomenclature arose when investigators studying bond formation during leukocyte rolling on endothelial ligands began to refer to formation of molecular bonds between the cell and the substrate as cell "tethering." These attachments include both molecular bonds (which may be a few tens of nanometers in dimension) and the associated cell surface structures. In this chapter, we restrict our definition of "tether" to describe membrane tethers, which consist of thin cylinders of bilayer membrane and can be many microns in length.

Then, we review tether formation from more complex cells, where the detailed mechanisms governing tether formation and properties remain to be identified with certainty.

III. TETHERS FORMED FROM BILAYER VESICLES

A. Bilayer Vesicle Equilibrium

The experiments that provide the clearest understanding of bilayer tether equilibrium involve the formation of tethers from phospholipid vesicles held in micropipettes, under conditions where the tethering force can be accurately measured (Fig. 2). In the simplest sense, tether equilibrium can be thought of as a tug of war between the aspiration pressure in the micropipette, which is trying to pull more membrane area into the pipette, and the force applied to the tether, which is trying to pull more membrane area onto the tether. The subtle point to recognize is that the balance in this tug of war must also account for the elastic energy stored in the tether as it moves from the cell body onto the tether. Thus, the force applied to the tether must not only work against the pressure in the pipette, it must also work to deform the

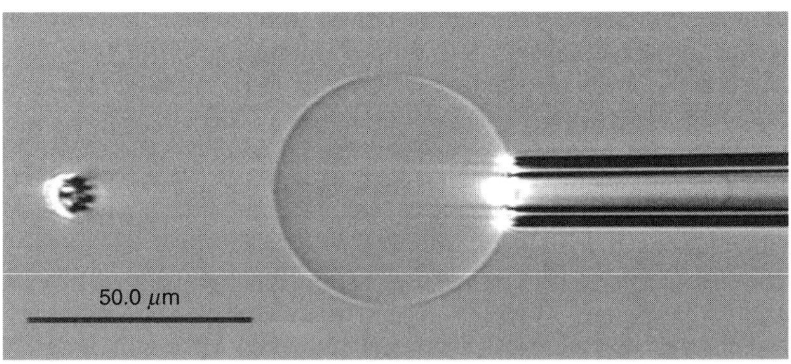

FIGURE 2 A phospholipid vesicle is aspirated into a micropipette, and a magnetic particle (left side of image) is attached to the vesicle via a tether. The edge of the vesicle projection in the pipette is visible near the edge of the image. By adjusting the pressure in the pipette, the tension in the membrane can be controlled. The magnetic force on the particle is adjusted and determined by controlling and measuring the current to an electromagnet located far to the left of the image. Measurements of the force required to maintain a tether at constant length as a function of the membrane tension can be used to determine the bending modulus of the membrane (k_c) (see Heinrich & Waugh, 1996).

membrane from a state of relatively low curvature on the cell body to the relatively high curvature of the tether. The property of the membrane bilayer that resists this deformation is the membrane bending stiffness.

Even in the simplest model systems, complexities in deriving the tether equilibrium arise because of complexities in the origins of the membrane bending stiffness. For the simplest form of the bending energy, there is a single elastic constant k_c (units of N m or J) that characterizes the resistance of the membrane to bending. In terms of the bending energy per unit area of the surface W_c (Helfrich, 1973),

$$W_c = \frac{k_c}{2}(\bar{c} - c_0)^2, \tag{1}$$

where $\bar{c}$ is the total curvature of the membrane and c_0 is the natural or "spontaneous" curvature. The natural curvature for all symmetric membranes is zero (flat) and for phospholipids that naturally form bilayers, the spontaneous curvature cannot be very large, or else the lipids would tend to form micelles rather than bilayers. Taking the spontaneous curvature to be zero, simple equilibrium relationships can be obtained (Bozic, Svetina, Zeks, & Waugh, 1992; Evans & Yeung, 1994; Hochmuth, Shao, Dai, & Sheetz, 1996; Waugh & Hochmuth, 1987):

$$f = \frac{2\pi k_c}{R_t}, \tag{2}$$

$$\tau_0 = \frac{k_c}{2R_t^2}, \tag{3}$$

where R_t is the tether radius, f is the force on the tether, and τ_0 is the tension (force per unit length) in the membrane far from the tether where curvature is small and bending contributions are negligible.

More complicated expressions are obtained when one takes into account the contributions to bending resistance that arise from the fact that the bilayer consists of two semi-independent layers, each of which has its own preferred surface area. If there is an imbalance between the natural areas of the two leaflets, then the membrane tends to curve. For example in a membrane with a symmetric chemical composition, if there is a greater number of molecules in the outer leaflet, the membrane would tend to curve outward, to relieve the crowding of molecules there. This property is nonlocal, because the molecules of the two leaflets can redistribute themselves laterally to relieve compression or expansion of the molecules in a local region. Thus, at equilibrium, the energy stored by this mechanism W_{nl} depends on the difference in the areas of the two leaflets integrated over the entire surface of the vesicle or cell:

$$W_{\mathrm{nl}} = \frac{k_{\mathrm{r}}}{2A_{\mathrm{ves}}}\left[\oint \bar{c}\,\mathrm{d}A - \oint \bar{c}_{\mathrm{i}}\,\mathrm{d}A\right]^2, \tag{4}$$

where A_{ves} is the area of the vesicular membrane, $\bar{c}$ is the total curvature at a point on the deformed surface, $\bar{c}_{\mathrm{i}}$ is the total curvature in the initial, resting shape, and k_{r} is a characteristic modulus that depends on the distance between the centers of the leaflets[2] h and the area compressibility of the two leaflets (Bozic et al., 1992). For a tethered cell or vesicle, a difference in the areas of the leaflets [the term in brackets in Eq. (4)] is generated because more surface area is pulled onto the outer leaflet than the inner leaflet of the tether. In this case the difference in area is approximately $2\pi hL_{\mathrm{t}}$, where L_{t} is the length of the tether. Note that in this case the energy depends on the square of the tether length, indicating that the tether force is expected to increase linearly as the tether length increases. Thus, a direct test of this mechanism (in a pure system) is to measure differences in the equilibrium tethering force at different tether lengths. These measurements reveal a relatively small contribution from nonlocal bending to the total force required for tether formation, typically less than 10% for tethers shorter than 100 μm from cell-sized membrane vesicles. (The force attributable to nonlocal bending resistance for a tether 100 μm in length from a vesicle with an area of 200 μm^2 would be approximately 3 pN, a small fraction of the 30–50 pN required to form tethers from cells.)

A further complication in the nonlocal bending contribution is that the energy stored by this mechanism can be dissipated by transporting molecules from one leaflet to another, to reduce the area strain in both leaflets. This phenomenon can be observed even in pure lipid systems (Raphael & Waugh, 1996), where nonlocal stresses appear to relax over times on the order of a few hundred seconds. In cell membranes, a relaxation of the force also appears to occur over periods of several hundred seconds (Hwang &

[2]Technically, h is the distance between the "neutral surfaces" of the two leaflets. In the mechanics of shells, the neutral surface is defined as the surface at which the mean force acts, and around which the bending moments are balanced. This is typically close to, but not actually at the center of the layer (or shell), and its exact position depends on the distribution of mechanical properties across the thickness of the leaflet. Therefore, we approximate h to be approximately half of the bilayer thickness. X-ray diffraction measurements suggest that the bilayer thickness is $\sim$5 nm, and we take $h = 2.5$–3.0 nm.

Waugh, 1997), but force relaxation in cells may involve more complex processes than a simple relaxation of nonlocal bending. Thus, nonlocal bending contributions are expected to relax when tethers are maintained for periods on the order of minutes, and, even initially, contribute a small fraction of the total resistance to tether formation.

B. Bilayer Vesicle Dynamics

Forming tethers at finite rates inevitably requires greater forces than those required to maintain a tether at a fixed length. The additional force arises from frictional dissipation of energy within the membrane as the tether forms as well as external forces generated as a result of the membrane movement. In pure lipid systems, there are two principal mechanisms that account for the internal membrane dissipation, first, the in-plane shear flow of the membrane as it moves from the cell body onto the tether, and second, the friction between the two leaflets as they slide past one another onto the tether (because more molecules flow onto the outer leaflet than the inner one) (Evans & Yeung, 1994; Hochmuth et al., 1996). The contributions to the force of these two contributions are

$$f_{\mathrm{vis}} = 4\pi\eta_{\mathrm{m}} V_{\mathrm{t}} + 2\pi b h^2 \ln(R_0/R_{\mathrm{t}}) V_{\mathrm{t}}, \tag{5}$$

where f_{vis} is the contribution to the force from viscous mechanisms within the membrane, η_{m} is the two-dimensional membrane viscosity for in-plane shear, b is the interleaflet drag coefficient, R_0 is the cell or vesicle radius, and V_{t} is the velocity of tether growth. The membrane viscosity coefficient, estimated from lateral mobility measurements, is thought to be $\sim 5 \times 10^{-10}$ N s/m in bilayer membrane systems (Waugh, 1982). Using this value, even for a tether growth velocity of as much as 100 μm/s, the contribution to the force is expected to be $\sim$0.5 pN, an even smaller contribution than is estimated for nonlocal bending. Contributions to the force from the interleaflet frictional stresses, however, can be more significant (Evans & Yeung, 1994). The interleaflet drag coefficient b depends on the lipid composition of the membrane, but ranges from $\sim 2 \times 10^8$ N s/m^3, for a typical phosphatidylcholine, to 12×10^8 N s/m^3 for sphingomyelin and cholesterol (Evans & Yeung). Taking an intermediate value of $b = 5 \times 10^8$ N s/m^3, and taking $h = 2.5$ nm, for a tether formation velocity of 100 μm/s, the contribution to the force is estimated to be $\sim$12 pN. Thus, at high rates of tether formation, the interleaflet friction can add significantly to the force.

IV. TETHERS FORMED FROM RED BLOOD CELLS

A. Red Blood Cell Equilibrium

After the pure phospholipid vesicle, the next simplest system in which to study tethers is the red blood cell. These cells lack an interior cytoskeleton, vastly simplifying the mechanical analysis of their deformation. Indeed, as was noted in the introduction, it was from red cells that the first experimentally formed tethers were observed (Hochmuth et al., 1973). A critical step in the advancement of our understanding of tether formation came with the development of experimental systems that enabled us to control or measure essentially all of the critical parameters needed to characterize the tether equilibrium (Heinrich & Waugh, 1996; Hochmuth, Wiles, Evans, & McCown, 1982; Hwang & Waugh, 1997) (e.g., see Fig. 3A). These experimental approaches,

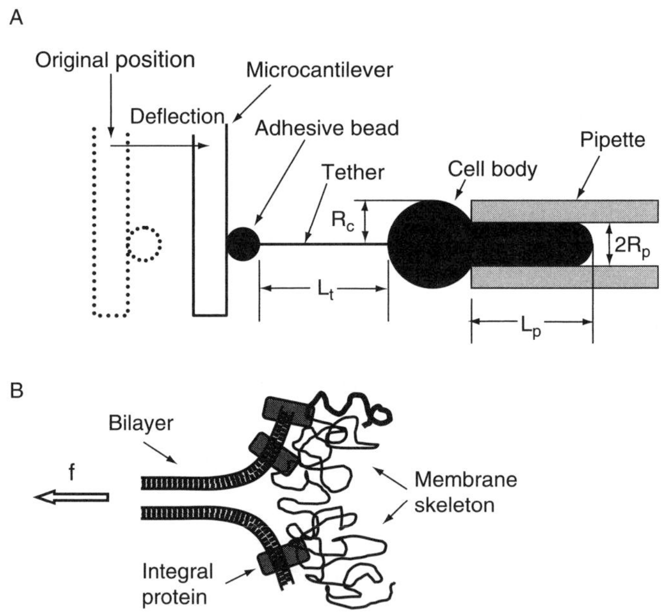

FIGURE 3 Schematic illustration of a tether formation measurement for a red blood cell. (A) The cell is held in a micropipette and attached to an adherent bead that is stuck to a thin glass fiber (microcantilever). The cell is withdrawn from the bead forming a tether between the body of the cell and the bead. The deflection of the cantilever provides a measure of the force, and the aspiration pressure in the pipette is used to control the membrane tension. (B) Schematic showing that tether formation from cells involves a lateral segregation of membrane components. Lipid bilayer is pulled from the cell surface, but the membrane skeleton and associated integral proteins remain on the cell body.

applied to both phospholipid vesicles and red cells, led to the confirmation of theoretical relationships describing tether equilibrium, and revealed the contribution to the work of tether formation due to the presence of the membrane-associated cytoskeleton. A critical aspect of tether formation from red blood cells is that it involves a separation of the membrane bilayer from the underlying membrane skeleton (Fig. 3B). This separation requires work to overcome the natural tendency of bilayer and skeleton to remain in contact with each other. This contribution has been named the bilayer–skeletal separation energy, W_{sk}, and enters the equilibrium relationships in the following way (Waugh, Mantalaris, Bauserman, Hwang, & Wu, 2001):

$$f^2 = 8\pi^2 k_c(\tau_0 + W_{sk}), \tag{6}$$

where τ_0 represents the mechanical tension in the membrane bilayer. Note that the first term in this expression applies to pure bilayer vesicles, and is obtained by solving Eqs. (2) and (3) to eliminate R_t. Equation (6) points to a simple and direct method for determining the separation energy W_{sk} (and k_c) experimentally. One needs simply to measure the force of tether equilibrium as a function of the bilayer membrane tension. This is accomplished by holding the cell in a micropipette under a controlled aspiration pressure and pulling the tether using a method that provides a measure of the tethering force, either with a microcantilever, magnet, or optical tweezers (Butler, Mohandas, & Waugh, 2008; Heinrich & Waugh, 1996; Ounkomol, Xie, Dayton, & Heinrich, 2009; Waugh et al., 2001). The membrane tension is controlled by adjusting the aspiration pressure in the micropipette. (A numerical calculation to determine the contribution to the membrane tension from the membrane skeleton may be needed in some circumstances in order to accurately determine the bilayer tension τ_0; (Butler et al.) A linear relationship is expected between the bilayer tension and the square of the tethering force. Experiments confirm this expectation, providing direct measures for both k_c and W_{sk}. For bilayer membranes, $k_c \approx 1.0 \times 10^{-19}$ J (Heinrich & Waugh) and for red blood cells, $k_c \approx 2.0 \times 10^{-19}$ J (Butler et al.). For normal mature human red cells, $W_{sk} \approx 70\ \mu J/m^2$ (Butler et al.; Hochmuth & Marcus, 2002; Hwang & Waugh, 1997; Waugh et al.), but it is smaller for red blood cells from patients with hemolytic anemia (Butler et al.; Waugh & LaCelle, 1980), in early-stage reticulocytes (Waugh et al.), and in normal red blood cells from mice (unpublished data).

While the origin of this additional energy contribution is not completely certain, both theoretical considerations and experimental measurements support the concept that it is the lateral segregation of membrane components during tether formation that accounts for this energy. Measurement of fluoresce intensity in tethers formed from red blood cells with fluorescently labeled integral membrane protein (band 3) reveal that there is significantly higher surface concentration of band 3 remaining on the cell body than appears in the tether (Butler et al., 2008). Thus, the boundary between the cell body and the tether acts like a semipermeable membrane, allowing lipids to pass onto the tether, but retaining integral proteins attached to, or entangled in, the membrane skeleton. This creates a difference between the chemical potential of the lipid on the tether and the lipid on the cell body that tends to pull the lipid of the tether back onto the cell body generating a force equal to (Waugh & Bauserman, 1995):

$$f = kT \ln\left(\frac{X_b}{X_t}\right) \frac{dn_\ell}{dL_t}, \tag{7}$$

where X_b is the mole fraction of lipid on the cell body and X_t is the mole fraction of lipid on the tether, kT is Boltzmann's constant times temperature, and dn_ℓ/dL_t is the number of lipid molecules added to the tether per unit length of tether. (The astute reader will note that the mole fraction of lipid is typically close to 1.0 in both regions, but the small difference in the concentration of proteins within the two regions accounts for a substantial energy difference.) The origin of this force is analogous that of osmotic pressure, and the concept has been popularized as the osmotic tension hypothesis (Dai & Sheetz, 1995). An estimate of the energy corresponding to this mechanism reveals that a difference in protein concentration of just 1.0% would amount to an energy of $\sim 160\ \mu J/m^2$, more than enough to account for the measured bilayer–skeletal separation energy. Further support for this concept comes from recent measurements showing that the energy of dissociation is zero in the absence of band 3 on the membrane, and increases when the concentration difference between the cell body and the tether was increased by immobilizing band 3 on the cell body (Butler et al., 2008).

B. Red Blood Cell Dynamics

Just as there is an additional contribution to the equilibrium tether force, there is an additional contribution to the tether force at increasing rates of formation arising from the presence of the membrane skeleton. In the simplest case, this takes the form (Hochmuth et al., 1996)

$$f_{vs} = 2\pi\eta_{eff} V_t, \tag{8}$$

where f_{vs} represents the force contribution from viscous forces arising from interactions with the skeleton, η_{eff} is an effective surface viscosity coefficient, and V_t is the tether growth rate. For red blood cells, the effective viscosity is approximately 30 × 10^{-6} N s/m (Hochmuth & Marcus, 2002; Hwang & Waugh, 1997). Equation (8) is an exact expression if η_{eff} is the intrinsic viscosity of the membrane bilayer that flows from the cell body onto the tether. The origin of the viscous dissipation, however, appears to be more complex than a simple surface viscosity. The main argument for this is that the rate at which proteins and lipids diffuse in the surface of the red cell membrane is much higher than would be predicted if the surface viscosity had a value of 30 × 10^{-6} N s/m. Given this value for membrane viscosity, the theoretical predictions of Saffman and Delbruck (1975) give an expected value for the lateral mobility of $\sim 10^{-16}$ $\mathrm{m^2/s}$, but measured values of mobility for lipids or mobile proteins on red cells are three orders of magnitude larger than this (Golan & Veatch, 1980). Thus, it appears unlikely that the internal viscous resistance of the membrane itself can account for the velocity-dependent force of tether formation.

Rate-dependent external forces on the flowing bilayer may arise from the presence of the adjacent cytoskeleton. Two principal mechanisms have been proposed as the physical basis for external contributions to the apparent viscous coefficient: the viscous drag of the membrane bilayer material as it flows over the cytoskeleton (Fig. 4A) (Hochmuth et al., 1996), and the drag caused by flow of membrane around integral proteins attached to the cytoskeleton (Fig. 4B) (Brochard-Wyart, Borghi, Cuvelier, & Nassoy, 2006). Thus, the total external contribution to η_{eff} can be written as a sum of an epitactic sliding friction, analogous to the interleaflet drag at the center of the bilayer (Hochmuth et al.), plus a contribution due to the flow of bilayer around stationary integral proteins (Brochard-Wyart et al.):

$$\eta_{\mathrm{eff}} = [b_{\mathrm{sc}} + b_{\mathrm{ip}}] R_{\mathrm{t}}^2 \ln (R_0 / R_{\mathrm{t}}). \tag{9}$$

The coefficient b_{sc} is an epitactic coupling coefficient (units of N $\mathrm{s/m^3}$)[3] analogous to the parameter b in Eq. (5) (and written as η_{sc} in Hochmuth et al., 1996), and the coefficient b_{ip} (same units) represents the drag contribution due to stationary integral membrane proteins (Brochard-Wyart et al., 2006):

$$b_{\mathrm{ip}} = \frac{4\pi\rho\eta_{\mathrm{bl}}}{\ln(\sqrt{1/\rho}/a)}, \tag{10}$$

where ρ is the surface density of immobile proteins, a is the protein radius, and η_{bl} is the surface viscosity of the bilayer. It is unclear which of these two contributions is more significant for tethers formed from cells, but it is

[3]Hochmuth et al. (1996) refer to this quantity as η_{sc}. In the present chapter, we reserve the symbol η for surface viscosity coefficients (units of N s/m) and use the symbol b (units of N $\mathrm{s/m^3}$) as proposed by Evans and Yeung (1994) for viscous drag coefficients at interfaces between membrane layers.

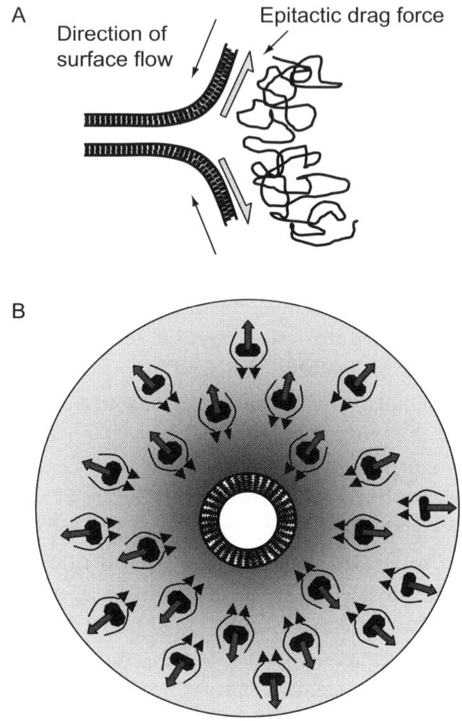

FIGURE 4 (A) Schematic illustration showing the epitactic drag forces acting on a bilayer membrane being dragged over the membrane skeleton to form the tether. (B) Cartoon showing a tether end-on (at the center) with membrane flowing around proteins (dark ovals) toward the tether base. Thin arrows represent flow of bilayer around the stationary integral proteins, and the gray block arrows indicate the radial reaction forces exerted by the proteins.

possible to estimate the expected contribution from integral proteins, based on knowledge of the composition of the red cell membrane. Taking $\rho = 1600\ \mu\text{m}^{-2}$, $\eta_{\text{bl}} = 3.7 \times 10^{-9}$ N s/m (higher than for pure lipid systems; Waugh, 1982), and setting the molecular dimension a to 5 nm, we obtain $b_{\text{ip}} \approx 5 \times 10^{7}$ N s/m^3. For a tether radius of 24 nm, this works out to an effective viscosity of $\sim 0.13 \times 10^{-6}$ N s/m, much smaller than what is measured. This calculation indicates that epitactic coupling between the bilayer and the underlying cytoskeleton may be the more dominant mechanism, although nonideal effects of molecular crowding [not included in the theory leading to Eq. (10)] could lead to higher resistance to flow around integral proteins than expected.

It is important to note that the effective viscosity arising from external cytoskeletal tractions scales with the square of the tether radius R_t, regardless of whether this drag results from integral proteins or an epitactic friction. In many published reports, the dynamic resistance to tether formation is expressed as a constant effective viscosity coefficient, in essence, making the assumption that the tether radius R_t is a constant. In this case there is a direct relationship between the intrinsic viscous properties (represented by b_{eff}) and the effective coefficient η_{eff} [Eq. (8)], and a linear relationship between tethering force and the velocity of tether formation is predicted (Hochmuth et al., 1996; Shao & Hochmuth, 1996). Experimental results for red cells appear to agree with this prediction (Hwang & Waugh, 1997), although the scatter in the data prevents a definitive test. If, on the other hand, one accounts for the expected change in tether radius with increasing force [Eq. (2)], a nonlinear prediction for the dependence of the force on the tether formation rate is obtained (Brochard-Wyart et al., 2006). Extension of the force balance relationships given in Eqs. (2) and (3) to include the bilayer–skeletal separation energy and the dynamic contribution leads to

$$R_t = \left(\frac{k_c}{2[\tau_0 + W_{sk} + b_{eff} R_t V_t \ln (R_0/R_t)]} \right)^{1/2}, \tag{11}$$

and this in turn leads to a prediction of a third-order dependence of force on tether formation velocity (Borghi & Brochard-Wyart, 2007; Brochard-Wyart et al., 2006):

$$f^3 - f f_0^2 = (2\pi)^3 2k_c^2 b_{eff} \ln (R_0/R_t) V_t, \tag{12}$$

where f_0 is the equilibrium force. In a recent report, Borghi and Brochard-Wyart (2007) show evidence that for individual cells, there is excellent agreement between the predictions of Eq. (12) and their experimental measurements, with values for b_{eff} ranging from 5×10^9 to 5.5×10^{10} N s/m^3. (Note that η_e reported by Borghi and Brochard-Wyart corresponds to b_{eff}/ρ in the present context. The value of ρ, the density of immobile proteins, is taken to be 500 μm^{-2}.) For large populations of cells, however, the theory appears to over-predict the increase in the velocity of tether formation with increasing force. This is illustrated in Fig. 5, where the original data of Hwang and Waugh (1997) are shown with theoretical predictions for different values of b_{eff}. For the left-hand curve, fit to the data where $f < 120$ pN, $b_{eff} = 4.2 \times 10^{10}$ N s/m^3, and for the right-hand curve, fit to all of the data, $b_{eff} = 9 \times 10^{10}$ N s/m^3. Note that these values are an order of magnitude larger than the interleaflet drag coefficient for pure bilayer membrane,

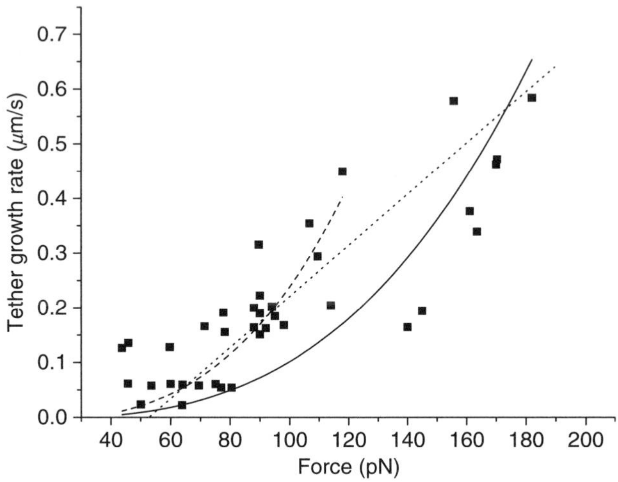

FIGURE 5 The original data of Hwang and Waugh (1997) (points) plotted with two theoretical predictions based on Eq. (12). For both curves, $k_c = 2 \times 10^{-19}$ J, $f_0 = 30$ pN, and $\ln(R_c/R_t) = 4.6$. The fits were almost insensitive to the value of the equilibrium force f_0, which was fixed at a value of 30 pN. A single parameter (b_{eff}) was varied to obtain the best fit to the data by nonlinear regression. For the solid curve (fit to all of the data), $b_{eff} = 9 \times 10^6$, for the dashed curve (fit to the data when $f < 120$ pN), $b_{eff} = 3.8 \times 10^6$. Linear regression to the data (dotted line) is shown for reference ($\eta_{eff} = 34 \times 10^{-6}$ N s/m).

indicating that the cytoskeletal contributions are dominant. The failure of the theory to accurately track the dependence of force on tether growth rate for the entire cell population is most probably due to the wide variability in properties across the red cell population. For practical calculations, the effective viscosity coefficient, neglecting any dependence on tether radius, may be the better choice when trying to predict the behavior of red cell populations with variable properties.

V. TETHERS FROM NEUTROPHILS AND OTHER CELLS

Of primary interest in the present context are the forces required to form tethers from neutrophils and other leukocytes. The seminal work on neutrophil tethers was by Shao and Hochmuth (1996), but it was not until Schmidtke and Diamond (2000) published observations of tether formation from neutrophils as they rolled over an adhesive substrate that the physiological significance of these observations was recognized.

A. Equilibrium in Complex Cells

A complication in interpreting tether formation measurements from neutrophils and other cells with extensive cytoskeletons is that the membrane tension is generally not known with certainty. In neutrophils, it is known that there is a contractile force resultant at the cell cortex that generates a force of 10–30 μN/m (Herant, Heinrich, & Dembo, 2005; Needham & Hochmuth, 1992; Tsai, Frank, & Waugh, 1994). Assuming that the membrane bilayer supports the resulting intracellular pressure, one can argue that the far field tension must be of this magnitude. The equilibrium tether force for neutrophils is similar to that measured for red blood cells, falling in the range of 30–45 pN. Applying Eq. (6), and taking values for $k_c = 2 \times 10^{-19}$ J and $\tau_0 = 20$ μN/m, we estimate a bilayer–skeletal separation energy of 40–100 μJ/m^2, in very close agreement with W_{sk} for red blood cells. This agreement is remarkable, considering that the red blood cell and the neutrophil have little in common from a structural perspective. It is interesting to speculate that this separation energy may be physiologically determined such that the stability of the bilayer–skeletal association is sufficient to maintain this association in the vasculature. Such a possibility is supported by observations that T-lymphocytes also exhibit a similar range of values for the equilibrium tethering force (Xu & Shao, 2005). These values are substantially greater than those measured for neuronal growth cones (Hochmuth et al., 1996), which do not need to withstand the forces within the vasculature, but similar to those measured for HEK cells, a transformed human cell line derived from kidney (Ermilov, Murdock, Qian, Brownell, & Anvari, 2007). Understanding the physical basis and regulation of the bilayer–skeletal separation energy in different cells and under different conditions remains a current challenge within the field.

B. Neutrophil Tether Dynamics

The dynamic behavior of tether formation from neutrophils is of particular interest because of their role in modulating cell rolling behavior. Early investigators, examining a limited range of tether formation rates ($<$10 μm/s), used a linear regression to measured values of force as a function of tether growth rate and obtained an effective viscosity coefficient of $\sim$1.6 $\times$ 10^{-6} N s/m (Shao & Hochmuth, 1996; Xu & Shao, 2005). More recently, Heinrich, Leung, and Evans (2005) examined tether formation rates over a much wider range (up to 200 μm/s) and found a substantially nonlinear relationship between force and tether growth rate. Their original interpretation of this was that the tethered membranes were exhibiting shear thinning behavior, and they proposed a power law dependence of growth rate on force. Subsequent analysis of these data by Brochard-Wyart et al. (2006) [see Eqs. (9)–(12)] revealed that the apparent shear thinning could be accounted for by recognizing that the radius of the tether is a function of the applied force. This is illustrated in Fig. 6, where the data of Heinrich et al. and Shao and Hochmuth are replotted and fit to Eq. (12). The agreement is excellent, revealing an interfacial drag coefficient $b_{\text{eff}} = 7.7 \times 10^8$ N s/m^3 for neutrophil membrane. Note that this value is substantially smaller than drag coefficients determined for red blood cells, even greater than the 20-fold difference in effective viscosities that has been reported. Thus, even though the threshold force for tether formation is similar for leukocytes and red blood cells, the dynamic resistance to tether formation from both neutrophils and T-lymphocytes is substantially less that for red cells. Interestingly, the value is remarkably close to that determined for pure bilayer membranes containing cholesterol. However, in the case of the lipid bilayer, the contribution to the force scales as the square of the bilayer thickness [see Eq. (5)], whereas the cytoskeletal contribution scales as the square of the tether radius. Thus, even

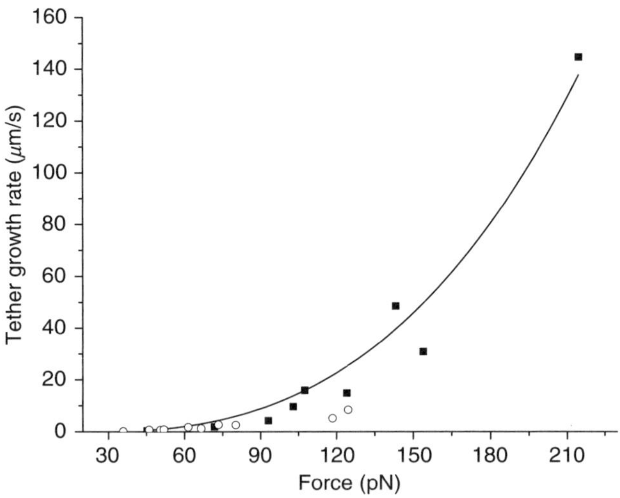

FIGURE 6 Illustration showing agreement between the theoretical prediction given in Eq. (12) and measurements of tethering force versus rate of tether formation. The data were replotted from Shao and Hochmuth (1996) (open circles) and from Heinrich et al. (2005) (solid squares). The equilibrium force f_0 was fixed at a value of 35 pN, and the parameter b_{eff} was determined by nonlinear least squares regression with b_{eff} as the only free parameter. From the fit, $b_{\text{eff}} = 7.7 \pm 0.4 \times 10^8$ N s/m^3.

though the coefficients are similar (possibly reflecting a similar underlying physical origin), the contribution of the cytoskeleton to the force is expected to be an order of magnitude greater than that of the bilayer.

C. *Other Cell Types*

Much of our understanding of the dynamics of tether formation was originally developed to interpret measurements of tethers formed from neuronal growth cones (Hochmuth et al., 1996). Tethers from these structures exhibit even less resistance to high rates of growth than leukocytes, exhibiting effective viscosities 10-fold lower than those estimated for neutrophils (Hochmuth et al.). Tethers pulled from outer hair cells appear to require larger forces to maintain fixed lengths, although results for a fully equilibrated tether are only reported for a single event ($f_0 = 60$ pN) (Li et al., 2002). Effective viscosities for these cells were also higher than for leukocytes, but not as large as for red blood cells: $\eta_{\text{eff}} = 2.4\text{–}5.6 \times 10^{-6}$ N s/m.

VI. IMPLICATIONS FOR CELL ADHESION IN THE VASCULATURE

An important first step in the response to inflammation is a rolling interaction between leukocytes and endothelium. These interactions are principally mediated by selectins and their ligands, but some integrins can also mediate rolling interactions. During this process, the forward and reverse kinetics of bond formation, and the changes in these rates under mechanical loading play a critical role in determining the speed and duration of rolling interactions, and this can have substantial effects on the ability of the cell to respond to the chemical signals presented by the endothelium. The underlying mechanisms of rolling interactions have been studied from a variety of perspectives, including atomic force microscopy to measure the force dependence of bond lifetimes (Fritz, Katopodis, Kolbinger, & Anselmetti, 1998; Hanley et al., 2003; Marshall et al., 2003), as well as flow channel experiments using both beads and various cell types, including neutrophils (Alon, Hammer, & Springer, 1995; Hanley et al.; Park et al., 2002). Understanding how the bonds between the cell and the substrate experience mechanical force is critical for understanding this process at a fundamental level. As first analyzed by Shao and colleagues (Shao, Ting-Beall, & Hochmuth, 1998), the formation of extended membrane strands from the cell surface can lead to a twofold decrease on the bond force, simply from the geometry of loading and changes in the moment arm by which force is applied to the bond. The most complete analysis of selectin-mediated rolling appeared in a trio of papers that examined P-selectin–PSGL-1 unbinding kinetics (including characterization of the unbinding from the underlying cytoskeleton), membrane tether formation and elongation, and culminating in a full computational analysis of the adhesive dynamics in fluid flow (Evans, Heinrich, Leung, & Kinoshita, 2005; Heinrich et al., 2005; King, Heinrich, Evans, & Hammer, 2005). In another study, the influence of microvillus deformability and the transition to membrane tether formation have been examined in detail in simulations, and show good agreement with experimental measurements made at low ($<200\ s^{-1}$) shear rates (Caputo & Hammer, 2005). The detachment of P-selectin from the cytoskeleton and the subsequent formation of membrane tethers as the cell rolled proved to be critical for accurately predicting cell behavior.

VII. CONCLUSION

Membrane tethers are unique structures that are formed from cell surfaces in response to point forces. Much of what we know about the physical properties of tethers and the structural characteristics that give them their unique properties has been learned from model systems and red blood cells (see Table I for a summary of coefficients characterizing tethers formed in different systems). In vesicle systems, tether formation provides an unusual opportunity to determine the intrinsic resistance of the membrane to bending, and to explore different mechanisms that are associated with controlling membrane curvature. In more complex cells, the equilibrium force of tether formation provides a measure of the energy of association between the membrane bilayer and the cytoskeletal structure of the cell, although in cells more complex than red blood cells, the origins of this association energy are difficult to define. The rate-dependent force of tether formation in simple bilayers is dominated by the epitactic friction at the interface between the two membrane leaflets. In red blood cells and more complex cells, the origins of the dynamic resistance are less clear, but clearly involve drag forces on the bilayer exerted by the cytoskeleton. The most likely mechanism appears to be epitactic friction between the bilayer and the skeleton, but a role for integral proteins in contributing to the drag forces is an alternative possibility. The formation of tethers during separation of adhesive cell–cell contacts can have a significant effect on the dynamics of separation. Thus, the physical characteristics of membrane tethers are of interest from a variety of perspectives, from defining the fundamental physical characteristics of bilayer membranes and their association with cytoskeleton, to understanding the physical mechanisms accounting for the dynamic interactions of circulating cells with endothelium in the vasculature.

TABLE I
Physical Constants of Tethers

Source	k_c (10^{-19} J)	W_{sk} (μJ/m^2)	b_{eff} (10^9 N s/m^3)	η_{eff} (μN s/m)
Vesicles (SOPC)[a]	1.0[b]	0.0[b]	0.2[c]	–
Vesicles (SM/chol)[d]	5.5[c]	0.0	1.2[c]	–
Red blood cells	2.0[e]	60–80[f]	5–55[g]	27–34[h]
Neutrophils	–	40–100[i]	0.8[j]	1.6[i]
T-lymphocytes	–	~85[k]		1.6[k]

[a]SOPC, stearoyl–oleoyl phosphatidylcholine.
[b]Heinrich and Waugh (1996).
[c]Evans and Yeung (1994) and Yeung (1994).
[d]SM/chol, sphingomyelin and cholesterol (1:1 mixture).
[e]Butler et al. (2008) and Hwang and Waugh (1997).
[f]Butler et al. (2008), Hochmuth and Marcus (2002), and Hwang and Waugh (1997).
[g]Borghi and Brochard-Wyart (2007).
[h]Hochmuth and Marcus (2002) and Hwang and Waugh (1997).
[i]Shao and Hochmuth (1996), Shao and Xu (2002), and Xu and Shao (2005).
[j]Brochard-Wyart et al. (2006), Heinrich et al. (2005), and Shao and Hochmuth (1996).
[k]Xu and Shao (2005).

VIII. FUTURE CHALLENGES

Significant challenges remain in understanding the role that the cytoskeleton may play during and after tether formation. The inability to precisely measure and control the far-field tension in cell surfaces remains a significant barrier to precise determination of bilayer–skeletal separation energies in complex cells, and while measurements of the force of tether formation can still reveal much about the tightness of association between bilayer and skeleton, a full understanding the mechanisms underlying the resistance to bilayer extraction from cell surfaces is yet to be realized. The ability to test theories about the origins of the bilayer–skeletal separation energy is limited by opportunity. There are few systems wherein membrane composition can be altered substantially and the parameters of tether formation controlled sufficiently well to draw firm conclusions about the mechanisms by which cells maintain their surface integrity.

Another feature of tether formation that remains to be fully understood is the growth of cytoskeletal structures within tethers. This phenomenon has been characterized in tethers formed from neutrophils (Zhelev, Alteraifi, & Hochmuth, 1996), and has been posed as a possible tool for understanding the initiation of cell protrusions and subsequent cell motility. But the biochemical requirements for cytoskeletal growth, and the consequences this growth may have for tether equilibrium, remain to be fully characterized. The fundamental properties of tethers that have been described in this chapter will serve as a solid foundation for exploring these phenomena, and should lead to an even more detailed understanding of the mechanical properties of cell surfaces.

References

Alon, R., Hammer, D. A., & Springer, T. A. (1995). Lifetime of the p-selectin—carbohydrate bond and its response to tensile force in hydrodynamic flow. *Nature, 374*, 539–542.

Borghi, N., & Brochard-Wyart, F. (2007). Tether extrusion from red blood cells: integral proteins unbinding from cytoskeleton. *Biophysical Journal, 93*(4), 1369–1379.

Bozic, B., Svetina, S., Zeks, B., & Waugh, R. E. (1992). Role of lamellar membrane structure in tether formation from bilayer vesicles. *Biophysical Journal, 61*, 963–973.

Brochard-Wyart, F., Borghi, N., Cuvelier, D., & Nassoy, P. (2006). Hydrodynamic narrowing of tubes extruded from cells. *Proceedings of the National Academy of Sciences of the United States of America, 103*(20), 7660–7663.

Butler, J., Mohandas, N., & Waugh, R. E. (2008). Integral protein linkage and the bilayer–skeletal separation energy in red blood cells. *Biophysical Journal, 95*(4), 1826–1836.

Caputo, K. E., & Hammer, D. A. (2005). Effect of microvillus deformability on leukocyte adhesion explored using adhesive dynamics simulations. *Biophysical Journal, 89*(1), 187–200.

Dai, J. W., & Sheetz, M. P. (1995). Mechanical properties of neuronal growth cone membranes studied by tether formation with laser optical tweezers. *Biophysical Journal, 68*, 988–996.

Ermilov, S. A., Murdock, D. R., Qian, F., Brownell, W. E., & Anvari, B. (2007). Studies of plasma membrane mechanics and plasma membrane–cytoskeleton interactions using optical tweezers and fluorescence imaging. *Journal of Biomechanics, 40*(2), 476–480.

Evans, E., Heinrich, V., Leung, A., & Kinoshita, K. (2005). Nano- to microscale dynamics of P-selectin detachment from leukocyte interfaces. I. Membrane separation from the cytoskeleton. *Biophysical Journal, 88*(3), 2288–2298.

Evans, E., & Yeung, A. (1994). Hidden dynamics in rapid changes of bilayer shape. *Chemistry and Physics of Lipids, 73*, 39–56.

Fritz, J., Katopodis, A. G., Kolbinger, F., & Anselmetti, D. (1998). Force-mediated kinetics of single P-selectin/ligand complexes observed by atomic force microscopy. *Proceedings of the National Academy of Sciences of the United States of America, 95*(21), 12283–12288.

Golan, D. E., & Veatch, W. (1980). Lateral mobility of band 3 in the human erythrocyte membrane studied by fluorescence photobleaching recovery: evidence for control by cytoskeletal interactions. *Proceedings of the National Academy of Sciences of the United States of America, 77*(5), 2537–2541.

Hanley, W., McCarty, O., Jadhav, S., Tseng, Y., Wirtz, D., & Konstantopoulos, K. (2003). Single molecule characterization of P-selectin/ligand binding. *Journal of Biological Chemistry, 278*(12), 10556–10561.

Heinrich, V., Leung, A., & Evans, E. (2005). Nano- to microscale dynamics of P-selectin detachment from leukocyte interfaces. II. Tether flow terminated by P-selectin dissociation from PSGL-1. *Biophysical Journal, 88*(3), 2299–2308.

Heinrich, V., & Waugh, R. E. (1996). A piconewton force transducer and its application to measurement of the bending stiffness of phospholipid membranes. *Annals of Biomedical Engineering, 24*, 595–605.

Helfrich, W. (1973). Elastic properties of lipid bilayers: theory and possible experiments. *Zeitschrift für Naturforschung C, 28*(11), 693–703.

Herant, M., Heinrich, V., & Dembo, M. (2005). Mechanics of neutrophil phagocytosis: behavior of the cortical tension. *Journal of Cell Science, 118*(Pt 9), 1789–1797.

Hochmuth, R. M., & Marcus, W. D. (2002). Membrane tethers formed from blood cells with available area and determination of their adhesion energy. *Biophysical Journal, 82*, 2964–2969.

Hochmuth, R. M., Mohandas, N., & Blackshear, P. L. (1973). Measurement of the elastic modulus for red cell membrane using a fluid mechanical technique. *Biophysical Journal, 13*, 747–762.

Hochmuth, R. M., Shao, J. Y., Dai, J., & Sheetz, M. P. (1996). Deformation and flow of membrane into tethers extracted from neuronal growth cones. *Biophysical Journal, 70*, 358–369.

Hochmuth, R. M., Wiles, H. C., Evans, E. A., & McCown, J. T. (1982). Extensional flow of erythrocyte membrane from cell body to elastic tether. II. Experiment. *Biophysical Journal, 39*, 83–89.

Hwang, W. C., & Waugh, R. E. (1997). Energy of dissociation of lipid bilayer from the membrane skeleton of red blood cells. *Biophysical Journal, 72*, 2669–2678.

King, M. R., Heinrich, V., Evans, E., & Hammer, D. A. (2005). Nano-to-micro scale dynamics of P-selectin detachment from leukocyte interfaces. III. Numerical simulation of tethering under flow. *Biophysical Journal, 88*(3), 1676–1683.

Li, Z., Anvari, B., Takashima, M., Brecht, P., Torres, J. H., & Brownell, W. E. (2002). Membrane tether formation from outer hair cells with optical tweezers. *Biophysical Journal, 82*(3), 1386–1395.

Marshall, B. T., Long, M., Piper, J. W., Yago, T., McEver, R. P., & Zhu, C. (2003). Direct observation of catch bonds involving cell-adhesion molecules. *Nature, 423*(6936), 190–193.

Needham, D., & Hochmuth, R. M. (1992). A sensitive measure of surface stress in the resting neutrophil. *Biophysical Journal, 61*, 1664–1670.

Ounkomol, C., Xie, H., Dayton, P. A., & Heinrich, V. (2009). Versatile horizontal force probe for mechanical tests on pipette-held cells, particles, and membrane capsules. *Biophysical Journal, 96*(3), 1218–1231.

Park, E. Y., Smith, M. J., Stropp, E. S., Snapp, K. R., DiVietro, J. A., Walker, W. F., et al. (2002). Comparison of PSGL-1 microbead and neutrophil rolling: microvillus elongation stabilizes P-selectin bond clusters. *Biophysical Journal, 82*(4), 1835–1847.

Raphael, R. M., & Waugh, R. E. (1996). Accelerated interleaflet transport of phosphatidylcholine molecules in membranes under deformation. *Biophysical Journal, 71*, 1374–1388.

Saffman, P. G., & Delbruck, M. (1975). Brownian motion in biological membranes. *Proceedings of the National Academy of Sciences of the United States of America, 72*(8), 3111–3113.

Schmidtke, D. W., & Diamond, S. L. (2000). Direct observation of membrane tethers formed during neutrophil attachment to platelets or P-selectin under physiological flow. *Journal of Cell Biology, 149*(3), 719–730.

Shao, J. Y., & Hochmuth, R. M. (1996). Micropipette suction for measuring piconewton forces of adhesion and tether formation from neutrophil membranes. *Biophysical Journal, 71*, 2892–2901.

Shao, J. Y., Ting-Beall, H. P., & Hochmuth, R. M. (1998). Static and dynamic lengths of neutrophil microvilli. *Proceedings of the National Academy of Sciences of the United States of America, 95*(12), 6797–6802.

Shao, J. Y., & Xu, J. (2002). A modified micropipette aspiration technique and its application to tether formation from human neutrophils. *Journal of Biomechanical Engineering, 124*, 388–396.

Tsai, M. A., Frank, R. S., & Waugh, R. E. (1994). Passive mechanical behavior of human neutrophils: effect of cytochalasin B. *Biophysical Journal, 66*, 2166–2172.

Waugh, R. E. (1982). Surface viscosity measurements from large bilayer vesicle tether formation. II. Experiments. *Biophysical Journal, 38*, 29–37.

Waugh, R. E., & Bauserman, R. G. (1995). Physical measurements of bilayer–skeletal separation forces. *Annals of Biomedical Engineering, 23*, 308–321.

Waugh, R. E., & Hochmuth, R. M. (1987). Mechanical equilibrium of thick hollow liquid membrane cylinders. *Biophysical Journal, 52*, 391–400.

Waugh, R. E., & LaCelle, P. L. (1980). Abnormalities in the membrane material properties of hereditary spherocytes. *Journal of Biomechanical Engineering, 102*, 240–246.

Waugh, R. E., Mantalaris, A., Bauserman, R. G., Hwang, W. C., & Wu, J. H. (2001). Membrane instability in late-stage erythropoiesis. *Blood, 97*(6), 1869–1875.

Xu, G., & Shao, J. Y. (2005). Double tether extraction from human neutrophils and its comparison with CD4+ T-lymphocytes. *Biophysical Journal, 88*(1), 661–669.

Yeung, A. K. C. (1994). *Mechanics of inter-monolayer coupling in fluid surfactant bilayers.* Doctoral Thesis, Department of Physics, University of British Columbia, Vancouver, BC.

Zhelev, D. V., Alteraifi, A. M., & Hochmuth, R. M. (1996). F-actin network formation in tethers and in pseudopods stimulated by chemoattractant. *Cell Motility and the Cytoskeleton, 35*, 331–344.

CHAPTER 2

Biomechanics of Leukocyte and Endothelial Cell Surface

Jin-Yu Shao

Department of Biomedical Engineering, Washington University, Saint Louis, Missouri 63011, USA

I. OVERVIEW

Human leukocytes, including neutrophils, lymphocytes, and monocytes, are the major immune cells within blood. Leukocytes travel around the human body through circulatory vessels that are lined with ECs. When they leave blood, leukocytes roll on ECs first. This is due, in part, to weak adhesive interactions mediated mainly by selectins on both leukocytes and ECs. This weak adhesion, coupled with the force and torque imposed on the leukocyte by blood flow, yields a consecutive series of jerky motion of the

Current Topics in Membranes, Volume 64 1063-5823/09 $35.00

DOI: 10.1016/S1063-5823(09)64002-3

leukocyte, often coined *rolling*. Neutrophil and lymphocyte rolling are essential for their immunological functions, whereas monocyte rolling is critical for its arrest on the vascular wall and invasion into the aortic intima. After their extravasation, neutrophils and lymphocytes migrate to the infection site to fight against invading microorganisms, whereas monocytes differentiate into macrophages that can form foam cells upon internalizing lipids, resulting in fatty streak lesions where more advanced lesions develop (Ross, 1999). Therefore, leukocyte rolling on the endothelium is the first critical step of leukocyte migration to sites of infection, injury, or atherosclerotic lesions.

Unstable rolling impairs effective neutrophil and lymphocyte recruitment to infected tissues and renders human beings defenseless against invading microorganisms, resulting in immunological diseases such as leukocyte adhesion deficiency syndrome (Serhan & Savill, 2005; Simon & Green, 2005; Smeeth et al., 2006; Springer, 1995). On the other hand, stable rolling is critical for monocyte infiltration into the aortic intima that initiates and exacerbates cardiovascular diseases like atherosclerosis (Ross, 1993, 1999). Elucidating the mechanisms that underlie leukocyte rolling can pave the way toward therapeutic amelioration of such maladies. However, the rolling of leukocytes is a complex dynamic process that is mediated concertedly by adhesion molecules expressed on leukocytes and ECs, shear stress due to blood flow, and mechanical properties of leukocytes and ECs. A complete understanding of leukocyte rolling requires thorough knowledge in all these areas. In this chapter, we review recent advances gained in biomechanics of leukocyte and endothelial cell surface, as well as their impact on leukocyte rolling.

II. INTRODUCTION

During leukocyte rolling, leukocyte adhesion to the endothelium is mainly mediated by a superfamily of receptors called *selectins*, including L-selectin (CD62L) on leukocytes and P- and E-selectin (CD62P and CD62E) on the endothelium (Springer, 1990). Selectins consist of an extracellular domain for ligand binding, a transmembrane domain, and a cytoplasmic tail. Many ligands have been identified for selectins, including P-selectin glycoprotein ligand 1 (PSGL-1) mucosal addressin cell adhesion molecule-1, and glycosylation-dependent cell adhesion molecule-1 (Panes & Granger, 1998). In addition, L-selectin can bind to both P- and E-selectin.

A leukocyte can roll for ~90 s and 270 μm before becoming adherent to ECs (Kunkel, Dunne, & Ley, 2000). Effective leukocyte rolling requires its adhesion to the endothelium, an optimal shear stress, and microvilli (Finger, Bruehl, Bainton, & Springer, 1996; Finger, Puri, et al., 1996; Lawrence,

Kansas, Kunkel, & Ley, 1997). Leukocyte microvilli are numerous membrane protrusions on its surface, which can be as long as 0.7 μm and as short as 0.05 μm with an average length of about 0.35 μm (Bruehl, Springer, & Bainton, 1996; Erlandsen, Hasslen, & Nelson, 1993; Shao, Ting-Beall, & Hochmuth, 1998; Von Andrian, Hasslen, Nelson, Erlandsen, & Butcher, 1995). L-selectin and PSGL-1 are both mainly distributed on leukocyte microvillus tips (Bruehl et al.; Erlandsen et al.; Moore et al., 1995). These microvilli are indispensable for leukocyte rolling on the endothelium because the rolling of leukocytes can be dramatically decreased if the microvillus number is reduced by cytochalasin B or hypotonic swelling (Finger, Bruehl, et al., 1996; Majstoravich et al., 2004).

Before adhering to the endothelium, a leukocyte translates and rotates simultaneously, but moves smoothly and freely in the bloodstream (Alon, Hammer, & Springer, 1995; Goldman, Cox, & Brenner, 1967). When a leukocyte touches an endothelial cell surface, the initial contact point is very likely the tip of a microvillus because of the richness of microvilli on the leukocyte surface (Bruehl et al., 1996; Erlandsen et al., 1993). If a bond forms between the microvillus and EC (Fig. 1A), a horizontal force and a clockwise torque will develop on the leukocyte because of the flowing blood. To balance this force and torque, force will also develop at the tip of the microvillus in two directions: along and perpendicular to its long axis. In other words, the microvillus will be bent and stretched simultaneously until the cell tumbles forward and forms another contact with the endothelium (Fig. 1B). The second contact is likely also made by a microvillus. At this point, the blood flow will cause the microvillus on the right (Fig. 1B) to be compressed and the microvillus on the left (Fig. 1B) to be pulled in two directions: along and perpendicular to its long axis. The force developed at the microvillus tip is also imposed on the EC. All these forces are imposed through membrane receptors, which are much smaller than the cells, so these forces can be considered point forces mechanically. In the following sections,

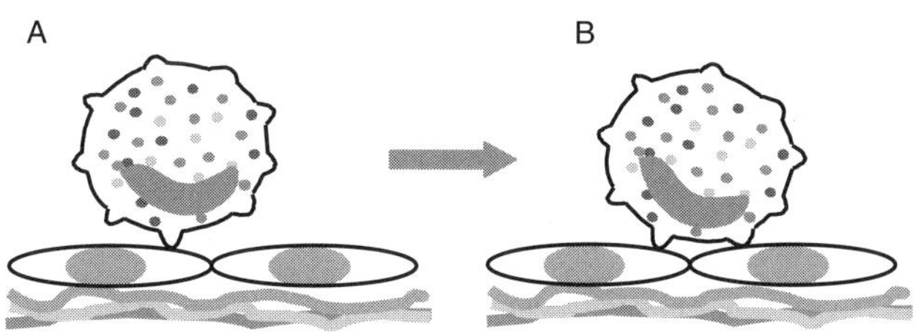

FIGURE 1 Leukocyte attachment to the endothelium (A) and rolling because of the blood flow (B).

we will discuss how cellular surfaces are deformed by these point forces and how these deformations affect leukocyte rolling stability, hence also its function and pathology.

III. SURFACE PROTRUSION AND COMPRESSION

When an increasing point pulling force is imposed perpendicularly on a cell surface, the cell would first extend like a solid as shown in Fig. 2. An appropriate term for describing this tent-like deformation is *surface extension*. However, *surface extension* has been historically used to describe surface dilation of a cell, so *surface protrusion* has been used instead to describe the type of deformation shown in Fig. 2. Correspondingly, cellular stiffness (force/deformation) observed during surface protrusion has been referred to as protrusional stiffness although extensional stiffness may be more appropriate. Once the pulling force shown in Fig. 2 reaches the crossover force (F_c), surface protrusion will transition to membrane tether extraction, a fluid-like deformation (Fig. 3), which will be dealt with in Section V and also in Waugh (2009).

A. Protrusional Stiffness

Because of blood flow, point pulling forces are exerted on leukocyte microvilli through selectins and their ligands during leukocyte rolling, so these microvilli are stretched. If a microvillus was pulled slowly with a small force, it extended like a spring with a stiffness of ~0.043 pN/nm (Shao et al., 1998). However, by pulling microvillus tips and nonmicrovillar regions of leukocytes with point forces, a recent study found that microvillus extension was actually surface protrusion due to both cytoskeletal and membranous deformation as shown in Fig. 2 (Xu & Shao, 2008). In other words, pulled microvilli were not lengthened themselves and their tips were displaced during surface protrusion because their

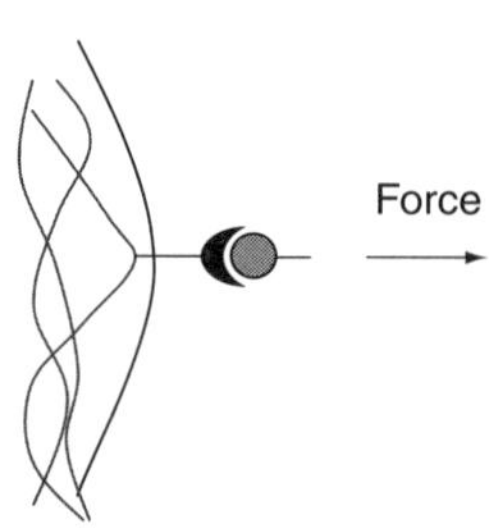

FIGURE 2 Tent-like solid deformation of a cellular surface when a point force is imposed via a receptor–ligand bond.

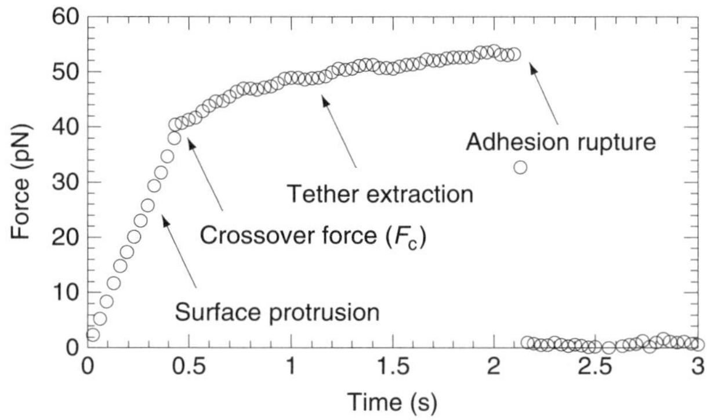

FIGURE 3 History of force when a cellular surface is stretched by a point force (adapted from Xu & Shao, 2008). No cell types have been found that deviate from this behavior under an increasing point force: start from a solid-like deformation and transition to a fluid-like deformation.

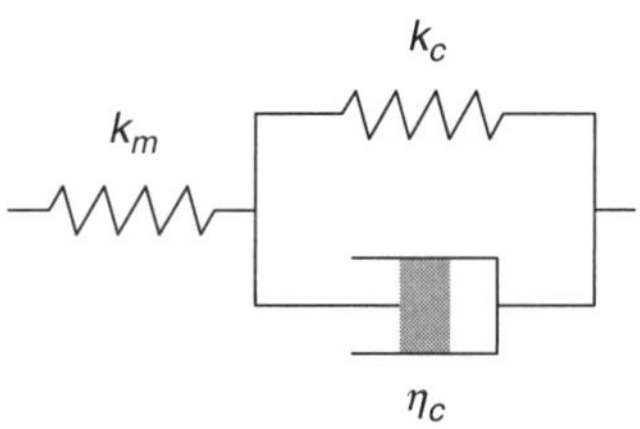

FIGURE 4 The three-parameter solid model for leukocyte and EC surface protrusion (Flugge, 1975).

other ends moved. Although the protrusional stiffness appeared to be constant in each pulling event, it was found to be dependent on the force-loading rate, increasing from ~0.05 to ~0.3 pN/nm when the force-loading rate increased from 20 to 38,000 pN/s (Evans, Heinrich, Leung, & Kinoshita, 2005; Xu & Shao). This shows that surface protrusion is a viscoelastic process, which can be clearly seen from its exhibition of relaxation and hysteresis (Xu & Shao). However, it is worthy of note that, at very high force-loading rates, the protrusional stiffness may be overestimated due to the viscous drag caused by fast probe or cell retraction. For human neutrophils, their surface protrusion is best described by the three-parameter solid model shown in Fig. 4 (Flugge, 1975), where $k_c = 0.043$ pN/nm, $\eta_c = 0.033$ pN s/nm, and $k_m = 0.12$ pN/nm (Xu & Shao). This model, whose parameters were obtained from two different studies with two different techniques, accurately predicted the behavior of neutrophils during relaxation and hysteresis. If neutrophil surface protrusion lasts for only a short period of time, it will appear to be linear elastic with a spring constant of k_m; if it lasts for a long period of time (longer than its characteristic time of η_c/k_c), it will appear to be linear elastic with a spring constant of $k_m k_c/(k_m + k_c)$. For leukocyte rolling, k_m is more relevant because of the short lifetime of each adhesive bond.

When a point pulling force was applied perpendicularly on an EC surface, the same force history as shown in Fig. 2 was observed, that is, surface protrusion was also observed before membrane tether extraction (Chen, 2008). For either suspended or attached human umbilical vein endothelial cells (HUVECs), their surface protrusions can be also described by the three-parameter solid model shown in Fig. 4, where $k_c = 0.15$ pN/nm, $\eta_c = 0.05$ pN s/nm, and $k_m = 0.87$ pN/nm, showing that HUVECs are much stiffer than leukocytes. These values were not affected by stimulating HUVECs with tumor necrosis factor-α (TNF-α) or interleukin-1β (IL-1β). For a Jurkat cell, which is an immortalized lymphocyte, $k_c = 0.017$ pN/nm, $\eta_c = 0.059$ pN s/nm, and $k_m = 0.276$ pN/nm (note that these parameters were converted from a slightly different model); for a Jurkat cell treated with higher Mg^{2+} concentration (5 mM Mg^{2+} with no Ca^{2+}), $k_c = 0.009$ pN/nm, $\eta_c = 0.06$ pN s/nm, and $k_m = 0.199$ pN/nm (Schmitz, Benoit, & Gottschalk, 2008). The decrease in k_c and k_m was very likely caused by the loss of microtubules and F-actin stress fibers (Prescott, Comerford, Magrath, Lamb, & Warn, 1988), not by any change in very late antigen-4 that was used as the force handle in this study. All aforementioned surface viscosities (η_c) are similar to each other, indicating similar mechanisms of energy dissipation in these cells. On the other hand, different initial elastic constants (k_m) may represent different densities of cytoskeletal components in them.

B. Crossover Force

The crossover force (F_c) is a critical parameter that dictates whether membrane tether extraction can occur. Over a wide range of force-loading rates (from 240 to 38,000 pN/s), F_c showed a behavior that is consistent with a weak chemical bond (Evans et al., 2005). However, this finding should not be interpreted as that F_c is governed by the receptor–cytoskeleton interaction alone. In fact, at small force-loading rates (20–400 pN/s), the relationship between F_c and the force-loading rate deviated from what was determined by Evans et al. (Fig. 5) (Xu & Shao, 2008), indicating different governing mechanisms at different force-loading rates. This is also true in ECs where the dependence of F_c on the force-loading rate showed a two-phase behavior,

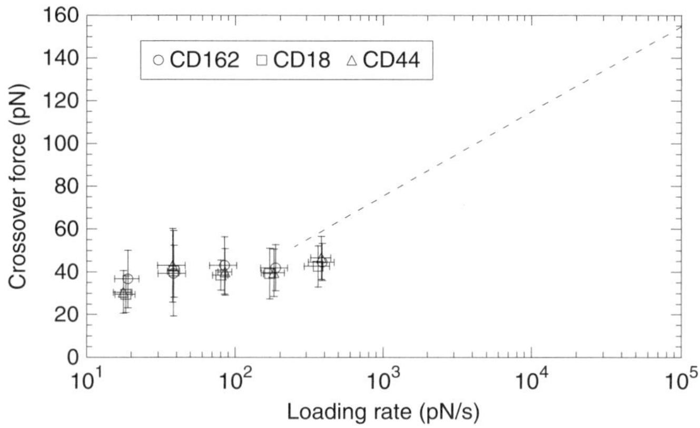

FIGURE 5 Dependence of the crossover force on the force-loading rate, measured with CD162, CD18, or CD44 as the force handle (adapted from Xu & Shao, 2008). The dashed line represents the measurements by Evans et al. (2005).

measured with the micropipette aspiration technique (Chen, 2008). In addition, the cytoplasmic tail of L-selectin contributed little to F_c at the force-loading rates of 10–4000 pN/s, so the receptor–cytoskeleton interaction does not play a dominant role in determining F_c when L-selectin is pulled (Yao & Shao, 2008b). Regardless of how F_c is governed, constant amount of deformation was obtained when the transition from surface protrusion to tether extraction occurred in ECs, showing that the transition was induced by the same structural cause (Chen).

Although it is still not completely clear what governs F_c, many factors may contribute, including receptor–cytoskeleton binding, membrane–cytoskeleton adhesion, and inherent membrane mechanical properties such as bending rigidity and tension. Membrane–cytoskeleton adhesion arises mainly due to some linker proteins like ezrin, moesin, α-actinin, and fimbrin, as well as some lipid molecules like phosphatidylinositol 4,5-biophosphate (PIP_2) that interact with the cytoskeleton (Berryman, Franck, & Bretscher, 1993; Bretscher, Reczek, & Berryman, 1997; Nebl, Oh, & Luna, 2000; Raucher et al., 2000), although one study found that direct interaction between F-actin and phospholipid was also possible (Le Bihan et al., 2005). In addition, the size of the membrane receptor where the pulling force is imposed may also contribute to F_c, which increased when more and more receptors were engaged in stretching a liposome surface (Koster, Cacciuto, Derenyi, Frenkel, & Dosterom, 2005).

C. Compressional Stiffness

During leukocyte rolling, blood flow also imposes compressive force to push the leukocyte against the EC. When a neutrophil was indented by a spherical force probe as used in the biomembrane force probe and the optical trap, its compressional stiffness was found to be very close to the protrusional stiffness, ranging from 0.054 to 0.07 pN/nm over a range of loading rates from ~0.1 to ~20 pN/s (Simon et al., 2007; Xu, 2006). These values are close to what can be estimated from the micropipette experiments (Girdhar, Chen, & Shao, 2007; Lomakina, Spillmann, King, & Waugh, 2004) and they are consistent with the model of neutrophil as a liquid drop with a cortical tension of ~0.025–0.035 pN/nm. The compressional stiffness of neutrophils decreased to 0.033 pN/nm after cytochalasin D treatment (Simon et al.) and increased to 0.25 pN/nm by 0.1 μM fMLP stimulation (Xu). For either suspended or attached HUVECs, their compressional stiffness (0.4–1 pN/nm over 10–1000 pN/s) was found to be consistently larger than their protrusional stiffness (Chen, 2008). At small force-loading rates, attached HUVECs were stiffer than suspended ones in compression, but this difference disappeared at high force-loading rates, indicating different arrangements of cytoskeletal components and possible effect of substrate rigidity.

IV. FLEXURAL STIFFNESS OF LEUKOCYTE MICROVILLI

When a leukocyte microvillus is subjected to a force tangential to the cell surface, it is going to be bent. Whereas many techniques can be easily applied to stretching leukocyte microvilli, none can be easily applied to imposing tangential forces because the force transducers will rotate as in the optical trap or cannot be maneuvered into a working position as in the micropipette aspiration technique. By coupling small fluorescent beads to the tips of microvilli and monitoring their thermal fluctuation, the flexural stiffness of microvilli was found to be 0.007 pN/nm for neutrophils and 0.004 pN/nm for lymphocytes (Yao & Shao, 2007, 2008a). These values should be only applicable to small deformation of microvilli since the microvillus tip was only displaced tens of nanometers during measurement. However, when a neutrophil microvillus was pressed down by a small bead for a large deformation of 500–1000 nm, a stiffness of 0.005 pN/nm was found, which is really close to 0.007 pN/nm (Simon et al., 2007). Therefore, it appears that the flexural stiffness of leukocyte microvilli is much smaller than its protrusional stiffness irrespective of small or large deformation. Besides, these measurements are consistent with the fact that a force of tens of piconewtons could easily press down microvilli as indicated by their natural length recovery after the force direction was reversed (Shao et al., 1998).

If conventional solid mechanics is applied to analyzing microvillus deformation, its flexural stiffness (K_F) is related to its protrusional stiffness (K_P) by

$$K_F = \frac{3R^2}{4L^2} K_P, \tag{1}$$

where R and L are the radius and length of the microvillus. For a K_P of 0.043 pN/nm, $K_F = 0.004$ pN/nm; for a K_P of 0.12 pN/nm, $K_F = 0.011$ pN/nm. Both estimates are not far off the measured values of 0.005 and 0.007 pN/nm. However, it should be recognized that this agreement may just be a coincidence because isotropy and homogeneity have to be assumed for microvilli and Eq. (1) cannot be applied otherwise. Interestingly, the flexural stiffness of an actin filament estimated using Eq. (1) and its longitudinal stiffness is also in very good agreement with its measured value (Howard, 2001), so it is tempting to think that necessity may be hidden behind this coincidence. If we assume that actin filaments are aligned inside microvilli along their longitudinal direction (Majstoravich et al., 2004), the flexural stiffness of microvilli may only represent a flexible connection between the microvillus and the cell body since it is much smaller than the flexural stiffness of even one actin filament with similar length. Therefore, it seems that leukocyte microvilli should be modeled mechanically as a rod with a ball-and-socket-like connection to the cytoskeleton.

V. MEMBRANE TETHER EXTRACTION

Once the pulling force exerted on a leukocyte or an EC exceeds the crossover force (F_c), a *membrane tether* is extracted. As shown in Fig. 6, a membrane tether is a cylindrical membrane tube with nanometer diameter. Figure 7 illustrates the terms used for describing single-, double-, and simultaneous tether extraction.

A. Single-Tether Extraction

When a membrane tether is extracted from a cell, cell membrane continues to flow from the cell body to the tether and the tether growth velocity (U_t) increases with the pulling force (F). Under physiological conditions, U_t falls in the range of 6–40 μm/s when neutrophils roll on P-selectin- or activated platelet-coated surfaces (Schmidtke & Diamond, 2000). This is consistent with the finding that neutrophils generally roll in a velocity range of 10–40 μm/s under postcapillary venular wall shear stress levels *in vitro* and *in vivo* (Jones, Smith, & McIntire, 1996). In a small range of U_t, F and U_t can be well described by the following equation (Shao & Hochmuth, 1996; Shao & Xu, 2002):

$$F = F_t + 2\pi\eta_{\text{eff}} U_t, \tag{2}$$

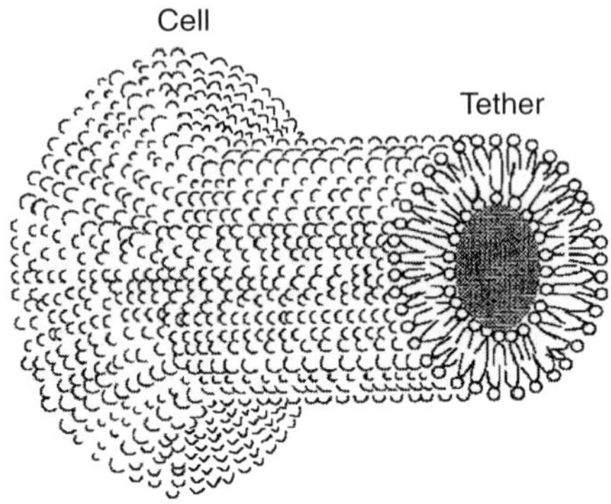

FIGURE 6 Schematic drawing of a membrane tether (cut in the middle of its length) extracted from a cell. Only a small part of the cell is shown. The diameter of the tether is typically tens of nanometers (adapted from Hochmuth et al., 1996).

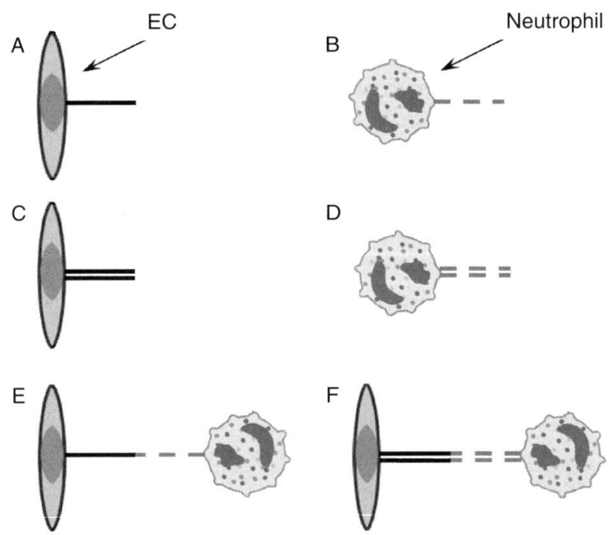

FIGURE 7 Six cases of tether extraction from ECs and neutrophils: single-tether extraction from an EC (A) or a neutrophil (B) alone; double-tether extraction from an EC (C) or a neutrophil (D) alone; and simultaneous single- (E) or double- (F) tether extraction from both ECs and neutrophils (adapted from Girdhar & Shao, 2007).

where F_t is the threshold force required for tether extraction to occur and η_{eff} is the effective viscosity. The range of U_t in which Eq. (2) is valid depends on the cell type and the magnitude of U_t. For ECs, this range is at least 20 μm/s starting at zero U_t.

The threshold force (F_t) and the crossover force (F_c) are inherently different because F_c includes the contribution from rupturing the receptor–cytoskeleton connection while F_t does not. In general, $F_t < F_c$ since receptor–cytoskeleton interaction is not expected to contribute to F_t, which is intrinsically determined by membrane tension, membrane bending, and membrane–cytoskeleton adhesion, whereas η_{eff} is determined by membrane viscosity, interbilayer slip, and membrane slip over the cytoskeleton (Evans & Yeung, 1994; Hochmuth, Shao, Dai, & Sheetz, 1996). The relationship shown in Eq. (2) is based on the finding that tethers lack any cytoskeletal support (Berk & Hochmuth, 1992; Raucher et al., 2000). It has been shown that Eq. (2) can be applied for single-tether extraction from many cell types, including neutrophils, lymphocytes, ECs, lipid vesicles, erythrocytes, neuronal growth cones, outer hair cells, HL-60, Chinese hamster ovary (CHO) cells, and K562 cells (Chen, Girdhar, & Shao, 2007; Dai & Sheetz, 1995; Evans & Yeung; Girdhar & Shao, 2004; Hochmuth et al.; Li et al., 2002; Marcus, McEver, & Zhu, 2004; Shao & Hochmuth, 1996; Waugh & Bauserman, 1995; Yeung, 1994).

For single-tether extraction from passive neutrophils, $F_t = 45$ pN and $\eta_{eff} = 1.8$ pN s/μm (Shao & Hochmuth, 1996). η_{eff} was decreased by raising temperature, whereas F_t was not affected much (Liu, Goergen, & Shao, 2007). For single-tether extraction from neutrophils activated with interleulin-8 or phorbol myristate acetate, $F_t = 86$ pN and $\eta_{eff} = 0.32$ pN s/μm (Shao & Xu, 2002). This implies that, if a neutrophil rolls on the endothelium at a constant velocity between 6 and 40 μm/s and becomes activated gradually, the pulling force required for tether extraction at this velocity will gradually decrease, hence stabilizing rolling. F_t and η_{eff} can also be affected by ethanol treatment or cholesterol depletion (Edmondson, Denney, & Diamond, 2005; Oh & Diamond, 2008). When a point force is exerted on an EC surface with the micropipette aspiration technique, a membrane tether can be extracted with $F_t = 50$ pN and $\eta_{eff} = 0.5$ pN s/μm (Chen et al., 2007; Girdhar & Shao, 2004). Stimulation of ECs with IL-1β and TNF-α did not change these parameters (Chen et al.), but cholesterol depletion or enrichment did (Sun et al., 2007).

B. Double-, Multiple-, and Simultaneous Tether Extraction

During the rolling of leukocytes on P-selectin-coated surfaces, single-bond adhesion events could only be achieved when the concentration of P-selectin was a few sites per μm^2 (Alon et al., 1995). *In vivo*, the stimulated endothelium likely has a higher P-selectin concentration on its surface in addition to other molecules that like E-selectin mediate leukocyte rolling (Panes & Granger,

1998; Rodgers, Camphausen, & Hammer, 2000). Therefore, double- or multiple-bond adhesion events are very likely during the rolling of leukocytes on the endothelium. If these bonds are not close to each other, double or multiple tethers can be extracted, as has been observed with the flow chamber technique, especially at high shear stress (Park et al., 2002; Ramachandran, Williams, Yago, Schmidtke, & McEver, 2004; Schmidtke & Diamond, 2000). For double- or multiple-tether extraction, the overall F–U_t relationship may not be simply the sum of all tethers because the competition for membrane materials might occur. Thus, understanding the relationship between F and U_t for double- or multiple-tether extraction should be beneficial to interpreting experimental findings and understanding leukocyte rolling.

For double-tether extraction from neutrophils or ECs, both F_t and η_{eff} doubled, so these tethers were independent of each other mechanically (Girdhar et al., 2007; Xu & Shao, 2005). Multiple-tether extraction from ECs was observed with the atomic force microscopy (Sun et al., 2005). Even when six tethers were ruptured one by one from the tip of an AFM cantilever, the force drop corresponding to each rupture event was almost the same, indicating that all of them were independent of each other mechanically. Therefore, double or multiple tethers during leukocyte rolling can be considered individually and independently. However, we do not expect that this conclusion holds forever with increasing tether numbers because of limited availability of membrane materials. Besides, tethers might coalesce when they become too close to each other structurally, which would significantly affect the F–U_t relationship.

During leukocyte rolling on the endothelium, the pulling force due to blood flow is equally exerted on both the leukocyte and the EC. As a result, the crossover forces of both cells may be overcome, then two tethers with the adhesive bond in the middle can be extracted (Fig. 8). If the crossover force of either the leukocyte or EC is appreciably larger than the other at comparable force-loading rates, simultaneous tether extraction would be very difficult. However, when a neutrophil or lymphocyte first adhered to an EC and was then separated from it, simultaneous tether extraction was observed (Girdhar & Shao, 2007), showing the two crossover forces can indeed be overcome. Because of the smaller effective viscosity for tether extraction from ECs, the tether from the EC grows faster than the one from the leukocyte. Consequently, the EC tether contributes more to the composite tether length during simultaneous tether extraction. Compared with tether extraction from either a leukocyte or an EC alone, simultaneous tether extraction further decreases the dislodging force of the adhesive bond, thus stabilizing rolling by increasing the adhesion lifetime even more (Yu & Shao, 2007).

FIGURE 8 Simultaneous tether extraction from a leukocyte and an endothelial cell with the adhesive bond in between (adapted from Girdhar & Shao, 2004).

C. Constitutive Equation

In a large range of U_t, F does not depend on U_t linearly any more and their relationship can be better described by a shear-thinning power-law constitutive relationship (Heinrich, Leung, & Evans, 2005):

$$F = \alpha U_t^{1-\beta}, \tag{3}$$

where α and β define the weak dependence of η_{eff} on the tether growth velocity U_t. For neutrophils, $\alpha = 60$ pN (s/μm)$^{0.25}$ and $\beta = 0.75$ (Heinrich et al., 2005). More recently, a theoretical analysis showed that $\beta = 2/3$ at high tether growth velocities where viscous dissipation dominates membrane tension and membrane–cytoskeleton adhesion (Brochard-Wyart, Borghi, Cuvelier, & Nassoy, 2006). Although this theoretical analysis provided a physical foundation for Eq. (3), there are other issues with this equation that remain unresolved. According to Eq. (3), F_t (the force at zero U_t) is equal to zero; besides, Eq. (3) is meaningless mathematically if U_t is negative. These are in direct conflict with our experimental result from ECs that $F_t \neq 0$ (Chen, 2008) and our simulation result that U_t may be negative during simultaneous tether extraction when leukocytes roll on the endothelium (unpublished observation). In addition, it has been shown theoretically (Evans & Yeung, 1994; Hochmuth et al., 1996) and experimentally (Heinrich et al.) that $F_t \neq 0$ because of membrane tension, membrane bending, and membrane–cytoskeleton adhesion. The relationship between F and U_t when U_t is negative describes how an extracted tether shortens when F is less than F_t. For small U_t, Brochard-Wyart et al. successfully predicted that $F_t \neq 0$, but their equation cannot be applied when U_t is negative. This is because, according to their model, $U_t = 0$ when $F = 0$, which is not true because an extracted tether shortens back to the cell body once F is removed (Girdhar & Shao, 2007).

Recently, a new constitutive equation for membrane tether extraction was proposed in the form of (Chen, 2008)

$$F = \alpha(U_t - U_0)^{\beta}, \tag{4}$$

where U_0 is the tether velocity when $F = 0$. According to Eq. (4),

$$F_t = \alpha(-U_0)^{\beta}. \tag{5}$$

Equation (4) captures all known characteristics of tether extraction, including nonlinearity, nonzero threshold force, and possible negative tether velocity. Its comparison with Eqs. (2) and (3) is shown in Fig. 9. For neutrophils and ECs, $\alpha = 63.76$ and 26.54 pN (s/μm)$^{\beta}$, $\beta = 0.24$ and 0.46, and $U_0 = -0.29$ and -5.11 μm/s, respectively (Chen, 2008). The threshold forces calculated with Eq. (5) are 47 and 56 pN, respectively, for neutrophils and ECs. It should be pointed out that, in regard to negative U_t, Eq. (4) can only be applied to the case when the pulling force on the tether is gradually decreased. Equations (2) and (3) do not appear to be dependent on the force-loading rate. Whether Eq. (4) will depend on the force-loading rate, especially when U_t is negative, remains to be seen.

D. Tether Retraction and Coalescence

Since simultaneous tethers can be extracted during leukocyte rolling and the force acting on the tethers decreases over time as the leukocyte rolls on the endothelium, the force may eventually fall below one of the threshold forces

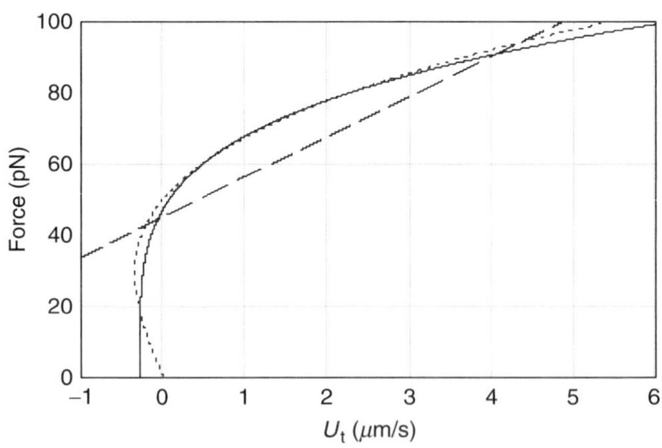

FIGURE 9 The constitutive relationship between the pulling force (F) and the tether growth velocity (U_t) for neutrophils. Solid line: Eq. (4); dotted line: Eq. (3); dashed line: Eq. (2).

for tether extraction, then the tether from that cell will retract and shorten. Although $F_t = 45$ and 50 pN for neutrophils and ECs, respectively, these values fluctuate for each individual cell, so the tether from the EC does not always have the larger F_t. When the threshold force for an EC is smaller than the one for a leukocyte because of heterogeneity, the tether from the leukocyte will retract first. If the force on the tether gradually decreased, the decrease in U_t can be described by Eq. (4) (unpublished observation). However, if the force on the tether is abruptly removed as in the case of adhesive bond rupture, tether retraction showed a two-phase behavior: first a fast retraction due to elastic energy stored in the tether, then a slow retraction due to membrane flow back to the cell body (Liu, 2008). This can be empirically described by

$$L_t = V_i t_0 e^{-t/t_0} - V_s t + L_0, \qquad (6)$$

where L_t, V_i, t_0, V_s, and L_0 are the tether length, the initial retraction velocity, the exponential decay time constant, the steady retraction velocity, and the residual length. V_i can be very large, reaching $\sim$100 μm/s when U_t was only $\sim$5 μm/s. In contrast, tethers extracted from liposomes retracted at more or less constant velocities after the force removal (Rossier et al., 2003).

While tethers extracted from lipid vesicles can join each other (coalescence) once the angle between them becomes small enough and their lengths become long enough (Cuvelier, Derényi, Bassereau, & Nassoy, 2005), tethers from neutrophils were rarely observed joining each other (Liu, 2008). This illustrates another major difference (other than tether retraction) between tether extraction from cells and liposomes. Tethers extracted from liposomes are mobile (Evans, Bowman, Leung, Needham, & Tirrell, 1996), whereas tethers extracted from live cells are not (Liu). This is in large part due to the membrane–cytoskeleton interaction in cells. Even when tethers were extracted tangentially to the cell surface with the microcantilever technique, the cell–tether junction did not move and the relationship between F and U_t did not change (Liu; Liu, Yu, Yao, & Shao, 2009). Therefore, even when tethers are not extracted perpendicularly to cell surfaces as occurs likely *in vivo*, Eq. (4) is still valid.

VI. IMPACT OF SURFACE PROTRUSION AND TETHER EXTRACTION ON LEUKOCYTE ROLLING

Since the existence of tethers during leukocyte rolling was predicted with the micropipette aspiration technique (Shao & Hochmuth, 1996; Shao et al., 1998), it has been confirmed several times with the flow chamber technique (Park et al., 2002; Ramachandran et al., 2004; Schmidtke & Diamond, 2000). The impact of surface protrusion and membrane tether extraction on leukocyte rolling has been studied both experimentally and numerically.

On P-selectin-coated surfaces, normal leukocytes rolled more stably than fixed leukocytes, which rolled more stably than PSGL-1-coated beads (Park et al.). Leukocytes can change their tether number and structure dynamically in response to varying shear stress as they roll (Ramachandran et al.). Treating leukocytes with cytochalasin can make tether extraction easier, rendering leukocyte rolling more stable and resistant to detachment (Yago et al., 2002). On the other hand, PSGL-1-coated microspheres rolled more stably on normal ECs than on fixed ECs (Yu, 2008). Therefore, surface protrusion and tether extraction of both leukocytes and ECs can definitely increase rolling stability.

On another front, simulation of leukocyte rolling on the endothelium has greatly improved our understanding of this complex process, especially the roles of various contributors to the rolling stability. Because the data on tether extraction from ECs has only become available recently, it has not been incorporated into many models to date, but surface protrusion and membrane tether extraction of leukocytes have been incorporated into quite a few models of leukocyte rolling. With numerical simulation, it was found that deformable microvilli definitely contribute to rolling stability (Pospieszalska & Ley, 2009; Yu & Shao, 2007). The deformable leukocyte surface is more important for rolling stability at high shear stress and high densities of P-selectin on substrates (Pawar, Jadhav, Eggleton, & Konstantopoulos, 2008). During later rolling stages, large contact area may develop between leukocytes and the endothelium. Large cellular deformation even occurs when cells collide with each other at venous shear rate and the resulted large contact area may be responsible for increased leukocyte arrest (Kadash, Lawrence, & Diamond, 2004). However, leukocyte rolling stability does not always translate into more effective leukocyte arrest (Oh & Diamond, 2008). Nevertheless, it has become clear that both surface protrusion and tether extraction facilitate leukocyte rolling on the endothelium by lengthening the lifetime of adhesive bonds that mediate rolling (Edmondson et al., 2005; Yu & Shao).

VII. CONCLUDING REMARKS

Although much has been learned about how surface protrusion and membrane tether extraction may affect leukocyte rolling on the endothelium, many questions remain unanswered. For example, although simultaneous tether extraction has been shown to exist with the micropipette aspiration technique (Girdhar & Shao, 2007), it has never been shown under flow conditions and it has not been incorporated in most models. What governs the transition from surface protrusion to membrane tether extraction? Under force, how do cytoplasmic tails of selectins and their ligands interact kinetically with the cytoskeleton? Can Eq. (4), the constitutive equation for tether extraction, be derived theoretically and how is it affected by activating agents like thrombin? Addressing these questions will provide us fundamental knowledge on leukocyte rolling, hence helping us understand leukocyte extravasation and its related diseases.

Acknowledgments

This work was supported by NIH grants R01 HL069947 and R21/R33 RR017014.

References

Alon, R., Hammer, D. A., & Springer, T. A. (1995). Lifetime of the P-selectin-carbohydrate bond and its response to tensile force in hydrodynamic flow. *Nature, 374*, 539–542.

Berk, D. A., & Hochmuth, R. M. (1992). Lateral mobility of integral proteins in red blood cell tethers. *Biophysical Journal, 61*, 9–18.

Berryman, M., Franck, Z., & Bretscher, A. (1993). Ezrin is concentrated in the apical microvilli of a wide variety of epithelial cells whereas moesin is found primarily in endothelial cells. *Journal of Cell Science, 105*, 1025–1043.

Bretscher, A., Reczek, D., & Berryman, M. (1997). Ezrin: A protein requiring conformational activation to link microfilaments to the plasma membrane in the assembly of cell surface structures. *Journal of Cell Science, 110*, 3011–3018.

Brochard-Wyart, F., Borghi, N., Cuvelier, D., & Nassoy, P. (2006). Hydrodynamic narrowing of tubes extruded from cells. *Proceedings of the National Academy of Sciences of the United States of America, 103*, 7660–7663.

Bruehl, R. E., Springer, T. A., & Bainton, D. F. (1996). Quantitation of L-selectin distribution on human leukocyte microvilli by immunogold labeling and electron microscopy. *Journal of Histochemistry and Cytochemistry, 44*, 835–844.

Chen, Y. (2008). *Mechanics of surface protrusion and tether extraction from endothelial cells.* Saint Louis, MO: Washington University.

Chen, Y., Girdhar, G., & Shao, J. Y. (2007). Single membrane tether extraction from adult and neonatal dermal microvascular endothelial cells. *American Journal of Physiology: Cell Physiology, 292*, C1272–C1279.

Cuvelier, D., Derényi, I., Bassereau, P., & Nassoy, P. (2005). Coalescence of membrane tethers: Experiments, theory, and applications. *Biophysical Journal, 88*, 2714–2726.

Dai, J., & Sheetz, M. P. (1995). Mechanical properties of neuronal growth cone membrane studied by tether formation with laser optical tweezers. *Biophysical Journal, 68*, 988–996.

Edmondson, K. E., Denney, W. S., & Diamond, S. L. (2005). Neutrophil–bead collision assay: Pharmacologically induced changes in membrane mechanics regulate the PSGL-1/P-selectin adhesion lifetime. *Biophysical Journal, 89*, 3603–3614.

Erlandsen, S. L., Hasslen, S. R., & Nelson, R. D. (1993). Detection and spatial distribution of the β_2 integrin (Mac-1) and L-selectin (LECAM-1) adherence receptors on human neutrophils by high-resolution field emission SEM. *Journal of Histochemistry and Cytochemistry, 41*, 327–333.

Evans, E., Bowman, H., Leung, A., Needham, D., & Tirrell, D. (1996). Biomembrane templates for nanoscale conduits and networks. *Science, 273*, 933–935.

Evans, E., Heinrich, V., Leung, A., & Kinoshita, K. (2005). Nano- to microscale dynamics of P-selectin detachment from leukocyte interfaces. I. Membrane separation from the cytoskeleton. *Biophysical Journal, 88*, 2288–2298.

Evans, E., & Yeung, A. (1994). Hidden dynamics in rapid changes of bilayer shape. *Chemistry and Physics of Lipids, 73*, 39–56.

Finger, E. B., Bruehl, R. E., Bainton, D. F., & Springer, T. A. (1996a). A differential role for cell shape in neutrophil tethering and rolling on endothelial selectins under flow. *Journal of Immunology, 157*, 5085–5096.

Finger, E. B., Puri, K. D., Alon, R., Lawrence, M. B., von Andrian, U. H., & Springer, T. A. (1996b). Adhesion through L-selectin requires a threshold hydrodynamic shear. *Nature, 379*, 266–269.

Flugge, W. (1975). *Viscoelasticity* (2nd Ed.). Berlin: Springer-Verlag.

Girdhar, G., Chen, Y., & Shao, J. Y. (2007). Double tether extraction from human umbilical vein and dermal microvascular endothelial cells. *Biophysical Journal, 92*, 1035–1045.

Girdhar, G., & Shao, J. Y. (2004). Membrane tether extraction from human umbilical vein endothelial cells and its implication in leukocyte rolling. *Biophysical Journal, 87*, 3561–3568.

Girdhar, G., & Shao, J. Y. (2007). Simultaneous tether extraction from endothelial cells and leukocytes: Observation, mechanics and significance. *Biophysical Journal, 93*, 4041–4052.

Goldman, A. J., Cox, R. G., & Brenner, H. (1967). Slow viscous motion of a sphere parallel to a plane wall. II. Couette flow. *Chemical Engineering Science, 22*, 653–660.

Heinrich, V., Leung, A., & Evans, E. (2005). Nano- to microscale dynamics of P-selectin detachment from leukocyte interfaces. II. Tether flow terminated by P-selectin dissociation from PSGL-1. *Biophysical Journal, 88*, 2299–2308.

Hochmuth, R. M., Shao, J. Y., Dai, J., & Sheetz, M. P. (1996). Deformation and flow of membrane into tethers extracted from neuronal growth cones. *Biophysical Journal, 70*, 358–369.

Howard, J. (2001). *Mechanics of motor proteins and the cytoskeleton.* Sunderland, MA: Sinauer Associates.

Jones, D. A., Smith, C. W., & McIntire, L. V. (1996). Leucocyte adhesion under flow conditions: Principles important in tissue engineering. *Biomaterials, 17*, 337–347.

Kadash, K. E., Lawrence, M. B., & Diamond, S. L. (2004). Neutrophil string formation: Hydrodynamic thresholding and cellular deformation during cell collisions. *Biophysical Journal, 86*, 4030–4039.

Koster, G., Cacciuto, A., Derenyi, I., Frenkel, D., & Dosterom, M. (2005). Force barriers for membrane tube formation. *Physical Review Letters, 94*, 068101.

Kunkel, E. J., Dunne, J. L., & Ley, K. (2000). Leukocyte arrest during cytokine-dependent inflammation *in vivo*. *Journal of Immunology, 164*, 3301–3308.

Lawrence, M. B., Kansas, G. S., Kunkel, E. J., & Ley, K. (1997). Threshold levels of fluid shear promote leukocyte adhesion through selectins (CD62L, P, E). *Journal of Cell Biology, 136*, 717–727.

Le Bihan, T., Pelletier, D., Tancrede, P., Heppell, B., Chauvet, J. P., & Gicquaud, C. R. (2005). Effect of the polar headgroup of phospholipids on their interaction with actin. *Journal of Colloid and Interface Science, 288*, 88–96.

Li, Z., Anvari, B., Takashima, M., Brecht, P., Torres, J. H., & Brownell, W. E. (2002). Membrane tether formation from outer hair cells with optical tweezers. *Biophysical Journal, 82*, 1386–1395.

Liu, B. (2008). *Mechanics of tether extraction, surface protrusion, and cortical tension of human neutrophils and extension of single DNA molecules.* Saint Louis, MO: Washington University.

Liu, B., Goergen, C. J., & Shao, J. Y. (2007). Effect of temperature on tether extraction, surface protrusion, and cortical tension of human neutrophils. *Biophysical Journal, 93*, 2923–2933.

Liu, B., Yu, Y., Yao, D. K., & Shao, J. Y. (2009). A direct micropipette-based calibration method for AFM cantilevers. *Review of Scientific Instruments, 80*, 065109.

Lomakina, E. B., Spillmann, C. M., King, M. R., & Waugh, R. E. (2004). Rheological analysis and measurement of neutrophil indentation. *Biophysical Journal, 87*, 4246–4258.

Majstoravich, S., Zhang, J., Nicholson-Dykstra, S., Linder, S., Friedrich, W., Siminovitch, K. A., et al. (2004). Lymphocyte microvilli are dynamic, actin-dependent structures that do not require Wiskott–Aldrich syndrome protein (WASp) for their morphology. *Blood, 104*, 1396–1403.

Marcus, W. D., McEver, R. P., & Zhu, C. (2004). Forces required to initiate membrane tether extrusion from cell surface depend on cell type but not on the surface molecule. *Mechanics & Chemistry of Biosystems, 1*, 245–251.

Moore, K. L., Patel, K. D., Bruehl, R. E., Li, F., Johnson, D. A., Lichenstein, H. S., et al. (1995). P-selectin glycoprotein ligand-1 mediates rolling of human neutrophils on P-selectin. *Journal of Cell Biology, 128*, 661–671.

Nebl, T., Oh, S. W., & Luna, E. J. (2000). Membrane cytoskeleton: PIP2 pulls the strings. *Current Biology, 10*, R351–R354.

Oh, H., & Diamond, S. L. (2008). Ethanol enhances neutrophil membrane tether growth and slows rolling on P-selectin but reduces capture from flow and firm arrest on IL-1-treated endothelium. *Journal of Immunology, 181*, 2472–2482.

Panes, J., & Granger, D. N. (1998). Leukocyte–endothelial cell interactions: Molecular mechanisms and implications in gastrointestinal disease. *Gastroenterology, 114*, 1066–1090.

Park, E. Y., Smith, M. J., Stropp, E. S., Snapp, K. R., DiVietro, J. A., Walker, W. F., et al. (2002). Comparison of PSGL-1 microbead and neutrophil rolling: Microvillus elongation stabilizes P-selectin bond clusters. *Biophysical Journal, 82*, 1835–1847.

Pawar, P., Jadhav, S., Eggleton, C. D., & Konstantopoulos, K. (2008). Roles of cell and microvillus deformation and receptor–ligand binding kinetics in cell rolling. *American Journal of Physiology: Heart and Circulatory Physiology, 295*, H1439–H1450.

Pospieszalska, M., & Ley, K. (2009). Dynamics of microvillus extension and tether formation in rolling leukocytes. *Cellular and Molecular Bioengineering, 2*, 207–217.

Prescott, A. R., Comerford, J. G., Magrath, R., Lamb, N. J., & Warn, R. M. (1988). Effects of elevated intracellular magnesium on cytoskeletal integrity. *Journal of Cell Science, 89*, 321–329.

Ramachandran, V., Williams, M., Yago, T., Schmidtke, D. W., & McEver, R. P. (2004). Dynamic alterations of membrane tethers stabilize leukocyte rolling on P-selectin. *Proceedings of the National Academy of Sciences of the United States of America, 101*, 13519–13524.

Raucher, D., Stauffer, T., Chen, W., Shen, K., Guo, S., York, J. D., et al. (2000). Phosphatidylinositol 4,5-bisphosphate functions as a second messenger that regulates cytoskeleton–plasma membrane adhesion. *Cell, 100*, 221–228.

Rodgers, S. D., Camphausen, R. T., & Hammer, D. A. (2000). Sialyl LewisX-mediated, PSGL-1-independent rolling adhesion on P-selectin. *Biophysical Journal, 79*, 694–706.

Ross, R. (1993). The pathogenesis of atherosclerosis: A perspective for the 1990s. *Nature, 362*, 801–809.

Ross, R. (1999). Atherosclerosis—An inflammatory disease. *The New England Journal of Medicine, 340*, 115–126.

Rossier, O., Cuvelier, D., Borghi, N., Puech, P. H., Derenyi, I., Buguin, A., et al. (2003). Giant vesicles under flows: extrusion and retraction of tubes. *Langmuir, 19*, 575–584.

Schmidtke, D. W., & Diamond, S. L. (2000). Direct observation of membrane tethers formed during neutrophil attachment to platelets or P-selectin under physiological flow. *Journal of Cell Biology, 149*, 719–729.

Schmitz, J., Benoit, M., & Gottschalk, K. E. (2008). The viscoelasticity of membrane tethers and its importance for cell adhesion. *Biophysical Journal, 95*(3), 1448–1459, (Abstract).

Serhan, C. N., & Savill, J. (2005). Resolution of inflammation: The beginning programs the end. *Nature Immunology, 6*, 1191–1197.

Shao, J. Y., & Hochmuth, R. M. (1996). Micropipette suction for measuring piconewton forces of adhesion and tether formation from neutrophil membranes. *Biophysical Journal, 71*, 2892–2901.

Shao, J. Y., Ting-Beall, H. P., & Hochmuth, R. M. (1998). Static and dynamic lengths of neutrophil microvilli. *Proceedings of the National Academy of Sciences of the United States of America, 95*, 6797–6802.

Shao, J. Y., & Xu, J. (2002). A modified micropipette aspiration technique and its application to tether formation from human neutrophils. *Journal of Biomechanical Engineering, 124*, 388–396.

Simon, S. I., & Green, C. E. (2005). Molecular mechanics and dynamics of leukocyte recruitment during inflammation. *Annual Review of Biomedical Engineering, 7*, 151–185.

Simon, S., Nyunt, T., Florine-Casteel, K., Ritchie, K., Ting-Beall, H., Evans, E., et al. (2007). Dynamics of neutrophil membrane compliance and microstructure probed with a micropipet-based piconewton force transducer. *Annals of Biomedical Engineering, 35*, 595–604.

Smeeth, L., Cook, C., Thomas, S., Hall, A. J., Hubbard, R., & Vallance, P. (2006). Risk of deep vein thrombosis and pulmonary embolism after acute infection in a community setting. *Lancet, 367*, 1075–1079.

Springer, T. A. (1990). Adhesion receptors of the immune system. *Nature, 346*, 425–434.

Springer, T. A. (1995). Traffic signals on endothelium for lymphocyte recirculation and leukocyte emigration. *Annual Review of Physiology, 57*, 827–872.

Sun, M. Z., Graham, J. S., Hegedus, B., Marga, F., Zhang, Y., Forgacs, G., et al. (2005). Multiple membrane tethers probed by atomic force microscopy. *Biophysical Journal, 89*, 4320–4329.

Sun, M., Northup, N., Marga, F., Huber, T., Byfield, F. J., Levitan, I., et al. (2007). The effect of cellular cholesterol on membrane–cytoskeleton adhesion. *Journal of Cell Science, 120*, 2223–2231.

Von Andrian, U. H., Hasslen, S. R., Nelson, R. D., Erlandsen, S. L., & Butcher, E. C. (1995). A central role for microvillous receptor presentation in leukocyte adhesion under flow. *Cell, 82*, 989–999.

Waugh, R. E. (2009). Membrane tethers. In K. Ley (Ed.), *Leukocyte adhesion*. Amsterdam: Elsevier.

Waugh, R. E., & Bauserman, R. G. (1995). Physical measurements of bilayer–skeletal separation forces. *Annals of Biomedical Engineering, 23*, 308–321.

Xu, G. (2006). *Mechanics of single microvilli and double tethers of human neutrophils and lymphocytes*. Saint Louis, MO: Washington University.

Xu, G., & Shao, J. Y. (2005). Double tether extraction from human neutrophils and its comparison with CD4+ T-lymphocytes. *Biophysical Journal, 88*, 661–669.

Xu, G., & Shao, J. Y. (2008). Human neutrophil surface protrusion under a point load: Location independence and viscoelasticity. *American Journal of Physiology: Cell Physiology, 295*, C1434–C1444.

Yago, T., Leppanen, A., Qiu, H., Marcus, W. D., Nollert, M. U., Zhu, C., et al. (2002). Distinct molecular and cellular contributions to stabilizing selectin-mediated rolling under flow. *Journal of Cell Biology, 158*, 787–799.

Yao, D. K., & Shao, J. Y. (2007). Flexibility of single microvilli on live neutrophils and lymphocytes. *Physical Review E, 76*, 021907.

Yao, D. K., & Shao, J. Y. (2008a). A novel technique of quantifying flexural stiffness of rod-like structures. *Cellular and Molecular Bioengineering, 1*, 75–83.

Yao, D. K., & Shao, J. Y. (2008b). *Effects of the cytoplasmic tail of L-selectin on crossover force and tether extraction* (p. P2.64). *The Abstracts of the 2008 Annual Fall Meeting of the Biomedical Engineering Society,* Saint Louis, MO.
Yeung, A. K. C. (1994). *Mechanics of inter-monolayer coupling in fluid surfactant bilayers.* Ph.D. Thesis, University of British Columbia.
Yu, Y. (2008). *Numerical and experimental study of tether extraction from endothelial cells.* Saint Louis, MO: Washington University.
Yu, Y., & Shao, J. Y. (2007). Simultaneous tether extraction contributes to neutrophil rolling stabilization: A model study. *Biophysical Journal, 92*, 418–429.

CHAPTER 3

The Cytoskeleton and Deformability of White Blood Cells

Damir B. Khismatullin
Department of Biomedical Engineering, Tulane University, New Orleans, Louisiana 70118, USA

I. OVERVIEW

White blood cells (WBCs), also known as leukocytes, migrate to sites of infection to destroy pathogenic microorganisms. The ability of WBCs to deform is essential for this function but it is also an important determinant

Current Topics in Membranes, Volume 64

1063-5823/09 $35.00
DOI: 10.1016/S1063-5823(09)64003-5

of healthy vasculature. This chapter analyzes the effects of leukocyte deformability on leukocyte–endothelial interactions, presents the evidence for the critical role of the cytoskeleton in bulk mechanical properties of leukocytes, summarizes recent advances in rheological measurements of leukocytes, and discusses pathologies associated with leukocyte activation and reduced deformability of these cells.

II. INTRODUCTION

WBCs (leukocytes) are the cellular elements of the immune system in humans and other animals. Their primary function is to patrol the body for potential sources of infection and destroy invading pathogens, such as viruses, bacteria, fungi, and parasitic microorganisms. Leukocytes derive from pluripotent hematopoietic stem cells in the bone marrow and are specialized into four morphologically different cell types of myeloid lineage (neutrophils, eosinophils, basophils, and monocytes) and lymphoid cells that include natural killer (NK) cells and T- and B-lymphocytes (Janeway, Travers, Walport, & Shlomchik, 2005).

Most of our knowledge about mechanical and adhesive properties of leukocytes comes from *in vitro* and *in vivo* investigations of mature neutrophils. These cells make up more than 50% of all leukocytes in human blood and can be easily isolated from blood for *in vitro* analysis (Quinn, DeLeo, & Bokoch, 2007). Neutrophils play a central role in acute inflammation, which is the physiological reaction of the body to infection or tissue injury. They circulate in blood but penetrate through the walls of postcapillary or small collecting venules when they contact vascular endothelial cells activated by proinflammatory mediators (cytokines, chemokines, and bacterial peptides). Then, neutrophils migrate to the extravascular sites of infection where they engulf bacteria or fungi and destroy these microorganisms by releasing toxic products located in their cytoplasmic granules. Both neutrophil adhesiveness and deformability are essential for trafficking of these cells to sites of infection. Neutrophil rolling and arrest (firm adhesion) on vascular endothelium are prerequisite steps for neutrophil transendothelial migration (diapedesis) (Ley, 1996). These events occur through binding of specific cell adhesion molecules expressed on the surface of neutrophils to their counterparts on endothelial cells provided the total binding force overcomes the hydrodynamic force on the neutrophil. These adhesion molecules include selectins (L-selectin on leukocytes, P- and E-selectins on endothelial cells) and their ligands (e.g., PSGL-1) as well the members of the integrin (β_2- and α_4-integrins) and immunoglobulin superfamilies (ICAM-1, VCAM-1) (Ley, 2002). Neutrophil diapedesis, then its chemotaxis (directed migration toward the source of infection), and finally phagocytosis are all associated with significant deformation of the cell. In addition, as we will review in this chapter, neutrophil deformability significantly contributes to its rolling and firm adhesion and is an important determinant of microvascular resistance to blood flow. All these indicate that understanding the leukocyte behavior requires knowledge of the leukocyte mechanical properties.

Active migration through and in tissues is a characteristic of all types of WBCs. Eosinophils, which, together with neutrophils and basophils, form a class of granulocytes, migrate to the extravascular sites to combat parasites and large foreign cells. Granulocytes play an important role in allergy and asthma. Basophils' granules contain histamine which is released when these cells encounter allergens (Janeway et al., 2005). Histamine in turn triggers the inflammatory response (which is in this case the allergic reaction) by upregulation of cell adhesion molecules for neutrophils and eosinophils (specifically, P-selectin) on the endothelial cell surface (Asako et al., 1994; Kubes & Kanwar, 1994; Yamaki et al., 1998). When fully matured, human neutrophils, eosinophils, and basophils have a multilobed nucleus and, therefore, are called polymorphonuclear cells (PMNs).

Monocytes are the largest leukocytes circulating in blood. They do not have cytoplasmic granules and their nucleus is not segmented. The migration of monocytes through a vascular wall is a part of the normal process that leads to differentiation of these cells into macrophages and inflammatory dendritic cells. However, if this migration and differentiation occurs in a large artery where the tunica intima may contain oxidized low-density lipoproteins (ox-LDL), macrophages engulf ox-LDL and transform into foam cells, leading to the development of an atherosclerotic plaque (Ross, 1999). It should be noted that macrophages are big phagocytes ($\sim$21 μm in diameter) that ingest cellular debris, foreign cells, and various pathogenic microorganisms in the tissues. Pathogens can be destroyed by fusion with a macrophage lysosome that contains toxic peroxides.

NK cells defend the body against tumor and virally infected cells. B- and T-lymphocytes are cells of the adaptive immune system. The B-lymphocytes differentiate into plasma cells which make antibodies against antigens. The majority of T-lymphocytes are helper T cells that activate other immune cells and cytotoxic T cells that, together with NK cells, combat viruses and tumors. Both B- and T-lymphocytes can also differentiate into long-lived memory cells that can recognize previously encountered antigens (Janeway et al., 2005). To patrol the body for foreign antigens, B and T cells recirculate between the cardiovascular and lymphatic systems. Lymphocyte homing, that is, the migration of these cells form blood to lymph nodes, preferentially occur at high endothelial venules (HEVs), which are postcapillary venules found in lymphoid tissues (Butcher & Picker, 1996; Girard & Springer, 1995).

Although lymphocyte homing is characterized by the same steps as neutrophil trafficking (specifically, rolling, firm adhesion, and diapedesis), there are differences in the type and role of cell adhesion molecules involved in these processes (Sackstein, 2005). Lymphocyte homing receptors include L-selectin (note that neutrophil L-selectin is important for neutrophil capture but not rolling), CD44, and integrins. These molecules interact with addressins expressed on HEVs (Berg et al., 1989). The differences in the structural organization between various types of WBCs make it questionable whether all aspects of leukocyte–endothelial cell interactions can be predicted from the properties of human neutrophils. We need a thorough biomechanical analysis of each major type of leukocytes to understand better leukocyte adhesion and migration in health and disease.

In this chapter, our goals are to understand the role of leukocyte deformation in leukocyte–endothelial cell interactions, analyze the relationship between leukocyte deformability and structure, review current progress in measurement and modeling of leukocyte rheology, and discuss pathologies associated with changes in the leukocyte mechanical properties.

III. PASSIVE DEFORMATION OF THE CELL CONTRIBUTES TO CELL ROLLING

When leukocytes under normal physiological conditions travel in the vessels with a diameter greater than their size, they have a roughly spherical shape with numerous surface projections (Fig. 1). In this passive, or quiescent, state they behave like droplets of a viscoelastic liquid and their surface folds provide excess surface area needed for protection of the plasma membrane during extensive deformation in capillaries (Evans & Yeung, 1989). Some of these folds are organized into 0.3–1.0 μm long F-actin-rich

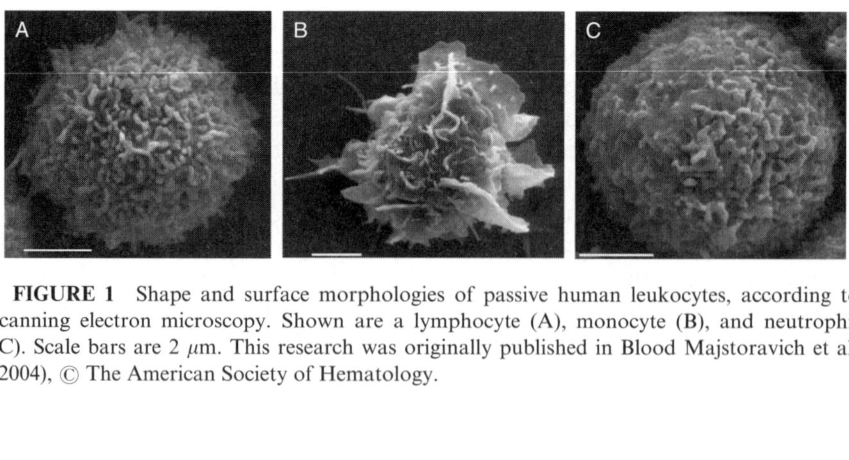

FIGURE 1 Shape and surface morphologies of passive human leukocytes, according to scanning electron microscopy. Shown are a lymphocyte (A), monocyte (B), and neutrophil (C). Scale bars are 2 μm. This research was originally published in Blood Majstoravich et al. (2004), © The American Society of Hematology.

projections, often called leukocyte microvilli. The microvilli of human neutrophils, shown in Fig. 1C, are well studied *in vitro* (Bruehl, Springer, & Bainton, 1996; Marcus & Hochmuth, 2002; Moore et al., 1995; Schmidtke & Diamond, 2000; Shao, Ting-Beall, & Hochmuth, 1998). It is established that L-selectin, P-selectin glycoprotein ligand-1 (PSGL-1), and other cell adhesion molecules that initiate contact between neutrophils and endothelial cells are clustered on the microvilli tips (Stein et al., 1999; Vestweber & Blanks, 1999; von Andrian, Hasslen, Nelson, Erlandsen, & Butcher, 1995). The importance of microvilli deformation in neutrophil tethering and rolling was shown in parallel-plate flow chamber assays (Finger, Bruehl, Bainton, & Springer, 1996; Moore et al.; Park et al., 2002; Ramachandran, Williams, Yago, Schmidtke, & McEver, 2004). The reader is referred to chapters by Shao and Waugh in this book for further discussion of this topic.

We again emphasize that neutrophil biology cannot be blindly applied to other types of leukocytes. The surface projections on lymphocytes and monocytes are significantly different from those on neutrophils, as demonstrated by Majstoravich et al. (2004). Lymphocyte microvilli are similar to those of epithelial cells (Alberts et al., 2002), that is, they are composed of parallel bundles of actin filaments crosslinked side-to-side by actin-binding proteins (Fig. 1A). On the contrary, neutrophil microvilli are membrane ruffles (Fig. 1C) that "contain a meshwork of short actin filaments cross-linked in an end-to-side manner" (Majstoravich et al.). Moreover, this group of researchers found that 84% of all surface structures in human monocytes are large ruffles with embedded epithelial-like microvilli (Fig. 1B).

Passive leukocytes pushed by red blood cells out of the centerline of postcapillary venules come in close contact with endothelial cells. Normal microvascular endothelium has a low density of selectins and other receptors that can bind to the corresponding ligands on the tips of leukocyte microvilli and is also coated with a surface layer called glycocalyx (Smith, Long, Damiano, & Ley, 2003; Vink & Duling, 1996) that reduces the probability of receptor–ligand binding (Zhao, Chien, & Weinbaum, 2001). However, the exposure of endothelium to proinflammatory mediators leads to upregulation of the expression level of selectins and shedding of the glycocalyx (Potter, Jiang, & Damiano, 2009). This allows the endothelium to capture leukocytes from blood flow. Most capture occurs where capillary merge to form postcapillary venules (Dunne, Ballantyne, Beaudet, & Ley, 2002; Schmid-Schonbein, Usami, Skalak, & Chien, 1980).

Here is our understanding of leukocyte tethering (capture) and rolling based on the results of our computational studies (Khismatullin & Truskey, 2004, 2005) and available experimental data. When a free-flowing leukocyte starts interacting with activated endothelium, it has an insufficient number of receptor–ligand bonds to overcome the drag force and torque induced by shear flow. (The drag force is the hydrodynamic force component in the flow direction.) It is, therefore, displaced in the flow direction and begins to rotate clockwise to balance this force and torque. With this motion, bonds are shifted to the trailing edge of the cell–substrate contact zone and stretched, thus creating a tensile force and countertorque on the cell. The tangential component of this bond force counteracts the drag force and thus leads to a decrease in the translational velocity of the cell. The normal component represents a force that pushes the cell closer to the substrate. The cell responds to the bond force by microvilli extension and local deformation of the cell at the trailing edge and a change in the pressure field in a thin, lubrication layer beneath the cell producing the lift force (normal component of the hydrodynamic force) (Abkarian & Viallat, 2005).

Receptor–ligand bonds are transient: they dissociate when reaching a critical length and time. As a result, microvilli at the trailing edge eventually detach from the surface. This reduces the countertorque and hence increases the speed of cell rotation, allowing the leukocyte to form more bonds at the leading edge. By this way, that is, through bond association and dissociation at the leading and trailing edges of the cell, respectively, leukocytes roll on the adhesive substrate. Rolling events can be determined from analysis of the cell trajectories: the rolling cells move in the flow direction with a velocity significantly less than the hydrodynamic velocity.

It is worth mentioning that the total bond force is a function of leukocyte deformability. The more deformable adherent cell has less stretched microvilli and bonds and can survive greater shear stresses. In addition, deformable leukocytes form a large, flat contact area with the substrate during tethering (Fig. 2) and thus have a large number of receptor–ligand bonds to support their rolling interactions with endothelial cells.

Neutrophils in postcapillary venules continuously deform during rolling, transitioning between ellipsoidal and tear drop shapes (Fig. 3). However, this important observation is largely ignored by research groups who focus upon *in vitro* investigations of WBCs (Chang & Hammer, 2000; Dooyoung & King, 2008; King & Hammer, 2001; King, Heinrich, Evans, & Hammer, 2005). Most *in vitro* rolling assays are done using a parallel-plate flow chamber which provides only a top (or bottom) view of the cells. It is extremely difficult to observe cell deformation and contact area changes from the top view. We should also note that the "middle of the road" approach in which microvillus extension and tether pulling gain acceptance but deformability of the cell body is neglected (Caputo & Hammer, 2005; Ramachandran et al., 2004; Yu & Shao, 2007) cannot fully describe leukocyte–endothelial cell interactions. Rolling neutrophils *in vivo* are flattened in the region of cell–substrate contact. Their large contact area, formed as a result of cell deformability, reduces the importance of microvillus

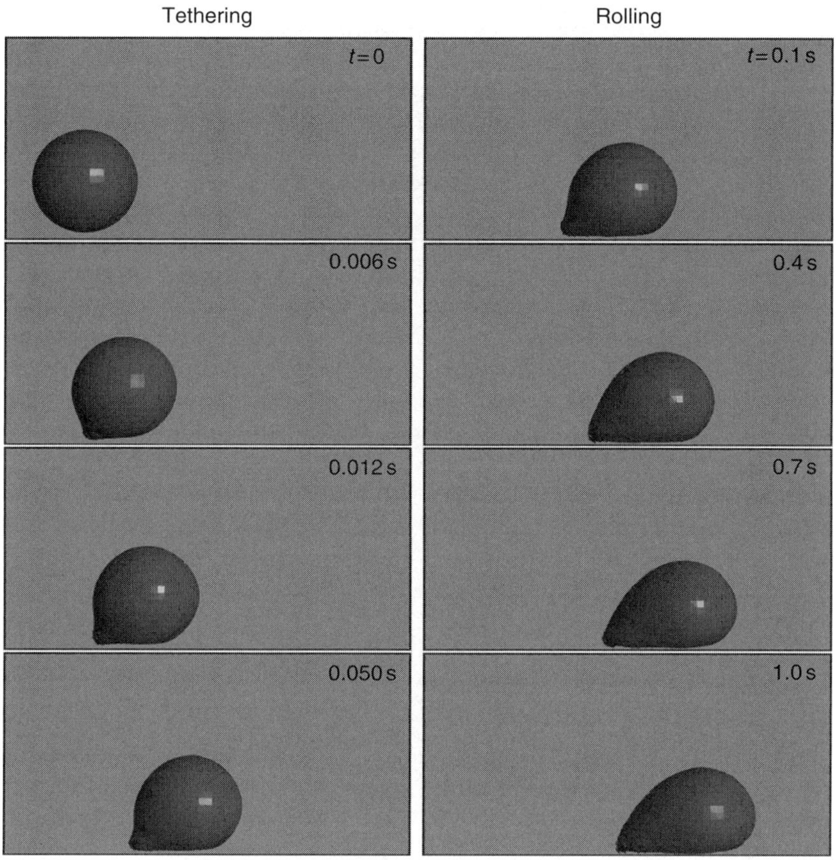

FIGURE 2 Snapshots showing shape changes of a deformable leukocyte (side view) in the tethering and rolling stages, according to numerical simulations. The initial radius and nuclear volume fraction of the cell are 3.82 μm and 25.9%, which corresponds to human monocyte. The cell is located in a 30-μm height rectangular microchannel coated with P-selectin. The wall shear stress is 2.5 dyn cm^{-2}. The viscosity of leukocyte cytoplasm is 50 poise. The nucleus-to-cytoplasm viscosity ratio is 2.5. The cell has 3112 microvilli of initial length 0.5 μm (not shown), distributed uniformly over the membrane. The surface densities of P-selectin and PSGL-1 are 320 sites μm^{-2}. The rolling velocity approached a steady-state value at 1 s (provided by D. Khismatullin, Tulane University, unpublished data).

mechanics, because the microvilli density at the trailing edge of the contact area (where all bonds are load-bearing) is large (Fig. 4) and the force per microvillus may be below the critical value for tether formation. Our computational studies in which both leukocyte and microvilli deformability are taken into account predict leukocyte deformation to a tear drop shape (Fig. 2) and the cell-rolling velocities.

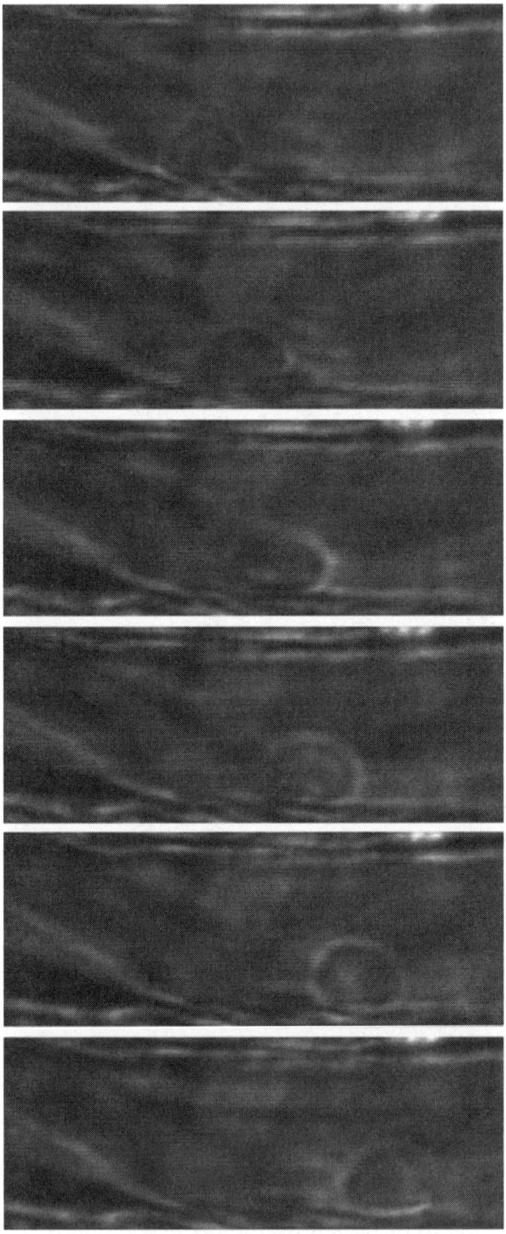

FIGURE 3 Videomicrograph frames showing trauma-induced neutrophil rolling along a postcapillary venule in the mouse cremaster muscle. The flow is from left to right. The cell continuously deforms during rolling (provided by D.R. Potter & E.R. Damiano, Boston University, unpublished data).

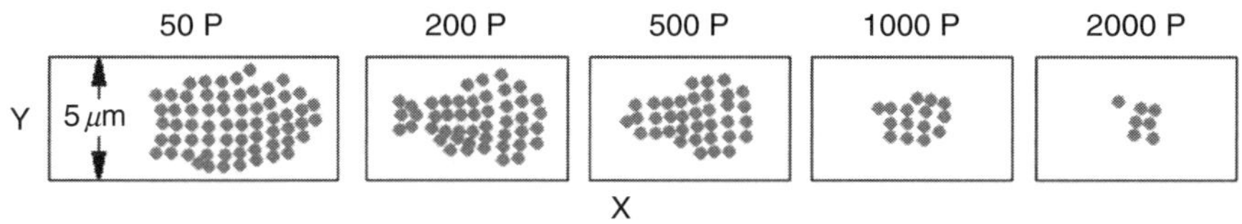

FIGURE 4 Effect of leukocyte deformability on the leukocyte–substrate contact area as a function of cytoplasmic viscosity in poise (P), according to numerical simulations. Shown are microvilli footprints (dots) of human monocyte of different cytoplasmic viscosity at 2 s after cell tethering to a P-selectin-coated surface in a rectangular microchannel. The wall shear stress is 0.25 dyn cm^{-2}. The P-selectin density is 145 sites μm^{-2}. Flow is from left to right. The number of microvilli footprints on the substrate decreases with an increase in cytoplasmic viscosity of the cell (provided by D. Khismatullin, Tulane University, unpublished data).

Cellular deformation is the main reason why WBCs are able to roll on vascular endothelium at relatively high values of the wall shear stress (WSS, 8–30 dyn cm^{-2}) (Damiano, Westheider, Tozeren, & Ley, 1996; Lipowsky, Scott, & Cartmell, 1996; Smith, Smith, Lawrence, & Ley, 2002; Yago et al., 2002). It also explains, in part, why the leukocyte-rolling velocity is relatively constant within a wide range of shear stresses (Firrell & Lipowsky, 1989; Lawrence & Springer, 1991, 1993). This hypothesis was confirmed in recent studies by Smith et al. and Yago et al. Smith et al. observed no increase in rolling velocity in L-selectin knockout mice (i.e., leukocyte rolling was mediated by P-selectin) when the plasma viscosity (and hence the wall shear stress) was doubled through hemodilution. The same trend was shown for neutrophils, but not for PSGL-1 beads, rolling on a P-selectin-coated substrate (170 sites μm^{-2}) in the parallel-plate flow chamber. The velocity of bead rolling on P-selectin reached the value of 30 μm s^{-1} at the WSS of 0.88 dyn cm^{-2}, which was six times higher than the rolling velocity of human neutrophils under the same flow conditions on the substrate coated with a lower density of P-selectin: 20 sites μm^{-2}.

Detachment assays with sPSGL-1- and 2-GSP-6- (a P-selectin ligand peptide) coated beads (200 sites μm^{-2}) (Yago et al., 2002) demonstrated no rolling of the beads on P-selectin if the WSS is higher than 3 dyn cm^{-2} (at a P-selectin density of 145 sites μm^{-2}) or 4 dyn cm^{-2} (at 486 sites μm^{-2}). Yago et al. also showed that human neutrophils roll on P-selectin with a velocity less than 10 μm s^{-1} at a WSS of up to 32 dyn cm^{-2}, while no rigid beads remain bound at the WSS of higher than 5 dyn cm^{-2} (Fig. 5). These data clearly suggest that the rolling dynamics of neutrophils and rigid beads are different.

Rinker, Prabhakar, and Truskey (2001) investigated the effect of WSS on Mono Mac 6 (MM6) cell (monocytic cell line) adhesion to TNF-α-activated HUVEC in a flow chamber. The wall shear rate (WSR) was fixed but the

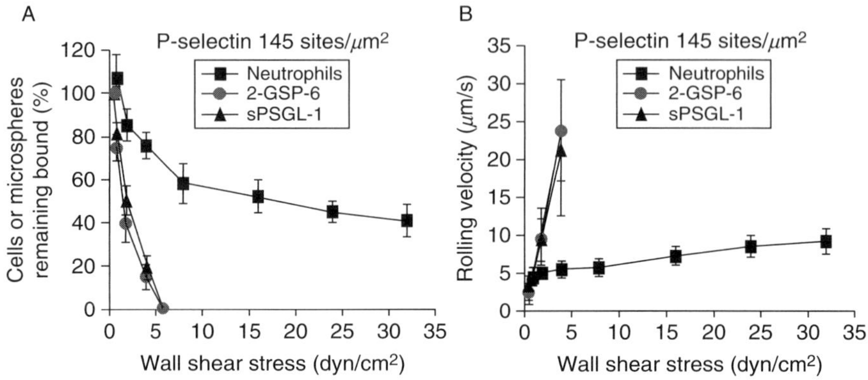

FIGURE 5 Detachment resistance (A) and mean rolling velocity (B) of human neutrophils and ligand-coupled microbeads on a P-selectin-coated surface in a parallel-plate flow chamber. Microbeads are coated either with 2-glycosulfopeptide (GSP)-6, which is similar in structure to NH_2-terminal P-selectin-binding domain of PSGL-1, or with a recombinant soluble form of PSGL-1. Neutrophils or microbeads first accumulate on P-selectin at 0.5 dyn cm^{-2}. The wall shear stress is then increased every 30 s. All microbeads detach from the surface at shear stresses greater than 5 dyn cm^{-2} while at least 40% of neutrophils support their rolling interactions with P-selectin at shear stresses as high as 30 dyn cm^{-2}. Reproduced from Yago et al., 2002. Originally published in Journal of Cell Biology. doi:10.1083/jcb.200204041.

buffer viscosity was elevated with dextran. They found that tethering frequencies increased and rolling velocities decreased with an increase in the WSS. Firrell and Lipowsky (1989) observed that rolling neutrophils in rat mesenteric venules elongated to 140% of their diameter when the WSR increased from 50 to 800 s^{-1}. Such deformation resulted in a 3.6-fold increase in the leukocyte-to-endothelium contact area.

In conclusion of this section, we would like to note that the passive mechanical properties of WBCs, including cellular viscosity and elasticity, are important determinants of both rolling fluxes and rolling velocities of these cells. As the leukocyte cytoplasmic viscosity increases (i.e., leukocyte deformability decreases), there is a corresponding decrease in contact area between the rolling leukocyte and substrate (Fig. 4) until a critical value of viscosity is reached at which the leukocyte detaches from the substrate during the tethering stage. The less the cytoplasmic viscosity of the cell, the larger its contact area with the substrate and the more stable the cell rolls. The elastic adaptation of the cell body to shear stresses leads to a physiologically relevant shape of rolling leukocytes in numerical simulations (Khismatullin & Truskey, 2005).

IV. INTEGRIN ACTIVATION AND CELL ARREST ARE DEPENDENT ON CELL DEFORMABILITY

The venular endothelium of wild-type mice is known to express P-selectin in response to surgical trauma resulted from animal preparation for intravital microscopy (Kunkel & Ley, 1996). Neutrophils roll on trauma-stimulated endothelium, as well as on a P-selectin-coated substrate *in vitro*, without activation and firm adhesion (Kunkel & Ley; Lawrence & Springer, 1991). In this experimental model, neutrophil rolling is accompanied by significant and time-dependent shape changes of the cell (Fig. 3). These shape changes are due to passive deformation that occurs before integrin binding and cell activation.

The mechanical response of WBCs becomes very complex when they encounter chemoattractants (chemokines and formyl peptides such as fMLP) (Ley, 2002) and/or certain receptors (e.g., E-selectin) present on cytokine-activated endothelium (Green, Pearson, Camphausen, Staunton, & Simon, 2004). Chemoattractants transmit their promigratory signals to leukocytes through binding G protein-coupled receptors expressed on the leukocyte surface (van Buul & Hordijk, 2004). Intracellular signaling induced by this binding leads to:

- Upregulation and conformational change of leukocyte integrins to a high-affinity state (Berger et al., 2002; Ginsberg, Partridge, & Shattil, 2005; Luo, Carman, & Springer, 2007; Shamri et al., 2005)
- Activation of the whole cell (Niggli, 2003b; Vicente-Manzanares & Sanchez-Madrid, 2004; Zhelev & Alteraifi, 2002)

These processes serve to arrest rolling leukocytes and prepare them for transendothelial migration.

Integrins are heterodimeric (i.e., consisting of two different chains: α and β) cell adhesion molecules that are present on many types of living cells. All leukocytes constitutively express integrins with β_2-subunit (CD18), which are localized to the cell body but not to leukocyte microvilli (Abitorabi, Pachynski, Ferrando, Tidswell, & Erle, 1997). β_2-integrins are not efficient in leukocyte tethering because they are of low affinity before activation and because they cannot form bonds with endothelial ligands until the cell surface is flattened enough in the cell–substrate contact zone. However, when activated, β_2-integrins mediate leukocyte arrest on endothelium (Ley, 2002). Note that attachment of microvilli in the leading edge of a flattened cell leads to their compression or reorients them closer to the cell body surface, thus decreasing the separation distance between the cell body surface and the substrate. The critical role of β_2-integrins including LFA-1 (CD11a/CD18), Mac-1 (CD11b/CD18), and p150,95 (CD11c/CD18) in the inflammatory response and, specifically, in leukocyte adherence to endothelium was recognized long ago (Anderson, Roswit, Holtzman, Hogg, & Van Eeden, 2001; Pohlman, Stanness, Beatty, Ochs, & Harlan, 1986; Springer, Thompson, Miller, Schmalstieg, & Anderson, 1984). It was established (Diamond et al., 1990; Makgoba et al., 1988) that LFA-1 and Mac-1 bind intercellular adhesion molecule-1 (ICAM-1) expressed on endothelial cells (another ligand for LFA-1 is ICAM-2; Staunton, Dustin, & Springer, 1989). Integrin p150,95 is largely expressed on monocytes/macrophages and dendritic cells and interacts with a counter-receptor on stimulated endothelium (Stacker & Springer, 1991). The structure and activation of integrins are reviewed in detail in Chapters 4 and 6.

Other types of integrins are also present on the surface of leukocytes. For example, $\alpha_4\beta_1$-integrin, also known as very late antigen-4 (VLA-4) or CD49d/CD29, is localized to the leukocyte microvilli (Abitorabi et al., 1997) and binds vascular cell adhesion molecule-1 (VCAM-1) (Pulido et al., 1991). Apart from its role in leukocyte firm adhesion (Burns, Issekutz, Yagita, & Issekutz, 2001; Feigelson et al., 2001; Hyduk & Cybulsky, 2009), this integrin mediates tethering and rolling of lymphocytes (Alon et al., 1995; Grabovsky et al., 2000) and may be involved in rolling of neutrophils (Reinhardt, Elliott, & Kubes, 1997) and eosinophils (Sriramarao, von Andrian, Butcher, Bourdon, & Broide, 1994) under certain conditions. Note that VLA-4 alone cannot support neutrophil rolling in a parallel-plate flow chamber at shear stresses greater than 2 dyn cm^{-2} (Reinhardt et al.) and is inefficient in capture of monocytes (Luscinskas et al., 1996). Many β_1-integrins interact with extracellular matrix proteins (Rosales & Juliano, 1995) and, therefore, they are important for leukocyte migration in the tissue (Werr, Xie, Hedqvist, Ruoslahti, & Lindbom, 1998).

We should mention several results indicating that passive deformation of leukocytes is an important determinant of leukocyte firm adhesion. First, Kitagawa et al. (1997) found that the filtration of human neutrophils through 3 μm pores elevates the expression level of Mac-1 (but not L-selectin). Such a filtration is associated with significant deformation of the cell. This group of researchers showed that neutrophil deformation causes an increase in intracellular Ca^{2+} concentration, and it is known that the release of Ca^{2+} from intracellular stores occurs after integrin activation and binding (Pettit & Hallett, 1996). There was also an increase in the F-actin content after filtration through 3 μm but not 5 μm pores, though there was no evidence of cellular polarization typical for migrating cells (Kitagawa et al.). As we will see later, actin polymerization leads to active deformation of the cell. Using a similar approach, Anderson et al. (2001) showed that mechanical deformation of human neutrophils enhances β_2-integrin-dependent leukocyte adhesion. Specifically, leukocytes filtered through 3–5 μm polycarbonate pores were more than two times more adhesive to ICAM-1 than prefiltered cells. The studies by Anderson et al. and Kitagawa et al. complement recent observations of leukocyte activation in narrow (1.5 and 2.5 μm in height) channels (Yap & Kamm, 2005a,b). Overall, these data show that WBCs are activated when they reach a certain critical level of deformation. Flattening of rolling leukocytes above this level, coupled with chemoattractant signaling, results in leukocyte firm adhesion. Note also that the passive deformation-induced activation of WBCs suggests that rheological techniques in which the sample cell is forced into large deformation (e.g., complete aspiration in a micropipette) cannot accurately determine passive mechanical properties of WBCs.

Second, Park and Shimaoka show, in Chapter 4, that LFA-1 activation occurs in two steps. The first step is the chemokine-induced "switchblade-like" opening that releases the integrin chains from the low-affinity bent state but keeps the headpiece closed. The second step is full extension of the molecule that exposes the high-affinity I-domain of the β-subunit. This extension requires intermediate binding of LFA-1 to ICAM-1 and occurs in response to the force imposed on LFA-1/ICAM-1 bonds by cell motion under shear flow. We discussed above the necessity of cell flattening for β_2-integrin interactions with endothelial immunoglobulins. If the mechanism described by Park and Shimaoka is correct, β_2-integrins cannot be fully activated without the prior passive deformation of the cell.

Third, Sheikh and Nash (1998) found that human neutrophils treated with cytochalasin B rolls on a platelet-coated surface in a parallel-plate flow chamber at 1 dyn cm^{-2} until reaching a critical level of cellular deformation. Then, they attach firmly to the substrate. Cytochalasins are fungal metabolites than bind to the barbed end of actin monomers and prevent actin polymerization and crosslinking (Cooper, 1987). Treatment of the cell with these metabolites makes the cell more deformable and disallows its activation. Using blocking antibodies, Sheikh and Nash showed that neutrophil rolling in their experimental system was mediated by P-selectin and the transition to firm adhesion did not involve β_2-integrins. This result indicates that passive deformation alone is able to induce firm adhesion of WBCs to a selectin-coated substrate.

One of the reasons why the role of cellular deformation in firm adhesion is not properly accounted for is the popular hypothesis that leukocyte arrest on activated endothelium is a rapid event (Alon, Grabovsky, & Feigelson, 2003). It is of course possible to make the transition from rolling to firm adhesion of the cell very fast by stimulating with a high concentration of chemoattractant. Does this approach reconstruct the conditions existing in the human body during the inflammatory response? Can we extend the rapid arrest hypothesis largely based on experiments with lymphocytes

(Alon et al.; Campbell et al., 1998) to other types of leukocytes? As reported by Kunkel, Dunne, and Ley (2000), neutrophils in wild-type mice gradually reduce their rolling velocity when interacting with tumor necrosis factor-α (TNF-α)-stimulated endothelium. On average, 86 s of contact time are required to firmly attach these cells. TNF-α is a well-studied inflammatory cytokine, which upregulates the expression of E-selectin, with a peak at 4 h after stimulation (Bevilacqua, Stengelin, Gimbrone, & Seed, 1989), and increases the surface densities of VCAM-1 and ICAM-1 on endothelial cells (McHale, Harari, Marshall, & Haskard, 1999). TNF-α induces neutrophil trafficking to sites of infection. It is largely produced by macrophages when they encounter pathogenic microorganisms (Gosset et al., 1991) but it can also be secreted by other WBCs (Hershkoviz et al., 1993; Jablonska et al., 2001; Yoshimura, Hara, Kaneko, & Kato, 1997). Using knockout mice, Kunkel et al. convincingly showed that the gradual transition to firm adhesion for neutrophils on TNF-α-stimulated endothelium is E-selectin dependent. Thus, if vascular endothelium expresses E-selectin, WBCs can gradually decrease their rolling velocity until reaching stationary adhesion. This allows them to form a large contact area with vascular endothelium and efficiently integrate signals from various chemoattractants present on the surface of endothelial cells.

V. FIRMLY ADHERENT CELLS EXPERIENCE ACTIVE DEFORMATION

Active leukocyte migration is the result of cytoskeleton remodeling, a highly dynamic process that can involve hundreds of different proteins. Due to its complexity, it is the least understood topic of leukocyte biology. Although we are able to identify key players in rolling and firm adhesion of WBCs, we still do not know how and when these cells switch to a migratory phenotype. Our knowledge of signaling pathways that lead to leukocyte activation is incomplete. Second messengers identified in chemoattractant-mediated signaling of WBCs include phosphoinositide 3-kinases (see the reviews by Curnock, Logan, & Ward, 2002; Stephens, Ellson, & Hawkins, 2002), Rho family GTPases such as Rac and Cdc42 (Makino, Glogauer, Bokoch, Chien, & Schmid-Schonbein, 2005; Niggli, 2003b; Sanchez-Madrid & del Pozo, 1999), the GTPase Rap1 (Shimonaka et al., 2003), the tyrosine kinase Syk (Schymeinsky, Then, & Walzog, 2005), sphingosine kinase (SphK) (Wang, Graeler, & Goetzl, 2005), cyclic AMP (Kaneider et al., 2002), cyclic ADP-ribose (Partida-Sanchez et al., 2007), calcium divalent cation Ca^{2+} (Cao et al., 1995), and nitric oxide (Cherla & Ganju, 2001; VanUffelen, de Koster, Van den Broek, VanSteveninck, & Elferink, 1996). Phosphoinositide 3-kinases (PI3-Ks) induce the phosphorylation of phosphatidylinositol lipids, for example, PtIns(3,4,5)P_2 or PIP_3, which are involved in leukocyte cytoskeleton remodeling (Sarraj et al., 2009). The role of PIP_3 and other phosphatidylinositol lipids in leukocyte motility was discussed in detail by Niggli and Zhelev and Alteraifi (2002). Rho GTPases regulate actin polymerization in leukocytes and other motile cells (Burridge & Wennerberg, 2004; Hall & Nobes, 2000; Makino et al.). Specifically, they activate the Arp2/3 complex (Glogauer, Hartwig, & Stossel, 2000; Higgs & Pollard, 2001), which is necessary for actin filament growth (see next section).

Signaling pathways associated with chemoattractant receptors are interconnected and the roles of second messengers are interrelated. For example, PIP_3 induced by the γ-isoform of phosphoinositide 3-kinase (PI3-Kγ) in neutrophils is involved in activation of Rac, while Rac itself can activate PI3-Kγ (Niggli, 2003b). The activation of Arp2/3 complex in neutrophils is synergistically mediated by phosphatidylinositols and Cdc42 (Higgs & Pollard, 2000). An added complication is that the transition of WBCs to the active state can be chemoattractant independent; for instance, it can occur as a result of cellular deformation (Yap & Kamm, 2005a), disruption of microtubules (Niggli, 2003a), and supplementation of the substrate by β-glucan (Harler & Reichner, 2001). In addition to that, the role of integrins in leukocyte motility is not well understood. Integrin activation and binding are not required for active motion of WBCs. Leukocytes migrate in a three-dimensional (3D) environment even when all integrins on their surface are ablated (Lammermann et al., 2008). In a micropipette system, neutrophils extend their projections toward the pipette filled with a chemoattractant (Zhelev, Alteraifi, & Chodniewicz, 2004). Since neutrophils are suspended in an aqueous solution, this active response occurs without integrin binding.

Nevertheless, priming and binding of leukocyte β_2-integrins are essential for strong attachment of leukocytes to endothelium, which is a prerequisite for transendothelial migration of these cells. Integrin binding to endothelial ligands provides traction to maintain leukocyte migration on two-dimensional (2D) adhesive substrates or squeeze the leukocyte between endothelial cells. There is also evidence about the supportive role of integrins in chemoattractant-induced activation of adherent leukocytes (Lofgren, Ng-Sikorski, Sjolander, & Andersson, 1993; Menegazzi et al., 1999; Zhu et al., 2008).

Leukocytes attached firmly to inflamed endothelium are characterized by mechanical properties very different from the cells circulating in normal blood. Passive cells are large and not deformable enough to pass through tiny gaps between vascular endothelial cells. They need to be activated to achieve this goal. Cellular activation is associated with substantial polymerization of active filaments in the cell cortex (the region beneath the plasma membrane full of actin). The formation of a 3D meshwork of short, branched actin filaments in this region (Ryder, Weinreb, & Niederman, 1984) makes activated leukocytes stiffer than passive ones. On the other hand, actin polymerization exerts a protrusive force on the plasma membrane, resulting in cell shape changes. Activated cells can, therefore, experience greater deformation than passive cells despite their increased stiffness. Such a behavior cannot be described by classical constitutive laws of continuum mechanics, thus requiring the development of new theory based on a combination of cellular biochemistry and biomechanics.

WBCs respond to actin polymerization by first developing large ruffles over the cell body and then transitioning to a polarized shape characterized by a thin, veil-like structure, called a lamellipodium, at the front of the cell and a contracted tail, known as an uropod, at the back (Eisenbach, 2004; Zhelev & Alteraifi, 2002). Adherent cells spread out on the substrate during polarization (Sengupta, Aranda-Espinoza, Smith, Janmey, & Hammer, 2006), reaching twice the diameter, in a lateral cross section, of circulating cells (Marschel & Schmid-Schonbein, 2002). Cell polarization is necessary for active migration of neutrophils and occurs whether the cell attaches to the substrate or is exposed to soluble chemokines in a 3D environment (Niggli, 2003b). Leukocytes migrate either in the direction of a chemoattractant gradient or in random directions if the chemoattractant concentration is uniform (Zigmond, Levitsky, & Kreel, 1981).

A cell migrating on 2D substrates moves by continual extension of finger-like projections (filopodia or microspikes) at the lamellipodium front and subsequent retraction of the uropod (Eisenbach, 2004). Note that the term *pseudopod* is often used to describe any temporary projections of an actively migrating leukocyte (Moazzam, DeLano, Zweifach, & Schmid-Schonbein, 1997; Schmid-Schonbein, 1986; Yap & Kamm, 2005b). During activation, the cell relocates integrins that can bind endothelial cell adhesion molecules and extracellular matrix proteins to the lamellipodium (Kiosses, Shattil, Pampori, & Schwartz, 2001). Binding of integrins to their ligands stabilizes (anchors) the filopodia and exerts traction on the rest of the cell (Eddy, Pierini, & Maxfield, 2002), thus propelling the cell forward. Uropod retraction is associated with disassembly of focal adhesion sites at the trailing edge of the cell, which occur due to tension on the plasma membrane created by actin–myosin network contraction (Cozens-Roberts, Lauffenburger, & Quinn, 1990). Recent experimental studies (Burridge & Wennerberg, 2004; Inoue et al., 2005; Smalley & Ley, 2005) also show that microtubules and Rho family GTPases are involved in the turnover of cell adhesion.

In the active state, leukocytes transmigrate across the endothelium and move toward the source of inflammation. Leukocyte diapedesis occurs via a paracellular route (between endothelial cells) (Hordijk, 2006), though there is some evidence that a transcellular route (through an endothelial cell) may be involved (Carman & Springer, 2004). Paracellular transmigration of leukocytes is regulated by complex signaling pathways which involve interactions between β_2-integrins and endothelial cell junction proteins, such as PECAM-1 (Muller & Randolph, 1999; Vaporciyan et al., 1993; Woodfin et al., 2009) and members of the JAM family (Dejana, Spagnuolo, & Bazzoni, 2001; Martin-Padura et al., 1998; Ostermann, Weber, Zernecke, Schroder, & Weber, 2002), as well as β_2-integrin binding to ICAM-1 (Rahman & Fazal, 2009) and ICAM-2 (Woodfin et al.). PECAM-1 also regulates transendothelial migration of leukocytes through interaction with its counterpart (PECAM-1 on leukocyte surface) upon stimulation of the endothelium by inflammatory mediators, such as IL-1β (O'Brien, Lim, Sun, & Albelda, 2003). In addition, the crosstalk between the activated leukocyte and endothelial cells leads to NF-κB-mediated expression of certain cytokines (IL-6 and IL-8) and upregulation of ICAM-1 (Naves et al., 2006) and can induce endothelial deformation that will facilitate leukocyte diapedesis (Kaplanski et al., 1994). The reader is referred to the review by Weber (2003) for more information about receptors involved in transendothelial migration.

VI. CYTOSKELETON IS THE SOURCE OF BULK MECHANICAL PROPERTIES OF WHITE BLOOD CELLS

From a mechanical perspective, a living cell is a dispersed mixture of deformable particles (organelles) embedded in a reactive viscoelastic fluid (cytosol and cytoskeleton) and enclosed within a semipermeable boundary (plasma membrane). A multiphase flow approach (Nigmatulin, 1990) combined with the interpenetrative reactive flow formalism (Dembo & Harlow, 1986) is the most appropriate (but never fully realized) way to describe the mechanical response of this system. Based on volume fraction, we can identify two components that play a critical role in deformability of WBCs: the cytoskeleton and the nucleus. The cytoskeleton is a network of actin filaments, intermediate filaments and microtubules immersed in the intracellular fluid called the cytosol. It ties together the plasma membrane and cell organelles, thus providing structural integrity of the cell, and is also the major source of cell viscoelasticity (Lenormand & Fredberg, 2006). Three major types of filamentous proteins that form the cytoskeleton (actin, tubulin, and a family of intermediate filament proteins), together with hundreds of accessory proteins that crosslink the filaments, connect the cytoskeleton to other components of the cell, and mediate its dynamic response, comprise about 60% of the cell dry mass (Alberts et al., 2002). The cytoskeleton rheology is determined by the properties of individual filaments (Gardel et al., 2004; Janmey, Euteneuer, Traub, & Schliwa, 1991; Pelletier, Gal,

Fournier, & Kilfoil, 2009), as well as the density and elasticity of crosslinkers (Gardel et al.; Head, Levine, & MacKintosh, 2003; Kim, Hwang, Lee, & Kamm, 2009; Levy & Shoseyov, 2004; Lieleg, Schmoller, Claessens, & Bausch, 2009) and the number of entanglements (Dichtl & Sackmann, 2002; Keller, Tharmann, Dichtl, Bausch, & Sackmann, 2003), which may lead to significant changes in deformability between different cell types and local variations of the material constants within the cell (Lau, Hoffman, Davies, Crocker, & Lubensky, 2003; Simon et al., 2004). Adding this to the fact that the cytoskeleton is a highly dynamic structure characterized by rapid reorganization of actin filaments and microtubules (Brangwynne, MacKintosh, & Weitz, 2007; Ding, Robinson, Behrens, & Vandre, 1995; Howard & Watts, 1994) makes understanding of the cytoskeleton functions extremely difficult. However, a lot is known about cytoskeletal dynamics by GFP fusion proteins (Gerisch & Muller-Taubenberger, 2003; Ludin & Matus, 1998), and structure by immunofluorescence studies (Becker & Gard, 2006; Franklin & Martin, 1980).

One of the functions of the cell nucleus is to protect DNA molecules from destruction by external mechanical stresses. This function is realized by surrounding the chromatin by a meshwork of intermediate filaments, known as lamins, attached from inside to a double membrane (nuclear envelope) permeable to macromolecules only in specific regions (nuclear pores) (Alberts et al., 2002). Four different types of lamins (A, B_1, B_2, and C) are present in the nuclear lamina of animal cells (Prokocimer et al., 2009), where they, in conjunction with lamin-associated proteins (Georgatos, Meier, & Simos, 1994), form a stable viscoelastic structure that makes the nucleus resistant to deformation. Stiffening of the cell nucleus occurs during differentiation of embryonic stem cells (Pajerowski, Dahl, Zhong, Sammak, & Discher, 2007). The nuclear cytoskeleton is not very reactive (otherwise, DNA protection would be compromised), but its deformation may result in signal transduction to the rest of the cell (Arnoczky, Lavagnino, Whallon, & Hoonjan, 2002). Fluid shear stresses and cytoplasmic deformation can also send signals to the nucleus through the cytoskeleton (Hu, Chen, Butler, & Wang, 2005). All leukocytes have lamin B_1 (Yabuki et al., 1999), which is critical for homeostasis and mammalian development, as evident from premature death of lamin B_1 mutant mice (Vergnes, Peterfy, Bergo, Young, & Reue, 2004). Human neutrophils but not mononuclear leukocytes lack lamins A and C, indicating that the downregulation of these proteins leads to the formation of a multilobed nucleus (Yabuki et al.).

Despite the facts that the nucleus takes from 18% to 44% of the volume in human leukocytes (Schmid-Schonbein, Shih, & Chien, 1980) and computational studies show that the deformability of this organelle influences leukocyte motion in capillaries (Kan, Shyy, Udaykumar, Vigneron, &

Tran-Son-Tay, 1999) and leukocyte–endothelial cell interactions (Khismatullin & Truskey, 2005; N'Dri, Shyy, & Tran-Son-Tay, 2003), there is only one inconclusive report about measurement of the mechanical properties of the leukocyte nucleus (Dong, Skalak, & Sung, 1991). Dong and coworkers used the micropipette aspiration and a standard viscoelastic solid model to measure deformability of the human lymphocyte nucleus. They found that the nucleus of this leukocyte type is characterized by a viscosity of 2323 poise when aspirated completely, which was much higher than the viscosity of neutrophil cytoplasm (50–650 poise). However, they did not provide data on the lymphocyte cytoplasmic viscosity. Measurements done with other cell types (chondrocytes, endothelial cells, Swiss 3T3 fibroblasts) also indicate that the nucleus is less deformable than the cytoplasm (Caille, Thoumine, Tardy, & Meister, 2002; Guilak, 1995; Guilak, Tedrow, & Burgkart, 2000; Ofek, Natoli, & Athanasiou, 2009; Tseng, Lee, Kole, Jiang, & Wirtz, 2004).

A. *Rheology of the Cell Cytoskeleton Is Well Described by Continuum Models*

The mechanical behavior of the cytoskeleton can be described in a number of ways. The first approach is the *tensegrity hypothesis* (Ingber, 2003a,b), which proposes that microtubules are compression struts connected to actin and intermediate filaments serving as tension cables. Based on the structural organization of microtubules and their large persistence length (see next paragraph) that makes them stiff filaments, this theory suggests that the microtubular network is able to resist the cellular prestress created by the networks of actin and intermediate filaments, thus stabilizing the cell shape. While the prestress, also known as cortical tension in leukocyte biomechanics, is certainly a feature of living cells (Evans & Yeung, 1989; Needham & Hochmuth, 1992; Tsai, Frank, & Waugh, 1993; Wang et al., 2002), the main proposition of this theory, microtubules as supporting elements, does not match well with dynamic properties of microtubules (cf. Section VI.B). In addition, current tensegrity-based computational models of cell mechanics consider only 24 tension cables and 6 compression struts (Canadas, Laurent, Oddou, Isabey, & Wendling, 2002; Stamenovic & Coughlin, 2000). This is clearly not enough to realistically describe the mechanical behavior of the cell cytoskeleton. Ingber (2003a) mentioned that the tensional prestress of the cell can also be balanced by focal adhesions. However, leukocytes do not form stress fibers (Yuruker & Niggli, 1992) or focal adhesions when they contact endothelial cells. With focal adhesions absent and microtubules dynamically unstable, there is a big "if" about application of the tensegrity approach to leukocyte biomechanics.

The second approach, known as a *biopolymer network*, is based on the theory of flexible polymers, that is, it considers the cytoskeleton as a polymer matrix made of semiflexible crosslinked filaments (Broedersz, Storm, & Mackintosh, 2009; Roy & Qi, 2008). A semiflexible polymer is a chain of rigid links connected at angles that can undergo temperature-driven fluctuations in the absence of external forces (Wiggins & Nelson, 2006). It is characterized by the number and length of links, the persistence length proportional to flexural rigidity if the filament is treated as a wormlike chain (links of infinitesimally small length), and by the end-to-end distance. The application of this theory to cytoskeletal proteins allowed to estimate the persistence length of actin filaments (12–20 μm) (Gittes, Mickey, Nettleton, & Howard, 1993; Isambert et al., 1995), microtubules (1–6 mm) (Felgner, Frank, & Schliwa, 1996; Gittes et al.), and leukocyte intermediate filaments such as vimentin (0.3–1 μm) (Mucke et al., 2004). The biopolymer network models were able to predict the rheological properties of structures reconstituted from purified actin (Broedersz et al.; Kim et al., 2009). However, the complexity of protein interactions and remodeling of the cytoskeleton make this approach very difficult to implement for the true cytoskeleton. No models of the leukocyte cytoskeleton based on a biopolymer network were yet proposed.

In the third approach, the cytoskeleton is considered as an *effective continuum*. This includes viscoelastic (Bathe, Shirai, Doerschuk, & Kamm, 2002; Dong, Skalak, Sung, Schmid-Schonbein, & Chien, 1988; Karcher et al., 2003; Khismatullin & Truskey, 2005; Schmid-Schonbein, Sung, Tozeren, Skalak, & Chien, 1981), poroelastic (Mitchison, Charras, & Mahadevan, 2008), and non-Newtonian models (Tsai et al., 1993) of the cell cytoplasm. It is important to emphasize that dilute and semidilute polymer networks characterized by a small number of intersections are essentially polymeric liquids, whose deformation is described by viscoelastic models (Grigorescu & Kulicke, 2000). Additionally, a concentrated polymer network with a large number of crosslinks, such as the cytoskeleton, can be modeled as a non-Newtonian material. For example, Fredberg and his coworkers (Lenormand, Millet, Fabry, Butler, & Fredberg, 2004; Maksym et al., 2000) as well as other research groups (Lau et al., 2003; Puig-De-Morales et al., 2001) convincingly showed that the cytoskeleton behaves as a power-law fluid. The power-law model was first suggested to describe human neutrophil rheology by Tsai et al. While discrete approaches may account for the microstructure of the cytoskeleton, it is currently infeasible to develop a computational model that can implement this microstructure. The computational algorithms based on continuum models are easy to develop and they have a more predictive power than the algorithms using discrete models. The continuum approach is also more flexible than the discrete one because it allows introducing time- and location-dependent active force field within the cell, using, for example the interpenetrative reactive flow formalism by Dembo and Harlow (1986), to describe active deformation of the cell.

We should also comment that it is inappropriate to model the leukocyte cytoskeleton as a 2D continuum (i.e., as a liquid capsule or a Newtonian drop with cortical tension), an approach that is still popular (Evans & Yeung, 1989; Jadhav, Eggleton, & Konstantopoulos, 2005; N'Dri et al., 2003; Pappu & Bagchi, 2008; Pawar, Jadhav, Eggleton, & Konstantopoulos, 2008). Although the network of actin filaments is located in the cell cortex, it is tightly connected to the networks of intermediate filaments and microtubules in all normal nucleated cells. Thus, the cytoskeleton, together with the cytosol, forms a 3D material which behaves as a viscoelastic or non-Newtonian fluid.

B. Microtubules Organize the Cell Interior

Microtubules are hollow fibers (14 and 25 nm in inner and outer diameters) made of a heterodimer protein called tubulin. They form a 3D network that originates at the centrosome (also known as the microtubule-organizing center, MTOC) located near the cell nucleus and radially propagates to the plasma membrane (Alberts et al., 2002). Aside from their critical role in cell division, microtubules together with their motor proteins (kinesins and dyneins) are responsible for positioning the organelles and moving cargo within the cell (Howard, 2001; Vaughan, 2005).

Although stiffer than other cytoskeletal filaments (see above), microtubules are dynamically unstable. They are characterized by structural polarity, that is, they are preferentially polymerized at so-called *plus end* that extends away from the centrosome (Bergen & Borisy, 1980). While growing toward the plasma membrane, they can rapidly shrink back to the centrosome (Cassimeris, Pryer, & Salmon, 1988; Hunyadi, Chretien, & Janosi, 2005; Mitchison & Kirschner, 1984). This dynamic instability is an indication that microtubules serve to sense but not support the cell against deformation. The network of microtubules relocates the cell organelles if it senses significant changes in the shape of the cell (Blocker, Griffiths, Olivo, Hyman, & Severin, 1998; Rothwell, Nath, & Wright, 1989; Sanchez-Madrid & Serrador, 2007; Stinchcombe, Majorovits, Bossi, Fuller, & Griffiths, 2006). The dynamic changes in the length of microtubules are especially important for the movement of the organelles in actively migrating cells including leukocytes (Chiplonkar, Vandre, & Robinson, 1992; Sanchez-Madrid & Serrador; Sumoza-Toledo & Santos-Argumedo, 2004). For the tensegrity hypothesis to work, the plus end of microtubules should be captured and stabilized in the cell cortex by special accessory proteins. This happens during cell migration on adhesive substrates (Kaverina, Rottner, & Small, 1998) but not in a 3D environment. Note that microtubule stabilization also occurs during cell division (Higuchi & Uhlmann, 2005; Kusch, Meyer, Snyder, & Barral, 2002).

It was recognized that the interactions between microtubules and intermediate filaments are mediated by conventional kinesin and cytoplasmic dynein (Gyoeva & Gelfand, 1991; Helfand, Mikami, Vallee, & Goldman, 2002; Huang et al., 1999; Prahlad, Yoon, Moir, Vale, & Goldman, 1998). The collapse of the vimentin network in fibroblasts after the microtubules were depolymerized by vinblastine (Soltys & Gupta, 1992) was probably the result of antegrade transport by kinesin. Dynein leads to the retrograde movement of vimentin filaments, as shown by Helfand et al. Chien and Sung (1984) studied the effect of colchicine (another microtubule-depolymerizing drug) on human neutrophil aspiration in a micropipette. They found that colchicine reduces deformability of these leukocytes and concluded that "the integrity of the microtubules plays a significant role in providing the viscoelastic resistance" of the cells. However, the observed effect could be the direct result of the microtubule depolymerization-associated collapse of the vimentin cytoskeleton in neutrophils. Another observation that illustrates strong interactions between vimentin filaments and microtubules is the coordinated movement of the centrosome and the vimentin network to the growing end of a migrating B-lymphocyte (Sumoza-Toledo & Santos-Argumedo, 2004).

C. Actin Filaments Control the Deformation of Actively Migrating Cells

Actin filaments (F-actin) are twisted strands of actin monomers (globular actin, G-actin) about 7 nm in diameter (Alberts et al., 2002). They are always present in the cell cortex, where they together with crosslinkers form a network called cortical cytoskeleton. The cortical cytoskeleton is an important determinant of the mechanical properties of both passive and active leukocytes. It is particularly responsible for the leukocyte tensional prestress (or cortical tension) (Zhelev & Hochmuth, 1995) and bending elasticity (Zhelev, Needham, & Hochmuth, 1994). Note that the plasma membrane is characterized by negligible surface tension due to its bilayer structure. As we already discussed in this chapter, F-actin is an important structural unit of leukocyte microvilli. The mechanical properties of purified actin solutions were extensively studied. It was established that F-actin diluted in an aqueous medium behaves as a viscoelastic material (Sato, Leimbach, Schwarz, & Pollard, 1985; Wachsstock, Schwarz, & Pollard, 1994). Its shear viscosity and elasticity increases when adding crosslinkers and actin gelation factors (Janssen et al., 1996; Wachsstock et al.; Zaner & Valberg, 1989) but decreases with shortening of the filaments by gelsolin (Janmey et al., 1994). The cortical cytoskeleton is certainly important for structural stability of the cell. However, the preferential distribution of active filaments at the cell periphery makes it questionable whether the cortical cytoskeleton is the only structural unit that protects the passive leukocyte against mechanical forces. There should be a 3D cytoskeletal structure in the cell interior that provides tensile strength to the cell.

More important is the role of actin filaments, and specifically, actin polymerization/depolymerization in the active deformation of migrating cells (cf. Section V). Unfortunately, we are very far from understanding of how biochemical reactions within the cell and associated actin polymerization produce mechanical forces that lead to cell migration. The simplest hypothesis for active force generation is *treadmilling*. Like microtubules, actin filaments are characterized by a molecular polarity, that is, they have the plus (barbed) end, which is favored for growth of the filament, and the minus (pointed) end, which is relatively inert and slow-growing (Alberts et al., 2002). F-actin polymerizes by adding its globular monomer (G-actin) to the plus end provided G-actin has ATP bound. According to the treadmilling hypothesis, the hydrolysis of the bound ATP and the resulting dissociation of actin monomers occur at the minus end. Thus, actin subunits treadmill from the barbed end to the pointed end, creating the directional movement of actin filaments (Higgs & Pollard, 2000). Unfortunately, the treadmilling velocity is very slow (0.04 $\mu m\ min^{-1}$) (Higgs & Pollard) and thus this hypothesis cannot explain fast motility of fish epidermal keratinocytes (30 $\mu m\ min^{-1}$) (Alberts et al.; Grimm, Verkhovsky, Mogilner, & Meister, 2003) and white blood cell (up to 20 $\mu m\ min^{-1}$) (Niggli, 2003b). Another popular hypothesis, *dendritic nucleation* of actin filaments, was proposed by Pollard and coworkers (Higgs & Pollard). Dendritic nucleation is mediated by Arp2/3 complex, a stable assembly of two actin-binding proteins (Machesky et al., 1999; May et al., 1999). This complex caps the pointed ends and then initiates the growth of daughter filaments at 70° to the barbed direction of the mother filament (Mullins, Heuser, & Pollard, 1998). The fast and directed growth of this branched structure is supported by several other actin-binding proteins, including profilin and thymosin-β_4, which bind to the barbed end and speed up filament elongation (De La Cruz et al., 2000; Leyton-Mange et al., 2006), and the capping protein, which limits the length of filaments directed away from the migration path (Cooper, 1987). It is well known that the Arp2/3 complex is activated in leukocytes as a result of chemoattractant receptor signaling (cf. Section V).

Several other models/hypotheses of active force generation in the cell were also suggested. In particular, the network swelling and polymerization force models of Herant, Marganski, and Dembo (2003) view cytoskeleton remodeling as an interpenetrating reactive flow (Dembo & Harlow, 1986) and take into account the active force field as an additional stress tensor in the Navier–Stokes equations. Here, the network swelling model includes interfilament forces (isotropic stress tensor), while the polymerization force model considers the active stresses resulting from interaction of the cytoskeletal network with the cell membrane activated by chemoattractant (so-called polymerization messengers). These models were developed to describe motility of human neutrophils toward a micropipette filled with chemoattractant (Herant et al.). The Brownian ratchet model of Peskin, Odell, and Oster (1993) deals with polymerizing filaments that exert the protrusive force on the membrane through rectifying Brownian motion. The treadsevering model of Dufort and Lumsden (1996) accounts for the capping of barbed ends and suggests that this capping speeds up treadmilling of actin filaments. The stochastic branching model of Carlsson (2001) is based on the dendritic nucleation hypothesis and assumes that the processes of monomer attachment/detachment, capping, and formation of daughter filaments are stochastic with probabilities determined by associated rate constants. Finally, the molecular motors model (actoclampin model) of Dickinson and coworkers (Moschakis, Murray, & Dickinson, 2006) suggests the "Lock, Load & Fire" (more precisely, "jumps between energy wells") mechanism for active force generation. Mogilner and Oster (1996, 2003) incorporated elasticity of actin filaments into the Brownian ratchet model (which assumed that filaments were stiff). The basic idea of this "elastic Brownian ratchet model" is that bending undulations of a filament permit intercalation of a monomer between the barbed end of the filament and the membrane.

D. Intermediate Filaments Contribute to the Deformability of the Cell

The continuous reorganization of actin filaments and microtubules in an animal cell leads to dynamic changes in the cell's mechanical properties. Hypothetically, there could be a situation where the cell temporarily lacks or has an insufficient number of these filaments to support its shape. If the cell were not protected by another cytoskeletal structure at this moment, it would be destroyed by extracellular mechanical stresses. This structure is a network of intermediate filaments. There are several reasons for that.

First, the building blocks of intermediate filaments are highly elongated fibrous molecules (note that tubulin and G-actin are globular molecules) that have three domains: amino-terminal head, carboxyl-terminal tail, and α-helical central rod domain. The central region contains so-called heptad repeats that are responsible for the formation of coiled-coil dimers from these molecules. An intermediate filament consists of tetrameric subunits formed by anti-parallel arrangement of dimers (head-to-tail). This organization keeps the filament nonpolarized and makes it capable of withstanding significant mechanical stresses (Alberts et al., 2002).

Second, the majority of intermediate filaments in the cell are in the fully polymerized state provided that the cell is not in mitosis (the nuclear lamina disassembles during cell division), the cell is not exposed to drugs that can depolymerize cytoskeletal filaments, and the cell is in a passive state. This makes the network of intermediate filaments more stable than microtubules and actin filaments.

Third, there are many more types of intermediate filament proteins than those of tubulin and actin, and this family of proteins is encoded by 65 genes in the human genome (Hesse, Magin, & Weber, 2001). Such diversity makes intermediate filaments more adaptable to a specific cell type. For example, leukocytes are motile cells that undergo significant deformation during migration. They need a dynamic cytoskeleton to perform this role. The major intermediate filament protein in the cytoplasm of these cells is vimentin, which is known to form a dynamic network that can quickly collapse to the nucleus during microtubule depolymerization (Gyoeva & Gelfand, 1991; Soltys & Gupta, 1992) or move with the centrosome to one of the poles of the migrating cell (Parysek & Eckert, 1984; Sumoza-Toledo & Santos-Argumedo, 2004). As demonstrated by Martys, Ho, Liem, and Gundersen (1999) and Yoon, Moir, Prahlad, and Goldman (1998), dynamic changes in the vimentin network structure are the result of the interaction of vimentin filaments with microtubules and actin filaments. Note that only short fibrils of vimentin move rapidly in the cytoplasm. Fully polymerized, long vimentin filaments are much less active because they provide structural support of the cell against mechanical stresses (Chou & Goldman, 2000; Helfand, Chang, & Goldman, 2004).

By contrast, intermediate filaments in the cells which primarily function to protect underlying tissue from external forces (e.g., keratin filaments in epithelial cells) are more inert than in motile cells. The keratin network in epithelial cells is not affected by microtubule-depolymerizing agents (Yoon et al., 2001) but it is associated with actin filaments through stress fibers (Helfand et al., 2004). Therefore, it provides a high resistance of epithelial cells to deformation and a strong adhesion of these cells to the substrate.

Fourth, intermediate filaments are distributed through the cytoplasm as well as the nucleus, forming a completely 3D structure inside the cell. Vimentin shows viscoelastic properties different from actin. The network of vimentin filaments hardens at high strains and thus becomes more resistant to breakage (Janmey et al., 1991). This is a critical property to maintain integrity of the cell.

The role of vimentin in endothelial cell deformability is well established (Helmke, 2005) and there are a few reports that show the importance of this protein in deformability of circulating lymphocytes (Brown, Hallam, Colucci-Guyon, & Shaw, 2001) and lymphocyte–endothelial cell interactions (Nieminen et al., 2006). However, its role in the biomechanics of granulocytes and monocytes was largely ignored in previous studies.

In conclusion of this section, we would like to propose that the basic level of the leukocyte deformability is determined by the network of intermediate filaments. Leukocytes are characterized by bulk non-Newtonian properties rather than the surface ones because of the 3D structure of this network, a direct linkage between actin and vimentin filaments (Esue, Carson, Tseng, & Wirtz, 2006), and tight interactions between vimentin filaments and microtubules.

VII. WHITE BLOOD CELL DEFORMABILITY CAN BE MEASURED BY SEVERAL RHEOLOGICAL TECHNIQUES

The ability of WBCs to deform is essential for their interactions with endothelial cells, for their diapedesis and chemotaxis, and, as will be reviewed in Section VIII, for health and life. The mechanical response of these cells, however, may be very different from that of polymeric droplets or other lifeless objects because they can switch to an active state, where they move and deform using their own chemical machinery. This raises the question of whether we can measure the intrinsic mechanical properties of leukocytes. The answer is a conditional yes. Yes, we have a number of *in vitro* techniques that allow us to track cell deformation under controlled conditions and we developed a number of mathematical models to extract the information about shear modulus, viscosity and other material constants of living cells. However, there is no guarantee that the properties measured by these techniques will describe living cells *in vivo*. Recent findings, for instance, indicate that leukocytes become hypo- or hyper-responsive, change sensitivity to cytokine levels, and show altered phosphorylation kinetics when taken from the *in vivo* environment (Krutzik, Hale, & Nolan, 2005). Besides changing the phenotype upon isolation and culture, WBCs can be activated when exposed to controlled conditions of an *in vitro* experiment. Without proper understanding of how the behavior of leukocytes changes when they become active, we cannot trustfully use *in vitro* approaches to measure leukocyte deformability. Thus, we are in urgent need to develop a realistic mathematical model that can predict both passive and active deformation of a specific type of the leukocyte. Having said that, we point out that the material constants of WBCs (mostly, human neutrophils) were almost exclusively measured by micropipette aspiration. Only recently, particle-tracking microrheology and atomic force microscopy have been used to study the rheology of these cells. Here, we give a short review of the rheological methods applied or potentially applicable to leukocytes.

A. Micropipette Aspiration

The basic idea of this method is to aspirate the cell into a glass pipette with diameter $2R_p$ smaller than the undeformed cell diameter $2R_{c0}$ (Fig. 6), measure the cell projection length L_p in the pipette at different time instances t and suction pressures ΔP, and determine the material constants of the cell by fitting a rheological model (constitutive relation) to the experimental curves of the projection length versus suction pressure and time (Fig. 7). In a typical design (Hochmuth, 2000), the pipette is connected through tubing to the water reservoir and immersed in the flow chamber with cells. Its movement is controlled by a pneumatic micromanipulator. The suction pressure, that is, the difference in the ambient pressure between the flow chamber P_0 and the pipette P_p, is regulated by the hydrostatic head. The aspiration experiment begins with first applying the suction pressure of desired magnitude via the pipette and then selecting a relatively round leukocyte with no *visual* sign of activation (in most studies, this sign was the presence of pseudopods). Bringing the mouth of the pipette into close proximity of the selected cell leads to rapid initial entry of the cell into the pipette (Fig. 6, top), followed by slow deformation and, at high suction pressures, complete entrance of the cell (Evans & Yeung, 1989; Needham & Hochmuth, 1990).

The following tests were performed to measure the material constants of aspirating cells: step aspiration (Chien, Schmid-Schonbein, Sung, Schmalzer, & Skalak, 1984; Dong et al., 1988; Schmid-Schonbein et al., 1981), complete aspiration (Evans & Yeung, 1989; Needham & Hochmuth, 1990; Tsai et al., 1993), and recovery (Dong et al., 1988, 1991; Evans & Kukan, 1984; Hochmuth, Ting-Beall, Beaty, Needham, & Tran-Son-Tay, 1993; Sung, Dong, Schmid-Schonbein, Chien, & Skalak, 1988; Tran-Son-Tay, Needham, Yeung, & Hochmuth, 1991). In the *step aspiration test*, the suction pressure is low but increases by a small amount every fixed time interval. The length of the cell projection into the pipette is measured as a function of time. These data are then "best-fitted" to a rheological model. So far, two rheological models were considered with this small deformation

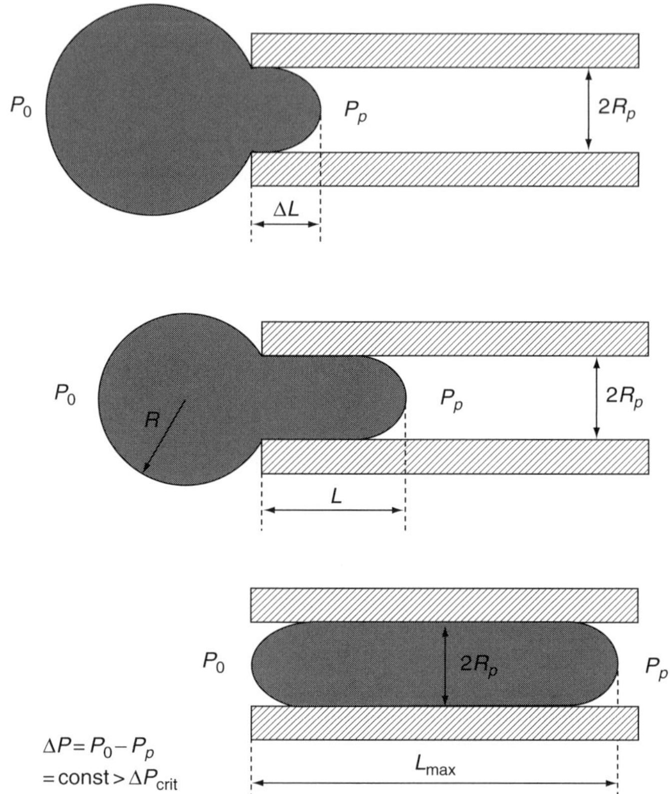

FIGURE 6 Schematics of complete aspiration of the leukocyte in a micropipette. The top picture shows an initial jump of the projection length associated with leukocyte elasticity (finite relaxation time) and rapid ultrafiltration of water in the cell. The middle and bottom pictures illustrate how the leukocyte flows into the pipette. After complete aspiration, the sausage-shaped cell (bottom) moves farther away from the pipette mouth.

experiment: standard viscoelastic solid (Chien et al.; Schmid-Schonbein et al.)

$$\frac{\mu_s}{k_2}\frac{\partial \tau}{\partial t} + \tau = 2k_1\gamma + 2\mu_s\left(1 + \frac{k_1}{k_2}\right)\dot{\gamma} \tag{1}$$

and Maxwell fluid (Dong et al., 1988):

$$\frac{\mu_s}{k_2}\frac{\partial \tau}{\partial t} + \tau = 2\mu_s\dot{\gamma}. \tag{2}$$

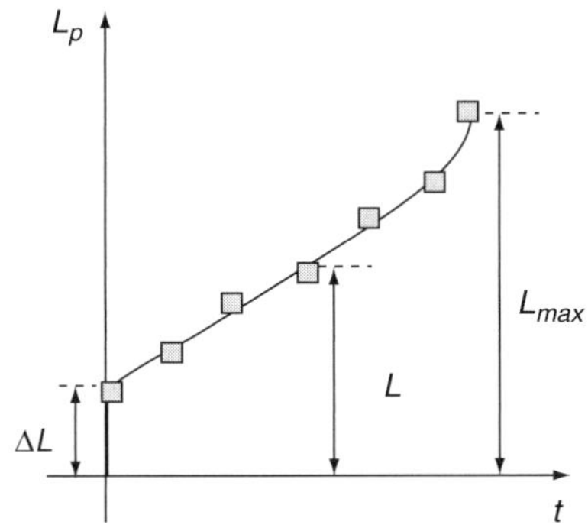

FIGURE 7 Illustration of how the projection length changes with time during complete aspiration of the leukocyte. After an initial jump ΔL, the projection length increases linearly with time, $L = L(t)$, until it reaches the maximal value L_{max} at which the cell completely enters the pipette.

Here τ, γ, $\dot{\gamma}$ are the shear stress, strain, and rate-of-strain tensors; k_1 the principal elastic element; k_2 the Maxwell fluid elastic element; and μ_s the shear viscosity. When applied to step aspiration data, the standard viscoelastic solid model predicts the following values for the material constants of human neutrophils (Schmid-Schonbein et al., 1981):

$$\begin{aligned} \mu_s &= 65 \pm 27\,\text{poise} = 6.5 \pm 2.7\,\text{Pa s}, \\ k_1 &= 137.5 \pm 59.5\,\text{dyn cm}^{-2}, \\ k_2 &= 368.5 \pm 173.0\,\text{dyn cm}^{-2}. \end{aligned} \qquad (3)$$

From Eq. (3), it follows that the relaxation time of neutrophil cytoplasm $\lambda = \mu_s/k_2 \approx 0.176\,\text{s}$. It is important to say that the rheological model used by Schmid-Schonbein et al. (1981) misses a factor of 2 in the elastic and viscous terms. For proper comparison with other rheological data, we decreased the values of material constants given in that paper by this factor.

Dong et al. (1988) modeled the neutrophil as a droplet of Maxwell fluid covered with the shell with isotropic tension T_0. The application of this model to step aspiration data gave the following values: $\mu_s \approx 300\,\text{poise}$, $k_2 \approx 285\,\text{dyn cm}^{-2}$, and $T_0 \approx 0.031\,\text{dyn cm}^{-2}$. This shows that measurements of the shear viscosity of the leukocyte cytoplasm are sensitive to the type of the rheological model.

From studies with human granulocytes (PMNs), it is known that these cells completely enter the pipette at suction pressures exceeding a critical value ΔP_{crit} but stabilize their shape after the projection length reaches a certain limit (less than the pipette diameter), if the pressure is less than critical (Evans & Yeung, 1989; Hochmuth, 2000; Needham & Hochmuth, 1992).

The existence of the critical suction pressure is evidence that granulocytes possess cortical tension and behave as a liquid. A simple approach to measure cortical tension is to aspirate the cell into pipettes of different radii, find the critical suction pressure as a function of the pipette radius, and, assuming that the cell is a liquid droplet, "best-fit" the Young–Laplace equation to the data. This approach gives values for the cortical tension in human granulocytes in the range from 0.016 to 0.035 dyn cm^{-1} (Evans & Yeung; Lomakina, Spillmann, King, & Waugh, 2004; Needham & Hochmuth; Simon et al., 2007; Tsai et al., 1993). However, measurements of the cortical tension are sensitive to the pipette radius due to bending elasticity and finite thickness of the cortex, as discussed by Zhelev et al. (1994).

In the *complete aspiration test*, the leukocyte is exposed to suction pressures above critical. The projection length, defined as a distance from the pipette mouth to the leading edge of the cell, is measured as a function of time from images of the aspirating cell at different pressures. The material constants of the cell can be found either by "best-fitting" a rheological model to the projection length versus time curves or by finding the slopes of the curves in the linear (i.e., Newtonian) region (cf. Fig. 7) and substituting these values in the expressions that describe aspiration of a Newtonian liquid droplet (Needham & Hochmuth, 1990; Tsai et al., 1993; Yeung & Evans, 1989). The cytoplasmic viscosity of human neutrophils measured by this test ranges from 1350 to 2100 poise, one order of magnitude higher than predictions of small deformation analysis. Most of these measurements are based on the Newtonian model of the leukocyte. The complete aspiration data were also analyzed by Tsai et al. using the power-law fluid model:

$$\tau = 2\mu_{\mathrm{app}}\dot{\gamma}, \mu_{\mathrm{app}} = \mu_{\mathrm{c}}\left(\frac{\dot{\gamma}_{\mathrm{m}}}{\dot{\gamma}_{\mathrm{c}}}\right)^{-b}, \tag{4}$$

with μ_{app} being the apparent viscosity of the leukocyte cytoplasm, μ_{m} the mean shear rate equal to the square root of the rate of viscous energy dissipation during the cell entry (not the same as the fluid shear rate), b the power-law exponent, and μ_{c} a characteristic viscosity determined at a characteristic shear rate $\dot{\gamma}_{\mathrm{c}}$. For the characteristic shear rate of 1 s^{-1}, this model gives the following material constants of human neutrophils: $\mu_{\mathrm{s}} \approx 1300$ poise, $b \approx 0.52$. Since the power-law exponent is positive, the apparent viscosity of the neutrophil cytoplasm decreases with an increase in the mean shear rate. This is a characteristic of a shear-thinning fluid.

It is, however, very questionable to claim that the power-law fluid model can predict all aspects of neutrophil deformation. For example, Tsai et al. (1993) were unable to fit this model to the initial portion of the projection length versus time curve. The only attempt to analyze the whole dynamics of neutrophil complete aspiration was done by Dong, Skalak, and Sung (1991). By fitting the Maxwell model to the data, this group of researchers found that the neutrophil cytoplasmic viscosity increases from 183 poise at the initial stage of cell aspiration to 2179 poise (12 times more) at the late stage (Dong et al.). Such dramatic changes in the mechanical properties cannot be a failure of the Maxwell model but rather the result of leukocyte activation during large deformation.

In the *recovery test*, the fully aspirated leukocyte is allowed to relax to its original spherical shape. This is achieved by expelling the cell rapidly out of the pipette. The material constants are found from matching the experimental recovery profiles and the corresponding predictions of a rheological model. When the neutrophil is modeled as a Newtonian liquid droplet with cortical tension, the recovery test after large deformation (aspiration into a small pipette) gives a cytoplasmic viscosity of about 1,517 poise (Tran-Son-Tay et al., 1991) but the recovery test after small deformation results in a much smaller value (600 poise) (Hochmuth et al., 1993). The cells aspirated in smaller pipettes show more irregularities in their shape during recovery (Fig. 3 in Zhelev & Hochmuth, 1994). This is additional evidence that neutrophils change their mechanical properties when undergoing large deformation in a pipette as a result of cellular activation.

Overall, micropipette aspiration introduced a lot of controversies in the field of leukocyte biomechanics. A large discrepancy in the reported values of the neutrophil cytoplasmic viscosity, even with the use of the same rheological models, questions the ability of this technique to measure passive mechanical properties of leukocytes. The active response of neutrophils to mechanical stresses is the most probable reason why current rheological models cannot fully capture all of the features of neutrophil aspiration (Drury & Dembo, 2001).

B. Atomic Force Microscopy

AFM belongs to a class of scanning probe microscopes that acquire the information about a sample through raster scanning the surface of the sample with a physical probe (tip) (Binnig, Quate, & Gerber, 1986). In AFM, the tip is connected to a cantilever that deflects when the tip is brought in close contact with the surface. The vertical deflection of the cantilever is measured by an optical beam deflection system in which the reflection of a laser beam from the cantilever surface is detected by an array of photodiodes (Meyer & Amer, 1988).

The AFM can operate in three modes (Lehenkari, Charras, Nesbitt, & Horton, 2000): (1) *contact mode* when the tip-to-sample distance is adjusted through a feedback mechanism (piezoelectric actuators) to maintain a

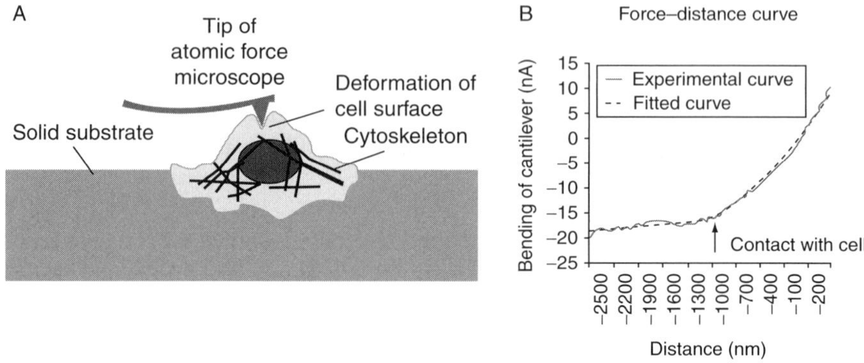

FIGURE 8 Illustration of the force mapping mode of AFM. (A) The stiff tip of the cantilever penetrates into the cell. (B) A typical curve of cantilever deflection (bending) versus sample height (distance). Such a curve can be fitted to a theoretical model based on contact mechanics to determine the material constants of the sample. Reproduced from Lehenkari et al. (2000). © Cambridge University Press.

constant force of tip–sample interaction; (2) *tapping mode* in which the cantilever vibrates in an acoustic or electromagnetic field, allowing to calculate a local height of the sample from changes in the amplitude of cantilever oscillation; and (3) *force mapping mode* (Fig. 8A) in which the tip penetrates into the sample and the cantilever deflection (i.e., force) is determined as a function of penetration depth (sample height). The first two modes are used for surface topography of living cells and the last mode is suitable for rheological measurements. Specifically, the experimental curves of cantilever deflection versus penetration depth (force–distance curves) (Fig. 8B) can be fit to the equation describing mechanical indentation of the sample. In most rheological studies using AFM (Costa, 2006; Radmacher, 2007; Vinckier & Semenza, 1998), this equation is based on the Hertz theory of contact mechanics (Timoshenko & Goodier, 1970). The Hertz theory is the simplest approach to describe contact of two elastic bodies. However, it does not consider the viscous properties of the sample. Only recently, Guilak and coworkers (Darling, Zauscher, Block, & Guilak, 2007; Darling, Zauscher, & Guilak, 2006) have extended the Hertz theory to viscoelastic bodies and derived equations that describe mechanical indentation of a viscoelastic cell. Besides the complex description of mechanical indentation, AFM is not suited for bulk rheological measurements. It could work well for spreading cells with small thickness but, due to inhomogeneity of the cell cytoplasm, AFM measurements of deformability of round cells such as passive leukocytes could be erroneous.

AFM was applied to measure elasticity of fibroblasts (Hutter et al., 2005; Martens & Radmacher, 2008; Rotsch, Jacobson, & Radmacher, 1999; Wu, Kuhn, & Moy, 1998), endothelial cells (Kang et al., 2008; Mathur, Truskey, &

Reichert, 2000; Sato et al., 2004), chondrocytes (Darling et al., 2006; Park, Costa, Ateshian, & Hong, 2009; Wozniak, Kawazoe, Tateishi, & Chen, 2009), and mesenchymal stem cells (Darling, Topel, Zauscher, Vail, & Guilak, 2008). However, rheological measurements of leukocytes with AFM are scarce. There exist only a few reports about using this technique to measure deformability of neutrophils and lymphocytes. Roca-Cusachs et al. (2006) have showed that the deformation of both passive nonadhered and activated adhered neutrophils by AFM is well described by the power-law fluid model. Hu, Wang, Zhao, Dong, and Cai (2009) have scanned the surface of passive and activated lymphocytes by AFM operated in the force mapping mode and extracted the Young's modulus of these cells by fitting the Hertz model to force–distance curves. The average Young's modulus of a resting lymphocyte was found to be 11.2 kPa, while activation of the lymphocyte increased its elasticity to about 19.7 kPa. These results support the hypothesis that leukocytes become significantly stiffer when activated. This group of researchers also showed that apoptosis leads to two- to threefold increase in lymphocyte deformability (7.1 kPa). Zhang, Wojcikiewicz, and Moy (2006) have used atomic force microscopy to characterize dynamic adhesion between T-lymphocytes (Jurkat E6-1 cell line) and endothelial cells (HUVEC). Specifically, they attached individual Jurkat cells to the AFM tip and then scanned the endothelial cell surface by AFM in the force mapping mode. The strength of lymphocyte–endothelial cell adhesion was determined from force–distance curves. One of the results of this work is that an increase in compression force (i.e., an increase in leukocyte deformation in the contact region) elevates lymphocyte adhesion to both resting and TNF-α-stimulated HUVEC.

C. *Particle-Tracking Microrheology*

The essence of microrheological methods is to measure the mechanical properties of living cells from the dynamics of small particles located/embedded in the cytoplasm or attached to the cell surface. There are two basic approaches in microrheology (MacKintosh & Schmidt, 1999): (1) passive tracking of particles, that is, measurement of thermal fluctuations in the coordinates of particles and (2) active tracking, that is, manipulation of particles with specific properties by external force fields. The term "particle-tracking microrheology" is usually applied to the methods that employ the first approach (Wirtz, 2009). In these methods, the motion of articles can be passively monitored by laser light scattering, as in the case of *laser-tracking microrheology* (LTM) (Gittes, Schnurr, Olmsted, MacKintosh, & Schmidt, 1997; Mason, Ganesan, vanZanten, Wirtz, & Kuo, 1997; Yamada, Wirtz, & Kuo, 2000) and *diffusing wave spectroscopy* (Mason &

Weitz, 1995), or can be quantified from analysis of live cell images obtained with the use of phase-contrast (Yap & Kamm, 2005a,b), differential interference contrast (Lau et al., 2003; Tseng, Kole, & Wirtz, 2002), or fluorescence microscopy (Tseng et al.).

LTM is based on the fact that any deviation of a particle, initially located at the laser focus, from the optical axis deflects energy away from the axis. This leads to imbalances between the forward-scattered signals at photodiode quadrants from which particle displacements can be determined (Fig. 9). LTM is essentially a single-particle-tracking method because it can capture the trajectories of no more than two particles at a time (Wirtz, 2009).

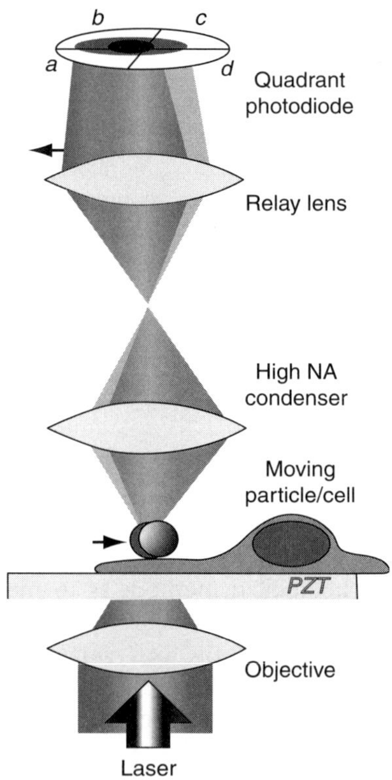

FIGURE 9 Schematics of the laser-tracking microrheology system. The particle either attached to the cell surface or present inside the cell (as an endogenous granule) is first centered in the beam of a low-power laser. Laser beam deflection due to particle motion is detected by a quadrant photodiode centered about the optical axis beyond a high NA condenser and relay lens. The photocurrent differences between opposing quadrant pairs are then converted into voltages that give particle coordinates in the x–y plane. Reprinted with permission from Macmillan Publishers Ltd. (Girard et al., 2004)

However, it is a high-resolution technique that can measure much smaller particle displacements over much shorter time scales than the existing multiple-particle-tracking methods.

Application of the theory of Brownian motion to the LTM data allows determining the bulk viscoelasticity of the cytoplasm if the tracked particle is an endogenous granule of the cell (Yamada et al., 2000) or the viscoelasticity of the cortical cytoskeleton if the particle is attached to the cell surface (Girard, Chaney, Delannoy, Kuo, & Robinson, 2004). This is done by first calculating the mean squared displacement (MSD) of the particle in the x–y plane as a function of the lag time τ:

$$\mathrm{MSD}(\tau) = \langle[\mathbf{r}(t+\tau) - \mathbf{r}(t)]^2\rangle = \langle[x(t+\tau) - x(t)]^2 + [y(t+\tau) - y(t)]^2\rangle, \tag{5}$$

where $\mathbf{r}(t) = (x(t), y(t))$ is the position vector of the particle in the x–y plane with an origin at the optical axis. In the general case, the next step should be to solve the generalized Langevin equation that describes the Brownian motion of the particle in a viscoelastic fluid. The cell sample is, however, not exposed to external forces in the LTM system and the interia of the cytoplasm is negligible due to the small size and high viscosity of the sample. This allows us to simplify the Langevin equation to the generalized Stokes–Einstein relation. If the fluid is isotropic and incompressible and the particle is spherical and rigid, this relation can be written in Laplace space as follows (Mason et al., 1997):

$$G_{\mathrm{L}}(s) = \frac{k_{\mathrm{B}}T}{\pi R s}\left[\frac{1}{\mathrm{MSD}_{\mathrm{L}}(s)}\right]. \tag{6}$$

Here s is the Laplace frequency, T the temperature, k_{B} the Boltzmann's constant, R the particle radius, $G_{\mathrm{L}}(s)$ the fluid viscoelastic spectrum, and $\mathrm{MSD}_{\mathrm{L}}(s)$ the unilateral Laplace transform of 2D $\mathrm{MSD}(\tau)$. Note that Eq. (6) is missing a factor of 2/3, present in the classical Stokes–Einstein formula, due to extension of $\mathrm{MSD}_{\mathrm{L}}(s)$ to 3D space. The last two steps in LTM is to take the inverse Laplace transform of $G_{\mathrm{L}}(s)$and then compute the unilateral Fourier transform of the resulting function. This gives us the complex shear modulus $G^*(\omega)$ of the material:

$$\begin{aligned} G^*(\omega) &= G'(\omega) + \mathrm{i}G''(\omega) = G_{\mathrm{a}}(\omega)\exp[\mathrm{i}\delta(\omega)], \\ G_{\mathrm{a}}(\omega) &= |G^*|, \delta(\omega) = \tan^{-1}[G''(\omega)/G'(\omega)], \end{aligned} \tag{7}$$

with ω being the angular frequency, $G'(\omega)$ the shear storage modulus (a measure of fluid elasticity in oscillatory experiments), and $G''(\omega)$ the shear loss modulus (a measure of fluid viscosity). By substituting $G'(\omega)$ and

$G''(\omega)$ into the general linear viscoelastic solid equation (Bird, Armstrong, & Hassager, 1987), we can find the particular form of the rheological model that describes the cell cytoplasm and determine the values for shear viscosity and elastic parameters (shear elasticity, relaxation time) of this material.

LTM was used to measure deformability of kidney epithelial cells (Massiera, Van Citters, Biancaniello, & Crocker, 2007; Yamada et al., 2000), fibroblasts (Jonas, Huang, Kamm, & So, 2008), and amoebas (Girard et al., 2004). These reports clearly show that living cells are characterized by bulk viscoelasticity due to cytoskeleton (Jonas et al.) and indicate that actin filaments in the cortex are not the sole contributors to the cell's mechanical properties (Yamada et al.). Although LTM is well suited to study granulocyte biomechanics due to the presence of endogenous particles in the cytoplasm, no studies have been reported on the application of this technique to leukocytes.

Microrheological methods based on analysis of live cell images deal with the Brownian diffusion of multiple particles and, therefore, this class of methods is often called *multiple-particle-tracking microrheology*. These techniques became very popular because of simple experimental implementation and the ability to measure spatial variations in mechanical properties within the cytoplasm. The bottlenecks of multiple-particle-tracking microrheology are low resolution as compared to LTM and the need for complicated theoretical analysis of the trajectories of multiple particles. Truly speaking, we have here multiple interactions between particles of similar size, a problem that is impossible to solve analytically, even with the use of Laplace or Fourier transforms. Another problem is that particles are distributed and move in 3D space. We are unable to fully capture the diffusion of these particles from 2D phase-contrast or fluorescence images and the use of the factor 3/2 in the Stokes–Einstein formula to account for 2D projections of 3D displacements of multiple particles could give more errors in determination of material constants than in laser-tracking microrheology. This problem can be resolved if particle tracking will be done using confocal microscopy. Such approach was used to measure viscoelasticity of emulsions (Moschakis et al., 2006) but not applied yet to living cells. Nevertheless, multiple-particle-tracking data were analyzed in most rheological studies with cells under the assumption that particles move in an isotropic incompressible fluid and do not interact (i.e., do not affect each other's motion) (Dangaria & Butler, 2007; Pai, Sundd, & Tees, 2008; Tseng et al., 2002; Yap & Kamm, 2005b). The theory used to determine the material constants was not different from that of single-particle tracking (Eqs. 5–7), except using the ensemble averaged MSD instead of the single-particle MSD in the Stokes–Einstein relation (Wirtz, 2009):

$$\mathrm{MSD}_{\mathrm{avg}}(\tau) = \frac{1}{N}\sum_{i=1}^{N} \mathrm{MSD}_i(\tau) = \frac{1}{N}\sum_{i=1}^{N} \langle [x_i(t+\tau) - x_i(t)]^2 + [y_i(t+\tau) - y_i(t)]^2 \rangle \tag{8}$$

(N is the number of particles). A more realistic theory called two-point microrheology was suggested few years ago by Crocker et al. (2000). Here, the correlated motion of pairs of tracer particles was taken into account. Lau et al. (2003) have applied two-point microrheology to mouse embryonic carcinoma cells (F9 cell line) and mouse macrophages (J774A.1 cell line) and, through comparison with convectional one-point microrheology data, showed that this theory is able to measure random stress fluctuations in the cytoplasm. This indicates that two-point microrheology is a step toward measurement of the active force field in living cells and thus understanding active cell deformation.

Besides the work by Lau et al. (2003), measurements of leukocyte deformability with multiple-particle-tracking microrheology were reported by Pai et al. (2008) and Yap and Kamm (2005b). As we already discussed in this chapter, Yap and Kamm aspirated human neutrophils into narrow rectangular PDMS channels (1.5 and 2.5 μm in height) and observed that neutrophils are activated at the threshold suction pressure. This indicates that neutrophils undergoing large, long-lasting deformation change their phenotype to a migratory one. Interestingly, their multiple-particle-tracking data (cytoplasmic granules were used as tracer particles) predict a significant decrease in both storage and loss moduli during the early stage of the neutrophil entry into the channel. This sudden fluidization of the neutrophil cytoplasm is associated with rupture of actin filaments or actin crosslinks, as evident from a recent computational study from the laboratory of Dr. Kamm (Kim et al., 2009). It occurs before the cell forms pseudopods. The neutrophil recovers its original deformability within a minute and becomes stiffer at later times due to cytoskeleton remodeling and activation (R.D. Kamm, personal communication). As seen in Fig. 10 in Yap and Kamm (2005b), the cells entering smaller channels at higher suction pressures recover their material constants faster. This work suggests that the 12-fold increase in the neutrophil cytoplasmic viscosity between the early and late stages of complete aspiration of the cell into a micropipette (Dong et al., 1988) is the result of rupture of the cortical cytoskeleton and subsequent remodeling and activation of the cell.

Pai et al. (2008) have also tracked cytoplasmic granules of human neutrophils and determined changes in neutrophil deformability during aspiration from the multiple-particle-tracking data. However, their channels were 5.0–7.5 μm diameter glass micropipettes coated with bovine serum albumin (BSA; prevents nonspecific adhesion) and some of them were physisorbed with ICAM-1. They found that the shear moduli of aspirating neutrophils show regional differences: the tail of the cell is the most deformable region (lowest moduli), while the central region is least deformable (highest moduli). Neutrophils exposed to ICAM-1 become 25–30% stiffer than the cells in

BSA-coated micropipettes. An increase in shear moduli for these cells was evident for each region of the cell body. This result is an indication that the presence of cell adhesion molecules on the endothelial cell surface coupled with large deformation of rolling leukocytes leads to rapid stiffening of the leukocyte cytoplasm and leukocyte activation.

There is "one rotten apple that spoils" *in vitro* studies of leukocyte deformation. Normal leukocytes are not activated, that is, they experience only minimal changes in the cytoskeleton structure, during their passage through pulmonary capillaries *in vivo* (Redenbach, English, & Hogg, 1997). Premature activation of these cells could lead to potentially fatal consequences. There could be several hypotheses to explain this discrepancy. One of them is that normal leukocytes *in vivo* are probably much more deformable than the cells taken out of the living body. They are able to pass through capillaries at pressure gradients less than the threshold value for cellular activation.

D. Optical Tweezers and Magnetic Twisting Cytometry

Active tracking methods employed to measure the mechanical properties of living cells can be divided into two categories: (1) those that use light to manipulate dielectric particles (optical tweezers-based microrheology) (Henon, Lenormand, Richert, & Gallet, 1999; Laurent, Henon, et al., 2002; Yanai, Butler, Suzuki, Sasaki, & Higuchi, 2004) and (2) those that impose magnetic fields to rotate or move ferromagnetic or paramagnetic particles (magnetic twisting cytometry and magnetic bead microrheometry) (Bausch, Ziemann, Boulbitch, Jacobson, & Sackmann, 1998; Fabry et al., 2001; Wang & Ingber, 1995).

In *optical tweezers*, dielectric particles are tracked actively by a focused laser beam. Such a particle when trapped in the laser focus experiences radiation pressure. The radiation pressure is the result of refraction of light by the object characterized by a refractive index different from the refractive index of the surrounding media. The radiation force on a particle can be split into two orthogonal components: the scattering force parallel to the direction of the laser beam and the gradient force. When the scattering force balances the force due to gravity, the particle located at the laser focus is levitated but can move out of the optical axis unless stabilized by the gradient force. The gradient force is the result of nonuniform distribution of light intensity across the laser beam. It points toward the high-intensity region of the beam, that is, the optical axis. Because of the gradient force, a particle displaced from the optical axis will be forced to return to its original location. Thus, it can be trapped in all three directions at appropriate intensity of the laser beam as a result of the scattering and gradient forces (Ashkin, 1997). Laser light is able to trap and move small endogenous particles in the cytoplasm (Valero, Nevian, Ho, & Lindau, 2008; Yanai et al., 1999, 2004) and can also capture and immobilize individual cells (Mannie, McConnell, Xie, & Li, 2005).

The standard approach to measure cell rheology by optical tweezers is to trap a small spherical particle (microbead) attached to the cell surface or present in the cell cytoplasm by a laser and then observe the bead motion by turning the laser off (Laurent, Henon, et al., 2002). The released bead motion will be dictated by the hydrodynamic force on the bead, which in turn depends on the mechanical properties of the cell. Alternatively, the position of the optical trap can be suddenly changed, allowing a particle (e.g., a cytoplasmic granule of a neutrophil) located in the original focal point to move in a viscoelastic cytoplasm (Yanai et al., 1999, 2004). The material constants of the cytoplasm can then be determined by fitting a specific rheological model to the granule displacement versus time curves. It should be noted that optical tweezers were not used and are not suitable to measure the passive mechanical properties of leukocytes. This is because the radiation force imposed on a trapped particle or cell leads to cellular activation. However, optical tweezers-based methods are useful to study the active behavior of leukocytes (Holm, Sundqvist, Oberg, & Magnusson, 1999; Mannie et al., 2005) and leukocyte interactions with adhesive surfaces (Anvari, Torres, & McIntyre, 2004; Snijder-Van As et al., 2009; Wang et al., 2006).

Magnetic twisting cytometry deals with ferromagnetic beads attached to the cell surface. It is well known that ferromagnetic magnetized particles dispersed in a fluid medium (e.g., the cell cytoplasm) respond to a uniform magnetic field like compass needles (Moller, Takenaka, Rust, Stahlhofen, & Heyder, 1997). Their dipoles rotate toward the direction of this field at a rate dependent on the mechanical properties of the medium, thus changing the remanent magnetic field in the direction of original magnetization of these particles. When applied to cells (Laurent, Henon, et al., 2002; Saito, Lai, Rogers, & Doerschuk, 2002; Wang & Ingber, 1995), magnetic twisting cytometry consists of the following steps (Fig. 10): (1) ferromagnetic beads are attached to or internalized by the cell located on the horizontal substrate; (2) the beads are magnetized in a horizontal direction by a brief magnetic pulse of high intensity; (3) a weak uniform "twisting" field $\mathbf{H}_{\mathrm{tw}}$ is applied orthogonally to the direction of original magnetization for about 1 min and, at the end of this procedure, the remanent magnetic field of the bead in the direction of original magnetization $B_{\mathrm{r}}(t) = B_{\mathrm{r}0}\cos\Omega(t)$ is measured by a magnetometer (here Ω is the angle between the direction of original magnetization and the direction of a uniform magnetic field); and (4) the twisting filed is turned off and the recovery of the remanent magnetic field is measured.

Stage 1: Bead attachment

Stage 2: Brief application of a strong external magnetic field

Stage 3: Bead twisting by a weak magnetic field applied perpendicular to the original field

Stage 4: Recovery, i.e., the twisting field is off

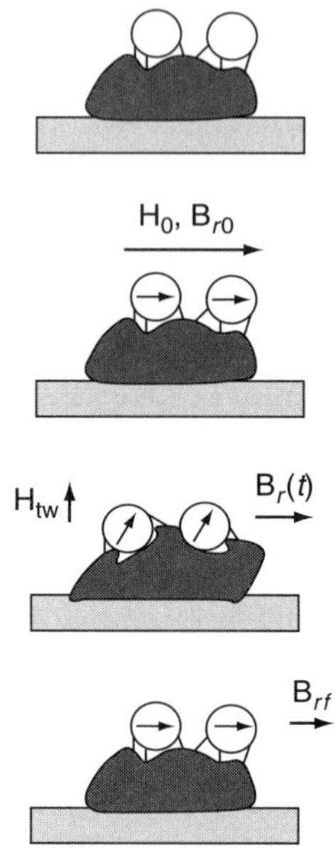

FIGURE 10 Manipulation of particles in magnetic twisting cytometry.

The time dependence of the remanent magnetic field gives information about the mechanical properties of the cell. Bead rotation is described by the equation of angular motion:

$$I\frac{\mathrm{d}^2\Omega}{\mathrm{d}t^2} = T_\mathrm{m} + T_\mathrm{h}, T_\mathrm{m} = \int_{\text{bead volume}} \mu_0(\mathbf{M}_\mathrm{r} \times \mathbf{H}_\mathrm{tw})\mathrm{d}V, T_\mathrm{h} = \int_{\text{bead surface area}} (\mathbf{F}_\mathrm{h} \times \mathbf{r})\mathrm{d}S, \tag{9}$$

where I is the moment of inertia, μ_0 the magnetic constant equal to $4\pi \times 10^{-7}\,\mathrm{N\,A^{-2}}$, $\mathbf{M}_\mathrm{r}$ the remanent magnetic moment per unit volume, $\mathbf{F}_\mathrm{h}$ the hydrodynamic force on the bead, $\mathbf{r}$ the radius vector from the center of mass of the bead to the bead surface, T_m the magnetic torque due to the twisting field, and T_h the hydrodynamic countertorque due to cell viscoelasticity. The angle Ω is determined as a function of time from measured values of the remanent magnetic field $B_\mathrm{r}(t)$ using the formula:

$$\Omega(t) = \cos^{-1}(B_\mathrm{r}/B_{\mathrm{r}0}). \tag{10}$$

The period of 1 min is selected in stage 3 to have an approximately linear dependence of the angle on time. Then, the inertial term in Eq. (9) is negligible and the equation of angular motion is reduced to the torque balance condition

$T_m + T_h = 0$. Using a specific rheological model (e.g., Kelvin–Voigt or Maxwell) and taking into account that the bead is spherical, this condition can be written as a differential equation for the angle Ω. Fitting this equation to the experimental data on the angle gives the material constants of the cell. In *oscillatory magnetic twisting cytometry* (Fabry et al., 2001; Massiera et al., 2007; Wang et al., 2002), the twisting magnetic field varies sinusoidally with time and is characterized by angular frequency ω, that is, $\mathbf{H}_{tw} = \mathbf{H}_{tw}\cos(\omega t)$. Then, the torque balance condition allows estimating the shear storage and loss moduli of the cell.

Magnetic twisting cytometry was used by Wang and Ingber (Lele et al., 2007; Wang & Ingber, 1995; Wang et al., 2002) as well as other researchers (Laurent, Canadas, et al., 2002) to measure the mechanical properties of human airway smooth muscle (HASM) cells and A549 human alveolar epithelial cells and to validate the tensegrity model of the cell. The oscillatory version of this technique was also extensively used by the group of Dr. Fredberg (Fabry et al., 2001; Lenormand & Fredberg, 2006; Lenormand et al., 2004; Maksym et al., 2000) to measure the shear moduli of HASM cells, from which they concluded that living cells are soft glassy materials that behave as power-law fluids. It should be noted, however, that the cells are inhomogeneous in their mechanical properties according to particle-tracking microrheology. Magnetic twisting of beads attached to the cell surface allows measuring the mechanical properties of the cortical cytoskeleton but the whole cell could behave differently from what was predicted by this technique. Nethertheless, Saito et al. (2002) measured deformability of rat neutrophils by magnetic twisting cytometry. They found that mature neutrophils isolated from the rat bone marrow are much stiffer than the neutrophils isolated from rat blood. Wang et al. (2001) have applied magnetic twisting cytometry to assess changes in the stiffness of human neutrophils and human pulmonary microvascular endothelial cells during their interactions. An important result of this work is that adhesion of neutrophils to TNF-α-treated endothelial cells increased the stiffness of both neutrophils and endothelial cells.

Magnetic bead microrheometry, also called magnetic tweezers, deals with paramagnetic beads (i.e., the particles are magnetized only when they are exposed to a magnetic field). In this method, a strong external magnetic field is briefly applied to the bead either bound to the cell surface (Bausch et al., 1998) or embedded into the cell cytoplasm (de Vries, Krenn, van Driel, & Kanger, 2005). Displacements of the bead before, during, and after the application of the magnetic field were determined by single-particle tracking, giving the creep response and relaxation curves. By fitting a rheological model to these curves it is possible to determine the mechanical properties of the cell.

VIII. REDUCED DEFORMABILITY OF WHITE BLOOD CELLS LEADS TO PATHOLOGIES

The volume fraction of leukocytes in normal human blood is less than 0.5%, and yet these cells, by virtue of their mechanical properties, significantly influence the flow of blood through the body. Leukocytes have a much more complex structure and, as a result, are larger in volume and less deformable than red blood cells and platelets. The diameter of circulating leukocytes ranges from 7 to 10 μm (Schmid-Schonbein, Shih, et al., 1980; Ting-Beall, Needham, & Hochmuth, 1993), generally exceeding the size of true capillaries (5–9 μm) (Ganong, 2001). Thus, WBCs cannot pass through the smallest vessels of the body without deformation. Moving more slowly than red blood cells (Ben-nun, 1996) and often temporarily trapped in capillaries (Nishiwaki, Ogura, Kimura, Kiryu, & Honda, 1995; Su et al., 2003), leukocytes significantly increase microvascular resistance to blood flow. Adding just 0.1% concentration of passive leukocytes to the suspension of red blood cells in plasma leads to a 22% increase in the vascular resistance to suspension flow in a rat skeletal muscle (Sutton & Schmid-Schonbein, 1992). Several mechanisms are involved in this effect: leukocyte plugging including slow entry into and long transit through capillaries (Bathe et al., 2002; Eppihimer & Lipowsky, 1996; Fenton, Wilson, & Cokelet, 1985; Harris & Skalak, 1993; House & Lipowsky, 1987; Warnke & Skalak, 1992), local changes in hematocrit due to redistribution of red blood cells on the upstream side of a slowly moving leukocyte (Helmke, Bremner, Zweifach, Skalak, & Schmid-Schonbein, 1997; Helmke, Sugihara-Seki, Skalak, & Schmid-Schonbein, 1998), and reduction in the effective venular diameter due to adherent leukocytes (House & Lipowsky).

It is well known from micropipette studies (Needham & Hochmuth, 1990; Yeung & Evans, 1989) that the rate of leukocyte entry into a vessel is inversely proportional to the cytoplasmic viscosity. Therefore, an increase in the viscosity (i.e., a decrease in the leukocyte deformability) increases the duration and incidence of leukocyte plugging (Harris & Skalak, 1993). This indicates that if the deformability of WBCs is decreased as compared to normal levels, tissue perfusion may be compromised (Ellis, Jagger, & Sharpe, 2005). Reduced deformability of circulating leukocytes is a consequence of untimely activation of these cells. It is a hallmark of many pathological conditions, such as acute and ventilator-induced lung injury (Choudhury, Wilson, Goddard, O'Dea, & Takata, 2004; Drost & MacNee, 2002; Suemori et al., 2009; Worthen, Schwab, Elson, & Downey, 1989), sepsis and posttraumatic shock (Nishino et al., 2005; Poschl, Ruef, & Linderkamp, 2005; Skoutelis, Kaleridis, Athanassiou, et al., 2000; Yodice, Astiz, Kurian, Lin, & Rackow, 1997), diabetes (Athanassiou, Matsouka,

Kaleridis, & Missirlis, 2000; Linderkamp, Ruef, Zilow, & Hoffmann, 1999), stroke (Mercuri, Ciuffetti, Robinson, & Toole, 1989; Vermes & Strik, 1988), ischemia–reperfusion injury (Harris & Skalak, 1996; Tillmanns et al., 1993), hypertension (Shen et al., 1995), chronic renal failure (Skoutelis, Kaleridis, Goumenos, et al., 2000), and HIV infection (Dadgostar et al., 2006).

The lungs are the most susceptible organ to leukocyte-mediated injury because of significant retention of neutrophils and other leukocytes in pulmonary capillaries. The pulmonary capillary bed is a complex interconnected network of short capillary segments in the alveolar wall (Doerschuk, 2001), while systemic capillaries are relatively long tubes characterized by a small number of intersections (Rhoades & Bell, 2008). The presence of numerous turns and intersections certainly reduces the leukocyte transit time. The average diameter of pulmonary capillaries is about 5 μm (Doerschuk; Doerschuk, Beyers, Coxson, Wiggs, & Hogg, 1993), which is smaller than the average size of systemic capillaries. Leukocytes should deform significantly to enter and pass through these vessels but the driving pressure in the pulmonary capillary beds is only 10 mmHg ($\sim$1300 Pa) (Rhoades & Bell), which is insufficient to quickly move the cells to larger vessels. As a result, leukocytes can be immobilized within pulmonary capillaries for several minutes (Lien et al., 1987; Yamaguchi et al., 1997). An additional reason why leukocytes play a critical role in lung injury is that pulmonary vessels are the sites of leukocyte transendothelial migration (Doerschuk). (Please note that leukocyte trafficking in systemic circulation occurs in much larger postcapillary venules.) Accumulated in these small interconnected vessels, leukocytes can be easily activated by various mediators present on endothelial cells (Doerschuk; Drost & MacNee, 2002; Sato, Hogg, English, & van Eeden, 2000; Skoutelis, Kaleridis, Gogos, et al., 2000; Worthen et al., 1989). Activated WBCs can damage the alveolar walls by releasing reactive oxygen species and proteolytic enzymes (Downey, Fialkow, & Fukushima, 1995) and penetrate into the alveolar space, leading to pulmonary dysfunction and airway collapse (Choudhury et al., 2004). Leukocyte activation resulting from pathophysiological changes in other organs leads to a large pool of leukocytes in pulmonary vessels, which can also result in lung injury (Nishino et al., 2005; Suemori et al., 2009; Xiao, Eppihimer, Willis, & Carden, 1997; Xiao, Eppihimer, Young, Nguyen, & Carden, 1997).

Neutrophil activation and an associated decrease in deformability of these cells are responsible for clinical signs of septic shock and diabetes. Sepsis, or systemic inflammatory response syndrome, is the exaggerated manifestation of the body's reaction to pathogenic microorganisms (Hansel & Dintzis, 2004). This pathological condition is often fatal and most commonly associated with the release of endotoxin (lipopolysaccharide, LPS) found in Gram-negative bacteria. LPS binds to its signaling receptor on macrophages, triggering the release of various chemoattractants for neutrophils (TNF-α, interleukins, platelet-activating factor) (Hansel & Dintzis), which significantly reduce neutrophil deformability (Skoutelis, Kaleridis, Gogos, et al., 2000). Rigid, activated neutrophils in patients with sepsis (Linderkamp, Ruef, Brenner, Gulbins, & Lang, 1998; Nishino et al., 2005; Poschl et al., 2005) can occlude capillaries and damage tissues in many organs. Decreased deformability of leukocytes is a feature of both type I and type II diabetes mellitus (Athanassiou et al., 2000; Linderkamp et al., 1999; Perrault, Bray, Didier, Ozaki, & Tran-Son-Tay, 2004). It is responsible for hemorheological abnormalities (Harris, Skalak, & Hatchell, 1994) which may cause diabetic microangiopathy (Miyamoto, Ogura, Kenmochi, & Honda, 1997). The exact reason why diabetic patients have a large number of activated leukocytes is not yet known.

Under normal physiological conditions, however, circulating WBCs are not activated even when they pass through smallest pulmonary capillaries (Redenbach et al., 1997). Another interesting result about the ability of neutrophils to maintain a low level of cellular activation was reported by the laboratory of Dr. Schmid-Schonbein. They observed retraction of pseudopods and a decrease in the apparent viscosity for leukocytes exposed to high shear stresses (Fukuda, Yasu, Predescu, & Schmid-Schonbein, 2000; Makino et al., 2005; Marschel & Schmid-Schonbein, 2002; Moazzam et al., 1997). Clearly, it would be important to elucidate how they can support their passive state despite significant deformation *in vivo* and what mechanisms are involved in switching these cells to a migratory phenotype.

IX. CONCLUDING REMARKS

The evidence presented in this chapter shows the importance of leukocyte deformability in inflammation and other pathological conditions but also emphasizes our limited knowledge about the biomechanical behavior of these cells. In the last 15 years, we have witnessed the development of rheological methods that could potentially provide deep insights in passive mechanical properties of leukocytes and could help to understand the interplay between molecular signaling and leukocyte deformation. The following questions need to be addressed: What role do intermediate filaments and microtubules play in leukocyte biomechanics? How to describe the forces that drive actin polymerization and active deformation of white blood cells? How do the conditions of *in vitro* experiments influence the mechanical response of leukocytes? And, can we extend *in vitro* rheological techniques to measure the mechanical properties of leukocytes *in vivo*?

Acknowledgments

The author thanks Klaus Ley, Roger Kamm, and George Truskey for helpful discussions and Edward Damiano for providing images of rolling leukocytes.

References

Abitorabi, M. A., Pachynski, R. K., Ferrando, R. E., Tidswell, M., & Erle, D. J. (1997). Presentation of integrins on leukocyte microvilli: A role for the extracellular domain in determining membrane localization. *Journal of Cell Biology, 139*(2), 563–571.

Abkarian, M., & Viallat, A. (2005). Dynamics of vesicles in a wall-bounded shear flow. *Biophysical Journal, 89*(2), 1055–1066.

Alberts, B., Johnson, A., Lewis, J., Raff, M., Roberts, K., & Walter, P. (2002). *Molecular biology of the cell* (4th ed.). New York, NY: Garland.

Alon, R., Grabovsky, V., & Feigelson, S. (2003). Chemokine induction of integrin adhesiveness on rolling and arrested leukocytes local signaling events or global stepwise activation? *Microcirculation, 10*(3–4), 297–311.

Alon, R., Kassner, P. D., Carr, M. W., Finger, E. B., Hemler, M. E., & Springer, T. A. (1995). The integrin VLA-4 supports tethering and rolling in flow on VCAM-1. *Journal of Cell Biology, 128*(6), 1243–1253.

Anderson, G. J., Roswit, W. T., Holtzman, M. J., Hogg, J. C., & Van Eeden, S. F. (2001). Effect of mechanical deformation of neutrophils on their CD18/ICAM-1-dependent adhesion. *Journal of Applied Physiology, 91*(3), 1084–1090.

Anvari, B., Torres, J. H., & McIntyre, B. W. (2004). Regulation of pseudopodia localization in lymphocytes through application of mechanical forces by optical tweezers. *Journal of Biomedical Optics, 9*(5), 865–872.

Arnoczky, S. P., Lavagnino, M., Whallon, J. H., & Hoonjan, A. (2002). *In situ* cell nucleus deformation in tendons under tensile load: A morphological analysis using confocal laser microscopy. *Journal of Orthopaedic Research, 20*(1), 29–35.

Asako, H., Kurose, I., Wolf, R., DeFrees, S., Zheng, Z. L., Phillips, M. L., et al. (1994). Role of H1 receptors and P-selectin in histamine-induced leukocyte rolling and adhesion in postcapillary venules. *Journal of Clinical Investigation, 93*(4), 1508–1515.

Ashkin, A. (1997). Optical trapping and manipulation of neutral particles using lasers. *Proceedings of the National Academy of Sciences of the United States of America, 94*(10), 4853–4860.

Athanassiou, G., Matsouka, P., Kaleridis, V., & Missirlis, Y. (2000). Deformability and filterability of white blood cell subpopulations. Evaluation of these parameters in the cell line HL-60 and in type II diabetes mellitus. *Clinical Hemorheology and Microcirculation, 22*(1), 35–43.

Bathe, M., Shirai, A., Doerschuk, C. M., & Kamm, R. D. (2002). Neutrophil transit times through pulmonary capillaries: The effects of capillary geometry and fMLP-stimulation. *Biophysical Journal, 83*(4), 1917–1933.

Bausch, A. R., Ziemann, F., Boulbitch, A. A., Jacobson, K., & Sackmann, E. (1998). Local measurements of viscoelastic parameters of adherent cell surfaces by magnetic bead microrheometry. *Biophysical Journal, 75*(4), 2038–2049.

Becker, B. E., & Gard, D. L. (2006). Visualization of the cytoskeleton in *Xenopus* oocytes and eggs by confocal immunofluorescence microscopy. *Methods in Molecular Biology, 322*, 69–86.

Ben-nun, J. (1996). Comparative flow velocity of erythrocytes and leukocytes in feline retinal capillaries. *Investigative Ophthalmology & Visual Science, 37*(9), 1854–1859.

Berg, E. L., Goldstein, L. A., Jutila, M. A., Nakache, M., Picker, L. J., Streeter, P. R., et al. (1989). Homing receptors and vascular addressins: Cell adhesion molecules that direct lymphocyte traffic. *Immunological Reviews, 108*, 5–18.

Bergen, L. G., & Borisy, G. G. (1980). Head-to-tail polymerization of microtubules *in vitro*. Electron microscope analysis of seeded assembly. *Journal of Cell Biology, 84*(1), 141–150.

Berger, M., Budhu, S., Lu, E., Li, Y., Loike, D., Silverstein, S. C., et al. (2002). Different G(i)-coupled chemoattractant receptors signal qualitatively different functions in human neutrophils. *Journal of Leukocyte Biology, 71*(5), 798–806.

Bevilacqua, M. P., Stengelin, S., Gimbrone , M. A., Jr., & Seed, B. (1989). Endothelial leukocyte adhesion molecule 1: An inducible receptor for neutrophils related to complement regulatory proteins and lectins. *Science, 243*(4895), 1160–1165.

Binnig, G., Quate, C. F., & Gerber, C. (1986). Atomic force microscope. *Physical Review Letters, 56*(9), 930–933.

Bird, R. B., Armstrong, R. C., & Hassager, O. (1987). *Dynamics of polymeric liquids, Vol. 1: Fluid mechanics New York, NY: Wiley.*

Blocker, A., Griffiths, G., Olivo, J. C., Hyman, A. A., & Severin, F. F. (1998). A role for microtubule dynamics in phagosome movement. *Journal of Cell Science, 111*(Pt. 3), 303–312.

Brangwynne, C. P., MacKintosh, F. C., & Weitz, D. A. (2007). Force fluctuations and polymerization dynamics of intracellular microtubules. *Proceedings of the National Academy of Sciences of the United States of America, 104*(41), 16128–16133.

Broedersz, C. P., Storm, C., & Mackintosh, F. C. (2009). Effective-medium approach for stiff polymer networks with flexible cross-links. *Physical Review E: Statistical, Nonlinear, and Soft Matter Physics, 79*(6 Pt. 1), 061914.

Brown, M. J., Hallam, J. A., Colucci-Guyon, E., & Shaw, S. (2001). Rigidity of circulating lymphocytes is primarily conferred by vimentin intermediate filaments. *Journal of Immunology, 166*(11), 6640–6646.

Bruehl, R. E., Springer, T. A., & Bainton, D. F. (1996). Quantitation of L-selectin distribution on human leukocyte microvilli by immunogold labeling and electron microscopy. *Journal of Histochemistry and Cytochemistry, 44*(8), 835–844.

Burns, J. A., Issekutz, T. B., Yagita, H., & Issekutz, A. C. (2001). The alpha 4 beta 1 (very late antigen (VLA)-4, CD49d/CD29) and alpha 5 beta 1 (VLA-5, CD49e/CD29) integrins mediate beta 2 (CD11/CD18) integrin-independent neutrophil recruitment to endotoxin-induced lung inflammation. *Journal of Immunology, 166*(7), 4644–4649.

Burridge, K., & Wennerberg, K. (2004). Rho and Rac take center stage. *Cell, 116*(2), 167–179.

Butcher, E. C., & Picker, L. J. (1996). Lymphocyte homing and homeostasis. *Science, 272*(5258), 60–66.

Caille, N., Thoumine, O., Tardy, Y., & Meister, J. J. (2002). Contribution of the nucleus to the mechanical properties of endothelial cells. *Journal of Biomechanics, 35*(2), 177–187.

Campbell, J. J., Hedrick, J., Zlotnik, A., Siani, M. A., Thompson, D. A., & Butcher, E. C. (1998). Chemokines and the arrest of lymphocytes rolling under flow conditions. *Science, 279*(5349), 381–384.

Canadas, P., Laurent, V. M., Oddou, C., Isabey, D., & Wendling, S. (2002). A cellular tensegrity model to analyse the structural viscoelasticity of the cytoskeleton. *Journal of Theoretical Biology, 218*(2), 155–173.

Cao, D., Mizukami, I. F., Garni-Wagner, B. A., Kindzelskii, A. L., Todd , R. F., III, Boxer, L. A., et al. (1995). Human urokinase-type plasminogen activator primes neutrophils for superoxide anion release. Possible roles of complement receptor type 3 and calcium. *Journal of Immunology, 154*(4), 1817–1829.

Caputo, K. E., & Hammer, D. A. (2005). Effect of microvillus deformability on leukocyte adhesion explored using adhesive dynamics simulations. *Biophysical Journal, 89*(1), 187–200.

Carlsson, A. E. (2001). Growth of branched actin networks against obstacles. *Biophysical Journal, 81*(4), 1907–1923.

Carman, C. V., & Springer, T. A. (2004). A transmigratory cup in leukocyte diapedesis both through individual vascular endothelial cells and between them. *Journal of Cell Biology, 167*(2), 377–388.

Cassimeris, L., Pryer, N. K., & Salmon, E. D. (1988). Real-time observations of microtubule dynamic instability in living cells. *Journal of Cell Biology, 107*(6 Pt. 1), 2223–2231.

Chang, K. C., & Hammer, D. A. (2000). Adhesive dynamics simulations of sialyl-Lewis(x)/E-selectin-mediated rolling in a cell-free system. *Biophysical Journal, 79*(4), 1891–1902.

Cherla, R. P., & Ganju, R. K. (2001). Stromal cell-derived factor 1 alpha-induced chemotaxis in T cells is mediated by nitric oxide signaling pathways. *Journal of Immunology, 166*(5), 3067–3074.

Chien, S., Schmid-Schonbein, G. W., Sung, K. L., Schmalzer, E. A., & Skalak, R. (1984). Viscoelastic properties of leukocytes. *Kroc Foundation Series, 16*, 19–51.

Chien, S., & Sung, K. L. (1984). Effect of colchicine on viscoelastic properties of neutrophils. *Biophysical Journal, 46*(3), 383–386.

Chiplonkar, J. M., Vandre, D. D., & Robinson, J. M. (1992). Stimulus-dependent relocation of the microtubule organizing center in human polymorphonuclear leukocytes. *Journal of Cell Science, 102*(Pt. 4), 723–728.

Chou, Y. H., & Goldman, R. D. (2000). Intermediate filaments on the move. *Journal of Cell Biology, 150*(3), F101–F106.

Choudhury, S., Wilson, M. R., Goddard, M. E., O'Dea, K. P., & Takata, M. (2004). Mechanisms of early pulmonary neutrophil sequestration in ventilator-induced lung injury in mice. *American Journal of Physiology: Lung Cellular and Molecular Physiology, 287*(5), L902–L910.

Cooper, J. A. (1987). Effects of cytochalasin and phalloidin on actin. *Journal of Cell Biology, 105* (4), 1473–1478.

Costa, K. D. (2006). Imaging and probing cell mechanical properties with the atomic force microscope. *Methods in Molecular Biology, 319*, 331–361.

Cozens-Roberts, C., Lauffenburger, D. A., & Quinn, J. A. (1990). Receptor-mediated cell attachment and detachment kinetics. I. Probabilistic model and analysis. *Biophysical Journal, 58*(4), 841–856.

Crocker, J. C., Valentine, M. T., Weeks, E. R., Gisler, T., Kaplan, P. D., Yodh, A. G., et al. (2000). Two-point microrheology of inhomogeneous soft materials. *Physical Review Letters, 85*(4), 888–891.

Curnock, A. P., Logan, M. K., & Ward, S. G. (2002). Chemokine signalling: Pivoting around multiple phosphoinositide 3-kinases. *Immunology, 105*(2), 125–136.

Dadgostar, H., Holland, G. N., Huang, X., Tufail, A., Kim, A., Fisher, T. C., et al. (2006). Hemorheologic abnormalities associated with HIV infection: *In vivo* assessment of retinal microvascular blood flow. *Investigative Ophthalmology & Visual Science, 47*(9), 3933–3938.

Damiano, E. R., Westheider, J., Tozeren, A., & Ley, K. (1996). Variation in the velocity, deformation, and adhesion energy density of leukocytes rolling within venules. *Circulation Research, 79*(6), 1122–1130.

Dangaria, J. H., & Butler, P. J. (2007). Macrorheology and adaptive microrheology of endothelial cells subjected to fluid shear stress. *American Journal of Physiology: Cell Physiology, 293*(5), C1568–C1575.

Darling, E. M., Topel, M., Zauscher, S., Vail, T. P., & Guilak, F. (2008). Viscoelastic properties of human mesenchymally-derived stem cells and primary osteoblasts, chondrocytes, and adipocytes. *Journal of Biomechanics, 41*(2), 454–464.

Darling, E. M., Zauscher, S., Block, J. A., & Guilak, F. (2007). A thin-layer model for viscoelastic, stress-relaxation testing of cells using atomic force microscopy: Do cell properties reflect metastatic potential? *Biophysical Journal, 92*(5), 1784–1791.

Darling, E. M., Zauscher, S., & Guilak, F. (2006). Viscoelastic properties of zonal articular chondrocytes measured by atomic force microscopy. *Osteoarthritis Cartilage, 14*(6), 571–579.

De La Cruz, E. M., Ostap, E. M., Brundage, R. A., Reddy, K. S., Sweeney, H. L., & Safer, D. (2000). Thymosin-beta(4) changes the conformation and dynamics of actin monomers. *Biophysical Journal, 78*(5), 2516–2527.

de Vries, A. H., Krenn, B. E., van Driel, R., & Kanger, J. S. (2005). Micro magnetic tweezers for nanomanipulation inside live cells. *Biophysical Journal, 88*(3), 2137–2144.

Dejana, E., Spagnuolo, R., & Bazzoni, G. (2001). Interendothelial junctions and their role in the control of angiogenesis, vascular permeability and leukocyte transmigration. *Thrombosis and Haemostasis, 86*(1), 308–315.

Dembo, M., & Harlow, F. (1986). Cell motion, contractile networks, and the physics of interpenetrating reactive flow. *Biophysical Journal, 50*(1), 109–121.

Diamond, M. S., Staunton, D. E., de Fougerolles, A. R., Stacker, S. A., Garcia-Aguilar, J., Hibbs, M. L., et al. (1990). ICAM-1 (CD54): A counter-receptor for Mac-1 (CD11b/CD18). *Journal of Cell Biology, 111*(6 Pt. 2), 3129–3139.

Dichtl, M. A., & Sackmann, E. (2002). Microrheometry of semiflexible actin networks through enforced single-filament reptation: Frictional coupling and heterogeneities in entangled networks. *Proceedings of the National Academy of Sciences of the United States of America, 99*(10), 6533–6538.

Ding, M., Robinson, J. M., Behrens, B. C., & Vandre, D. D. (1995). The microtubule cytoskeleton in human phagocytic leukocytes is a highly dynamic structure. *European Journal of Cell Biology, 66*(3), 234–245.

Doerschuk, C. M. (2001). Mechanisms of leukocyte sequestration in inflamed lungs. *Microcirculation, 8*(2), 71–88.

Doerschuk, C. M., Beyers, N., Coxson, H. O., Wiggs, B., & Hogg, J. C. (1993). Comparison of neutrophil and capillary diameters and their relation to neutrophil sequestration in the lung. *Journal of Applied Physiology, 74*(6), 3040–3045.

Dong, C., Skalak, R., & Sung, K. L. (1991). Cytoplasmic rheology of passive neutrophils. *Biorheology, 28*(6), 557–567.

Dong, C., Skalak, R., Sung, K. L., Schmid-Schonbein, G. W., & Chien, S. (1988). Passive deformation analysis of human leukocytes. *Journal of Biomechanical Engineering, 110*(1), 27–36.

Dooyoung, L., & King, M. R. (2008). Microcontact printing of P-selectin increases the rate of neutrophil recruitment under shear flow. *Biotechnology Progress, 24*(5), 1052–1059.

Downey, G. P., Fialkow, L., & Fukushima, T. (1995). Initial interaction of leukocytes within the microvasculature: Deformability, adhesion, and transmigration. *New Horizons, 3*(2), 219–228.

Drost, E. M., & MacNee, W. (2002). Potential role of IL-8, platelet-activating factor and TNF-alpha in the sequestration of neutrophils in the lung: Effects on neutrophil deformability, adhesion receptor expression, and chemotaxis. *European Journal of Immunology, 32*(2), 393–403.

Drury, J. L., & Dembo, M. (2001). Aspiration of human neutrophils: Effects of shear thinning and cortical dissipation. *Biophysical Journal, 81*(6), 3166–3177.

Dufort, P. A., & Lumsden, C. J. (1996). How profilin/barbed-end synergy controls actin polymerization: A kinetic model of the ATP hydrolysis circuit. *Cell Motility and the Cytoskeleton, 35*(4), 309–330.

Dunne, J. L., Ballantyne, C. M., Beaudet, A. L., & Ley, K. (2002). Control of leukocyte rolling velocity in TNF-alpha-induced inflammation by LFA-1 and Mac-1. *Blood, 99*(1), 336–341.

Eddy, R. J., Pierini, L. M., & Maxfield, F. R. (2002). Microtubule asymmetry during neutrophil polarization and migration. *Molecular Biology of the Cell, 13*(12), 4470–4483.

Eisenbach, M. (Ed.), (2004). *Chemotaxis.* London: Imperial College Press.

Ellis, C. G., Jagger, J., & Sharpe, M. (2005). The microcirculation as a functional system. *Critical Care, 9* (Suppl. 4), S3–S8.

Eppihimer, M. J., & Lipowsky, H. H. (1996). Effects of leukocyte-capillary plugging on the resistance to flow in the microvasculature of cremaster muscle for normal and activated leukocytes. *Microvascular Research, 51*(2), 187–201.

Esue, O., Carson, A. A., Tseng, Y., & Wirtz, D. (2006). A direct interaction between actin and vimentin filaments mediated by the tail domain of vimentin. *Journal of Biological Chemistry, 281*(41), 30393–30399.

Evans, E., & Kukan, B. (1984). Passive material behavior of granulocytes based on large deformation and recovery after deformation tests. *Blood, 64*(5), 1028–1035.

Evans, E., & Yeung, A. (1989). Apparent viscosity and cortical tension of blood granulocytes determined by micropipet aspiration. *Biophysical Journal, 56*(1), 151–160.

Fabry, B., Maksym, G. N., Butler, J. P., Glogauer, M., Navajas, D., & Fredberg, J. J. (2001). Scaling the microrheology of living cells. *Physical Review Letters, 87*(14), 148102.

Feigelson, S. W., Grabovsky, V., Winter, E., Chen, L. L., Pepinsky, R. B., Yednock, T., et al. (2001). The Src kinase p56(lck) up-regulates VLA-4 integrin affinity. Implications for rapid spontaneous and chemokine-triggered T cell adhesion to VCAM-1 and fibronectin. *Journal of Biological Chemistry, 276*(17), 13891–13901.

Felgner, H., Frank, R., & Schliwa, M. (1996). Flexural rigidity of microtubules measured with the use of optical tweezers. *Journal of Cell Science, 109*(Pt. 2), 509–516.

Fenton, B. M., Wilson, D. W., & Cokelet, G. R. (1985). Analysis of the effects of measured white blood cell entrance times on hemodynamics in a computer model of a microvascular bed. *Pflügers Archiv, 403*(4), 396–401.

Finger, E. B., Bruehl, R. E., Bainton, D. F., & Springer, T. A. (1996). A differential role for cell shape in neutrophil tethering and rolling on endothelial selectins under flow. *Journal of Immunology, 157*(11), 5085–5096.

Firrell, J. C., & Lipowsky, H. H. (1989). Leukocyte margination and deformation in mesenteric venules of rat. *American Journal of Physiology, 256*(6 Pt. 2), H1667–H1674.

Franklin, R. M., & Martin, M. T. (1980). A novel method to improve immunofluorescence in histological specimens. *European Journal of Cell Biology, 21*(1), 134–138.

Fukuda, S., Yasu, T., Predescu, D. N., & Schmid-Schonbein, G. W. (2000). Mechanisms for regulation of fluid shear stress response in circulating leukocytes. *Circulation Research, 86*(1), E13–E18.

Ganong, W. F. (2001). *Review of medical physiology.* New York, NY: McGraw-Hill.

Gardel, M. L., Shin, J. H., MacKintosh, F. C., Mahadevan, L., Matsudaira, P., & Weitz, D. A. (2004). Elastic behavior of cross-linked and bundled actin networks. *Science, 304*(5675), 1301–1305.

Georgatos, S. D., Meier, J., & Simos, G. (1994). Lamins and lamin-associated proteins. *Current Opinion in Cell Biology, 6*(3), 347–353.

Gerisch, G., & Muller-Taubenberger, A. (2003). GFP-fusion proteins as fluorescent reporters to study organelle and cytoskeleton dynamics in chemotaxis and phagocytosis. *Methods in Enzymology, 361*, 320–337.

Ginsberg, M. H., Partridge, A., & Shattil, S. J. (2005). Integrin regulation. *Current Opinion in Cell Biology, 17*(5), 509–516.

Girard, K. D., Chaney, C., Delannoy, M., Kuo, S. C., & Robinson, D. N. (2004). Dynacortin contributes to cortical viscoelasticity and helps define the shape changes of cytokinesis. *EMBO Journal, 23*(7), 1536–1546.

Girard, J. P., & Springer, T. A. (1995). High endothelial venules (HEVs): Specialized endothelium for lymphocyte migration. *Immunology Today, 16*(9), 449–457.
Gittes, F., Mickey, B., Nettleton, J., & Howard, J. (1993). Flexural rigidity of microtubules and actin filaments measured from thermal fluctuations in shape. *Journal of Cell Biology, 120*(4), 923–934.
Gittes, F., Schnurr, B., Olmsted, P. D., MacKintosh, F. C., & Schmidt, C. F. (1997). Microscopic viscoelasticity: Shear moduli of soft materials determined from thermal fluctuations. *Physical Review Letters, 79*(17), 3286–3289.
Glogauer, M., Hartwig, J., & Stossel, T. (2000). Two pathways through Cdc42 couple the N-formyl receptor to actin nucleation in permeabilized human neutrophils. *Journal of Cell Biology, 150*(4), 785–796.
Gosset, P., Perez, T., Lassalle, P., Duquesnoy, B., Farre, J. M., Tonnel, A. B., et al. (1991). Increased TNF-alpha secretion by alveolar macrophages from patients with rheumatoid arthritis. *American Review of Respiratory Diseases, 143*(3), 593–597.
Grabovsky, V., Feigelson, S., Chen, C., Bleijs, D. A., Peled, A., Cinamon, G., et al. (2000). Subsecond induction of alpha4 integrin clustering by immobilized chemokines stimulates leukocyte tethering and rolling on endothelial vascular cell adhesion molecule 1 under flow conditions. *Journal of Experimental Medicine, 192*(4), 495–506.
Green, C. E., Pearson, D. N., Camphausen, R. T., Staunton, D. E., & Simon, S. I. (2004). Shear-dependent capping of L-selectin and P-selectin glycoprotein ligand 1 by E-selectin signals activation of high-avidity beta2-integrin on neutrophils. *Journal of Immunology, 172*(12), 7780–7790.
Grigorescu, G., & Kulicke, W. M. (2000). Prediction of viscoelastic properties and shear stability of polymers in solution. *Advances in Polymer Science, 152*, 1–40.
Grimm, H. P., Verkhovsky, A. B., Mogilner, A., & Meister, J. J. (2003). Analysis of actin dynamics at the leading edge of crawling cells: Implications for the shape of keratocyte lamellipodia. *European Biophysics Journal, 32*(6), 563–577.
Guilak, F. (1995). Compression-induced changes in the shape and volume of the chondrocyte nucleus. *Journal of Biomechanics, 28*(12), 1529–1541.
Guilak, F., Tedrow, J. R., & Burgkart, R. (2000). Viscoelastic properties of the cell nucleus. *Biochemical and Biophysical Research Communications, 269*(3), 781–786.
Gyoeva, F. K., & Gelfand, V. I. (1991). Coalignment of vimentin intermediate filaments with microtubules depends on kinesin. *Nature, 353*(6343), 445–448.
Hall, A., & Nobes, C. D. (2000). Rho GTPases: Molecular switches that control the organization and dynamics of the actin cytoskeleton. *Philosophical Transactions of the Royal Society of London Series B: Biological Sciences, 355*(1399), 965–970.
Hansel, D. E., & Dintzis, R. Z. (2004). *Lippincott's pocket pathology*. Philadelphia, PA: Lippincott Williams & Wilkins.
Harler, M. B., & Reichner, J. (2001). Increased neutrophil motility by beta-glucan in the absence of chemoattractant. *Shock, 16*(6), 419–424.
Harris, A. G., & Skalak, T. C. (1993). Leukocyte cytoskeletal structure determines capillary plugging and network resistance. *American Journal of Physiology, 265*(5 Pt. 2), H1670–H1675.
Harris, A. G., & Skalak, T. C. (1996). Effects of leukocyte capillary plugging in skeletal muscle ischemia–reperfusion injury. *American Journal of Physiology, 271*(6 Pt. 2), H2653–H2660.
Harris, A. G., Skalak, T. C., & Hatchell, D. L. (1994). Leukocyte-capillary plugging and network resistance are increased in skeletal muscle of rats with streptozotocin-induced hyperglycemia. *International Journal of Microcirculation: Clinical and Experimental, 14*(3), 159–166.
Head, D. A., Levine, A. J., & MacKintosh, F. C. (2003). Deformation of cross-linked semiflexible polymer networks. *Physical Review Letters, 91*(10), 108102.

Helfand, B. T., Chang, L., & Goldman, R. D. (2004). Intermediate filaments are dynamic and motile elements of cellular architecture. *Journal of Cell Science, 117*(Pt. 2), 133–141.

Helfand, B. T., Mikami, A., Vallee, R. B., & Goldman, R. D. (2002). A requirement for cytoplasmic dynein and dynactin in intermediate filament network assembly and organization. *Journal of Cell Biology, 157*(5), 795–806.

Helmke, B. P. (2005). Molecular control of cytoskeletal mechanics by hemodynamic forces. *Physiology (Bethesda), 20*, 43–53.

Helmke, B. P., Bremner, S. N., Zweifach, B. W., Skalak, R., & Schmid-Schonbein, G. W. (1997). Mechanisms for increased blood flow resistance due to leukocytes. *American Journal of Physiology, 273*(6 Pt. 2), H2884–H2890.

Helmke, B. P., Sugihara-Seki, M., Skalak, R., & Schmid-Schonbein, G. W. (1998). A mechanism for erythrocyte-mediated elevation of apparent viscosity by leukocytes *in vivo* without adhesion to the endothelium. *Biorheology, 35*(6), 437–448.

Henon, S., Lenormand, G., Richert, A., & Gallet, F. (1999). A new determination of the shear modulus of the human erythrocyte membrane using optical tweezers. *Biophysical Journal, 76*(2), 1145–1151.

Herant, M., Marganski, W. A., & Dembo, M. (2003). The mechanics of neutrophils: Synthetic modeling of three experiments. *Biophysical Journal, 84*(5), 3389–3413.

Hershkoviz, R., Cahalon, L., Gilat, D., Miron, S., Miller, A., & Lider, O. (1993). Physically damaged extracellular matrix induces TNF-alpha secretion by interacting resting CD4+ T cells and macrophages. *Scandinavian Journal of Immunology, 37*(1), 111–115.

Hesse, M., Magin, T. M., & Weber, K. (2001). Genes for intermediate filament proteins and the draft sequence of the human genome: Novel keratin genes and a surprisingly high number of pseudogenes related to keratin genes 8 and 18. *Journal of Cell Science, 114*(Pt. 14), 2569–2575.

Higgs, H. N., & Pollard, T. D. (2000). Activation by Cdc42 and PIP(2) of Wiskott–Aldrich syndrome protein (WASp) stimulates actin nucleation by Arp2/3 complex. *Journal of Cell Biology, 150*(6), 1311–1320.

Higgs, H. N., & Pollard, T. D. (2001). Regulation of actin filament network formation through ARP2/3 complex: Activation by a diverse array of proteins. *Annual Review of Biochemistry, 70*, 649–676.

Higuchi, T., & Uhlmann, F. (2005). Stabilization of microtubule dynamics at anaphase onset promotes chromosome segregation. *Nature, 433*(7022), 171–176.

Hochmuth, R. M. (2000). Micropipette aspiration of living cells. *Journal of Biomechanics, 33*(1), 15–22.

Hochmuth, R. M., Ting-Beall, H. P., Beaty, B. B., Needham, D., & Tran-Son-Tay, R. (1993). Viscosity of passive human neutrophils undergoing small deformations. *Biophysical Journal, 64*(5), 1596–1601.

Holm, A., Sundqvist, T., Oberg, A., & Magnusson, K. E. (1999). Mechanical manipulation of polymorphonuclear leukocyte plasma membranes with optical tweezers causes influx of extracellular calcium through membrane channels. *Medical & Biological Engineering & Computing, 37*(3), 410–412.

Hordijk, P. L. (2006). Endothelial signalling events during leukocyte transmigration. *FEBS Journal, 273*(19), 4408–4415.

House, S. D., & Lipowsky, H. H. (1987). Leukocyte–endothelium adhesion: Microhemodynamics in mesentery of the cat. *Microvascular Research, 34*(3), 363–379.

Howard, J. (2001). *Mechanics of motor proteins and the cytoskeleton.* Sunderland, MA: Sinauer Associates.

Howard, T. H., & Watts, R. G. (1994). Actin polymerization and leukocyte function. *Current Opinion in Hematology, 1*(1), 61–68.

Hu, S., Chen, J., Butler, J. P., & Wang, N. (2005). Prestress mediates force propagation into the nucleus. *Biochemical and Biophysical Research Communications, 329*(2), 423–428.

Hu, M., Wang, J., Zhao, H., Dong, S., & Cai, J. (2009). Nanostructure and nanomechanics analysis of lymphocyte using AFM: From resting, activated to apoptosis. *Journal of Biomechanics, 42*(10), 1513–1519.

Huang, J. D., Brady, S. T., Richards, B. W., Stenolen, D., Resau, J. H., Copeland, N. G., et al. (1999). Direct interaction of microtubule- and actin-based transport motors. *Nature, 397* (6716), 267–270.

Hunyadi, V., Chretien, D., & Janosi, I. M. (2005). Mechanical stress induced mechanism of microtubule catastrophes. *Journal of Molecular Biology, 348*(4), 927–938.

Hutter, J. L., Chen, J., Wan, W. K., Uniyal, S., Leabu, M., & Chan, B. M. (2005). Atomic force microscopy investigation of the dependence of cellular elastic moduli on glutaraldehyde fixation. *Journal of Microscopy, 219*(Pt. 2), 61–68.

Hyduk, S. J., & Cybulsky, M. I. (2009). Role of alpha4beta1 integrins in chemokine-induced monocyte arrest under conditions of shear stress. *Microcirculation, 16*(1), 17–30.

Ingber, D. E. (2003a). Tensegrity I. Cell structure and hierarchical systems biology. *Journal of Cell Science, 116*(Pt. 7), 1157–1173.

Ingber, D. E. (2003b). Tensegrity II. How structural networks influence cellular information processing networks. *Journal of Cell Science, 116*(Pt. 8), 1397–1408.

Inoue, A., Watanabe, T., Tominaga, K., Tsugawa, K., Nishio, K., Takahashi, K. P., et al. (2005). Association of hnRNP S1 proteins with vimentin intermediate filaments in migrating cells. *Journal of Cell Science, 118*(Pt. 10), 2303–2311.

Isambert, H., Venier, P., Maggs, A. C., Fattoum, A., Kassab, R., Pantaloni, D., et al. (1995). Flexibility of actin filaments derived from thermal fluctuations. Effect of bound nucleotide, phalloidin, and muscle regulatory proteins. *Journal of Biological Chemistry, 270*(19), 11437–11444.

Jablonska, E., Kiluk, M., Markiewicz, W., Piotrowski, L., Grabowska, Z., & Jablonski, J. (2001). TNF-alpha, IL-6 and their soluble receptor serum levels and secretion by neutrophils in cancer patients. *Archivum Immunologia et Therapiae Experimentalis (Warsz), 49*(1), 63–69.

Jadhav, S., Eggleton, C. D., & Konstantopoulos, K. (2005). A 3-D computational model predicts that cell deformation affects selectin-mediated leukocyte rolling. *Biophysical Journal, 88*(1), 96–104.

Janeway, C. A., Travers, P., Walport, M., & Shlomchik, M. J. (2005). *Immunobiology: The immune system in health and disease* (6th ed.). New York, NY: Garland.

Janmey, P. A., Euteneuer, U., Traub, P., & Schliwa, M. (1991). Viscoelastic properties of vimentin compared with other filamentous biopolymer networks. *Journal of Cell Biology, 113*(1), 155–160.

Janmey, P. A., Hvidt, S., Kas, J., Lerche, D., Maggs, A., Sackmann, E., et al. (1994). The mechanical properties of actin gels. Elastic modulus and filament motions. *Journal of Biological Chemistry, 269*(51), 32503–32513.

Janssen, K. P., Eichinger, L., Janmey, P. A., Noegel, A. A., Schliwa, M., Witke, W., et al. (1996). Viscoelastic properties of F-actin solutions in the presence of normal and mutated actin-binding proteins. *Archives of Biochemistry and Biophysics, 325*(2), 183–189.

Jonas, M., Huang, H., Kamm, R. D., & So, P. T. (2008). Fast fluorescence laser tracking microrheometry. II. Quantitative studies of cytoskeletal mechanotransduction. *Biophysical Journal, 95*(2), 895–909.

Kan, H. C., Shyy, W., Udaykumar, H. S., Vigneron, P., & Tran-Son-Tay, R. (1999). Effects of nucleus on leukocyte recovery. *Annals of Biomedical Engineering, 27*(5), 648–655.

Kaneider, N. C., Egger, P., Wiedermann, F. J., Ritter, M., Woll, E., & Wiedermann, C. J. (2002). Involvement of cyclic adenosine monophosphate-dependent protein kinase A and pertussis toxin-sensitive G proteins in the migratory response of human CD14+ mononuclear cells to katacalcin. *Journal of Bone and Mineral Research, 17*(10), 1872–1882.

Kang, I., Panneerselvam, D., Panoskaltsis, V. P., Eppell, S. J., Marchant, R. E., & Doerschuk, C. M. (2008). Changes in the hyperelastic properties of endothelial cells induced by tumor necrosis factor-alpha. *Biophysical Journal, 94*(8), 3273–3285.

Kaplanski, G., Farnarier, C., Benoliel, A. M., Foa, C., Kaplanski, S., & Bongrand, P. (1994). A novel role for E- and P-selectins: Shape control of endothelial cell monolayers. *Journal of Cell Science, 107*(Pt. 9), 2449–2457.

Karcher, H., Lammerding, J., Huang, H., Lee, R. T., Kamm, R. D., & Kaazempur-Mofrad, M. R. (2003). A three-dimensional viscoelastic model for cell deformation with experimental verification. *Biophysical Journal, 85*(5), 3336–3349.

Kaverina, I., Rottner, K., & Small, J. V. (1998). Targeting, capture, and stabilization of microtubules at early focal adhesions. *Journal of Cell Biology, 142*(1), 181–190.

Keller, M., Tharmann, R., Dichtl, M. A., Bausch, A. R., & Sackmann, E. (2003). Slow filament dynamics and viscoelasticity in entangled and active actin networks. *Philosophical Transactions of the Royal Society A: Mathematical, Physical and Engineering Sciences, 361*(1805), 699–711 discussion 711–712.

Khismatullin, D. B., & Truskey, G. A. (2004). A 3D numerical study of the effect of channel height on leukocyte deformation and adhesion in parallel-plate flow chambers. *Microvascular Research, 68*(3), 188–202.

Khismatullin, D., & Truskey, G. (2005). Three-dimensional numerical simulation of receptor-mediated leukocyte adhesion to surfaces: Effects of cell deformability and viscoelasticity. *Physics of Fluids, 17*(3), 031505.

Kim, T., Hwang, W., Lee, H., & Kamm, R. D. (2009). Computational analysis of viscoelastic properties of crosslinked actin networks. *PLoS Computational Biology, 5*(7), e1000439.

King, M. R., & Hammer, D. A. (2001). Multiparticle adhesive dynamics. Interactions between stably rolling cells. *Biophysical Journal, 81*(2), 799–813.

King, M. R., Heinrich, V., Evans, E., & Hammer, D. A. (2005). Nano-to-micro scale dynamics of P-selectin detachment from leukocyte interfaces. III. Numerical simulation of tethering under flow. *Biophysical Journal, 88*(3), 1676–1683.

Kiosses, W. B., Shattil, S. J., Pampori, N., & Schwartz, M. A. (2001). Rac recruits high-affinity integrin alphavbeta3 to lamellipodia in endothelial cell migration. *Nature Cell Biology, 3*(3), 316–320.

Kitagawa, Y., Van Eeden, S. F., Redenbach, D. M., Daya, M., Walker, B. A., Klut, M. E., et al. (1997). Effect of mechanical deformation on structure and function of polymorphonuclear leukocytes. *Journal of Applied Physiology, 82*(5), 1397–1405.

Krutzik, P. O., Hale, M. B., & Nolan, G. P. (2005). Characterization of the murine immunological signaling network with phosphospecific flow cytometry. *Journal of Immunology, 175*(4), 2366–2373.

Kubes, P., & Kanwar, S. (1994). Histamine induces leukocyte rolling in post-capillary venules. A P-selectin-mediated event. *Journal of Immunology, 152*(7), 3570–3577.

Kunkel, E. J., Dunne, J. L., & Ley, K. (2000). Leukocyte arrest during cytokine-dependent inflammation *in vivo*. *Journal of Immunology, 164*(6), 3301–3308.

Kunkel, E. J., & Ley, K. (1996). Distinct phenotype of E-selectin-deficient mice. E-selectin is required for slow leukocyte rolling *in vivo*. *Circulation Research, 79*(6), 1196–1204.

Kusch, J., Meyer, A., Snyder, M. P., & Barral, Y. (2002). Microtubule capture by the cleavage apparatus is required for proper spindle positioning in yeast. *Genes & Development, 16*(13), 1627–1639.

Lammermann, T., Bader, B. L., Monkley, S. J., Worbs, T., Wedlich-Soldner, R., Hirsch, K., et al. (2008). Rapid leukocyte migration by integrin-independent flowing and squeezing. *Nature, 453*(7191), 51–55.

Lau, A. W., Hoffman, B. D., Davies, A., Crocker, J. C., & Lubensky, T. C. (2003). Microrheology, stress fluctuations, and active behavior of living cells. *Physical Review Letters, 91*(19), 198101.

Laurent, V. M., Canadas, P., Fodil, R., Planus, E., Asnacios, A., Wendling, S., et al. (2002a). Tensegrity behaviour of cortical and cytosolic cytoskeletal components in twisted living adherent cells. *Acta Biotheoretica, 50*(4), 331–356.

Laurent, V. M., Henon, S., Planus, E., Fodil, R., Balland, M., Isabey, D., et al. (2002b). Assessment of mechanical properties of adherent living cells by bead micromanipulation: Comparison of magnetic twisting cytometry vs optical tweezers. *Journal of Biomechanical Engineering, 124*(4), 408–421.

Lawrence, M. B., & Springer, T. A. (1991). Leukocytes roll on a selectin at physiologic flow rates: Distinction from and prerequisite for adhesion through integrins. *Cell, 65*(5), 859–873.

Lawrence, M. B., & Springer, T. A. (1993). Neutrophils roll on E-selectin. *Journal of Immunology, 151*(11), 6338–6346.

Lehenkari, P. P., Charras, G. T., Nesbitt, S. A., & Horton, M. A. (2000). New technologies in scanning probe microscopy for studying molecular interactions in cells. *Expert Reviews in Molecular Medicine, 2*(2), 1–19.

Lele, T. P., Sero, J. E., Matthews, B. D., Kumar, S., Xia, S., Montoya-Zavala, M., et al. (2007). Tools to study cell mechanics and mechanotransduction. *Methods in Cell Biology, 83*, 443–472.

Lenormand, G., & Fredberg, J. J. (2006). Deformability, dynamics, and remodeling of cytoskeleton of the adherent living cell. *Biorheology, 43*(1), 1–30.

Lenormand, G., Millet, E., Fabry, B., Butler, J. P., & Fredberg, J. J. (2004). Linearity and time-scale invariance of the creep function in living cells. *Journal of the Royal Society Interface, 1*(1), 91–97.

Levy, I., & Shoseyov, O. (2004). Cross bridging proteins in nature and their utilization in bio- and nanotechnology. *Current Protein & Peptide Science, 5*(1), 33–49.

Ley, K. (1996). Molecular mechanisms of leukocyte recruitment in the inflammatory process. *Cardiovascular Research, 32*(4), 733–742.

Ley, K. (2002). Integration of inflammatory signals by rolling neutrophils. *Immunological Reviews, 186*, 8–18.

Leyton-Mange, J., Yang, S., Hoskins, M. H., Kunz, R. F., Zahn, J. D., & Dong, C. (2006). Design of a side-view particle imaging velocimetry flow system for cell–substrate adhesion studies. *Journal of Biomechanical Engineering, 128*(2), 271–278.

Lieleg, O., Schmoller, K. M., Claessens, M. M., & Bausch, A. R. (2009). Cytoskeletal polymer networks: Viscoelastic properties are determined by the microscopic interaction potential of cross-links. *Biophysical Journal, 96*(11), 4725–4732.

Lien, D. C., Wagner , W. W., Jr., Capen, R. L., Haslett, C., Hanson, W. L., Hofmeister, S. E., et al. (1987). Physiological neutrophil sequestration in the lung: Visual evidence for localization in capillaries. *Journal of Applied Physiology, 62*(3), 1236–1243.

Linderkamp, O., Ruef, P., Brenner, B., Gulbins, E., & Lang, F. (1998). Passive deformability of mature, immature, and active neutrophils in healthy and septicemic neonates. *Pediatric Research, 44*(6), 946–950.

Linderkamp, O., Ruef, P., Zilow, E. P., & Hoffmann, G. F. (1999). Impaired deformability of erythrocytes and neutrophils in children with newly diagnosed insulin-dependent diabetes mellitus. *Diabetologia, 42*(7), 865–869.

Lipowsky, H. H., Scott, D. A., & Cartmell, J. S. (1996). Leukocyte rolling velocity and its relation to leukocyte–endothelium adhesion and cell deformability. *American Journal of Physiology, 270*(4 Pt. 2), H1371–H1380.

Lofgren, R., Ng-Sikorski, J., Sjolander, A., & Andersson, T. (1993). Beta 2 integrin engagement triggers actin polymerization and phosphatidylinositol trisphosphate formation in non-adherent human neutrophils. *Journal of Cell Biology, 123*(6 Pt. 1), 1597–1605.

Lomakina, E. B., Spillmann, C. M., King, M. R., & Waugh, R. E. (2004). Rheological analysis and measurement of neutrophil indentation. *Biophysical Journal, 87* (6), 4246–4258.

Ludin, B., & Matus, A. (1998). GFP illuminates the cytoskeleton. *Trends in Cell Biology, 8*(2), 72–77.

Luo, B. H., Carman, C. V., & Springer, T. A. (2007). Structural basis of integrin regulation and signaling. *Annual Review of Immunology, 25*, 619–647.

Luscinskas, F. W., Ding, H., Tan, P., Cumming, D., Tedder, T. F., & Gerritsen, M. E. (1996). L- and P-selectins, but not CD49d (VLA-4) integrins, mediate monocyte initial attachment to TNF-alpha-activated vascular endothelium under flow *in vitro*. *Journal of Immunology, 157*(1), 326–335.

Machesky, L. M., Mullins, R. D., Higgs, H. N., Kaiser, D. A., Blanchoin, L., May, R. C., et al. (1999). Scar, a WASp-related protein, activates nucleation of actin filaments by the Arp2/3 complex. *Proceedings of the National Academy of Sciences of the United States of America, 96*(7), 3739–3744.

MacKintosh, F. C., & Schmidt, C. F. (1999). Microrheology. *Current Opinion in Colloid & Interface Science, 4*(4), 300–307.

Majstoravich, S., Zhang, J., Nicholson-Dykstra, S., Linder, S., Friedrich, W., Siminovitch, K. A., et al. (2004). Lymphocyte microvilli are dynamic, actin-dependent structures that do not require Wiskott–Aldrich syndrome protein (WASp) for their morphology. *Blood, 104*(5), 1396–1403.

Makgoba, M. W., Sanders, M. E., Ginther Luce, G. E., Dustin, M. L., Springer, T. A., Clark, E. A., et al. (1988). ICAM-1 a ligand for LFA-1-dependent adhesion of B, T and myeloid cells. *Nature, 331*(6151), 86–88.

Makino, A., Glogauer, M., Bokoch, G. M., Chien, S., & Schmid-Schonbein, G. W. (2005). Control of neutrophil pseudopods by fluid shear: Role of Rho family GTPases. *American Journal of Physiology: Cell Physiology, 288*(4), C863–C871.

Maksym, G. N., Fabry, B., Butler, J. P., Navajas, D., Tschumperlin, D. J., Laporte, J. D., et al. (2000). Mechanical properties of cultured human airway smooth muscle cells from 0.05 to 0.4 Hz. *Journal of Applied Physiology, 89*(4), 1619–1632.

Mannie, M. D., McConnell, T. J., Xie, C., & Li, Y. Q. (2005). Activation-dependent phases of T cells distinguished by use of optical tweezers and near infrared Raman spectroscopy. *Journal of Immunological Methods, 297*(1–2), 53–60.

Marcus, W. D., & Hochmuth, R. M. (2002). Experimental studies of membrane tethers formed from human neutrophils. *Annals of Biomedical Engineering, 30*(10), 1273–1280.

Marschel, P., & Schmid-Schonbein, G. W. (2002). Control of fluid shear response in circulating leukocytes by integrins. *Annals of Biomedical Engineering, 30*(3), 333–343.

Martens, J. C., & Radmacher, M. (2008). Softening of the actin cytoskeleton by inhibition of myosin II. *Pflügers Archiv, 456*(1), 95–100.

Martin-Padura, I., Lostaglio, S., Schneemann, M., Williams, L., Romano, M., Fruscella, P., et al. (1998). Junctional adhesion molecule, a novel member of the immunoglobulin superfamily that distributes at intercellular junctions and modulates monocyte transmigration. *Journal of Cell Biology, 142*(1), 117–127.

Martys, J. L., Ho, C. L., Liem, R. K., & Gundersen, G. G. (1999). Intermediate filaments in motion: Observations of intermediate filaments in cells using green fluorescent protein-vimentin. *Molecular Biology of the Cell, 10*(5), 1289–1295.

Mason, T. G., Ganesan, K., vanZanten, J. H., Wirtz, D., & Kuo, S. C. (1997). Particle tracking microrheology of complex fluids. *Physical Review Letters, 79*(17), 3282–3285.

Mason, T. G., & Weitz, D. A. (1995). Optical measurements of frequency-dependent linear viscoelastic moduli of complex fluids. *Physical Review Letters, 74*(7), 1250–1253.

Massiera, G., Van Citters, K. M., Biancaniello, P. L., & Crocker, J. C. (2007). Mechanics of single cells: Rheology, time dependence, and fluctuations. *Biophysical Journal, 93*(10), 3703–3713.

Mathur, A. B., Truskey, G. A., & Reichert, W. M. (2000). Atomic force and total internal reflection fluorescence microscopy for the study of force transmission in endothelial cells. *Biophysical Journal, 78*(4), 1725–1735.

May, R. C., Hall, M. E., Higgs, H. N., Pollard, T. D., Chakraborty, T., Wehland, J., et al. (1999). The Arp2/3 complex is essential for the actin-based motility of Listeria monocytogenes. *Current Biology, 9*(14), 759–762.

McHale, J. F., Harari, O. A., Marshall, D., & Haskard, D. O. (1999). Vascular endothelial cell expression of ICAM-1 and VCAM-1 at the onset of eliciting contact hypersensitivity in mice: Evidence for a dominant role of TNF-alpha. *Journal of Immunology, 162*(3), 1648–1655.

Menegazzi, R., Busetto, S., Decleva, E., Cramer, R., Dri, P., & Patriarca, P. (1999). Triggering of chloride ion efflux from human neutrophils as a novel function of leukocyte beta 2 integrins: Relationship with spreading and activation of the respiratory burst. *Journal of Immunology, 162*(1), 423–434.

Mercuri, M., Ciuffetti, G., Robinson, M., & Toole, J. (1989). Blood cell rheology in acute cerebral infarction. *Stroke, 20*(7), 959–962.

Meyer, G., & Amer, N. M. (1988). Novel optical approach to atomic force microscopy. *Applied Physics Letters, 53*(12), 1045–1047.

Mitchison, T. J., Charras, G. T., & Mahadevan, L. (2008). Implications of a poroelastic cytoplasm for the dynamics of animal cell shape. *Seminars in Cell and Developmental Biology, 19*(3), 215–223.

Mitchison, T., & Kirschner, M. (1984). Dynamic instability of microtubule growth. *Nature, 312*(5991), 237–242.

Miyamoto, K., Ogura, Y., Kenmochi, S., & Honda, Y. (1997). Role of leukocytes in diabetic microcirculatory disturbances. *Microvascular Research, 54*(1), 43–48.

Moazzam, F., DeLano, F. A., Zweifach, B. W., & Schmid-Schonbein, G. W. (1997). The leukocyte response to fluid stress. *Proceedings of the National Academy of Sciences of the United States of America, 94*(10), 5338–5343.

Mogilner, A., & Oster, G. (1996). Cell motility driven by actin polymerization. *Biophysical Journal, 71*(6), 3030–3045.

Mogilner, A., & Oster, G. (2003). Force generation by actin polymerization II: The elastic ratchet and tethered filaments. *Biophysical Journal, 84*(3), 1591–1605.

Moller, W., Takenaka, S., Rust, M., Stahlhofen, W., & Heyder, J. (1997). Probing mechanical properties of living cells by magnetopneumography. *Journal of Aerosol Medicine, 10*(3), 173–186.

Moore, K. L., Patel, K. D., Bruehl, R. E., Li, F., Johnson, D. A., Lichenstein, H. S., et al. (1995). P-selectin glycoprotein ligand-1 mediates rolling of human neutrophils on P-selectin. *Journal of Cell Biology, 128*(4), 661–671.

Moschakis, T., Murray, B. S., & Dickinson, E. (2006). Particle tracking using confocal microscopy to probe the microrheology in a phase-separating emulsion containing nonadsorbing polysaccharide. *Langmuir, 22*(10), 4710–4719.

Mucke, N., Kreplak, L., Kirmse, R., Wedig, T., Herrmann, H., Aebi, U., et al. (2004). Assessing the flexibility of intermediate filaments by atomic force microscopy. *Journal of Molecular Biology, 335*(5), 1241–1250.

Muller, W. A., & Randolph, G. J. (1999). Migration of leukocytes across endothelium and beyond: Molecules involved in the transmigration and fate of monocytes. *Journal of Leukocyte Biology, 66*(5), 698–704.

Mullins, R. D., Heuser, J. A., & Pollard, T. D. (1998). The interaction of Arp2/3 complex with actin: Nucleation, high affinity pointed end capping, and formation of branching networks of filaments. *Proceedings of the National Academy of Sciences of the United States of America, 95*(11), 6181–6186.

N'Dri, N. A., Shyy, W., & Tran-Son-Tay, R. (2003). Computational modeling of cell adhesion and movement using a continuum-kinetics approach. *Biophysical Journal, 85*(4), 2273–2286.

Naves, R., Reyes, L. I., Rosemblatt, M., Jacobelli, S., Gonzalez, A., & Bono, M. R. (2006). Lymphoid B cells induce NF-kappaB activation in high endothelial cells from human tonsils. *International Immunology, 18*(2), 259–267.

Needham, D., & Hochmuth, R. M. (1990). Rapid flow of passive neutrophils into a 4 microns pipet and measurement of cytoplasmic viscosity. *Journal of Biomechanical Engineering, 112*(3), 269–276.

Needham, D., & Hochmuth, R. M. (1992). A sensitive measure of surface stress in the resting neutrophil. *Biophysical Journal, 61*(6), 1664–1670.

Nieminen, M., Henttinen, T., Merinen, M., Marttila-Ichihara, F., Eriksson, J. E., & Jalkanen, S. (2006). Vimentin function in lymphocyte adhesion and transcellular migration. *Nature Cell Biology, 8*(2), 156–162.

Niggli, V. (2003a). Microtubule-disruption-induced and chemotactic-peptide-induced migration of human neutrophils: Implications for differential sets of signalling pathways. *Journal of Cell Science, 116*(Pt. 5), 813–822.

Niggli, V. (2003b). Signaling to migration in neutrophils: Importance of localized pathways. *International Journal of Biochemistry & Cell Biology, 35*(12), 1619–1638.

Nigmatulin, R. I. (1990). *Dynamics of multiphase media (Vols. 1 and 2)*. Washington, DC: Hemisphere.

Nishino, M., Tanaka, H., Ogura, H., Inoue, Y., Koh, T., Fujita, K., et al. (2005). Serial changes in leukocyte deformability and whole blood rheology in patients with sepsis or trauma. *Journal of Trauma, 59*(6), 1425–1431.

Nishiwaki, H., Ogura, Y., Kimura, H., Kiryu, J., & Honda, Y. (1995). Quantitative evaluation of leukocyte dynamics in retinal microcirculation. *Investigative Ophthalmology & Visual Science, 36*(1), 123–130.

O'Brien, C. D., Lim, P., Sun, J., & Albelda, S. M. (2003). PECAM-1-dependent neutrophil transmigration is independent of monolayer PECAM-1 signaling or localization. *Blood, 101*(7), 2816–2825.

Ofek, G., Natoli, R. M., & Athanasiou, K. A. (2009). In situ mechanical properties of the chondrocyte cytoplasm and nucleus. *Journal of Biomechanics, 42*(7), 873–877.

Ostermann, G., Weber, K. S., Zernecke, A., Schroder, A., & Weber, C. (2002). JAM-1 is a ligand of the beta(2) integrin LFA-1 involved in transendothelial migration of leukocytes. *Nature Immunology, 3*(2), 151–158.

Pai, A., Sundd, P., & Tees, D. F. (2008). In situ microrheological determination of neutrophil stiffening following adhesion in a model capillary. *Annals of Biomedical Engineering, 36*(4), 596–603.

Pajerowski, J. D., Dahl, K. N., Zhong, F. L., Sammak, P. J., & Discher, D. E. (2007). Physical plasticity of the nucleus in stem cell differentiation. *Proceedings of the National Academy of Sciences of the United States of America, 104*(40), 15619–15624.

Pappu, V., & Bagchi, P. (2008). 3D computational modeling and simulation of leukocyte rolling adhesion and deformation. *Computers in Biology and Medicine, 38*(6), 738–753.
Park, S., Costa, K. D., Ateshian, G. A., & Hong, K. S. (2009). Mechanical properties of bovine articular cartilage under microscale indentation loading from atomic force microscopy. *Proceedings of the Institution of Mechanical Engineers, Part H: Journal of Engineering in Medicine, 223*(3), 339–347.
Park, E. Y., Smith, M. J., Stropp, E. S., Snapp, K. R., DiVietro, J. A., Walker, W. F., et al. (2002). Comparison of PSGL-1 microbead and neutrophil rolling: Microvillus elongation stabilizes P-selectin bond clusters. *Biophysical Journal, 82*(4), 1835–1847.
Partida-Sanchez, S., Gasser, A., Fliegert, R., Siebrands, C. C., Dammermann, W., Shi, G., et al. (2007). Chemotaxis of mouse bone marrow neutrophils and dendritic cells is controlled by ADP-ribose, the major product generated by the CD38 enzyme reaction. *Journal of Immunology, 179*(11), 7827–7839.
Parysek, L. M., & Eckert, B. S. (1984). Vimentin filaments in spreading, randomly locomoting, and f-met-leu-phe-treated neutrophils. *Cell and Tissue Research, 235*(3), 575–581.
Pawar, P., Jadhav, S., Eggleton, C. D., & Konstantopoulos, K. (2008). Roles of cell and microvillus deformation and receptor–ligand binding kinetics in cell rolling. *American Journal of Physiology: Heart and Circulatory Physiology, 295*(4), H1439–H1450.
Pelletier, V., Gal, N., Fournier, P., & Kilfoil, M. L. (2009). Microrheology of microtubule solutions and actin–microtubule composite networks. *Physical Review Letters, 102*(18), 188303.
Perrault, C. M., Bray, E. J., Didier, N., Ozaki, C. K., & Tran-Son-Tay, R. (2004). Altered rheology of lymphocytes in the diabetic mouse. *Diabetologia, 47*(10), 1722–1726.
Peskin, C. S., Odell, G. M., & Oster, G. F. (1993). Cellular motions and thermal fluctuations: The Brownian ratchet. *Biophysical Journal, 65*(1), 316–324.
Pettit, E. J., & Hallett, M. B. (1996). Localised and global cytosolic Ca2+ changes in neutrophils during engagement of Cd11b/CD18 integrin visualised using confocal laser scanning reconstruction. *Journal of Cell Science, 109*(Pt. 7), 1689–1694.
Pohlman, T. H., Stanness, K. A., Beatty, P. G., Ochs, H. D., & Harlan, J. M. (1986). An endothelial cell surface factor(s) induced *in vitro* by lipopolysaccharide, interleukin 1, and tumor necrosis factor-alpha increases neutrophil adherence by a CDw18-dependent mechanism. *Journal of Immunology, 136*(12), 4548–4553.
Poschl, J. M., Ruef, P., & Linderkamp, O. (2005). Deformability of passive and activated neutrophils in children with Gram-negative septicemia. *Scandinavian Journal of Clinical and Laboratory Investigation, 65*(4), 333–339.
Potter, D. R., Jiang, J., & Damiano, E. R. (2009). The recovery time course of the endothelial cell glycocalyx *in vivo* and its implications *in vitro*. *Circulation Research, 104*(11), 1318–1325.
Prahlad, V., Yoon, M., Moir, R. D., Vale, R. D., & Goldman, R. D. (1998). Rapid movements of vimentin on microtubule tracks: Kinesin-dependent assembly of intermediate filament networks. *Journal of Cell Biology, 143*(1), 159–170.
Prokocimer, M., Davidovich, M., Nissim-Rafinia, M., Wiesel-Motiuk, N., Bar, D., Barkan, R., et al. (2009). Nuclear lamins: Key regulators of nuclear structure and activities. *Journal of Cellular and Molecular Medicine, [Epub ahead of print].*
Puig-De-Morales, M., Grabulosa, M., Alcaraz, J., Mullol, J., Maksym, G. N., Fredberg, J. J., et al. (2001). Measurement of cell microrheology by magnetic twisting cytometry with frequency domain demodulation. *Journal of Applied Physiology, 91*(3), 1152–1159.
Pulido, R., Elices, M. J., Campanero, M. R., Osborn, L., Schiffer, S., Garcia-Pardo, A., et al. (1991). Functional evidence for three distinct and independently inhibitable adhesion activities mediated by the human integrin VLA-4. Correlation with distinct alpha 4 epitopes. *Journal of Biological Chemistry, 266*(16), 10241–10245.

Quinn, M. T., DeLeo, F. R., & Bokoch, G. M. (Eds.), (2007). *Neutrophil methods and protocols*. New York, NY: Springer-Verlag.

Radmacher, M. (2007). Studying the mechanics of cellular processes by atomic force microscopy. *Methods in Cell Biology, 83*, 347–372.

Rahman, A., & Fazal, F. (2009). Hug tightly and say goodbye: Role of endothelial ICAM-1 in leukocyte transmigration. *Antioxidants & Redox Signaling, 11*(4), 823–839.

Ramachandran, V., Williams, M., Yago, T., Schmidtke, D. W., & McEver, R. P. (2004). Dynamic alterations of membrane tethers stabilize leukocyte rolling on P-selectin. *Proceedings of the National Academy of Sciences of the United States of America, 101*(37), 13519–13524.

Redenbach, D. M., English, D., & Hogg, J. C. (1997). The nature of leukocyte shape changes in the pulmonary capillaries. *American Journal of Physiology, 273*(4 Pt. 1), L733–L740.

Reinhardt, P. H., Elliott, J. F., & Kubes, P. (1997). Neutrophils can adhere via alpha4beta1-integrin under flow conditions. *Blood, 89*(10), 3837–3846.

Rhoades, R. A., & Bell, D. R. (2008). *Medical physiology: Principles for clinical medicine*. Philadelphia, PA: Lippincott Williams & Wilkins.

Rinker, K. D., Prabhakar, V., & Truskey, G. A. (2001). Effect of contact time and force on monocyte adhesion to vascular endothelium. *Biophysical Journal, 80*(4), 1722–1732.

Roca-Cusachs, P., Almendros, I., Sunyer, R., Gavara, N., Farre, R., & Navajas, D. (2006). Rheology of passive and adhesion-activated neutrophils probed by atomic force microscopy. *Biophysical Journal, 91*(9), 3508–3518.

Rosales, C., & Juliano, R. L. (1995). Signal transduction by cell adhesion receptors in leukocytes. *Journal of Leukocyte Biology, 57*(2), 189–198.

Ross, R. (1999). Atherosclerosis—An inflammatory disease. *New England Journal of Medicine, 340*(2), 115–126.

Rothwell, S. W., Nath, J., & Wright, D. G. (1989). Interactions of cytoplasmic granules with microtubules in human neutrophils. *Journal of Cell Biology, 108*(6), 2313–2326.

Rotsch, C., Jacobson, K., & Radmacher, M. (1999). Dimensional and mechanical dynamics of active and stable edges in motile fibroblasts investigated by using atomic force microscopy. *Proceedings of the National Academy of Sciences of the United States of America, 96*(3), 921–926.

Roy, S., & Qi, H. J. (2008). Micromechanical model for elasticity of the cell cytoskeleton. *Physical Review E: Statistical, Nonlinear, and Soft Matter Physics, 77*(6 Pt. 1), 061916.

Ryder, M. I., Weinreb, R. N., & Niederman, R. (1984). The organization of actin filaments in human polymorphonuclear leukocytes. *Anatomical Record, 209*(1), 7–20.

Sackstein, R. (2005). The lymphocyte homing receptors: Gatekeepers of the multistep paradigm. *Current Opinion in Hematology, 12*(6), 444–450.

Saito, H., Lai, J., Rogers, R., & Doerschuk, C. M. (2002). Mechanical properties of rat bone marrow and circulating neutrophils and their responses to inflammatory mediators. *Blood, 99*(6), 2207–2213.

Sanchez-Madrid, F., & del Pozo, M. A. (1999). Leukocyte polarization in cell migration and immune interactions. *EMBO Journal, 18*(3), 501–511.

Sanchez-Madrid, F., & Serrador, J. M. (2007). Mitochondrial redistribution: Adding new players to the chemotaxis game. *Trends in Immunology, 28*(5), 193–196.

Sarraj, B., Massberg, S., Li, Y., Kasorn, A., Subramanian, K., Loison, F., et al. (2009). Myeloid-specific deletion of tumor suppressor PTEN augments neutrophil transendothelial migration during inflammation. *Journal of Immunology, 182*(11), 7190–7200.

Sato, Y., Hogg, J. C., English, D., & van Eeden, S. F. (2000). Endothelin-1 changes polymorphonuclear leukocytes' deformability and CD11b expression and promotes their retention in the lung. *American Journal of Respiratory Cell and Molecular Biology, 23*(3), 404–410.

Sato, H., Kataoka, N., Kajiya, F., Katano, M., Takigawa, T., & Masuda, T. (2004). Kinetic study on the elastic change of vascular endothelial cells on collagen matrices by atomic force microscopy. *Colloids and Surfaces B: Biointerfaces, 34*(2), 141–146.

Sato, M., Leimbach, G., Schwarz, W. H., & Pollard, T. D. (1985). Mechanical properties of actin. *Journal of Biological Chemistry, 260*(14), 8585–8592.

Schmid-Schonbein, G. W. (1986). Rheology of leukocytes. In R. Skalak & S. Chien (Eds.), *Handbook of bioengineering* (pp. 13.11–13.25). New York, NY: McGraw-Hill.

Schmid-Schonbein, G. W., Shih, Y. Y., & Chien, S. (1980a). Morphometry of human leukocytes. *Blood, 56*(5), 866–875.

Schmid-Schonbein, G. W., Sung, K. L., Tozeren, H., Skalak, R., & Chien, S. (1981). Passive mechanical properties of human leukocytes. *Biophysical Journal, 36*(1), 243–256.

Schmid-Schonbein, G. W., Usami, S., Skalak, R., & Chien, S. (1980b). The interaction of leukocytes and erythrocytes in capillary and postcapillary vessels. *Microvascular Research, 19*(1), 45–70.

Schmidtke, D. W., & Diamond, S. L. (2000). Direct observation of membrane tethers formed during neutrophil attachment to platelets or P-selectin under physiological flow. *Journal of Cell Biology, 149*(3), 719–730.

Schymeinsky, J., Then, C., & Walzog, B. (2005). The non-receptor tyrosine kinase Syk regulates lamellipodium formation and site-directed migration of human leukocytes. *Journal of Cellular Physiology, 204*(2), 614–622.

Sengupta, K., Aranda-Espinoza, H., Smith, L., Janmey, P., & Hammer, D. (2006). Spreading of neutrophils: From activation to migration. *Biophysical Journal, 91*(12), 4638–4648.

Shamri, R., Grabovsky, V., Gauguet, J. M., Feigelson, S., Manevich, E., Kolanus, W., et al. (2005). Lymphocyte arrest requires instantaneous induction of an extended LFA-1 conformation mediated by endothelium-bound chemokines. *Nature Immunology, 6*(5), 497–506.

Shao, J. Y., Ting-Beall, H. P., & Hochmuth, R. M. (1998). Static and dynamic lengths of neutrophil microvilli. *Proceedings of the National Academy of Sciences of the United States of America, 95*(12), 6797–6802.

Sheikh, S., & Nash, G. B. (1998). Treatment of neutrophils with cytochalasins converts rolling to stationary adhesion on P-selectin. *Journal of Cellular Physiology, 174*(2), 206–216.

Shen, K., Sung, K. L., Whittemore, D. E., DeLano, F. A., Zweifach, B. W., & Schmid-Schonbein, G. W. (1995). Properties of circulating leukocytes in spontaneously hypertensive rats. *Biochemistry and Cell Biology, 73*(7–8), 491–500.

Shimonaka, M., Katagiri, K., Nakayama, T., Fujita, N., Tsuruo, T., Yoshie, O., et al. (2003). Rap1 translates chemokine signals to integrin activation, cell polarization, and motility across vascular endothelium under flow. *Journal of Cell Biology, 161*(2), 417–427.

Simon, A., Cohen-Bouhacina, T., Aime, J. P., Porte, M. C., Amedee, J., & Baquey, C. (2004). Heterogeneous cell mechanical properties: An atomic force microscopy study. *Cellular and Molecular Biology (Noisy-le-Grand), 50*(3), 255–266.

Simon, S. I., Nyunt, T., Florine-Casteel, K., Ritchie, K., Ting-Beall, H. P., Evans, E., et al. (2007). Dynamics of neutrophil membrane compliance and microstructure probed with a micropipet-based piconewton force transducer. *Annals of Biomedical Engineering, 35*(4), 595–604.

Skoutelis, A. T., Kaleridis, V., Athanassiou, G. M., Kokkinis, K. I., Missirlis, Y. F., & Bassaris, H. P. (2000a). Neutrophil deformability in patients with sepsis, septic shock, and adult respiratory distress syndrome. *Critical Care Medicine, 28*(7), 2355–2359.

Skoutelis, A. T., Kaleridis, V. E., Gogos, C. A., Athanassiou, G. M., Missirlis, Y. F., & Bassaris, H. P. (2000b). Effect of cytokines and colony-stimulating factors on passive polymorphonuclear leukocyte deformability *in vitro*. *Cytokine, 12*(11), 1737–1740.

Skoutelis, A. T., Kaleridis, V. E., Goumenos, D. S., Athanassiou, G. M., Missirlis, Y. F., Vlachojannis, J. G., et al. (2000c). Polymorphonuclear leukocyte rigidity is defective in patients with chronic renal failure. *Nephrology Dialysis Transplantation, 15*(11), 1788–1793.
Smalley, D. M., & Ley, K. (2005). L-selectin: Mechanisms and physiological significance of ectodomain cleavage. *Journal of Cellular and Molecular Medicine, 9*(2), 255–266.
Smith, M. L., Long, D. S., Damiano, E. R., & Ley, K. (2003). Near-wall micro-PIV reveals a hydrodynamically relevant endothelial surface layer in venules *in vivo*. *Biophysical Journal, 85*(1), 637–645.
Smith, M. L., Smith, M. J., Lawrence, M. B., & Ley, K. (2002). Viscosity-independent velocity of neutrophils rolling on p-selectin *in vitro* or *in vivo*. *Microcirculation, 9*(6), 523–536.
Snijder-Van As, M. I., Rieger, B., Joosten, B., Subramaniam, V., Figdor, C. G., & Kanger, J. S. (2009). A hybrid total internal reflection fluorescence and optical tweezers microscope to study cell adhesion and membrane protein dynamics of single living cells. *Journal of Microscopy, 233*(1), 84–92.
Soltys, B. J., & Gupta, R. S. (1992). Interrelationships of endoplasmic reticulum, mitochondria, intermediate filaments, and microtubules—A quadruple fluorescence labeling study. *Biochemistry and Cell Biology, 70*(10–11), 1174–1186.
Springer, T. A., Thompson, W. S., Miller, L. J., Schmalstieg, F. C., & Anderson, D. C. (1984). Inherited deficiency of the Mac-1, LFA-1, p150, 95 glycoprotein family and its molecular basis. *Journal of Experimental Medicine, 160*(6), 1901–1918.
Sriramarao, P., von Andrian, U. H., Butcher, E. C., Bourdon, M. A., & Broide, D. H. (1994). L-selectin and very late antigen-4 integrin promote eosinophil rolling at physiological shear rates *in vivo*. *Journal of Immunology, 153*(9), 4238–4246.
Stacker, S. A., & Springer, T. A. (1991). Leukocyte integrin P150, 95 (CD11c/CD18) functions as an adhesion molecule binding to a counter-receptor on stimulated endothelium. *Journal of Immunology, 146*(2), 648–655.
Stamenovic, D., & Coughlin, M. F. (2000). A quantitative model of cellular elasticity based on tensegrity. *Journal of Biomechanical Engineering, 122*(1), 39–43.
Staunton, D. E., Dustin, M. L., & Springer, T. A. (1989). Functional cloning of ICAM-2, a cell adhesion ligand for LFA-1 homologous to ICAM-1. *Nature, 339*(6219), 61–64.
Stein, J. V., Cheng, G., Stockton, B. M., Fors, B. P., Butcher, E. C., & von Andrian, U. H. (1999). L-selectin-mediated leukocyte adhesion *in vivo*: Microvillous distribution determines tethering efficiency, but not rolling velocity. *Journal of Experimental Medicine, 189*(1), 37–50.
Stephens, L., Ellson, C., & Hawkins, P. (2002). Roles of PI3Ks in leukocyte chemotaxis and phagocytosis. *Current Opinion in Cell Biology, 14*(2), 203–213.
Stinchcombe, J. C., Majorovits, E., Bossi, G., Fuller, S., & Griffiths, G. M. (2006). Centrosome polarization delivers secretory granules to the immunological synapse. *Nature, 443*(7110), 462–465.
Su, M., West, C. A., Young, A. J., He, C., Konerding, M. A., & Mentzer, S. J. (2003). Dynamic deformation of migratory efferent lymph-derived cells "trapped" in the inflammatory microcirculation. *Journal of Cellular Physiology, 194*(1), 54–62.
Suemori, T., Morimatsu, H., Mizobuchi, S., Morita, K., Katanosaka, Y., Mohri, S., et al. (2009). Impairment of leukocyte deformability in patients undergoing esophagectomy. *Clinical Hemorheology and Microcirculation, 41*(2), 127–136.
Sumoza-Toledo, A., & Santos-Argumedo, L. (2004). The spreading of B lymphocytes induced by CD44 cross-linking requires actin, tubulin, and vimentin rearrangements. *Journal of Leukocyte Biology, 75*(2), 233–239.
Sung, K. L., Dong, C., Schmid-Schonbein, G. W., Chien, S., & Skalak, R. (1988). Leukocyte relaxation properties. *Biophysical Journal, 54*(2), 331–336.

Sutton, D. W., & Schmid-Schonbein, G. W. (1992). Elevation of organ resistance due to leukocyte perfusion. *American Journal of Physiology, 262*(6 Pt. 2), H1646–H1650.
Tillmanns, H., Neumann, F. J., Tiefenbacher, C., Dorigo, O., Parekh, N., Waas, W., et al. (1993). Activation of neutrophils in the microvasculature of the ischaemic and reperfused myocardium. *European Heart Journal, 14*(Suppl. I), 82–86.
Timoshenko, S. P., & Goodier, J. N. (1970). *Theory of elasticity.* London: McGraw-Hill.
Ting-Beall, H. P., Needham, D., & Hochmuth, R. M. (1993). Volume and osmotic properties of human neutrophils. *Blood, 81*(10), 2774–2780.
Tran-Son-Tay, R., Needham, D., Yeung, A., & Hochmuth, R. M. (1991). Time-dependent recovery of passive neutrophils after large deformation. *Biophysical Journal, 60*(4), 856–866.
Tsai, M. A., Frank, R. S., & Waugh, R. E. (1993). Passive mechanical behavior of human neutrophils: Power-law fluid. *Biophysical Journal, 65*(5), 2078–2088.
Tseng, Y., Kole, T. P., & Wirtz, D. (2002). Micromechanical mapping of live cells by multiple-particle-tracking microrheology. *Biophysical Journal, 83*(6), 3162–3176.
Tseng, Y., Lee, J. S., Kole, T. P., Jiang, I., & Wirtz, D. (2004). Micro-organization and visco-elasticity of the interphase nucleus revealed by particle nanotracking. *Journal of Cell Science, 117*(Pt. 10), 2159–2167.
Valero, V., Nevian, T., Ho, D., & Lindau, M. (2008). Tethering forces of secretory granules measured with optical tweezers. *Biophysical Journal, 95*(10), 4972–4978.
van Buul, J. D., & Hordijk, P. L. (2004). Signaling in leukocyte transendothelial migration. *Arteriosclerosis, Thrombosis, and Vascular Biology, 24*(5), 824–833.
VanUffelen, B. E., de Koster, B. M., Van den Broek, P. J., VanSteveninck, J., & Elferink, J. G. (1996). Modulation of neutrophil migration by exogenous gaseous nitric oxide. *Journal of Leukocyte Biology, 60*(1), 94–100.
Vaporciyan, A. A., DeLisser, H. M., Yan, H. C., Mendiguren, I. I., Thom, S. R., Jones, M. L., et al. (1993). Involvement of platelet–endothelial cell adhesion molecule-1 in neutrophil recruitment *in vivo*. *Science, 262*(5139), 1580–1582.
Vaughan, K. T. (2005). Microtubule plus ends, motors, and traffic of Golgi membranes. *Biochimica et Biophysica Acta, 1744*(3), 316–324.
Vergnes, L., Peterfy, M., Bergo, M. O., Young, S. G., & Reue, K. (2004). Lamin B1 is required for mouse development and nuclear integrity. *Proceedings of the National Academy of Sciences of the United States of America, 101*(28), 10428–10433.
Vermes, I., & Strik, F. (1988). Altered leukocyte rheology in patients with chronic cerebrovascular disease. *Stroke, 19*(5), 631–633.
Vestweber, D., & Blanks, J. E. (1999). Mechanisms that regulate the function of the selectins and their ligands. *Physiological Reviews, 79*(1), 181–213.
Vicente-Manzanares, M., & Sanchez-Madrid, F. (2004). Role of the cytoskeleton during leukocyte responses. *Nature Reviews Immunology, 4*(2), 110–122.
Vinckier, A., & Semenza, G. (1998). Measuring elasticity of biological materials by atomic force microscopy. *FEBS Letters, 430*(1–2), 12–16.
Vink, H., & Duling, B. R. (1996). Identification of distinct luminal domains for macromolecules, erythrocytes, and leukocytes within mammalian capillaries. *Circulation Research, 79*(3), 581–589.
von Andrian, U. H., Hasslen, S. R., Nelson, R. D., Erlandsen, S. L., & Butcher, E. C. (1995). A central role for microvillous receptor presentation in leukocyte adhesion under flow. *Cell, 82*(6), 989–999.
Wachsstock, D. H., Schwarz, W. H., & Pollard, T. D. (1994). Cross-linker dynamics determine the mechanical properties of actin gels. *Biophysical Journal, 66*(3 Pt. 1), 801–809.

Wang, Q., Chiang, E. T., Lim, M., Lai, J., Rogers, R., Janmey, P. A., et al. (2001). Changes in the biomechanical properties of neutrophils and endothelial cells during adhesion. *Blood, 97*(3), 660–668.

Wang, S. K., Chiu, J. J., Lee, M. R., Chou, S. C., Chen, L. J., & Hwang, N. H. (2006). Leukocyte–endothelium interaction: Measurement by laser tweezers force spectroscopy. *Cardiovascular Engineering, 6*(3), 111–117.

Wang, W., Graeler, M. H., & Goetzl, E. J. (2005). Type 4 sphingosine 1-phosphate G protein-coupled receptor (S1P4) transduces S1P effects on T cell proliferation and cytokine secretion without signaling migration. *FASEB Journal, 19*(12), 1731–1733.

Wang, N., & Ingber, D. E. (1995). Probing transmembrane mechanical coupling and cytomechanics using magnetic twisting cytometry. *Biochemistry and Cell Biology, 73*(7–8), 327–335.

Wang, N., Tolic-Norrelykke, I. M., Chen, J., Mijailovich, S. M., Butler, J. P., Fredberg, J. J., et al. (2002). Cell prestress. I. Stiffness and prestress are closely associated in adherent contractile cells. *American Journal of Physiology: Cell Physiology, 282*(3), C606–C616.

Warnke, K. C., & Skalak, T. C. (1992). Leukocyte plugging *in vivo* in skeletal muscle arteriolar trees. *American Journal of Physiology, 262*(4 Pt. 2), H1149–H1155.

Weber, C. (2003). Novel mechanistic concepts for the control of leukocyte transmigration: Specialization of integrins, chemokines, and junctional molecules. *Journal of Molecular Medicine, 81*(1), 4–19.

Werr, J., Xie, X., Hedqvist, P., Ruoslahti, E., & Lindbom, L. (1998). beta1 integrins are critically involved in neutrophil locomotion in extravascular tissue *In vivo*. *Journal of Experimental Medicine, 187*(12), 2091–2096.

Wiggins, P. A., & Nelson, P. C. (2006). Generalized theory of semiflexible polymers. *Physical Review E: Statistical, Nonlinear, and Soft Matter Physics, 73*(3 Pt. 1), 031906.

Wirtz, D. (2009). Particle-tracking microrheology of living cells: Principles and applications. *Annual Review of Biophysics, 38*, 301–326.

Woodfin, A., Voisin, M. B., Imhof, B. A., Dejana, E., Engelhardt, B., & Nourshargh, S. (2009). Endothelial cell activation leads to neutrophil transmigration as supported by the sequential roles of ICAM-2, JAM-A, and PECAM-1. *Blood, 113*(24), 6246–6257.

Worthen, G. S., Schwab , B., III, Elson, E. L., & Downey, G. P. (1989). Mechanics of stimulated neutrophils: Cell stiffening induces retention in capillaries. *Science, 245*(4914), 183–186.

Wozniak, M. J., Kawazoe, N., Tateishi, T., & Chen, G. (2009). Monitoring of mechanical properties of serially passaged bovine articular chondrocytes by atomic force microscopy. *Micron, 40*(8), 870–875.

Wu, H. W., Kuhn, T., & Moy, V. T. (1998). Mechanical properties of L929 cells measured by atomic force microscopy: Effects of anticytoskeletal drugs and membrane crosslinking. *Scanning, 20*(5), 389–397.

Xiao, F., Eppihimer, M. J., Willis, B. H., & Carden, D. L. (1997a). Complement-mediated lung injury and neutrophil retention after intestinal ischemia–reperfusion. *Journal of Applied Physiology, 82*(5), 1459–1465.

Xiao, F., Eppihimer, M. J., Young, J. A., Nguyen, K., & Carden, D. L. (1997b). Lung neutrophil retention and injury after intestinal ischemia/reperfusion. *Microcirculation, 4*(3), 359–367.

Yabuki, M., Miyake, T., Doi, Y., Fujiwara, T., Hamazaki, K., Yoshioka, T., et al. (1999). Role of nuclear lamins in nuclear segmentation of human neutrophils. *Physiological Chemistry & Physics & Medical NMR, 31*(2), 77–84.

Yago, T., Leppanen, A., Qiu, H., Marcus, W. D., Nollert, M. U., Zhu, C., et al. (2002). Distinct molecular and cellular contributions to stabilizing selectin-mediated rolling under flow. *Journal of Cell Biology, 158*(4), 787–799.

Yamada, S., Wirtz, D., & Kuo, S. C. (2000). Mechanics of living cells measured by laser tracking microrheology. *Biophysical Journal, 78*(4), 1736–1747.

Yamaguchi, K., Nishio, K., Sato, N., Tsumura, H., Ichihara, A., Kudo, H., et al. (1997). Leukocyte kinetics in the pulmonary microcirculation: Observations using real-time confocal luminescence microscopy coupled with high-speed video analysis. *Laboratory Investigation, 76*(6), 809–822.

Yamaki, K., Thorlacius, H., Xie, X., Lindbom, L., Hedqvist, P., & Raud, J. (1998). Characteristics of histamine-induced leukocyte rolling in the undisturbed microcirculation of the rat mesentery. *British Journal of Pharmacology, 123*(3), 390–399.

Yanai, M., Butler, J. P., Suzuki, T., Kanda, A., Kurachi, M., Tashiro, H., et al. (1999). Intracellular elasticity and viscosity in the body, leading, and trailing regions of locomoting neutrophils. *American Journal of Physiology, 277*(3 Pt. 1), C432–C440.

Yanai, M., Butler, J. P., Suzuki, T., Sasaki, H., & Higuchi, H. (2004). Regional rheological differences in locomoting neutrophils. *American Journal of Physiology: Cell Physiology, 287*(3), C603–C611.

Yap, B., & Kamm, R. D. (2005a). Cytoskeletal remodeling and cellular activation during deformation of neutrophils into narrow channels. *Journal of Applied Physiology, 99*(6), 2323–2330.

Yap, B., & Kamm, R. D. (2005b). Mechanical deformation of neutrophils into narrow channels induces pseudopod projection and changes in biomechanical properties. *Journal of Applied Physiology, 98*(5), 1930–1939.

Yeung, A., & Evans, E. (1989). Cortical shell–liquid core model for passive flow of liquid-like spherical cells into micropipets. *Biophysical Journal, 56*(1), 139–149.

Yodice, P. C., Astiz, M. E., Kurian, B. M., Lin, R. Y., & Rackow, E. C. (1997). Neutrophil rheologic changes in septic shock. *American Journal of Respiratory and Critical Care Medicine, 155*(1), 38–42.

Yoon, M., Moir, R. D., Prahlad, V., & Goldman, R. D. (1998). Motile properties of vimentin intermediate filament networks in living cells. *Journal of Cell Biology, 143*(1), 147–157.

Yoon, K. H., Yoon, M., Moir, R. D., Khuon, S., Flitney, F. W., & Goldman, R. D. (2001). Insights into the dynamic properties of keratin intermediate filaments in living epithelial cells. *Journal of Cell Biology, 153*(3), 503–516.

Yoshimura, A., Hara, Y., Kaneko, T., & Kato, I. (1997). Secretion of IL-1 beta, TNF-alpha, IL-8 and IL-1ra by human polymorphonuclear leukocytes in response to lipopolysaccharides from periodontopathic bacteria. *Journal of Periodontal Research, 32*(3), 279–286.

Yu, Y., & Shao, J. Y. (2007). Simultaneous tether extraction contributes to neutrophil rolling stabilization: A model study. *Biophysical Journal, 92*(2), 418–429.

Yuruker, B., & Niggli, V. (1992). Alpha-actinin and vinculin in human neutrophils: Reorganization during adhesion and relation to the actin network. *Journal of Cell Science, 101*(Pt. 2), 403–414.

Zaner, K. S., & Valberg, P. A. (1989). Viscoelasticity of F-actin measured with magnetic microparticles. *Journal of Cell Biology, 109*(5), 2233–2243.

Zhang, X., Wojcikiewicz, E. P., & Moy, V. T. (2006). Dynamic adhesion of T lymphocytes to endothelial cells revealed by atomic force microscopy. *Experimental Biology and Medicine (Maywood), 231*(8), 1306–1312.

Zhao, Y. H., Chien, S., & Weinbaum, S. (2001). Dynamic contact forces on leukocyte microvilli and their penetration of the endothelial glycocalyx. *Biophysical Journal, 80*(3), 1124–1140.

Zhelev, D. V., & Alteraifi, A. (2002). Signaling in the motility responses of the human neutrophil. *Annals of Biomedical Engineering, 30*(3), 356–370.

Zhelev, D. V., Alteraifi, A. M., & Chodniewicz, D. (2004). Controlled pseudopod extension of human neutrophils stimulated with different chemoattractants. *Biophysical Journal, 87*(1), 688–695.

Zhelev, D. V., & Hochmuth, R. M. (1994). Human neutrophils under mechanical stress. In V. C. Mow, F. Guilak, R. Tran-Son-Tay & R. M. Hochmuth (Eds.), *Cell mechanics and cellular engineering* (pp. 3–20). New York, NY: Springer-Verlag.

Zhelev, D. V., & Hochmuth, R. M. (1995). Mechanically stimulated cytoskeleton rearrangement and cortical contraction in human neutrophils. *Biophysical Journal, 68*(5), 2004–2014.

Zhelev, D. V., Needham, D., & Hochmuth, R. M. (1994). Role of the membrane cortex in neutrophil deformation in small pipets. *Biophysical Journal, 67*(2), 696–705.

Zhu, X., Boetticher, E., Wang, L., Duan, Y., Learoyd, J., & Leff, A. R. (2008). Proline-rich tyrosine kinase 2 regulates spreading and migration of eosinophils after beta2-integrin adhesion. *American Journal of Respiratory Cell and Molecular Biology, 39*(3), 263–269.

Zigmond, S. H., Levitsky, H. I., & Kreel, B. J. (1981). Cell polarity: An examination of its behavioral expression and its consequences for polymorphonuclear leukocyte chemotaxis. *Journal of Cell Biology, 89*(3), 585–592.

SECTION 2

ADHESION MOLECULES

CHAPTER 4

Activation of Leukocyte Integrins

Eun Jeong Park and Motomu Shimaoka
Immune Disease Institute,
Program in Cellular and Molecular Medicine at Children's Hospital Boston, and
Department of Anesthesia, Harvard Medical School, Boston, Massachusetts 02115, USA

I. OVERVIEW

Integrins are the largest family of α/β-heterodimeric cell adhesion molecules that mediate cation-dependent adhesive interactions in a wide range of cell-to-cell and cell-to-matrix interactions. Integrin-mediated leukocyte adhesion and migration constitute the essential component in immune system, as exemplified by a group of genetic disorders leukocyte adhesion deficiencies (LAD) type I and III. The most unique feature of integrins is that they transmit the bidirectional signals across the plasma membrane. In response to internal cues (e.g., activation of chemokine receptor and T-cell receptor that elicit intracellular signaling cascades) that impinge on the integrin cytoplasmic domains, the integrin extracellular part undergoes the

Current Topics in Membranes, Volume 64 1063-5823/09 $35.00

DOI: 10.1016/S1063-5823(09)64004-7

global conformational changes, rapidly upregulating the ability to bind ligand. Conversely, in response to external cues (e.g., ligand binding and shear force) applied to the extracellular part, the integrin cytoplasmic domains change their configuration, thereby altering the ability to interact with intracellular signaling molecules that regulate gene expression and cell metabolism. This unique ability of integrins stems from the regulated conformational dynamism between the low-affinity bent to the high-affinity extended integrin. The point of equilibrium of the integrin conformational dynamism is determined by a balance of the internal and external cues as well as the activation threshold set by the intra- and interdomain interactions within the integrin molecule.

II. LEUKOCYTE INTEGRINS

Integrins are the foremost family of cell adhesion molecules containing noncovalently associated α- and β-subunits (Askari, Buckley, Mould, & Humphries, 2009; Hynes, 2002; Luo, Carman, & Springer, 2007; Shimaoka, Takagi, & Springer, 2002). Integrins mediate cell–cell, cell–extracellular matrix, and cell–pathogen interactions over a wide range of biological contexts. Not only do integrins support force-resistant stable firm adhesion, but also they are involved in dynamic adhesive interactions observed in cellular polarization and cell migration. Integrin-dependent physiological processes include tissue morphogenesis, inflammation, wound healing, and regulation of cell growth and differentiation. To date, 19 different integrin α-subunits and 8 different β-subunits have been reported in vertebrates, forming at least 25 $\alpha\beta$-heterodimers and perhaps revealing the integrins to be the most structurally and functionally diverse family of cell adhesion molecules yet known.

The adhesive and signaling activities of integrins are vital to many of the cell–cell and cell–extracellular matrix interactions involved in immune responses (Evans et al., 2009). Integrins on leukocytes play a critical role in their adhesive interactions with endothelial cells during migration to lymphoid organs and extravasation to sites of inflammation. The most unique feature of integrins is their ability to transmit bidirectional transmembrane signaling (Luo et al., 2007). Intracellular signaling pathways that are activated by other receptors (e.g., receptors coupled to G proteins or tyrosine kinases) impinge on integrin cytoplasmic domains and enhance the activity of the extracellular headpiece for ligand binding (inside-out signaling). In contrast, the binding of ligand to extracellular domains initiates intracellular signaling via conformational changes of the cytoplasmic domains (outside-in signaling). Such bidirectional signaling supports the dynamic and reversible transformation of integrins between nonadhesive and adhesive states, and thereby plays a critical role in the regulation of cell migration, an essential component of immune function.

β_2-Integrins ($\alpha_L\beta_2$ or LFA-1; $\alpha_M\beta_2$ or Mac-1; $\alpha_X\beta_2$ or p150,95; and $\alpha_D\beta_2$), β_7-integrins ($\alpha_4\beta_7$ and $\alpha_E\beta_7$) and a subset of β_1-integrins ($\alpha_4\beta_1$) are the major integrins expressed on leukocytes (Kellermann, Dell, Hunt, & Shimizu, 2002), although other integrins are also expressed in certain leukocyte subsets and/or under certain conditions (e.g., chronic inflammation, during the course of development). Integrin $\alpha_L\beta_2$ or leukocyte function-associated antigen-1 (LFA-1) is the predominant integrin present in lymphocytes. $\alpha_L\beta_2$ binds to IgSF cell-surface molecules (ICAMs) on endothelial cells. By acting in concert with $\alpha_4\beta_1$ and $\alpha_4\beta_7$, $\alpha_L\beta_2$ plays a pivotal role in firm adhesion to, and migration through, endothelial cells during normal lymphocyte recirculation, as well as in response to inflammatory signals (Dustin & Springer, 1999). $\alpha_L\beta_2$ is also required for a wide variety of cell–cell interactions including those of T cells with antigen-presenting cells, B cells with T cells, and NK cells with target cells. Moreover, $\alpha_L\beta_2$ acts as a co-stimulatory molecule in virtually all T-cell responses, in keeping with its role in forming the immunological synapse (Grakoui et al., 1999). $\alpha_4\beta_7$ and $\alpha_E\beta_7$ bind to mucosal addressin cell adhesion molecule-1 (MAdCAM-1) and E-cadherin, respectively (Wagner & Muller, 1998). MAdCAM-1 is constitutively expressed in high-endothelial venules (HEVs) in gut-associated lymphoid tissue (GALT) in adults, thus supporting mucosal tissue-specific homing by $\alpha_4\beta_7$ (Berlin et al., 1993). While most integrins predominantly support firm adhesion, $\alpha_4\beta_7$ mediates both rolling (when it is in a low-affinity state) and firm adhesion (when it is in a high-affinity state) (Berlin et al.; Chen, Salas, & Springer, 2003). Although $\alpha_L\beta_2$ can support rolling on its own under certain *in vitro* experimental conditions (Knorr & Dustin, 1997; Salas, Shimaoka, Chen, Carman, & Springer, 2002; Sigal et al., 2000), $\alpha_L\beta_2$ rather participates in the stabilization of selectin-mediated rolling *in vivo* (Henderson et al., 2001). $\alpha_4\beta_1$ also supports both rolling and firm adhesion on VCAM-1 (Alon et al., 1995).

III. PATHOLOGY OF INTEGRIN FUNCTION DEFICIENCY

The physiologic importance of $\alpha_L\beta_2$ and its related β_2-integrins is illustrated by a rare genetic disorder, leukocyte adhesion deficiency type I (LAD-I). LAD-I is caused by loss-of-function mutations in the β_2-integrin subunit that result in the absence, or severely reduced expression, of all β_2-integrin heterodimers on the cell surface of leukocytes (Anderson & Springer, 1987; Fischer, Lisowska-Grospierre, Anderson, & Springer, 1988). The typical clinical manifestations of LAD-I patients include delayed separation of the umbilical cord, lack of pus formation at the sites of infection, and increased numbers of neutrophils in circulation. LAD-I patients suffer from recurrent and often life-threatening bacterial infections and from impaired wound healing, since β_2-integrins are important for host defenses against microorganisms. Neutrophils from LAD-I patients showed a markedly reduced capacity to adhere to endothelial cells and to migrate to sites of inflammation. LAD-1 lymphocytes exhibited impaired function in antigen- and mitogen-induced proliferation, antibody-dependent killing, and T-cell-dependent antibody production.

LAD-III (a.k.a. LAD-I variant) demonstrates the physiological importance of integrin activation (Etzioni, 2009). LAD-III patients manifest not only an increased susceptibility to bacterial infections, such as occurs in LAD-I, but also to platelet dysfunction. The latter can be observed in Glanzmann's thrombasthenia, where one finds a lack of integrin $\alpha_{IIb}\beta_3$-expression or functionality. In contrast to LAD-I, leukocytes isolated from LAD-III patients express normal levels of integrins on leukocytes. Despite normal levels of integrin expression, these same LAD-III leukocytes failed to upregulate $\alpha_L\beta_2$-, $\alpha_M\beta_2$-, and $\alpha_4\beta_1$-adhesiveness in response to chemokine and/or the other chemoattractants that activate G protein-coupled receptor (GPCR) signaling. A genetic defect in kindelin-3 has been identified as a cause of LAD-III (Malinin et al., 2009; Mory et al., 2008; Svensson et al., 2009). Kindlin-3 is a cytoskeletal protein that activates integrins by binding to the integrin β-cytoplasmic tails. In kindlin-3$^{-/-}$ chimeric mice, defects in β_1- and β_2-integrin-mediated leukocyte adhesion were observed, supporting the critical role of kindlin-3 in integrin activation in leukocytes (Moser, Bauer, et al., 2009).

IV. PATHOLOGY UNDERLYING THE ABERRANT INTEGRIN REGULATION

Aberrantly increased expression of $\alpha_L\beta_2$ has been shown to induce autoimmune diseases in a mouse model (Giblin & Kelly, 2001; Yusuf-Makagiansar, Anderson, Yakovleva, Murray, & Siahaan, 2002). Adaptive transfer of a T-cell clone overexpressing $\alpha_L\beta_2$ may induce lupus-like autoimmune symptoms including alveolitis and nephritis (Yung et al., 1996; Yung, Ray, Mo, & Chen, 2003).

Naïve lymphocytes from healthy individuals predominantly express a low-affinity integrin conformation. Physiologic stimulation induces only a transient upregulation of integrin adhesiveness. In a variety of inflammatory diseases, aberrantly upregulated integrin affinity and/or adhesiveness has been amply demonstrated. Lymphocytes in inflamed salivary glands of Sjögren's syndrome patients showed increased expression of a high-affinity form of $\alpha_L\beta_2$ (Cauli, Yanni, Pitzalis, Challacombe, & Panayi, 1995). Monocytes from patients with acute myocardial infarction, which involves inflammatory responses as part of its pathogenesis, were shown to express elevated levels of the high-affinity forms of $\alpha_L\beta_2$ and $\alpha_M\beta_2$ (May et al., 2002). Furthermore, blood cancer cells have been shown to exhibit aberrantly enhanced $\alpha_L\beta_2$- and $\alpha_4\beta_1$-affinity and/or adhesiveness (Hideshima, Mitsiades, Tonon, Richardson, & Anderson, 2007; Peng, Liu, Andrei, Xiao, & Lam, 2008; Tanaka, 1999).

As is well known, integrins on leukocytes are valid therapeutic targets for treating autoimmune and inflammatory disorders. Efalizumab (Raptiva^{TM}), a humanized antibody to $\alpha_L\beta_2$, was approved by the FDA for the treatment of psoriasis (Cather & Menter, 2003; Lebwohl et al., 2003). $\alpha_L\beta_2$ plays pivotal roles in at least two important processes underlying the pathogenesis of T-cell-mediated autoimmunity, thereby rendering it an excellent therapeutic target for anti-inflammation therapy. $\alpha_L\beta_2$ participates in the interaction of circulating T cells with endothelial cells, which supports not only the migration of T cells to lymphoid organs and to sites of inflammation, but also the interaction of migratory T cells with antigen-presenting and target cells, which facilitates proliferation, cytokine production, and CTL activity (de Fougerolles, 2003).

Natalizumab (Tysabri^{TM}), a humanized antibody to the α_4-integrin, was approved for the treatment of multiple sclerosis and Crohn's diseases (Miller et al., 2003). This stems from the fact that α_4-integrins ($\alpha_4\beta_1$ and $\alpha_4\beta_7$) help facilitate the accumulation of inflammatory leukocytes to both the inflamed brain and inflamed gut. Although Efalizumab and Natalizumab have proven effective in clinical use, one must be aware of potential immune suppression brought about by the inhibition of $\alpha_L\beta_2$- and α_4-integrins, as exemplified by both LAD-I and LAD-III. Indeed, several cases of progressive multifocal leukoencephalopathy (PML) have been reported in patients treated with Efalizumab (Sterry et al., 2009) and Natalizumab (Van Assche et al., 2005). PML is believed to be caused by the reactivation of latent JC virus due to the iatrogenic immune suppression associated with the inhibition of $\alpha_L\beta_2$- and α_4-integrins.

V. STRUCTURES OF INTEGRIN HETERODIMERS AND INTEGRIN DOMAINS

Integrins are α/β-heterodimeric membrane proteins characterized by a complex multidomain organization. It is believed that there is a fundamental mechanism underlying the regulation of integrin activation and conformational changes common to most integrins (although there remain some important differences specific to particular integrins). Here, we use $\alpha_L\beta_2$ as a model to provide readers with an overview of integrin structure and conformational changes. Each integrin subunit is a type I transmembrane protein composed of a large ectodomain, a single transmembrane part, and a short cytoplasmic tail (except for the β_4-subunit, which contains a long cytoplasmic tail). The integrin α_L-subunit ectodomain (>940 residues) contains five domains and the β_2-subunit (~640 residues) contains eight domains. The N-terminals of the α- and β-subunits associate with each other to form a globular ligand-binding headpiece, which is connected to the plasma membrane via the leg pieces.

VI. CONFORMATIONAL CHANGES IN THE α AND β I-DOMAINS

About half (α_L, α_M, α_X, α_D, α_E, α_1, α_2, α_{10}, and α_{11}) of the α-subunits contain a von Willebrand factor-type A domain of ~200 amino acids, referred to as an α inserted (I)-domain (Humphries, 2000; Shimaoka et al., 2002). The ligand-binding capacity of $\alpha_L\beta_2$ is contained solely within the I-domain, whereas other domains play regulatory roles. The I-domain adopts an α/β-Rossmann fold with a metal ion-dependent adhesion site (MIDAS) on the top of the domain, whereas its C- and N-terminal connections are on the distal bottom face (Huang, Zang, Takagi, & Springer, 2000; Lee, Rieu, Arnaout, & Liddington, 1995; Shimaoka et al., 2002; Xiong et al., 2001). Divalent cations are universally required for integrins to bind ligands. Metals bind to the MIDAS in the integrin, and coordinate to a Glu or Asp residue in the ligand. This interaction through the MIDAS plays a central role in ligand recognition. The ability of the I-domain to bind ligand is controlled by conformational changes; the affinity of the I-domain for its ligand is enhanced by a downward axial displacement of its C-terminal helix, which is conformationally linked to alterations of the MIDAS loops and Mg^{2+} coordination (Huth et al., 2000; Shimaoka et al., 2001; Shimaoka, Xiao, et al., 2003; Vorup-Jensen, Ostermeier, Shimaoka, Hommel, & Springer, 2003). Compared to the default, low-affinity conformation, downward displacements by one and two turns of the helix lead to intermediate- and high-affinity conformations with ~500 and 10,000-fold increases in affinity, respectively (Shimaoka, Xiao, et al.). Conversely, the binding of ligand to the MIDAS of the I-domain induces conformational changes by stabilizing the high-affinity conformation. These changes include rearrangements in metal coordination in the MIDAS, as well as backbone movements in the loops surrounding the MIDAS, which are linked to a downward axial displacement of the C-terminal helix (Emsley, Knight, Farndale, Barnes, & Liddington, 2000; Lee et al.; Shimaoka, Xiao, et al.).

All integrin β-subunits contain another von Willebrand factor-type A domain, the β I-domain (Xiong et al., 2001, 2002). The function of the β I-domain is regulated by conformational changes similar to those observed in the α I-domain, in which a downward movement of the C-terminal α-helix allosterically alters the geometry of the MIDAS and increases the affinity for ligand (Luo, Takagi, & Springer, 2004; Shimaoka et al., 2002; Takagi & Springer, 2002; Yang, Shimaoka, Chen, & Springer, 2004). In contrast to the α I-domain, the β I-domain contains two additional metal-binding site LIMBS (a.k.a. SyMBS) and ADMIDAS at either side of the MIDAS. The metal coordinations at these sites are interconnected and coupled to the movement of the C-terminal helix (Chen et al., 2003; Valdramidou, Humphries, & Mould, 2008; Xiao, Takagi, Wang, Coller, & Springer, 2004; Zhu et al., 2008). The β I-domain associates with the α_L β-propeller domain at the linkage to the I-domain, forming a globular ligand-binding integrin head. The polypeptide linker between the C-terminus of the I-domain and β-sheet 3 of the β-propeller domain, including the α_L-residue Glu-310, is important in I-domain activation (Huth et al., 2000; Yang, Shimaoka, Salas, Takagi, & Springer, 2004). I-domain activation is thought to be induced by a downward pull on the C-terminal α-helix or linker (Salas et al., 2002; Shimaoka et al.). In fact, it has been postulated that the universally conserved residue Glu-310 in the I-domain linker is an intrinsic ligand, and that binding of the activated β I-domain to this ligand induces the downward movement of the α I-domain C-terminal α-helix, which upregulates ligand affinity at the α I-domain MIDAS (Alonso, Essafi, Xiong, Stehle, & Arnaout, 2002; Shimaoka et al.; Takagi & Springer; Yang, Shimaoka, Salas, et al.).

Both the α I-domain and the β I-domain are targets for allosteric small-molecule antagonists (Shimaoka & Springer, 2003). One class of small molecules, termed α I-allosteric antagonists, binds underneath the C-terminal α-helix of the α_L I-domain (Kallen et al., 1999; Last-Barney et al., 2001; Liu et al., 2001). By interfering with the downward movement of the C-terminal helix, the antagonists stabilize the low-affinity I-domain conformation, thereby allosterically inhibiting ligand binding. Another class of small molecules, designated as an α/β-I-like allosteric antagonist, binds to the MIDAS of the β_2 I-domain and a portion of the α_L β-propeller domain, thus preventing the I-like domain from interacting with the I-domain C-terminal linker. Consequently, the α I-domain is isolated from the β I-domain and is left in a default low-affinity conformation (Shimaoka, Salas, Yang, Weitz-Schmidt, & Springer, 2003a). Whereas the α I-allosteric antagonists favor the bent conformation containing the stabilized low-affinity I-domain, the α/β I-like allosteric antagonists induce the extended conformation containing the default low-affinity I-domain (Shimaoka, Salas, et al.).

VII. GLOBAL CONFORMATIONAL CHANGES

Integrin cytoplasmic tails, which lack enzymatic activities or actin-binding capabilities, regulate integrin activation by interacting with cytoplasmic effector and adaptor proteins (Moser, Legate, Zent, & Fassler, 2009b). The α- and β-integrin cytoplasmic tails are thought to associate with each other at the membrane-proximal regions, functioning as a "clasp" that restrains integrins in a default low-affinity conformation. Critical interactions at the association interface between the α/β-cytoplasmic domains have been revealed by structural investigations using NMR (Vinogradova et al., 2002). The arginine residue in the conserved GFFKR sequence at the membrane-proximal region of the α-subunit forms a putative salt bridge with the β-subunit. This putative membrane-proximal salt bridge constitutes a critical interaction, one which contributes to clasping the α/β-cytoplasmic domains and thereby restraining integrin activation.

This α/β-cytoplasmic association stabilizes the extracellular domains in a latent bent conformation, in which the headpiece is folded back to the leg piece (Luo et al., 2007; Nishida et al., 2006; Takagi, Petre, Walz, & Springer, 2002). In the bent conformation, the I-domain adopts a low-affinity state and is pointed toward the membrane, positioning the binding domain in an unfavorable orientation for any potential interaction with ligand on opposing cells (Xiong et al., 2001; Zhu et al., 2008). It is widely believed that the bent conformation represents a low-affinity ligand incompetent form. Activation of other receptors—including TCR and chemokine receptors—initiates an intracellular signaling cascade that eventually impinges upon the integrin cytoplasmic tails. Binding to the integrin cytoplasmic domains of signaling molecules triggers a dissociation of the integrin cytoplasmic tails (Harburger, Bouaouina, & Calderwood, 2009; Moser, Legate, et al., 2009b; Wegener et al., 2007). Talin as well as kindlins are the major integrin tail-interacting proteins that have been shown to trigger the separation of the tails and subsequent upregulation of integrin adhesiveness (Moser, Legate, et al.). The interaction of the talin head, a FREM domain, with the β-integrin tail involves two key steps (Wegener et al.). First, the talin head binds to the membrane-proximal NPxY motif that is located at the distal part of the β-integrin tail. Then, while maintaining the first interaction with the distal part, the talin head interacts with the membrane-proximal helical part of the β-tail, thereby separating the association of the α/β-integrin tails.

Separation of the α- and β-subunit cytoplasmic tails has been linked to a separation of the transmembrane domains from the membrane-proximal segments of the extracellular domains, resulting in a destabilization of the interface between the headpiece and tailpiece. This induces a switchblade-like opening, leading to an extended conformation, which reorients the ligand-binding face in the I-domains toward the opposing cells (Askari et al., 2009; Luo et al., 2007).

In the extended conformation, two different conformations of the headpiece, termed closed and open, can be observed (Nishida et al., 2006; Takagi et al., 2002; Xiao et al., 2004). In the bent conformation, only the closed conformation of the headpiece is present. Therefore, an extension facilitates the adoption of an open headpiece conformation, which corresponds to the ligand-bound and high-affinity conformation. The closed headpiece is converted to an open headpiece by the outward movement of the β-hybrid domain. This outward movement is coupled to a downward shift of the β I-domain's C-terminal α-helix, which triggers the conversion of the β-MIDAS into an open configuration (Xiao et al.). The open configuration of the β I-domain MIDAS is thought to bind to the C-terminal portion of the α I-domain. This intramolecular and intersubunit domain–domain interaction leads to a downward shift of the α I-domain's C-terminal helix, converting it to the ligand-competent open I-domain conformation, which binds to ligand ICAM-1.

Structural studies using crystallography (Xiao et al., 2004; Xiong et al., 2001; Zhu et al., 2008) and electron microscopy (Nishida et al., 2006; Takagi et al., 2002) have revealed multiple conformations in solution. However, it is thought that integrins on the cell surface are not fixed in a particular conformation, but rather in an equilibrium between the bent conformation (containing a low-affinity I-domain) and either the extended conformation (with closed headpiece containing a low- to intermediate-affinity I-domain) or the extended conformation (with the open headpiece containing a high-affinity I-domain). This equilibrium is affected by the presence of activating intracellular factors and the concentration of extracellular ligands. Activation by signals within the cell (inside-out signaling) induces both a separation of the cytoplasmic and transmembrane domains and a straightening of the extracellular part, thereby stabilizing the extended form (shifting the equilibrium to the right, as shown in Figure 1). Conversely, the binding of extracellular ligands stabilizes the extended conformation and therefore enhances the separation of integrin transmembrane domains and cytoplasmic tails, which transmits signals to the cytoplasm (outside-in signaling). The global conformational changes between the bent and the extended conformations described above function as a kind of allosteric machinery for transmitting bidirectional transmembrane signals (Hynes, 2002; Shimaoka & Springer, 2003).

VIII. INTEGRIN ACTIVATION IN LEUKOCYTE–ENDOTHELIAL INTERACTIONS

The dynamic regulation of $\alpha_L\beta_2$-activation and conformation in response to physiologic cues is of critically importance to controlling the adhesive interactions of lymphocytes with endothelial cells. The recruitment process

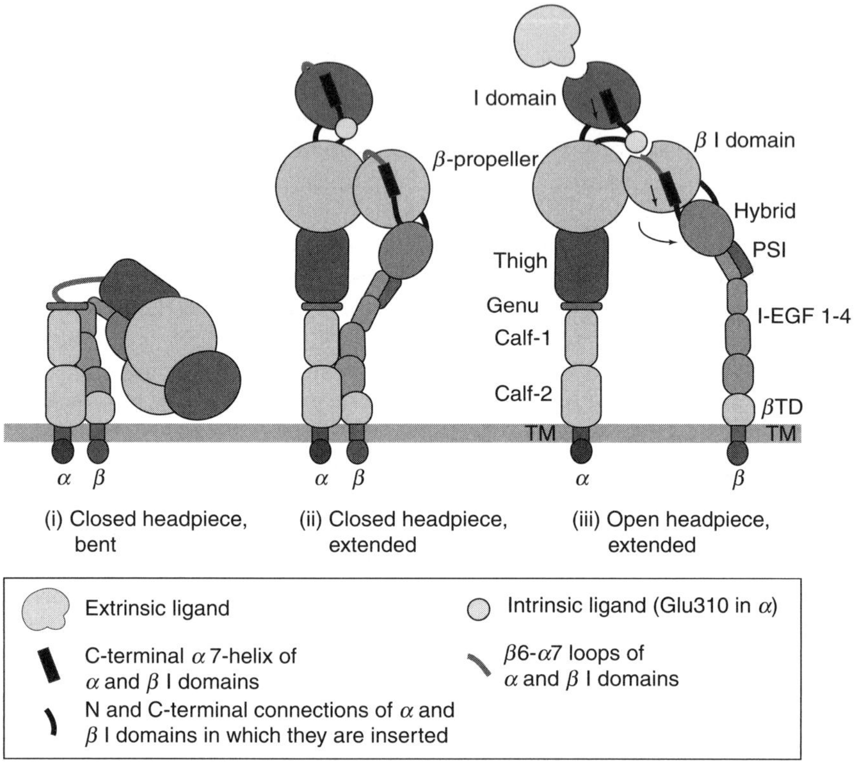

FIGURE 1 Integrin domains and activation-dependent conformational changes. Graphics show the conformational transition between the bent form containing the closed headpiece (with low-affinity I-domain) (left), the intermediate to extended form containing the closed headpiece (with the low- to intermediate-affinity I-domain) (middle), and the extended form containing the open headpiece (with the high-affinity I-domain). (See Color Insert.)

drawing lymphocytes from the blood stream to lymph node endothelial cells is initiated by the interactions of L-selectin with its ligands, which leads to their rolling along HEVs under shear flow (Butcher & Picker, 1996; Springer, 1994). During rolling, lymphocytes encounter surface-bound chemokines, which activate the GPCRs. This, in turn, leads to $\alpha_L\beta_2$-affinity upregulation, which mediates their firm arrest to endothelial cells under shear stress. Endothelial cell-bound chemokines are thought to induce a switchblade-like opening that converts a default bent form to an extended form. However, chemokine activation *per se* is not strong enough to induce the outward movement of the β-hybrid domain, which is required for inducing the open headpiece. The inability of chemokine to induce the outward movement of the β-hybrid domain was also shown in $\alpha_4\beta_1$ (Chigaev et al., 2009). The extended form with the closed headpiece (containing a low- to intermediate-affinity α I-domain) represents the chemokine-primed state of $\alpha_L\beta_2$. The interaction of this primed $\alpha_L\beta_2$ with ICAM-1 on endothelial cells under shear stress is thought to impose a force that induces the outward movement of the hybrid domain, thereby converting $\alpha_L\beta_2$ into a fully active extended form with an open headpiece (Shamri et al., 2005). In this way, internal cues (e.g., signaling elicited by chemokine activation, which leads to talin and kindlin interactions with integrin cytoplasmic tails) and external cues (e.g., shear stress and interaction with ligand that impact the integrin ectodomain) cooperatively regulate $\alpha_L\beta_2$-conformations, thereby supporting the firm adhesion of lymphocytes to endothelial cells.

Integrin-mediated cell adhesion can be enhanced through affinity upregulation and/or valency upregulation (a.k.a. clustering) (Kim, Carman, Yang, Salas, & Springer, 2004). Chemokine activation primarily induces the conformational changes of integrin that lead to affinity upregulation, whereas clustering is induced after binding to multivalent ligands (e.g., ICAM-1 on endothelial cells), thereby strengthening cell adhesion (Kim et al.).

IX. SPATIOTEMPORAL REGULATION OF INTEGRIN ACTIVATION

Coordinated and directional cell migration requires properly controlled spatiotemporal regulation of integrin activation and conformations. Cycles of adhesion by activated integrins at the leading edge, as well as deadhesion (detachment) by deactivated integrins at the trailing edge, are thought to be the underlying mechanisms that drive cells forward. In migrating endothelial cells, as is apparent in the wound healing process, the high-affinity form of integrin $\alpha_V\beta_3$ is preferentially localized at the leading edge (Kiosses, Shattil, Pampori, & Schwartz, 2001). However, lymphocytes, which migrate much faster than endothelial cells, appear to use a different mode of regulation (Smith et al., 2005). Lymphoblasts migrating on ICAM-1 substrates exhibited a distinct distribution pattern composed of different $\alpha_L\beta_2$-conformations (Stanley et al., 2008). At the protruding lamellipodia of the leading edge, intermediate-affinity $\alpha_L\beta_2$ is localized, whereas at the middle region, termed the focal zone, high-affinity $\alpha_L\beta_2$ becomes concentrated. The cytoskeletal proteins α-actinin-1 and talin were shown to associate with intermediate-affinity and high-affinity $\alpha_L\beta_2$, respectively, potentially regulating the spatial distributions of specific $\alpha_L\beta_2$-conformations. At the trailing edge of lymphocytes migrating on ICAM-1 substrates, inactive $\alpha_L\beta_2$ was found to selectively localize at the trailing edge, where it was associated with myosin heavy chain IIA, which would provide the necessary force to retract the uropod (Morin et al., 2008). Deadhesion at the trailing edge (uropod) critically influences lymphocyte migration. Lymphocytes isolated from genetically engineered mice that express constitutively activated integrins (Table I) exhibited perturbed lymphocyte migration pathways due to their inability to efficiently retract the uropod. Shear stress impacts the spatial regulation of integrin activation, thereby giving rise to a distinct distribution pattern of specific $\alpha_L\beta_2$-conformation, compared to that observed in cells migrating in the absence of shear stress. In shear- and chemokine-driven migration across the endothelial cell monolayer, lymphocytes exhibited numerous millipede-like filopodia underneath the cell body (Shulman et al., 2009). High-affinity $\alpha_L\beta_2$ is localized at these filopodias, in which it potentially helps facilitate crawling on, and transmigration across, an endothelial monolayer.

X. THE ROLE OF INTEGRINS IN THE INTERSTITIAL MIGRATION OF LEUKOCYTES

After completing transmigration, lymphocytes enter the interstitial space of lymph nodes, where they become highly motile (Bousso & Robey, 2003; Mempel, Henrickson, & Von Andrian, 2004; Miller, Wei, Parker, &

TABLE I

Perturbed Migrations of Knock-in Lymphocytes Engineered to Express Integrins Constitutively Activated Integrins

Integrins	Sites targeted	Migratory phenotypes of lymphocytes	References
α_L	Deletion of GFFKR motif (*Lfa-1$^{d/d}$*) in α_L-cytoplasmic domain	Constitutive activation of LFA-1-mediated cell adhesion, impaired deadhesion, and defected cell migration Impaired lymphocyte activation and neutrophil recruitment in peritonitis	Semmrich et al. (2005)
β_7	Mutation in the ADMIDAS of the β_7 I-domain	Increased firm adhesion of lymphocytes to Peyer's patch venules Reduced lymphocyte migration to the gut and colitis progression	Park et al. (2007)
α_4	Mutation in the α_4-cytoplasmic GFFKR motif	Suppressed migration of T cells on VCAM-1 and MAdCAM-1 substrates Reduced gut homing of lymphocytes	Imai et al. (2008)
α_L	Mutation in α_L I-domain	Suppressed migration of T cells on ICAM-1 substrates Perturbed diapedesis of T cells across lymph node high-endothelial venules	Park et al. (unpublished data)

Cahalan, 2002), crawling through a mesh-like network of branching collagen fibers ensheathed by fibroblastic reticular cells (FRCs) (Bajenoff et al., 2006; Katakai, Hara, Sugai, Gonda, & Shimizu, 2004). While migrating within the interstitial space, lymphocytes and other leukocytes utilize amoeboid-like cell migration mechanisms, which involve completely different modes of migration from those used for migratory purposes in either 2D or within endothelial lumen (Friedl & Weigelin, 2008). Amoeboid-like migration allows leukocytes to move fast (up to 30 μm/min), making this mode of migration suitable for searching and scanning the interstitial 3D microenvironment. Moreover, this mode of cell migration does not require strong adhesive interactions to the tissue. Instead of generating a driving force by strong adhesive interactions, amoeboid-like cell migration utilizes a combination of actin networks to form protrusive flows that power locomotion, in tandem with actomyosin contractions of the trailing edge, in order to squeeze the rigid nucleus through narrow spaces (Friedl & Weigelin).

It has been shown that integrins are likely to be a dispensable component of the amoeboid-like cell migration observed in the interstitial space. Dendritic cells that lack all integrins exhibited an interstitial cell migration pattern comparable to that of wild-type cells. β_2-Integrin-deficient T cells lacking $\alpha_L\beta_2$ suffered only moderately impaired interstitial motilities (Woolf et al., 2007). The lymph node interstitial space is thought to express sufficient levels of both chemokines and ICAM-1. The inability of $\alpha_L\beta_2$ to engage in ICAM-1 during interstitial migration is probably due to the lack of external cues (e.g., shear stress) that are required to induce conformational conversion to high-affinity conformation (Woolf et al.).

XI. CONCLUDING REMARKS

Activation-dependent global conformational changes between the low-affinity bent and high-affinity extended form constitute the structural basis of bidirectional transmembrane signaling through integrins. The ability to dynamically regulate their adhesiveness enables integrins to play a pivotal role in governing leukocyte movements in immune reactions. Integrins critically influence a wide range of leukocyte migratory and adhesive behaviors including migration on and across endothelial cells, as well as formation of stable T-cell contact with dendritic cells, except for the amoeboid-like migration observed in the interstitial microenvironment. A close cooperation of internal and external cues is required to induce high-affinity $\alpha_L\beta_2$. Activation and deactivation of integrins must be spatiotemporarily regulated to promote efficient cell migration.

References

Alon, R., Kassner, P. D., Carr, M. W., Finger, E. B., Hemler, M. E., & Springer, T. A. (1995). The integrin VLA-4 supports tethering and rolling in flow on VCAM-1. *Journal of Cell Biology, 128*, 1243–1253.

Alonso, J. L., Essafi, M., Xiong, J. P., Stehle, T., & Arnaout, M. A. (2002). Does the integrin aA domain act as a ligand for its bA domain? *Current Biology, 12*, R340–R342.

Anderson, D. C., & Springer, T. A. (1987). Leukocyte adhesion deficiency: An inherited defect in the Mac-1, LFA-1, and p150,95 glycoproteins. *Annual Review of Medicine, 38*, 175–194.

Askari, J. A., Buckley, P. A., Mould, A. P., & Humphries, M. J. (2009). Linking integrin conformation to function. *Journal of Cell Science, 122*, 165–170.

Bajenoff, M., Egen, J. G., Koo, L. Y., Laugier, J. P., Brau, F., Glaichenhaus, N., et al. (2006). Stromal cell networks regulate lymphocyte entry, migration, and territoriality in lymph nodes. *Immunity, 25*, 989–1001.

Berlin, C., Berg, E. L., Briskin, M. J., Andrew, D. P., Kilshaw, P. J., Holzmann, B., et al. (1993). a4b7 integrin mediates lymphocyte binding to the mucosal vascular address in MAdCAM-1. *Cell, 74*, 185–195.

Bousso, P., & Robey, E. (2003). Dynamics of CD8+ T cell priming by dendritic cells in intact lymph nodes. *Nature Immunology, 4*, 579–585.

Butcher, E. C., & Picker, L. J. (1996). Lymphocyte homing and homeostasis. *Science, 272*, 60–66.

Cather, J. C., & Menter, A. (2003). Modulating T cell responses for the treatment of psoriasis: A focus on efalizumab. *Expert Opinion on Biological Therapy, 3*, 361–370.

Cauli, A., Yanni, G., Pitzalis, C., Challacombe, S., & Panayi, G. S. (1995). Cytokine and adhesion molecule expression in the minor salivary glands of patients with Sjogren's syndrome and chronic sialoadenitis. *Annals of the Rheumatic Diseases, 54*, 209–215.

Chen, J. F., Salas, A., & Springer, T. A. (2003). Bistable regulation of integrin adhesiveness by a bipolar metal ion cluster. *Nature Structural Biology, 10*, 995–1001.

Chigaev, A., Waller, A., Amit, O., Halip, L., Bologa, C. G., & Sklar, L. A. (2009). Real-time analysis of conformation-sensitive antibody binding provides new insights into integrin conformational regulation. *Journal of Biological Chemistry, 284*, 14337–14346.

de Fougerolles, A. R. (2003). Integrins in immune and inflammatory diseases. In D. Gullberg (Ed.), *I domains in integrins* (pp. 165–177). Georgetown, TX: Plenum Publishers.

Dustin, M. L., & Springer, T. A. (1999). Lymphocyte function associated-1 (LFA-1, CD11a/CD18). In T. Kreis & R. Vale (Eds.), *Guidebook to the extracellular matrix and adhesion proteins* (pp. 228–232). New York, NY: Sambrook and Tooze.

Emsley, J., Knight, C. G., Farndale, R. W., Barnes, M. J., & Liddington, R. C. (2000). Structural basis of collagen recognition by integrin a2b1. *Cell, 101*, 47–56.

Etzioni, A. (2009). Defects in the leukocyte adhesion cascade. *Clinical Reviews in Allergy & Immunology,* [*Epub ahead of print*].

Evans, R., Patzak, I., Svensson, L., De Filippo, K., Jones, K., McDowall, A., et al. (2009). Integrins in immunity. *Journal of Cell Science, 122*, 215–225.

Fischer, A., Lisowska-Grospierre, B., Anderson, D. C., & Springer, T. A. (1988). The leukocyte adhesion deficiency: Molecular basis and functional consequences. *Immunodeficiency Reviews, 1*, 39–54.

Friedl, P., & Weigelin, B. (2008). Interstitial leukocyte migration and immune function. *Nature Immunology, 9*, 960–969.

Giblin, P. A., & Kelly, T. A. (2001). Antagonists of b2 integrin-mediated cell adhesion. *Annual Reports in Medicinal Chemistry, 36*, 181–190.

Grakoui, A., Bromley, S. K., Sumen, C., Davis, M. M., Shaw, A. S., Allen, P. M., et al. (1999). The immunological synapse: A molecular machine controlling T cell activation. *Science, 285*, 221–227.

Harburger, D. S., Bouaouina, M., & Calderwood, D. A. (2009). Kindlin-1 and -2 directly bind the C-terminal region of beta integrin cytoplasmic tails and exert integrin-specific activation effects. *Journal of Biological Chemistry, 284*, 11485–11497.

Henderson, R. B., Lim, L. H., Tessier, P. A., Gavins, F. N., Mathies, M., Perretti, M., et al. (2001). The use of lymphocyte function-associated antigen (LFA)-1-deficient mice to determine the role of LFA-1, Mac-1, and a4 integrin in the inflammatory response of neutrophils. *Journal of Experimental Medicine, 194*, 219–226.

Hideshima, T., Mitsiades, C., Tonon, G., Richardson, P. G., & Anderson, K. C. (2007). Understanding multiple myeloma pathogenesis in the bone marrow to identify new therapeutic targets. *Nature Reviews Cancer, 7*, 585–598.

Huang, C., Zang, Q., Takagi, J., & Springer, T. A. (2000). Structural and functional studies with antibodies to the integrin b2 subunit: A model for the I-like domain. *Journal of Biological Chemistry, 275*, 21514–21524.

Humphries, M. J. (2000). Integrin structure. *Biochemical Society Transactions, 28*, 311–339.

Huth, J. R., Olejniczak, E. T., Mendoza, R., Liang, H., Harris, E. A., Lupher, M .L., Jr. (2000). NMR and mutagenesis evidence for an I domain allosteric site that regulates lymphocyte function-associated antigen 1 ligand binding. *Proceedings of the National Academy of Sciences of the United States of America, 97*, 5231–5236.

Hynes, R. O. (2002). Integrins: Bi-directional, allosteric, signalling machines. *Cell, 110*, 673–687.

Imai, Y., Park, E. J., Peer, D., Peixoto, A., Cheng, G., Carman, C. V., et al. (2008). Genetic perturbation of the putative cytoplasmic membrane-proximal salt bridge aberrantly activates alpha4 integrins. *Blood, 112*, 5007–5015.

Kallen, J., Welzenbach, K., Ramage, P., Geyl, D., Kriwacki, R., Legge, G., et al. (1999). Structural basis for LFA-1 inhibition upon lovastatin binding to the CD11a I-domain. *Journal of Molecular Biology, 292*, 1–9.

Katakai, T., Hara, T., Sugai, M., Gonda, H., & Shimizu, A. (2004). Lymph node fibroblastic reticular cells construct the stromal reticulum via contact with lymphocytes. *Journal of Experimental Medicine, 200*, 783–795.

Kellermann, S. A., Dell, C. L., Hunt, S. W., III, & Shimizu, Y. (2002). Genetic analysis of integrin activation in T lymphocytes. *Immunological Reviews, 186*, 172–188.

Kim, M., Carman, C. V., Yang, W., Salas, A., & Springer, T. A. (2004). The primacy of affinity over clustering in regulation of adhesiveness of the integrin aLb2. *Journal of Cell Biology, 167*, 1241–1253.

Kiosses, W. B., Shattil, S. J., Pampori, N., & Schwartz, M. A. (2001). Rac recruits high affinity integrin aVb3 to lamellipodia in endothelial cell migration. *Nature Cell Biology, 3*, 316–320.

Knorr, R., & Dustin, M. L. (1997). The lymphocyte function-associated antigen 1 I domain is a transient binding module for intercellular adhesion molecule (ICAM)-1 and ICAM-1 in hydrodynamic flow. *Journal of Experimental Medicine, 186*, 719–730.

Last-Barney, K., Davidson, W., Cardozo, M., Frye, L. L., Grygon, C. A., Hopkins, J. L., et al. (2001). Binding site elucidation of hydantoin-based antagonists of LFA-1 using multidisciplinary technologies: Evidence for the allosteric inhibition of a protein-protein interaction. *Journal of the American Chemical Society, 123*, 5643–5650.

Lebwohl, M., Tyring, S. K., Hamilton, T. K., Toth, D., Glazer, S., Tawfik, N. H., et al. (2003). A novel targeted T-cell modulator, efalizumab, for plaque psoriasis. *New England Journal of Medicine, 349*, 2004–2013.

Lee, J.-O., Rieu, P., Arnaout, M. A., & Liddington, R. (1995). Crystal structure of the A domain from the a subunit of integrin CR3 (CD11b/CD18). *Cell, 80*, 631–638.

Liu, G., Huth, J. R., Olejniczak, E. T., Mendoza, R., DeVries, P., Leitza, S., et al. (2001). Novel *p*-arylthio cinnamides as antagonists of leukocyte function-associated antigen-1/intracellular adhesion molecule-1 interaction. 2. Mechanism of inhibition and structure-based improvement of pharmaceutical properties. *Journal of Medicinal Chemistry, 44*, 1202–1210.

Luo, B. H., Carman, C. V., & Springer, T. A. (2007). Structural basis of integrin regulation and signaling. *Annual Review of Immunology, 25*, 619–647.

Luo, B.-H., Takagi, J., & Springer, T. A. (2004). Locking the b3 integrin I-like domain into high and low affinity conformations with disulfides. *Journal of Biological Chemistry, 279*, 10215–10221.

Malinin, N. L., Zhang, L., Choi, J., Ciocea, A., Razorenova, O., Ma, Y. Q., et al. (2009). A point mutation in KINDLIN3 ablates activation of three integrin subfamilies in humans. *Nature Medicine, 15*, 313–318.

May, A. E., Schmidt, R., Kanse, S. M., Chavakis, T., Stephens, R. W., Schomig, A., et al. (2002). Urokinase receptor surface expression regulates monocyte adhesion in acute myocardial infarction. *Blood, 100*, 3611–3617.

Mempel, T. R., Henrickson, S. E., & Von Andrian, U. H. (2004). T-cell priming by dendritic cells in lymph nodes occurs in three distinct phases. *Nature, 427*, 154–159.

Miller, D. H., Khan, O. A., Sheremata, W. A., Blumhardt, L. D., Rice, G. P., Libonati, M. A., et al. (2003). A controlled trial of natalizumab for relapsing multiple sclerosis. *New England Journal of Medicine, 348*, 15–23.

Miller, M. J., Wei, S. H., Parker, I., & Cahalan, M. D. (2002). Two-photon imaging of lymphocyte motility and antigen response in intact lymph node. *Science, 296*, 1869–1873.

Morin, N. A., Oakes, P. W., Hyun, Y. M., Lee, D., Chin, E. Y., King, M. R., et al. (2008). Nonmuscle myosin heavy chain IIA mediates integrin LFA-1 de-adhesion during T lymphocyte migration. *Journal of Experimental Medicine, 205*, 195–205.

Mory, A., Feigelson, S. W., Yarali, N., Kilic, S. S., Bayhan, G. I., Gershoni-Baruch, R., et al. (2008). Kindlin-3: A new gene involved in the pathogenesis of LAD-III. *Blood, 112*, 2591.

Moser, M., Bauer, M., Schmid, S., Ruppert, R., Schmidt, S., Sixt, M., et al. (2009a). Kindlin-3 is required for beta2 integrin-mediated leukocyte adhesion to endothelial cells. *Nature Medicine, 15*, 300–305.

Moser, M., Legate, K. R., Zent, R., & Fassler, R. (2009b). The tail of integrins, talin, and kindlins. *Science, 324*, 895–899.

Nishida, N., Xie, C., Shimaoka, M., Cheng, Y., Walz, T., & Springer, T. A. (2006). Activation of leukocyte beta2 integrins by conversion from bent to extended conformations. *Immunity, 25*, 583–594.

Park, E. J., Mora, J. R., Carman, C. V., Chen, J., Sasaki, Y., Cheng, G., et al. (2007). Aberrant activation of integrin alpha4beta7 suppresses lymphocyte migration to the gut. *Journal of Clinical Investigation, 117*, 2526–2538.

Peng, L., Liu, R., Andrei, M., Xiao, W., & Lam, K. S. (2008). *In vivo* optical imaging of human lymphoma xenograft using a library-derived peptidomimetic against alpha4beta1 integrin. *Molecular Cancer Therapeutics, 7*, 432–437.

Salas, A., Shimaoka, M., Chen, S., Carman, C. V., & Springer, T. A. (2002). Transition from rolling to firm adhesion is regulated by the conformation of the I domain of the integrin LFA-1. *Journal of Biological Chemistry, 277*, 50255–50262.

Semmrich, M., Smith, A., Feterowski, C., Beer, S., Engelhardt, B., Busch, D. H., et al. (2005). Importance of integrin LFA-1 deactivation for the generation of immune responses. *Journal of Experimental Medicine, 201*, 1987–1998.

Shamri, R., Grabovsky, V., Gauguet, J. M., Feigelson, S., Manevich, E., Kolanus, W., et al. (2005). Lymphocyte arrest requires instantaneous induction of an extended LFA-1 conformation mediated by endothelium-bound chemokines. *Nature Immunology, 6*, 497–506.

Shimaoka, M., Lu, C., Palframan, R., von Andrian, U. H., Takagi, J., & Springer, T. A. (2001). Reversibly locking a protein fold in an active conformation with a disulfide bond: Integrin aL I domains with high affinity and antagonist activity *in vivo*. *Proceedings of the National Academy of Sciences of the United States of America, 98*, 6009–6014.

Shimaoka, M., Salas, A., Yang, W., Weitz-Schmidt, G., & Springer, T. A. (2003a). Small molecule integrin antagonists that bind to the b2 subunit I-like domain and activate signals in one direction and block them in another. *Immunity, 19*, 391–402.

Shimaoka, M., & Springer, T. A. (2003). Therapeutic antagonists and the conformational regulation of integrin structure and function. *Nature Reviews Drug Discovery, 2*, 703–716.

Shimaoka, M., Takagi, J., & Springer, T. A. (2002). Conformational regulation of integrin structure and function. *Annual Review of Biophysics and Biomolecular Structure, 31*, 485–516.

Shimaoka, M., Xiao, T., Liu, J.-H., Yang, Y., Dong, Y., Jun, C.-D., et al. (2003b). Structures of the aL I domain and its complex with ICAM-1 reveal a shape-shifting pathway for integrin regulation. *Cell, 112*, 99–111.

Shulman, Z., Shinder, V., Klein, E., Grabovsky, V., Yeger, O., Geron, E., et al. (2009). Lymphocyte crawling and transendothelial migration require chemokine triggering of high-affinity LFA-1 integrin. *Immunity, 30*, 384–396.

Sigal, A., Bleijs, D. A., Grabovsky, V., van Vliet, S. J., Dwir, O., Figdor, C. G., et al. (2000). The LFA-1 integrin supports rolling adhesions on ICAM-1 under physiological shear flow in a permissive cellular environment. *Journal of Immunology, 165*, 442–452.

Smith, A., Carrasco, Y. R., Stanley, P., Kieffer, N., Batista, F. D., & Hogg, N. (2005). A talin-dependent LFA-1 focal zone is formed by rapidly migrating T lymphocytes. *Journal of Cell Biology, 170*, 141–151.

Springer, T. A. (1994). Traffic signals for lymphocyte recirculation and leukocyte emigration: The multi-step paradigm. *Cell, 76*, 301–314.

Stanley, P., Smith, A., McDowall, A., Nicol, A., Zicha, D., & Hogg, N. (2008). Intermediate-affinity LFA-1 binds alpha-actinin-1 to control migration at the leading edge of the T cell. *EMBO Journal, 27*, 62–75.

Sterry, W., Bagot, M., Ferrandiz, C., Kragballe, K., Papp, K., & Stingl, G. (2009). Immunosuppressive therapy in dermatology and PML. *Journal der Deutschen Dermatologischen Gesellschaft, 7*, 5.

Svensson, L., Howarth, K., McDowall, A., Patzak, I., Evans, R., Ussar, S., et al. (2009). Leukocyte adhesion deficiency-III is caused by mutations in KINDLIN3 affecting integrin activation. *Nature Medicine, 15*, 306–312.

Takagi, J., Petre, B. M., Walz, T., & Springer, T. A. (2002). Global conformational rearrangements in integrin extracellular domains in outside-in and inside-out signaling. *Cell, 110*, 599–611.

Takagi, J., & Springer, T. A. (2002). Integrin activation and structural rearrangement. *Immunological Reviews, 186*, 141–163.

Tanaka, Y. (1999). Activation of leukocyte function-associated antigen-1 on adult T-cell leukemia cells. *Leukemia & Lymphoma, 36*, 15–23.

Valdramidou, D., Humphries, M. J., & Mould, A. P. (2008). Distinct roles of beta1 metal ion-dependent adhesion site (MIDAS), adjacent to MIDAS (ADMIDAS), and ligand-associated metal-binding site (LIMBS) cation-binding sites in ligand recognition by integrin alpha2-beta1. *Journal of Biological Chemistry, 283*, 32704–32714.

Van Assche, G., Van Ranst, M., Sciot, R., Dubois, B., Vermeire, S., Noman, M., et al. (2005). Progressive multifocal leukoencephalopathy after natalizumab therapy for Crohn's disease. *New England Journal of Medicine, 353*, 362–368.

Vinogradova, O., Velyvis, A., Velyviene, A., Hu, B., Haas, T. A., Plow, E. F., et al. (2002). A structural mechanism of integrin $a_{IIb}b_3$ "inside-out" activation as regulated by its cytoplasmic face. *Cell, 110*, 587–597.

Vorup-Jensen, T., Ostermeier, C., Shimaoka, M., Hommel, U., & Springer, T. A. (2003). Structure and allosteric regulation of the aXb2 integrin I domain. *Proceedings of the National Academy of Sciences of the United States of America, 100*, 1873–1878.

Wagner, N., & Muller, W. (1998). Functions of a4- and b7-integrins in hematopoiesis, lymphocyte trafficking and organ development. *Current Topics in Microbiology and Immunology, 231*, 23–32.

Wegener, K. L., Partridge, A. W., Han, J., Pickford, A. R., Liddington, R. C., Ginsberg, M. H., et al. (2007). Structural basis of integrin activation by talin. *Cell, 128*, 171–182.

Woolf, E., Grigorova, I., Sagiv, A., Grabovsky, V., Feigelson, S. W., Shulman, Z., et al. (2007). Lymph node chemokines promote sustained T lymphocyte motility without triggering stable integrin adhesiveness in the absence of shear forces. *Nature Immunology, 8*, 1076–1085.

Xiao, T., Takagi, J., Wang, J.-h., Coller, B. S., & Springer, T. A. (2004). Structural basis for allostery in integrins and binding of ligand-mimetic therapeutics to the platelet receptor for fibrinogen. *Nature, 432*, 59–67.

Xiong, J.-P., Stehle, T., Diefenbach, B., Zhang, R., Dunker, R., Scott, D. L., et al. (2001). Crystal structure of the extracellular segment of integrin aVb3. *Science, 294*, 339–345.

Xiong, J. P., Stehle, T., Zhang, R., Joachimiak, A., Frech, M., Goodman, S. L., et al. (2002). Crystal structure of the extracellular segment of integrin aVb3 in complex with an Arg–Gly–Asp ligand. *Science, 296*, 151–155.

Yang, W., Shimaoka, M., Chen, J. F., & Springer, T. A. (2004a). Activation of integrin b subunit I-like domains by one-turn C-terminal a-helix deletions. *Proceedings of the National Academy of Sciences of the United States of America, 101*, 2333–2338.

Yang, W., Shimaoka, M., Salas, A., Takagi, J., & Springer, T. A. (2004b). Inter-subunit signal transmission in integrins by a receptor-like interaction with a pull spring. *Proceedings of the National Academy of Sciences of the United States of America, 101*, 2906–2911.

Yung, R., Powers, D., Johnson, K., Amento, E., Carr, D., Laing, T., et al. (1996). Mechanisms of drug-induced lupus. II. T cells overexpressing lymphocyte function-associated antigen 1 become autoreactive and cause a lupuslike disease in syngeneic mice. *Journal of Clinical Investigation, 97*, 2866–2871.

Yung, R. L., Ray, D., Mo, R. R., & Chen, J. (2003). T cell integrin overexpression as a model of murine autoimmunity. *Biological Procedures Online, 5*, 211–221.

Yusuf-Makagiansar, H., Anderson, M. E., Yakovleva, T. V., Murray, J. S., & Siahaan, T. J. (2002). Inhibition of LFA-1/ICAM-1 and VLA-4/VCAM-1 as a therapeutic approach to inflammation and autoimmune diseases. *Medicinal Research Reviews, 22*, 146–167.

Zhu, J., Luo, B. H., Xiao, T., Zhang, C., Nishida, N., & Springer, T. A. (2008). Structure of a complete integrin ectodomain in a physiologic resting state and activation and deactivation by applied forces. *Molecular Cell, 32*, 849–861.

CHAPTER 5

Cytoskeletal Interactions with Leukocyte and Endothelial Cell Adhesion Molecules

Fredrick M. Pavalko
Department of Cellular and Integrative Physiology,
Indiana University School of Medicine, Indianapolis, Indiana 46202, USA

Current Topics in Membranes, Volume 64 1063-5823/09 $35.00

DOI: 10.1016/S1063-5823(09)64005-9

I. OVERVIEW

Each of the major classes of cell adhesion molecules are necessary for leukocyte trafficking. Selectins, integrins, and immunoglobulin superfamily members all play important roles in cellular immune responses and each have cytoplasmic tails that interact with cytoskeletal proteins. These interactions between cell adhesion molecule cytoplasmic tails and cytoskeletal proteins are essential for immune cell rolling, adhesion, migration, and diapedesis. This chapter describes the progress made to date in characterizing these interactions and focuses on those processes most essential to leukocyte rolling and adhesion.

II. INTRODUCTION

The major families of cell adhesion molecules involved in mediating leukocyte rolling, adhesion and migration (integrins, selectins, and immunoglobulin superfamily) each associate with, and are regulated by interactions with the cytoskeleton. Transendothelial migration requires participation of both leukocyte and endothelial cell adhesion molecules (reviewed recently in Abram & Lowell, 2009; Luo, Carman, & Springer, 2007; van Buul & Hordijk, 2009; Zarbock & Ley, 2008, 2009). This chapter provides an overview of the cytoskeletal interactions that occur between leukocyte and endothelial cell adhesion molecules and discusses how those interactions are thought to regulate adhesive functions that are important for leukocyte trafficking under conditions of fluid shear stress (recently reviewed in Alon & Ley, 2008; Simon, Sarantos, Green, & Schaff, 2009). Although this chapter will be limited primarily to discussion of cell adhesion molecules in leukocytes and endothelial cells that are involved in cellular immunity and are regulated through interaction with the actin cytoskeleton, recent reports do suggest a role for endothelial microtubules and kinesins in promoting diapedesis (Mamdouh, Kreitzer, & Muller, 2008).

III. INTEGRIN INTERACTIONS WITH THE CYTOSKELETON

Integrins are a large and diverse family of cell surface adhesion molecules that play vital functional roles in essentially all tissues in the body (Hynes, 2002; Hynes et al., 2002). Their name derives from the fact that they physically "integrate" the outside and inside of cells. Integrins are critically important in mediating leukocyte activation and migration in the immune system (Abram & Lowell, 2009; Hogg, Laschinger, Giles, & McDowall, 2003;

Kinashi, 2005; Luo et al., 2007). They are capable of bidirectional signaling across the plasma membrane through processes termed outside-in and inside-out signaling (Ginsberg, Partridge, & Shattil, 2005). These signaling mechanisms involve conformational changes in the extracellular and intracellular domains of integrins which affect the repertoire of extracellular matrix and cell surface receptors that integrins bind on the outside of cells (reviewed in Arnaout, Goodman, & Xiong, 2007). At the cytoplasmic face of the membrane, integrin cytoplasmic tails are regulated by associations with the actin cytoskeleton indirectly via binding to various cytoskeleton-associated linker proteins (Fig. 1). Importantly, these cytoskeletal associations with integrins further modify extracellular ligand interactions by modulating ligand affinity and clustering of integrins within the plasma membrane (Abram & Lowell), and defects in cytoplasmic signals are generally associated with defective leukocyte integrin function (Barreiro, de la Fuente, Mittelbrunn, & Sanchez-Madrid, 2007; Hogg et al., 2002; McDowall et al., 2003). Recently, shedding of leukocyte integrins has been proposed to play a regulatory role in leukocyte detachment after transendothelial migration and in regulating integrin-dependent outside-in signaling (Evans et al., 2006).

Approximately 24 distinct integrins have been identified that are comprised of a single α-chain and a single β-chain that combine to form a type I transmembrane glycoprotein heterodimer with a short cytoplasmic tail. There are 18 different α-chain subunits and 8 different β-chain subunits

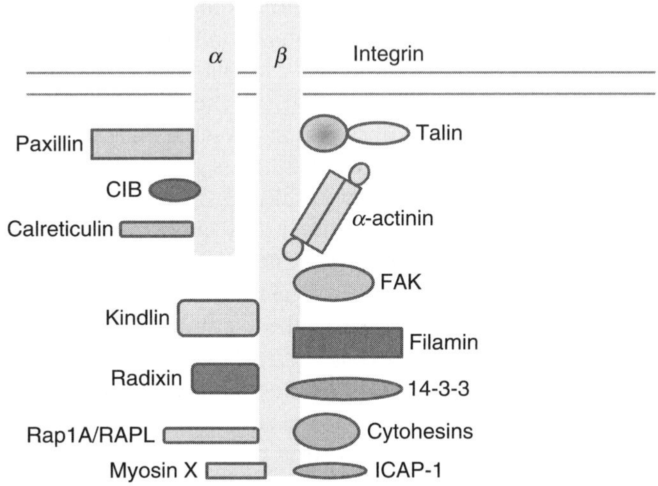

FIGURE 1 Cytoskeletal linker protein interactions with integrins.

which can assemble in a limited number of combinations that result in integrins with diverse ligand recognition specificities. Approximately 13 different integrins have been identified on leukocytes that are assembled from 4 different β-subunits (β_1, β_2, β_3, and β_7) and 12 different α-subunits (α_1, α_2, α_4, α_5, α_6, α_9, α_L, α_M, α_X, α_D, α_{IIb}, and α_V). The β_2- and β_7-subunits are unique in that their expression is limited to leukocytes. The cytoplasmic tails of integrin α- and β-subunits on leukocytes that mediate cytoskeletal interactions are relatively short; the α-subunits range from 20 to 58 residues, while the β-subunits tails range from 46 to 51 residues. Integrin cytoplasmic tails possess no intrinsic kinase activity, nor do they appear to be capable of binding directly to actin filaments. While the different integrin β-cytoplasmic tails share extensive homology, the α-subunit cytoplasmic domains share little sequence similarity other than the membrane proximal GFFKR sequence common to all α-subunit tails (Abram & Lowell, 2009).

Because of analysis of both the amino acid sequence of integrin subunits and from electron microscopy studies (Garcia-Alvarez et al., 2003), the general structure of integrins has been appreciated for some time. Integrin heterodimers consist of a relatively large globular extracellular domain which binds ligand and is formed from interactions of the α- and β-subunits which rest on a pair of legs or stalks that represent the transmembrane and cytoplasmic tails of each subunit. Knowledge gained from analyses of X-ray crystal structures and NMR data of integrin heterodimers have greatly enhanced our understanding of integrin regulation (Qu & Leahy, 1995; Takagi, Strokovich, Springer, & Walz, 2003; Vinogradova, Haas, Plow, & Qin, 2000; Vinogradova et al., 2002). These studies strongly suggest that information leading to conformational changes in integrin structure is transmitted to the extracellular ligand-binding domains of integrins through alterations in the transmembrane and cytoplasmic domains. A general picture has emerged which indicates that, in their inactive or low-affinity state, integrin extracellular domains exist in a V-shaped (or bent) conformation in which the ligand-binding globular heads are positioned close to the membrane, near the stalks. In this low-affinity or resting state, the integrin α- and β-cytoplasmic domains are positioned very close together, separated by only a few angstroms. The arginine residue in the highly conserved membrane proximal GFFKR motif in integrin α-subunits mediates a charge interaction with an aspartate residue in the β-subunit (Vinogradova et al.). This site is generally referred to as the "hinge" region and mutations of the arginine residue in the GFFKR motif of the α-chain default to a high-affinity conformation (Semmrich et al., 2005). *In vivo*, following the generation of cytoplasmic signals elicited from activation of G protein-coupled chemokine receptors (Constantin, 2008), cytokine receptors or stimulation of the T-cell receptor (TCR) on leukocytes (inside-out signals), the transmembrane and cytoplasmic domains are thought to become destabilized or spatially separated from each other in the hinge region (Kim, Carman, & Springer, 2003; Zhu et al., 2007). This results in propagation of a conformational signal through the transmembrane and stalk domains that extends or opens the globular head in a way that modulates integrin conformation to increase its affinity for ligand (reviewed in detail in Abram & Lowell, 2009; Moser, Legate, Zent, & Fassler, 2009).

Regulation of integrin affinity and avidity for ligand plays a particularly important role in modulating the response of leukocytes in the immune system under conditions of fluid shear stress (Alon & Ley, 2008; Evans et al., 2009). In the bent conformation, low-affinity or resting integrins on leukocytes are unable to bind strongly to ligand receptors such as ICAM-1 or VCAM-1 on endothelial cells. Numerous studies suggest that the integrin cytoplasmic tails, through their interactions with cytoskeletal and other cytoplasmic regulatory proteins, control regulation of integrin ligand-binding activity (Legate & Fassler, 2009). Inside-out signaling is characterized by the ability of signals generated within the cell that cause conformational changes in the integrin which lead to increased ligand affinity and clustering of integrin heterodimers in the membrane. At least two "active" states of integrins have been identified and characterized as intermediate- and high-affinity binding states (Smith et al., 2007). Regulation appears to be primarily mediated via integrin cytoplasmic tails and has been reviewed recently (Moser, Legate, et al., 2009). Mutations within the transmembrane or cytoplasmic sequences of α- and β-subunits have been identified which destabilize the close association between the subunits and activate integrins and cause increased ligand binding (Gottschalk, 2005; Hughes et al., 1996; Schneider & Engelman, 2004). Similarly, mutations or substitutions of the cytoplasmic tails with sequences that promote stabilization of the cytoplasmic tails in a tightly coupled conformation prevent activation of integrins (Partridge, Liu, Kim, Bowie, & Ginsberg, 2005). It has been widely speculated that differences in the strength of binding of different cytoskeletal β-tail-binding proteins regulate the ability of integrins to mediate processes vital to leukocyte function including adhesion, locomotion over and migration across endothelial and epithelial cell surfaces (Cairo, Mirchev, & Golan, 2006). Phosphorylation of the integrin cytoplasmic domains also plays a vital role in the dynamic regulation of integrins in cells (Fagerholm, Hilden, Nurmi, & Gahmberg, 2005). Therefore, abundant evidence indicates that interactions with structural and regulatory proteins that interact with the cytoskeleton play a key role in modulating the conformation of integrin cytoplasmic domains, and thus, influence integrin ligand-binding activity (Bolomini-Vittori et al., 2009).

IV. INTEGRIN CYTOPLASMIC DOMAIN-BINDING PROTEINS IN LEUKOCYTES

A. Talin, α-Actinin, Filamin, and 14-3-3 Proteins

Three cytoskeleton-associated proteins, talin, α-actinin, and filamin, are among the most well characterized integrin β-subunit-associated proteins. The first interaction between integrin cytoplasmic domains and a cytoplasmic protein was discovered by Horwitz, Duggan, Buck, Beckerle, and Burridge (1986). This group identified an association between the cytoskeletal-associated protein talin and the integrin β_1-subunit tail. Subsequently, talin has been shown to bind to other β-tails, including the β_2-subunit (Nayal, Webb, & Horwitz, 2004; Sampath, Gallagher, & Pavalko, 1998) and appears to be a central regulator of integrin activation in leukocytes (Ratnikov, Partridge, & Ginsberg, 2005).

Talin consists of a globular head domain that can be separated from a large extended rod domain by a calpain cleavage site (Critchley, 2009). The head domain of talin contains the binding region (located within FERM domain of the head) for integrin cytoplasmic domains and, when expressed independently from the talin rod domain, results in spatial separation of the α- and β-tails and promotes ligand binding (Moser, Legate, et al., 2009; Tadokoro et al., 2003). Activation of integrins requires the β-turn at NPX (Y/F) motifs conserved in integrin β-cytoplasmic domains (Calderwood et al., 2002; Wegener et al., 2007). In addition, siRNA knockdown of talin in leukocytes inhibits the ability of inside-out signals generated via T-cell receptor to upregulate LFA-1 (α_L/β_2) affinity for ICAM-1 (Simonson, Franco, & Huttenlocher, 2006).

One model of talin function in leukocytes with experimental support suggests that full length talin constitutively associates with the β_2-integrin subunit in resting cells and that upon activation of inside-out signals, talin is cleaved by calpain (Sampath et al., 1998). This may release actin cytoskeleton-mediated lateral restraints on integrin mobility within the plane of the membrane that allows integrins to form higher affinity clusters. Following clustering, binding of another cytoskeleton-associated protein that also can bind directly to integrin β-subunit tails, α-actinin (Pavalko & LaRoche, 1993), may then stabilize the active conformation and also promote modulation of integrin affinity states. Recent studies by Hogg and coworkers (Stanley et al., 2008) also support the notion that α-actinin may bind to β_2-integrins as a secondary event following integrin activation as a result of a conformational change in the cytoplasmic tail which is consistent with the idea that α-actinin binding stabilizes the extended high-affinity extracellular structure. Indeed, these authors showed that knockdown of α-actinin using siRNA or expression of a dominant-negative form of α-actinin that binds the integrin β-chain cytoplasmic tail but lacks the actin-binding domain, inhibits T-cell migration. The studies provide compelling evidence for a level of integrin regulation in which two active conformations of LFA-1 exist (intermediate- and high-affinity conformations) in T-lymphocytes that provide important biological regulation of lymphocyte motility. Talin has also been shown to be capable of undergoing conformational changes that do not require cleavage of talin but can result in exposure of the FERM domain, and a second integrin-binding site (IBS2) has been identified near the C-terminal end of the talin rod (Gingras et al., 2009).

Filamin is another actin-binding protein that can also associate directly with integrin β-subunit tails including leukocyte β_1-, β_2-, and β_7-integrins (Calderwood et al., 2001; Sharma, Ezzell, & Arnaout, 1995). Filamin association with integrin tails appears to negatively regulate integrin activity and may maintain integrins in an inactive conformation in resting leukocytes (Takala et al., 2008). Compelling evidence supports the notion that regulation of integrin–filamin interactions control cell migration. Calderwood et al. found that increased filamin binding suppresses random cell migration and that phosphorylation of the β_1- and β_7-tails appears to regulate filamin binding. Other recent studies with β_2-integrins investigated the binding of the β_2-integrin cytoplasmic domain to filamin, talin, and the molecular adaptor protein 14-3-3 (Takala et al.). These studies supported a role for phosphorylation of Thr758 phosphorylation in inhibition of filamin binding, but found that Thr758 phosphorylation is required for binding of 14-3-3 protein. Thus, phosphorylation of β_2 on Thr758 led to impairment of filamin interaction and binding of 14-3-3, suggesting that phosphorylation may act as a molecular switch between filamin and 14-3-3. Although phosphorylation did not directly regulate talin binding to the integrin β_2-tail, 14-3-3 was able to outcompete talin for binding to phosphorylated β_2-integrin suggesting a role in regulating talin–β_2 interaction. Physiologically, it may be very significant that phosphorylation of β_2-integrins at Thr758 may regulate leukocyte function. Mutations at this site resulted in defective T-cell adhesion and spreading, most likely by preventing the dissociation of filamin and thus blocking the association of 14-3-3 proteins to the β_2-tail. Filamin knockdown with siRNA also promotes leukocyte adhesion to ICAM-1, and filamin negatively regulates talin binding and blocks talin-dependent integrin activation (Takala et al.). Thus, filamin association with integrin cytoplasmic tails appears to play an inhibitory role in regulating leukocyte adhesion.

B. Cytohesins

Members of the cytohesin family, a subfamily of guanine nucleotide exchange factors (GEFs) for adenosine diphosphate (ADP)-ribosylation factor (ARF) GTPases, have recently received considerable attention for their ability to regulate leukocyte integrin function (Geiger et al., 2000; Kolanus, 2007; Mor, Dustin, & Philips, 2007; Nagel, Zeitlmann, et al., 1998; Quast et al., 2009; Weber et al., 2001). Cytohesin-1 has been shown to be a regulator of β_2-integrin-mediated activation of α_L/β_2-integrin in T cells and in recruitment of monocyte binding to endothelial cells under physiological fluid shear stress conditions (Weber et al.). Kolanus et al. (1996), using the β_2-cytoplasmic domain as "bait," identified cytohesin-1 as a novel regulator of leukocyte integrin activation. Binding occurs between residues in the membrane proximal portion of the β_2-cytoplasmic domain and the Sec7 domain of cytohesin. The interaction between cytohesin-1 and the β_2-tail appears to increase the avidity of leukocyte integrins. Importantly, efforts to rescue β_2-deficient cells with mutated β_2-integrin lacking the cytohesin-binding domain resulted in normal surface expression of α_L/β_2-integrin in transfected cells; however, these cells were unable to adhere to ICAM-1 surfaces (Kolanus). Thus, deletion of the cytohesin-1-binding region in β_2-integrins prevented activation of the integrin. Interestingly, although cytohesins possess GEF activity, the GEF activity of cytohesin-1 is not required for leukocyte integrin activation (as measured by binding of an activation state specific antibody) but is required for efficient cell adhesion and cell spreading (Geiger et al.). Thus, it seems that simple binding of cytohesin-1 can regulate cytohesin-dependent inside-out signaling and this binding may regulate the downstream activation of the small GTPase Rho and influence leukocyte adhesion, spreading and migration, as it has been shown to do in dendritic cells derived from monocytes (Quast et al.). Also of note, the presence of the pleckstrin homology (PH) domain is necessary for cytohesins to promote integrin activation, and expression of the isolated PH domain alone can function as a dominant negative inhibitor of α_L/β_2-integrin binding to ICAM-1 (Nagel, Schilcher, Zeitlmann, & Kolanus, 1998).

C. Rap1A

The small GTPase, Rap1A, appears to be necessary for TCR-mediated inside-out integrin activation (Sebzda, Bracke, Tugal, Hogg, & Cantrell, 2002). Furthermore, expression of a constitutive mutant of Rap1A is sufficient to increase TCR-mediated upregulation of LFA-1 affinity. Although Rap1A has been reported to antagonize Ras signaling, thereby rendering

T cells immunologically unresponsive (Boussiotis, Freeman, Berezovskaya, Barber, & Nadler, 1997; Li et al., 2005), constitutive expression of Rap1A in mice T-lymphocytes was shown to enhance TCR-mediated responses (Sebzda et al.). Rap1A has been suggested to be a direct modulator of integrin affinity for ligand in Jurkat T cells that were expressing constitutively active Rap1A (Katagiri et al., 2000; Reedquist et al., 2000). The Rap1A effector protein, RAPL, has been shown to associate with LFA-1 and become relocated to the leading edge of migrating lymphocytes suggesting that RAPL regulates lymphocyte adhesion through modulation of the spatial distribution of LFA-1 (Katagiri, Maeda, Shimonaka, & Kinashi, 2003). However, it is unclear whether Rap1A directly affects the activation state of leukocyte integrins or whether increased cell adhesion occurs through avidity modulation. Confocal microscopy of α_L/β_2-integrin on activated T cells revealed that Rap1A promoted clustering of integrins (Kinashi & Katagiri, 2004). In either case, the mechanism of action of Rap1A appears to be via strong activation of β_1- and β_2-integrins through inside-out modulation of integrins.

D. Paxillin

The cytoskeletal/signaling adaptor protein paxillin (reviewed in Deakin & Turner, 2008) interacts directly with the cytoplasmic domain of the integrin α_4-subunit (Liu et al., 1999). This interaction is critical for α_4/β_1-integrin-dependent cell adhesion under shear flow conditions (Rose, Han, & Ginsberg, 2002). The interaction of paxillin with α_4-integrin is also required for α_4/β_1-dependent leukocyte migration and requires regulation by the small GTPase Rac, FAK/Pyk2, and talin binding (Rose, Alon, & Ginsberg, 2007). Disruption of paxillin binding by a mutation in the α_4-tail (Y991A) reduced talin association with α_4/β_1 and decreased anchorage to the actin (Alon et al., 2005). In Jurkat T cells, this mutation suppressed α_4/β_1-mediated capture and adhesion strengthening under fluid shear stress conditions. These studies nicely demonstrated that cytoskeletal anchorage of integrins mechanically stabilizes adhesion in cells adhering under physiological levels of fluid shear. Paxillin may also regulate the formation of a ternary complex of proteins including a novel interaction between the α_4-integrin cytoplasmic domain and 14-3-3ζ (Deakin et al., 2009). Formation of this complex depends on serine phosphorylation of S978 in the α_4-tail which is distinct from the site (S988) which regulates paxillin binding (S988). It has been proposed that this interaction with α_4/β_1-integrin cytoplasmic tails could provide an attractive target for intervention in integrin-mediated leukocyte pathologies (Cantor, Ginsberg, & Rose, 2008).

E. Kindlins

Kindlins (consisting of Kindlin-1, -2, and -3) are a relatively recently identified family of evolutionarily conserved cytoplasmic proteins that regulate cell–matrix adhesion and bind to β-integrin cytoplasmic tails directly and cooperate with talin in integrin activation (Harburger, Bouaouina, & Calderwood, 2009; Larjava, Plow, & Wu, 2008). Recently kindlin-3, which is expressed in hematopoietic cells, was shown to bind directly to the integrin β_2-cytoplasmic tail via the distal NXXF motif, and mutations in kindling-3 cause LAD-III (Svensson et al., 2009). Kindlin-3 was previously shown to active β_1- and β_3-integrins in platelets and is an essential element for platelet integrin activation in hemostasis and thrombosis (Moser, Nieswandt, Ussar, Pozgajova, & Fassler, 2008). Mutations in kindlin-1 (which is expressed in epithelial cells) cause a skin blistering disease in humans (Kindler syndrome). Kindlin-2 is more widely expressed and deletion of this gene *in vivo* results in severe defects in β_1- and β_3-integrin functions and embryonic lethality. Kindlin-3 appears to be essential for activation of β_2-integrins in PMNs and in the absence of kindlin-3, PMNs failed to spread on β_2-integrin ligands including ICAM-1 and iC3b (Moser, Bauer, et al., 2009; Svensson et al.). *In vivo* studies with kindlin-3$^{-/-}$ mice indicate that loss of kindlin-3 results in leukocyte adhesion deficiency (LAD) and may explain the phenotype of LAD-III in mice (Moser, Bauer, et al.) and in human patients (Malinin et al., 2009). The mechanisms through which kindlin-3 regulates β_2-integrin function in blood cells could provide valuable insights into regulation of leukocyte adhesion and migration under conditions of dynamic blood flow. The physiologic roles of several other FERM domain-containing proteins that also bind to integrin β-subunit tails including radixin, ICAP-1, myosin X, Dok-1, and Numb remain to be determined (Calderwood et al., 2002, 2003; Chang, Wong, Smith, & Liu, 1997; Clemmons & Maile, 2005; Nishimura & Kaibuchi, 2007; Sousa & Cheney, 2005; Zhang et al., 2004).

V. SELECTIN INTERACTIONS WITH THE CYTOSKELETON

The selectin family of cell adhesion molecules consists of three closely related members (L-selectin, E-selectin, and P-selectin). L-selectin is expressed on the surface of all leukocytes and is essential for leukocyte rolling at sites of inflammation. P-selectin is present in endothelial cells and platelets and is rapidly mobilized to the cell surface from sites of intracellular storage in Weibel–Palade bodies or α-granules, respectively. E-selectin is expressed on endothelial cells following stimulation by cytokines, including TNF-α and IL-1β. All three selectins are structurally similar, consisting of an

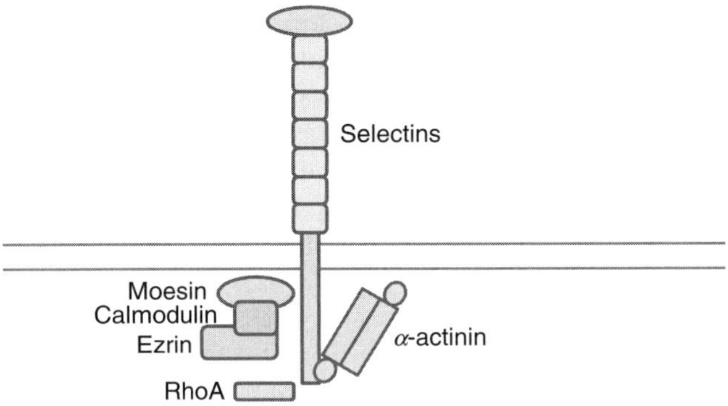

FIGURE 2 Cytoskeletal linker protein interactions with selectins.

extracellular region comprised of C-type lectin, EGF-like domains, complement-binding domains, a transmembrane domain and a short cytoplasmic tail. Cytoskeletal interactions with the cytoplasmic tails of selectins (Fig. 2) have been described and play important roles in regulating selectin function (Gonzalez-Amaro & Sanchez-Madrid, 1999). In addition to their roles in leukocytes, endothelial cells, and platelets, selectins also play an important function in cancer metastasis (Paschos, Canovas, & Bird, 2009).

A. L-Selectin

An essential role for the L-selectin cytoplasmic domain was shown in regulation of leukocyte adhesion to the endothelium that was independent of ligand recognition (Kansas, Ley, Munro, & Tedder, 1993). An indirect link between L-selectin and the actin cytoskeleton was shown to occur via the actin-binding protein α-actinin which interacts with the COOH-terminal 11 amino acids of the L-selectin tail (Pavalko et al., 1995). A role for the L-selectin tail in regulation of cytoskeletal organization has been shown following stimulation of L-selectin which induced an increase in actin filament polymerization (Brenner et al., 1997). L-selectin-directed actin remodeling was mediated by a Ras- and Rac2-regulated pathway that was inhibited by transient transfection of inhibitory N17Ras or suppression of Rac2 protein expression. The functional significance of the L-selectin tail was demonstrated by *in vitro* studies in which the COOH-terminal 11 residues which prevent association with α-actinin resulted in leukocytes that could mediate binding under physiological fluid flow conditions, but could not convert this initial tethering into rolling (Dwir, Kansas, & Alon, 2001).

Simon et al. (1999) have shown that intracellular signals mediated by L-selectin enhance neutrophil binding at sites of inflammation through a multistep process that increases cell stiffness and alters adhesiveness of β_2-integrins. ERM proteins also interact directly with L-selectin tails through a process in which moesin and ezrin bind to a similar region of the cytoplasmic tail of L-selectin, although binding of moesin is dependent upon activation of protein kinase C (PKC), while ezrin did not require PKC activation for binding (Ivetic, Deka, Ridley, & Ager, 2002). Thus, ezrin and moesin may be independently regulated but are required for correct positioning of L-selectin on leukocyte microvilli and are important both for leukocyte tethering and L-selectin shedding (Ivetic et al., 2004).

The lateral mobility of L-selectin is also regulated by interactions with the actin cytoskeleton that in turn fortifies leukocyte tethering (Mattila, Green, Schaff, Simon, & Walcheck, 2005). A recent report has shown that the L-selectin cytoplasmic tail could form a heterotrimeric complex with ERM proteins and calmodulin that interact simultaneously *in vivo* and promotes clustering of neighboring L-selectin tails (Killock et al., 2009). Together, these results suggest L-selectin can signal to the actin cytoskeleton via Ras and Rac2 and that this activity is physiologically important for leukocyte adhesion (Evans et al., 1999). The functional significance of the L-selectin cytoplasmic tail and its regulation by calmodulin is highlighted by characterization of novel intracellular signaling events that occurs as a consequence of L-selectin clustering via the cytoplasmic domain which promotes the transition from leukocyte rolling to arrest (Killock et al.).

B. E-Selectin

E-selectin, which is expressed on activated vascular endothelium, promotes leukocyte rolling and stable adhesion of leukocytes at sites of inflammation (Wiese, Barthel, & Dimitroff, 2009). Adhesion of leukocytes to cytokine-activated human umbilical vein endothelial cells (HUVEC) has been shown to induce linkage between E-selectin and the actin cytoskeleton through the E-selectin cytoplasmic domain via an outside-in signaling mechanism (Yoshida et al., 1996). This association does not appear to be mediated directly via α-actinin, as is the case with L-selectin (Kansas & Pavalko, 1996), but appears to be important in leukocyte adhesion to the endothelium. Using a series of cytoplasmic domain mutants of E-selectin expressed in COS-7 cells, Yoshida, Szente, Kiely, Rosenzweig, and Gimbrone (1998) found further evidence for a transmembrane signaling function for E-selectin–cytoskeletal interactions during leukocyte–endothelial adhesion. These investigators showed that E-selectin is constitutively phosphorylated in activated endothelial cells and undergoes dephosphorylation following leukocyte adhesion through a process triggered by engagement of E-selectin at the cell surface. Setiadi and McEver (2008) showed that deleting the cytoplasmic domain of E-selectin or disrupting clathrin-coated pits blocked internalization of E-selectin and inhibited E-selectin-mediated neutrophil rolling under flow. Thus, current evidence suggests that the cytoplasmic domain of E-selectin regulates its function through cytoskeletal interactions mediated by phosphorylation events.

C. P-Selectin and PSGL-1

Under conditions of fluid shear, leukocytes roll on P-selectin that is expressed on the surfaces of endothelial cells after histamine or thrombin stimulation. The ability of P-selectin to promote leukocyte rolling is enhanced by the clustering of P-selectin in clathrin-coated pits and subsequent internalization. RhoA seems to be involved in this process and inhibitors of RhoA or its effector, Rho kinase, prevent thrombin from increasing the expression and adhesive activity of P-selectin in endothelial cells (Setiadi & McEver, 2003). The investigators found that the cytoplasmic domain of P-selectin was required for Rho-mediated increases in P-selectin adhesion. Although relatively little information is available on the mechanisms of potential cytoskeletal interactions with the P-selectin cytoplasmic tail, the predominant ligand for P-selectin, P-selectin glycoprotein ligand-1 (PSGL-1) in leukocyte adhesion during inflammatory responses is clearly regulated by cytoskeletal associations. Snapp, Heitzig, and Kansas (2002) found that attachment of PSGL-1 to the leukocyte cytoskeleton is essential for leukocyte rolling on P-selectin. Using cells expressing either wild-type PSGL-1 or truncated PSGL-1 in which only four cytoplasmic residues were retained, these investigators found that rolling was almost completely absent in cells lacking the PSGL-1 cytoplasmic domain even at low shear stress. This impaired rolling was not due to the inability to bind P-selectin, or to alterations in subcellular localization since both wild-type and truncated PSGL-1 had similar surface distributions. Furthermore, in cells expressing endogenous PSGL-1, cytoskeletal poisons caused a dose-dependent decrease in adhesion. The PSGL-1 cytoplasmic domain interacted with the ezrin/radixin/moesin (ERM) protein moesin and ezrin, but not with other ERM proteins or several other cytoskeletal linker proteins to mediate an interaction with Syk (Urzainqui et al., 2002). Signals transduced by PSGL-1 can also promote β_2-integrin-mediated neutrophil adhesion to ICAM-1, and this process depends on an intact actin cytoskeleton (Wang, Cheng, & Ba, 2006).

Other recent studies suggest that the PSGL-1 cytoplasmic domain, while dispensable for leukocyte rolling on P-selectin, is required to activate β_2-integrins and alter the speed of rolling on ICAM-1 (Miner et al., 2008).

Takai, Kitano, Terawaki, Maesaki, and Hakoshima (2007) have recently found that upon activation, ERM proteins interact with the cytoplasmic tail of PSGL-1 and mediate redistribution of PSGL-1 on polarized cell surfaces. This cytoskeletal interaction with PSGL-1 is needed to facilitate binding to target molecules and therefore may play a physiologic role in inflammation. ERM proteins interact with a short binding motif, Motif-1, conserved in cytoplasmic tails of ICAMs; however, PSGL-1 lacks this motif.

VI. IMMUNOGLOBULIN SUPERFAMILY INTERACTIONS WITH THE CYTOSKELETON

Immunoglobulin superfamily (IgSF) members, including the intercellular adhesion molecule (ICAM) subfamily, vascular cell adhesion molecule-1 (VCAM-1), platelet endothelial cell adhesion molecule (PECAM), and junctional adhesion molecules (JAMs), play important roles in the inflammatory response. ICAM-1 serves as an important ligand for some leukocyte integrins. ICAM-1 was the first IgSF member for which an interaction to the cytoskeleton was characterized. Carpen, Pallai, Staunton, and Springer (1992) discovered that α-actinin bound directly to the ICAM-1 cytoplasmic tail; subsequently, ICAM-2 was also found to bind via its cytoplasmic domain to α-actinin (Heiska et al., 1996). Studies by VandenBerg, Reid, Edwards, and Davis (2004) evaluated the general role of the actin cytoskeleton TNF-α-induced cell adhesion molecule function. They found that a functional actin cytoskeleton was important for ICAM-1 and VCAM-1 expression (Fig. 3).

A. ICAMs

Cytoskeletal linkages to IgSF members during inflammation also appear to be necessary for transendothelial migration and migration into surrounding tissues (Millan et al., 2006). Translocation of ICAM-1 to specialized membrane domains called caveolae that are rich in filamentous actin are also necessary to promote adhesion and diapedesis. ICAM-1, but not ICAM-2, rapidly stimulates signaling responses involving RhoA (Thompson, Randi, & Ridley, 2002). RhoA association with ICAM-1 also appears to play a role in regulating ICAM-1 function via the cytoplasmic domain. Ligand engagement of ICAM-1 activates the small GTPase RhoA and promotes formation of actin stress fibers in endothelial cells. Thompson et al. reported that although

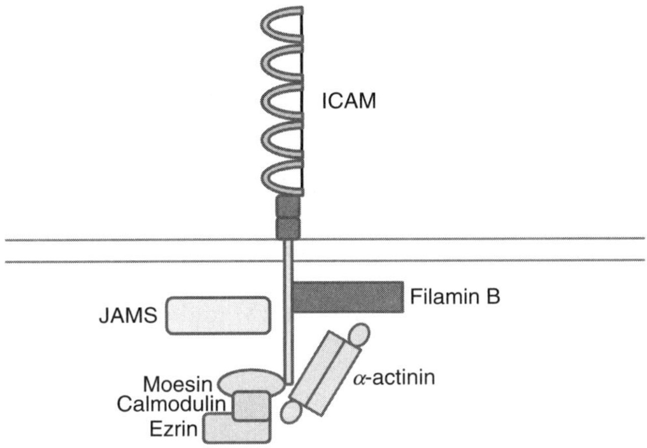

FIGURE 3 Cytoskeletal linker protein interactions with ICAMs.

ICAM-1 and ICAM-2 both localize with moesin in apical microvilli, only ICAM-1 colocalized with moesin after crosslinking by ligand. ICAM-2 did not activate RhoA or alter actin cytoskeletal organization, unlike ICAM-1 and only ICAM-1 stimulated transcription of the early response gene c-fos. Because ICAM-1 activation also upregulates the expression of the small GTPase RhoA, this suggests that a positive feedback pathway may be involved in ICAM-1 signaling.

Recently, filamin B was identified as a novel binding partner for ICAM-1 (Kanters et al., 2008). Clustering of ICAM-1 promoted the ICAM-1–filamin B interaction and, using siRNA to downregulate filamin B expression, these investigators found that filamin B is required for the lateral mobility of ICAM-1 and for ICAM-1-induced transmigration of leukocytes. When filamin B expression was reduced in endothelial cells it resulted in reduced recruitment of ICAM-1 to endothelial docking structures, reduced firm adhesion of the leukocytes to the endothelium, and inhibited transendothelial migration.

B. PECAM-1

PECAM-1 is an IgSF member found on the surface of leukocytes, endothelial cells, and platelets that has a large external domain consisting of Ig-like loops, a 19-residue transmembrane domain, and a relatively large, 118-amino acid cytoplasmic tail. The tail has several phosphorylation sites and may mediate interactions with the cytoskeleton. Following activation, PECAM-1 becomes highly phosphorylated and associates with the cytoskeleton of activated but not resting platelets (Newman et al., 1992). PECAM-1 interaction with the platelet cytoskeleton promotes movement within the plane of the membrane and may regulate the ability of PECAM-1 to localize to the intercellular borders of endothelial cells once cell-cell contact has been achieved (Newman, 1994). Extracellular binding of PECAM-1 promotes activation of integrins, suggesting that PECAM-1 may regulate integrin-mediating transendothelial migration, although the details of the implicated role of PECAM-actin cytoskeletal interactions and RhoA activation in this signaling process are unclear (Garnacho et al., 2008). Recent data actually suggest that microtubules may be involved (Mamdouh et al., 2008).

C. VCAM-1

Like ICAM-1, VCAM-1 serves as a ligand on vascular cells for some leukocyte integrins (i.e., VLA-4) and has a short cytoplasmic tail whose interactions with cytoskeletal proteins affect its function (Pozo, de Nicolas, Egido, & Gonzalez-Cabrero, 2006; VandenBerg et al., 2004). Leukocyte adhesion to endothelial cells under either static or fluid flow conditions induces clustering of VCAM-1 and ICAM-1with moesin and ezrin in actin-rich membrane structures that anchor leukocytes and contain additional cytoskeletal proteins including α-actinin, vinculin, and VASP (Barreiro et al., 2002).

D. HepaCAM

Moh, Zhang, Luo, Lee, and Shen (2005) recently identified a novel IgSF molecule, hepaCAM, that promotes cell–ECM adhesion and motility that is mediated by direct interactions with the actin cytoskeleton. Mutants of hepaCAM with deletions in the N- and C-terminal domains were used to map the actin-binding region as well as to evaluate the effect of the domains on the biological function of hepaCAM. Their results suggest that the extracellular and cytoplasmic domains of hepaCAM are required for stable physical association with the actin cytoskeleton and for modulating hepaCAM-mediated cell adhesion and motility (Moh, Tian, Zhang, Lee, & Shen, 2009).

E. JAMs

Several members of IgSF known as junctional cell adhesion molecules (JAMs) are also known to play important roles in leukocyte trafficking (Mandell & Parkos, 2005). They both function in barrier formation in endothelial cells and are expressed on leukocytes and platelets (Smith, 2008), and their function appears to require an intact cytoskeleton (Puthenedam, Williams, Lakshmi, & Balakrishnan, 2007). The molecular mechanisms through which JAMs interact with the actin cytoskeleton remain to be determined but may involve Rho family GTPases (Bruewer, Hopkins, Hobert, Nusrat, & Madara, 2004).

VII. CONCLUSIONS

The cytoplasmic domains of essentially all classes of cell adhesion molecules involved in leukocyte inflammatory responses play key regulatory roles in mediating cellular responses. Inside-out signaling processes that occur following leukocyte and endothelial cell activation are critical for proper regulation of leukocyte rolling, adhesion, migration, and diapedesis. Considerable progress has been made during the past 20 years in identifying cytoskeletal-binding partners for cell adhesion molecules. Further testing of the specific mechanisms through which cytoskeletal interactions with cell adhesion molecule cytoplasmic domains regulate adhesion and motility will be important from both basic science and clinical intervention (Cantor et al., 2008) perspectives.

Acknowledgements

The author thanks Ms. Rita O'Riley for editorial reading of the manuscript. This work is supported by NIH AR052682.

References

Abram, C. L., & Lowell, C. A. (2009). The ins and outs of leukocyte integrin signaling. *Annual Review of Immunology, 27*, 339–362.

Alon, R., Feigelson, S. W., Manevich, E., Rose, D. M., Schmitz, J., Overby, D. R., et al. (2005). Alpha4beta1-dependent adhesion strengthening under mechanical strain is regulated by paxillin association with the alpha4-cytoplasmic domain. *Journal of Cell Biology, 171*, 1073–1084.

Alon, R., & Ley, K. (2008). Cells on the run: Shear-regulated integrin activation in leukocyte rolling and arrest on endothelial cells. *Current Opinion in Cell Biology, 20*, 525–532.

Arnaout, M. A., Goodman, S. L., & Xiong, J. P. (2007). Structure and mechanics of integrin-based cell adhesion. *Current Opinion in Cell Biology, 19*, 495–507.

Barreiro, O., de la Fuente, H., Mittelbrunn, M., & Sanchez-Madrid, F. (2007). Functional insights on the polarized redistribution of leukocyte integrins and their ligands during leukocyte migration and immune interactions. *Immunological Reviews, 218*, 147–164.

Barreiro, O., Yanez-Mo, M., Serrador, J. M., Montoya, M. C., Vicente-Manzanares, M., Tejedor, R., et al. (2002). Dynamic interaction of VCAM-1 and ICAM-1 with moesin and ezrin in a novel endothelial docking structure for adherent leukocytes. *Journal of Cell Biology, 157*, 1233–1245.

Bolomini-Vittori, M., Montresor, A., Giagulli, C., Staunton, D., Rossi, B., Martinello, M., et al. (2009). Regulation of conformer-specific activation of the integrin LFA-1 by a chemokine-triggered Rho signaling module. *Nature Immunology, 10*, 185–194.

Boussiotis, V. A., Freeman, G. J., Berezovskaya, A., Barber, D. L., & Nadler, L. M. (1997). Maintenance of human T cell anergy: Blocking of IL-2 gene transcription by activated Rap1. *Science, 278*, 124–128.

Brenner, B., Gulbins, E., Busch, G. L., Koppenhoefer, U., Lang, F., & Linderkamp, O. (1997). L-selectin regulates actin polymerisation via activation of the small G-protein Rac2. *Biochemical and Biophysical Research Communications, 231*, 802–807.

Bruewer, M., Hopkins, A. M., Hobert, M. E., Nusrat, A., & Madara, J. L. (2004). RhoA, Rac1, and Cdc42 exert distinct effects on epithelial barrier via selective structural and biochemical modulation of junctional proteins and F-actin. *American Journal of Physiology: Cell Physiology, 287*, C327–C335.

Cairo, C. W., Mirchev, R., & Golan, D. E. (2006). Cytoskeletal regulation couples LFA-1 conformational changes to receptor lateral mobility and clustering. *Immunity, 25*, 297–308.

Calderwood, D. A., Fujioka, Y., de Pereda, J. M., Garcia-Alvarez, B., Nakamoto, T., Margolis, B., et al. (2003). Integrin beta cytoplasmic domain interactions with phosphotyrosine-binding domains: A structural prototype for diversity in integrin signaling. *Proceedings of the National Academy of Sciences of the United States of America, 100*, 2272–2277.

Calderwood, D. A., Huttenlocher, A., Kiosses, W. B., Rose, D. M., Woodside, D. G., Schwartz, M. A., et al. (2001). Increased filamin binding to beta-integrin cytoplasmic domains inhibits cell migration. *Nature Cell Biology, 3*, 1060–1068.

Calderwood, D. A., Yan, B., de Pereda, J. M., Alvarez, B. G., Fujioka, Y., Liddington, R. C., et al. (2002). The phosphotyrosine binding-like domain of talin activates integrins. *Journal of Biological Chemistry, 277*, 21749–21758.

Cantor, J. M., Ginsberg, M. H., & Rose, D. M. (2008). Integrin-associated proteins as potential therapeutic targets. *Immunological Reviews, 223*, 236–251.

Carpen, O., Pallai, P., Staunton, D., & Springer, T. A. (1992). Association of intercellular adhesion molecule-1 (ICAM-1) with actin-containing cytoskeleton and alpha-actinin. *Journal of Cell Biology, 118*, 1223–1234.

Chang, D. D., Wong, C., Smith, H., & Liu, J. (1997). ICAP-1, a novel beta1 integrin cytoplasmic domain-associated protein, binds to a conserved and functionally important NPXY sequence motif of beta1 integrin. *Journal of Cell Biology, 138*, 1149–1157.

Clemmons, D. R., & Maile, L. A. (2005). Interaction between insulin-like growth factor-I receptor and alphaVbeta3 integrin linked signaling pathways: Cellular responses to changes in multiple signaling inputs. *Molecular Endocrinology, 19*, 1–11.

Constantin, G. (2008). Chemokine signaling and integrin activation in lymphocyte migration into the inflamed brain. *Journal of Neuroimmunology, 198*, 20–26.

Critchley, D. R. (2009). Biochemical and structural properties of the integrin-associated cytoskeletal protein talin. *Annual Review of Biophysics, 38*, 235–254.

Deakin, N. O., Bass, M. D., Warwood, S., Schoelermann, J., Mostafavi-Pour, Z., Knight, D., et al. (2009). An integrin-{alpha}4-14-3-3{zeta}-paxillin ternary complex mediates localised Cdc42 activity and accelerates cell migration. *Journal of Cell Science, 122*, 1654–1664.

Deakin, N. O., & Turner, C. E. (2008). Paxillin comes of age. *Journal of Cell Science, 121*, 2435–2444.

Dwir, O., Kansas, G. S., & Alon, R. (2001). Cytoplasmic anchorage of L-selectin controls leukocyte capture and rolling by increasing the mechanical stability of the selectin tether. *Journal of Cell Biology, 155*, 145–156.

Evans, B. J., McDowall, A., Taylor, P. C., Hogg, N., Haskard, D. O., & Landis, R. C. (2006). Shedding of lymphocyte function-associated antigen-1 (LFA-1) in a human inflammatory response. *Blood, 107*, 3593–3599.

Evans, R., Patzak, I., Svensson, L., De Filippo, K., Jones, K., McDowall, A., et al. (2009). Integrins in immunity. *Journal of Cell Science, 122*, 215–225.

Evans, S. S., Schleider, D. M., Bowman, L. A., Francis, M. L., Kansas, G. S., & Black, J. D. (1999). Dynamic association of L-selectin with the lymphocyte cytoskeletal matrix. *Journal of Immunology, 162*, 3615–3624.

Fagerholm, S. C., Hilden, T. J., Nurmi, S. M., & Gahmberg, C. G. (2005). Specific integrin alpha and beta chain phosphorylations regulate LFA-1 activation through affinity-dependent and -independent mechanisms. *Journal of Cell Biology, 171*, 705–715.

Garcia-Alvarez, B., de Pereda, J. M., Calderwood, D. A., Ulmer, T. S., Critchley, D., Campbell, I. D., et al. (2003). Structural determinants of integrin recognition by talin. *Molecular Cell, 11*, 49–58.

Garnacho, C., Shuvaev, V., Thomas, A., McKenna, L., Sun, J., Koval, M., et al. (2008). RhoA activation and actin reorganization involved in endothelial CAM-mediated endocytosis of anti-PECAM carriers: Critical role for tyrosine 686 in the cytoplasmic tail of PECAM-1. *Blood, 111*, 3024–3033.

Geiger, C., Nagel, W., Boehm, T., van Kooyk, Y., Figdor, C. G., Kremmer, E., et al. (2000). Cytohesin-1 regulates beta-2 integrin-mediated adhesion through both ARF-GEF function and interaction with LFA-1. *EMBO Journal, 19*, 2525–2536.

Gingras, A. R., Ziegler, W. H., Bobkov, A. A., Joyce, M. G., Fasci, D., Himmel, M., et al. (2009). Structural determinants of integrin binding to the talin rod. *Journal of Biological Chemistry, 284*, 8866–8876.

Ginsberg, M. H., Partridge, A., & Shattil, S. J. (2005). Integrin regulation. *Current Opinion in Cell Biology, 17*, 509–516.

Gonzalez-Amaro, R., & Sanchez-Madrid, F. (1999). Cell adhesion molecules: Selectins and integrins. *Critical Reviews in Immunology, 19*, 389–429.

Gottschalk, K. E. (2005). A coiled-coil structure of the alphaIIbbeta3 integrin transmembrane and cytoplasmic domains in its resting state. *Structure, 13*, 703–712.

Harburger, D. S., Bouaouina, M., & Calderwood, D. A. (2009). Kindlin-1 and -2 directly bind the C-terminal region of beta integrin cytoplasmic tails and exert integrin-specific activation effects. *Journal of Biological Chemistry, 284*, 11485–11497.

Heiska, L., Kantor, C., Parr, T., Critchley, D. R., Vilja, P., Gahmberg, C. G., et al. (1996). Binding of the cytoplasmic domain of intercellular adhesion molecule-2 (ICAM-2) to alpha-actinin. *Journal of Biological Chemistry, 271*, 26214–26219.

Hogg, N., Henderson, R., Leitinger, B., McDowall, A., Porter, J., & Stanley, P. (2002). Mechanisms contributing to the activity of integrins on leukocytes. *Immunological Reviews, 186*, 164–171.

Hogg, N., Laschinger, M., Giles, K., & McDowall, A. (2003). T-cell integrins: More than just sticking points. *Journal of Cell Science, 116*, 4695–4705.

Horwitz, A., Duggan, K., Buck, C., Beckerle, M. C., & Burridge, K. (1986). Interaction of plasma membrane fibronectin receptor with talin—A transmembrane linkage. *Nature, 320*, 531–533.

Hughes, P. E., Diaz-Gonzalez, F., Leong, L., Wu, C., McDonald, J. A., Shattil, S. J., et al. (1996). Breaking the integrin hinge. A defined structural constraint regulates integrin signaling. *Journal of Biological Chemistry, 271*, 6571–6574.

Hynes, R. O. (2002). Integrins: Bidirectional, allosteric signaling machines. *Cell, 110*, 673–687.

Hynes, R. O., Lively, J. C., McCarty, J. H., Taverna, D., Francis, S. E., Hodivala-Dilke, K., et al. (2002). The diverse roles of integrins and their ligands in angiogenesis. *Cold Spring Harbor Symposia on Quantitative Biology, 67*, 143–153.

Ivetic, A., Deka, J., Ridley, A., & Ager, A. (2002). The cytoplasmic tail of L-selectin interacts with members of the ezrin–radixin–moesin (ERM) family of proteins: Cell activation-dependent binding of moesin but not ezrin. *Journal of Biological Chemistry, 277*, 2321–2329.

Ivetic, A., Florey, O., Deka, J., Haskard, D. O., Ager, A., & Ridley, A. J. (2004). Mutagenesis of the ezrin-radixin-moesin binding domain of L-selectin tail affects shedding, microvillar positioning, and leukocyte tethering. *Journal of Biological Chemistry, 279*, 33263–33272.

Kansas, G. S., Ley, K., Munro, J. M., & Tedder, T. F. (1993). Regulation of leukocyte rolling and adhesion to high endothelial venules through the cytoplasmic domain of L-selectin. *Journal of Experimental Medicine, 177*, 833–838.

Kansas, G. S., & Pavalko, F. M. (1996). The cytoplasmic domains of E- and P-selectin do not constitutively interact with alpha-actinin and are not essential for leukocyte adhesion. *Journal of Immunology, 157*, 321–325.

Kanters, E., van Rijssel, J., Hensbergen, P. J., Hondius, D., Mul, F. P., Deelder, A. M., et al. (2008). Filamin B mediates ICAM-1-driven leukocyte transendothelial migration. *Journal of Biological Chemistry, 283*, 31830–31839.

Katagiri, K., Hattori, M., Minato, N., Irie, S., Takatsu, K., & Kinashi, T. (2000). Rap1 is a potent activation signal for leukocyte function-associated antigen 1 distinct from protein kinase C and phosphatidylinositol-3-OH kinase. *Molecular Cell Biology, 20*, 1956–1969.

Katagiri, K., Maeda, A., Shimonaka, M., & Kinashi, T. (2003). RAPL, a Rap1-binding molecule that mediates Rap1-induced adhesion through spatial regulation of LFA-1. *Nature Immunology, 4*, 741–748.

Killock, D. J., Parsons, M., Zarrouk, M., Ameer-Beg, S. M., Ridley, A. J., Haskard, D. O., et al. (2009). *In vitro* and *in vivo* characterization of molecular interactions between calmodulin, ezrin/radixin/moesin, and L-selectin. *Journal of Biological Chemistry, 284*, 8833–8845.

Kim, M., Carman, C. V., & Springer, T. A. (2003). Bidirectional transmembrane signaling by cytoplasmic domain separation in integrins. *Science, 301*, 1720–1725.

Kinashi, T. (2005). Intracellular signalling controlling integrin activation in lymphocytes. *Nature Reviews Immunology, 5*, 546–559.

Kinashi, T., & Katagiri, K. (2004). Regulation of lymphocyte adhesion and migration by the small GTPase Rap1 and its effector molecule, RAPL. *Immunology Letters, 93*, 1–5.

Kolanus, W. (2007). Guanine nucleotide exchange factors of the cytohesin family and their roles in signal transduction. *Immunological Reviews, 218*, 102–113.

Kolanus, W., Nagel, W., Schiller, B., Zeitlmann, L., Godar, S., Stockinger, H., et al. (1996). Alpha L beta 2 integrin/LFA-1 binding to ICAM-1 induced by cytohesin-1, a cytoplasmic regulatory molecule. *Cell, 86*, 233–242.

Larjava, H., Plow, E. F., & Wu, C. (2008). Kindlins: Essential regulators of integrin signalling and cell–matrix adhesion. *EMBO Reports, 9*, 1203–1208.

Legate, K. R., & Fassler, R. (2009). Mechanisms that regulate adaptor binding to beta-integrin cytoplasmic tails. *Journal of Cell Science, 122*, 187–198.

Li, L., Greenwald, R. J., Lafuente, E. M., Tzachanis, D., Berezovskaya, A., Freeman, G. J., et al. (2005). Rap1-GTP is a negative regulator of Th cell function and promotes the generation of CD4+CD103+ regulatory T cells *in vivo*. *Journal of Immunology, 175*, 3133–3139.

Liu, S., Thomas, S. M., Woodside, D. G., Rose, D. M., Kiosses, W. B., Pfaff, M., et al. (1999). Binding of paxillin to alpha4 integrins modifies integrin-dependent biological responses. *Nature, 402*, 676–681.

Luo, B. H., Carman, C. V., & Springer, T. A. (2007). Structural basis of integrin regulation and signaling. *Annual Review of Immunology, 25*, 619–647.

Malinin, N. L., Zhang, L., Choi, J., Ciocea, A., Razorenova, O., Ma, Y. Q., et al. (2009). A point mutation in KINDLIN3 ablates activation of three integrin subfamilies in humans. *Nature Medicine, 15*, 313–318.

Mamdouh, Z., Kreitzer, G. E., & Muller, W. A. (2008). Leukocyte transmigration requires kinesin-mediated microtubule-dependent membrane trafficking from the lateral border recycling compartment. *Journal of Experimental Medicine, 205*, 951–966.

Mandell, K. J., & Parkos, C. A. (2005). The JAM family of proteins. *Advanced Drug Delivery Reviews, 57*, 857–867.

Mattila, P. E., Green, C. E., Schaff, U., Simon, S. I., & Walcheck, B. (2005). Cytoskeletal interactions regulate inducible L-selectin clustering. *American Journal of Physiology: Cell Physiology, 289*, C323–C332.

McDowall, A., Inwald, D., Leitinger, B., Jones, A., Liesner, R., Klein, N., et al. (2003). A novel form of integrin dysfunction involving beta1, beta2, and beta3 integrins. *Journal of Clinical Investigation, 111*, 51–60.

Millan, J., Hewlett, L., Glyn, M., Toomre, D., Clark, P., & Ridley, A. J. (2006). Lymphocyte transcellular migration occurs through recruitment of endothelial ICAM-1 to caveola- and F-actin-rich domains. *Nature Cell Biology, 8*, 113–123.

Miner, J. J., Xia, L., Yago, T., Kappelmayer, J., Liu, Z., Klopocki, A. G., et al. (2008). Separable requirements for cytoplasmic domain of PSGL-1 in leukocyte rolling and signaling under flow. *Blood, 112*, 2035–2045.

Moh, M. C., Tian, Q., Zhang, T., Lee, L. H., & Shen, S. (2009). The immunoglobulin-like cell adhesion molecule hepaCAM modulates cell adhesion and motility through direct interaction with the actin cytoskeleton. *Journal of Cell Physiology, 219*, 382–391.

Moh, M. C., Zhang, C., Luo, C., Lee, L. H., & Shen, S. (2005). Structural and functional analyses of a novel ig-like cell adhesion molecule, hepaCAM, in the human breast carcinoma MCF7 cells. *Journal of Biological Chemistry, 280*, 27366–27374.

Mor, A., Dustin, M. L., & Philips, M. R. (2007). Small GTPases and LFA-1 reciprocally modulate adhesion and signaling. *Immunological Reviews, 218*, 114–125.

Moser, M., Bauer, M., Schmid, S., Ruppert, R., Schmidt, S., Sixt, M., et al. (2009a). Kindlin-3 is required for beta2 integrin-mediated leukocyte adhesion to endothelial cells. *Nature Medicine, 15*, 300–305.

Moser, M., Legate, K. R., Zent, R., & Fassler, R. (2009b). The tail of integrins, talin, and kindlins. *Science, 324*, 895–899.

Moser, M., Nieswandt, B., Ussar, S., Pozgajova, M., & Fassler, R. (2008). Kindlin-3 is essential for integrin activation and platelet aggregation. *Nature Medicine, 14*, 325–330.

Nagel, W., Schilcher, P., Zeitlmann, L., & Kolanus, W. (1998a). The PH domain and the polybasic c domain of cytohesin-1 cooperate specifically in plasma membrane association and cellular function. *Molecular Biology of the Cell, 9*, 1981–1994.

Nagel, W., Zeitlmann, L., Schilcher, P., Geiger, C., Kolanus, J., & Kolanus, W. (1998b). Phosphoinositide 3-OH kinase activates the beta2 integrin adhesion pathway and induces membrane recruitment of cytohesin-1. *Journal of Biological Chemistry, 273*, 14853–14861.

Nayal, A., Webb, D. J., & Horwitz, A. F. (2004). Talin: An emerging focal point of adhesion dynamics. *Current Opinion in Cell Biology, 16*, 94–98.

Newman, P. J. (1994). The role of PECAM-1 in vascular cell biology. *Annals of the New York Academy of Sciences, 714*, 165–174.

Newman, P. J., Hillery, C. A., Albrecht, R., Parise, L. V., Berndt, M. C., Mazurov, A. V., et al. (1992). Activation-dependent changes in human platelet PECAM-1: Phosphorylation, cytoskeletal association, and surface membrane redistribution. *Journal of Cell Biology, 119*, 239–246.

Nishimura, T., & Kaibuchi, K. (2007). Numb controls integrin endocytosis for directional cell migration with aPKC and PAR-3. *Developmental Cell, 13*, 15–28.

Partridge, A. W., Liu, S., Kim, S., Bowie, J. U., & Ginsberg, M. H. (2005). Transmembrane domain helix packing stabilizes integrin alphaIIbbeta3 in the low affinity state. *Journal of Biological Chemistry, 280*, 7294–7300.

Paschos, K. A., Canovas, D., & Bird, N. C. (2009). The role of cell adhesion molecules in the progression of colorectal cancer and the development of liver metastasis. *Cellular Signalling, 21*, 665–674.

Pavalko, F. M., & LaRoche, S. M. (1993). Activation of human neutrophils induces an interaction between the integrin beta 2-subunit (CD18) and the actin binding protein alpha-actinin. *Journal of Immunology, 151*, 3795–3807.

Pavalko, F. M., Walker, D. M., Graham, L., Goheen, M., Doerschuk, C. M., & Kansas, G. S. (1995). The cytoplasmic domain of L-selectin interacts with cytoskeletal proteins via alpha-actinin: Receptor positioning in microvilli does not require interaction with alpha-actinin. *Journal of Cell Biology, 129*, 1155–1164.

Pozo, M., de Nicolas, R., Egido, J., & Gonzalez-Cabrero, J. (2006). Simvastatin inhibits the migration and adhesion of monocytic cells and disorganizes the cytoskeleton of activated endothelial cells. *European Journal of Pharmacology, 548*, 53–63.

Puthenedam, M., Williams, P. H., Lakshmi, B. S., & Balakrishnan, A. (2007). Modulation of tight junction barrier function by outer membrane proteins of enteropathogenic *Escherichia coli*: Role of F-actin and junctional adhesion molecule-1. *Cell Biology International, 31*, 836–844.

Qu, A., & Leahy, D. J. (1995). Crystal structure of the I-domain from the CD11a/CD18 (LFA-1, alpha L beta 2) integrin. *Proceedings of the National Academy of Sciences of the United States of America, 92*, 10277–10281.

Quast, T., Tappertzhofen, B., Schild, C., Grell, J., Czeloth, N., Forster, R., et al. (2009). Cytohesin-1 controls the activation of RhoA and modulates integrin-dependent adhesion and migration of dendritic cells. *Blood, 113*, 5801–5810.

Ratnikov, B. I., Partridge, A. W., & Ginsberg, M. H. (2005). Integrin activation by talin. *Journal of Thrombosis and Haemostasis, 3*, 1783–1790.

Reedquist, K. A., Ross, E., Koop, E. A., Wolthuis, R. M., Zwartkruis, F. J., van Kooyk, Y., et al. (2000). The small GTPase, Rap1, mediates CD31-induced integrin adhesion. *Journal of Cell Biology, 148*, 1151–1158.

Rose, D. M., Alon, R., & Ginsberg, M. H. (2007). Integrin modulation and signaling in leukocyte adhesion and migration. *Immunological Reviews, 218*, 126–134.

Rose, D. M., Han, J., & Ginsberg, M. H. (2002). Alpha4 integrins and the immune response. *Immunological Reviews, 186*, 118–124.

Sampath, R., Gallagher, P. J., & Pavalko, F. M. (1998). Cytoskeletal interactions with the leukocyte integrin beta2 cytoplasmic tail. activation-dependent regulation of associations with talin and alpha-actinin [In Process Citation]. *Journal of Biological Chemistry, 273*, 33588–33594.

Schneider, D., & Engelman, D. M. (2004). Involvement of transmembrane domain interactions in signal transduction by alpha/beta integrins. *Journal of Biological Chemistry, 279*, 9840–9846.

Sebzda, E., Bracke, M., Tugal, T., Hogg, N., & Cantrell, D. A. (2002). Rap1A positively regulates T cells via integrin activation rather than inhibiting lymphocyte signaling. *Nature Immunology, 3*, 251–258.

Semmrich, M., Smith, A., Feterowski, C., Beer, S., Engelhardt, B., Busch, D. H., et al. (2005). Importance of integrin LFA-1 deactivation for the generation of immune responses. *Journal of Experimental Medicine, 201*, 1987–1998.

Setiadi, H., & McEver, R. P. (2003). Signal-dependent distribution of cell surface P-selectin in clathrin-coated pits affects leukocyte rolling under flow. *Journal of Cell Biology, 163*, 1385–1395.

Setiadi, H., & McEver, R. P. (2008). Clustering endothelial E-selectin in clathrin-coated pits and lipid rafts enhances leukocyte adhesion under flow. *Blood, 111*, 1989–1998.

Sharma, C. P., Ezzell, R. M., & Arnaout, M. A. (1995). Direct interaction of filamin (ABP-280) with the beta 2-integrin subunit CD18. *Journal of Immunology, 154*, 3461–3470.

Simon, S. I., Cherapanov, V., Nadra, I., Waddell, T. K., Seo, S. M., Wang, Q., et al. (1999). Signaling functions of L-selectin in neutrophils: Alterations in the cytoskeleton and colocalization with CD18. *Journal of Immunology, 163*, 2891–2901.

Simon, S. I., Sarantos, M. R., Green, C. E., & Schaff, U. Y. (2009). Leucocyte recruitment under fluid shear: Mechanical and molecular regulation within the inflammatory synapse. *Clinical and Experimental Pharmacology & Physiology, 36*, 217–224.

Simonson, W. T., Franco, S. J., & Huttenlocher, A. (2006). Talin1 regulates TCR-mediated LFA-1 function. *Journal of Immunology, 177*, 7707–7714.

Smith, C. W. (2008). 3. Adhesion molecules and receptors. *Journal of Allergy and Clinical Immunology, 121*, S375–S379 quiz S414.

Smith, A., Stanley, P., Jones, K., Svensson, L., McDowall, A., & Hogg, N. (2007). The role of the integrin LFA-1 in T-lymphocyte migration. *Immunological Reviews, 218*, 135–146.

Snapp, K. R., Heitzig, C. E., & Kansas, G. S. (2002). Attachment of the PSGL-1 cytoplasmic domain to the actin cytoskeleton is essential for leukocyte rolling on P-selectin. *Blood, 99*, 4494–4502.

Sousa, A. D., & Cheney, R. E. (2005). Myosin-X: A molecular motor at the cell's fingertips. *Trends in Cell Biology, 15*, 533–539.

Stanley, P., Smith, A., McDowall, A., Nicol, A., Zicha, D., & Hogg, N. (2008). Intermediate-affinity LFA-1 binds alpha-actinin-1 to control migration at the leading edge of the T cell. *EMBO Journal, 27*, 62–75.

Svensson, L., Howarth, K., McDowall, A., Patzak, I., Evans, R., Ussar, S., et al. (2009). Leukocyte adhesion deficiency-III is caused by mutations in KINDLIN3 affecting integrin activation. *Nature Medicine, 15*, 306–312.

Tadokoro, S., Shattil, S. J., Eto, K., Tai, V., Liddington, R. C., de Pereda, J. M., et al. (2003). Talin binding to integrin beta tails: A final common step in integrin activation. *Science, 302*, 103–106.

Takagi, J., Strokovich, K., Springer, T. A., & Walz, T. (2003). Structure of integrin alpha5beta1 in complex with fibronectin. *EMBO Journal, 22*, 4607–4615.

Takai, Y., Kitano, K., Terawaki, S., Maesaki, R., & Hakoshima, T. (2007). Structural basis of PSGL-1 binding to ERM proteins. *Genes to Cells, 12*, 1329–1338.

Takala, H., Nurminen, E., Nurmi, S. M., Aatonen, M., Strandin, T., Takatalo, M., et al. (2008). Beta2 integrin phosphorylation on Thr758 acts as a molecular switch to regulate 14-3-3 and filamin binding. *Blood, 112*, 1853–1862.

Thompson, P. W., Randi, A. M., & Ridley, A. J. (2002). Intercellular adhesion molecule (ICAM)-1, but not ICAM-2, activates RhoA and stimulates c-fos and rhoA transcription in endothelial cells. *Journal of Immunology, 169*, 1007–1013.

Urzainqui, A., Serrador, J. M., Viedma, F., Yanez-Mo, M., Rodriguez, A., Corbi, A. L., et al. (2002). ITAM-based interaction of ERM proteins with Syk mediates signaling by the leukocyte adhesion receptor PSGL-1. *Immunity, 17*, 401–412.

van Buul, J. D., & Hordijk, P. L. (2009). Endothelial adapter proteins in leukocyte transmigration. *Thrombosis and Haemostasis, 101*, 649–655.

VandenBerg, E., Reid, M. D., Edwards, J. D., & Davis, H. W. (2004). The role of the cytoskeleton in cellular adhesion molecule expression in tumor necrosis factor-stimulated endothelial cells. *Journal of Cell Biochemistry, 91*, 926–937.

Vinogradova, O., Haas, T., Plow, E. F., & Qin, J. (2000). A structural basis for integrin activation by the cytoplasmic tail of the alpha IIb-subunit. *Proceedings of the National Academy of Sciences of the United States of America, 97*, 1450–1455.

Vinogradova, O., Velyvis, A., Velyviene, A., Hu, B., Haas, T., Plow, E., et al. (2002). A structural mechanism of integrin alpha(IIb)beta(3) "inside-out" activation as regulated by its cytoplasmic face. *Cell, 110*, 587–597.

Wang, X. G., Cheng, Y. P., & Ba, X. Q. (2006). Engagement of PSGL-1 enhances beta(2)-integrin-involved adhesion of neutrophils to recombinant ICAM-1. *Acta Pharmacologica Sinica, 27*, 617–622.

Weber, K. S., Weber, C., Ostermann, G., Dierks, H., Nagel, W., & Kolanus, W. (2001). Cytohesin-1 is a dynamic regulator of distinct LFA-1 functions in leukocyte arrest and transmigration triggered by chemokines. *Current Biology, 11*, 1969–1974.

Wegener, K. L., Partridge, A. W., Han, J., Pickford, A. R., Liddington, R. C., Ginsberg, M. H., et al. (2007). Structural basis of integrin activation by talin. *Cell, 128*, 171–182.

Wiese, G., Barthel, S. R., & Dimitroff, C. J. (2009). Analysis of physiologic E-selectin-mediated leukocyte rolling on microvascular endothelium. *Journal of Visualized Experiments, 24*, .

Yoshida, M., Szente, B. E., Kiely, J. M., Rosenzweig, A., & Gimbrone, M. A., Jr. (1998). Phosphorylation of the cytoplasmic domain of E-selectin is regulated during leukocyte–endothelial adhesion. *Journal of Immunology, 161*, 933–941.

Yoshida, M., Westlin, W. F., Wang, N., Ingber, D. E., Rosenzweig, A., Resnick, N., et al. (1996). Leukocyte adhesion to vascular endothelium induces E-selectin linkage to the actin cytoskeleton. *Journal of Cell Biology, 133*, 445–455.

Zarbock, A., & Ley, K. (2008). Mechanisms and consequences of neutrophil interaction with the endothelium. *American Journal of Pathology, 172*, 1–7.

Zarbock, A., & Ley, K. (2009). Neutrophil adhesion and activation under flow. *Microcirculation, 16*, 31–42.

Zhang, H., Berg, J. S., Li, Z., Wang, Y., Lang, P., Sousa, A. D., et al. (2004). Myosin-X provides a motor-based link between integrins and the cytoskeleton. *Nature Cell Biology, 6*, 523–531.

Zhu, J., Carman, C. V., Kim, M., Shimaoka, M., Springer, T. A., & Luo, B. H. (2007). Requirement of alpha and beta subunit transmembrane helix separation for integrin outside-in signaling. *Blood, 110*, 2475–2483.

CHAPTER 6

Membrane–Cytoskeletal Platforms for Rapid Chemokine Signaling to Integrins

Ronen Alon

Department of Immunology, The Weizmann Institute of Science, Rehovot 76100, Israel

Current Topics in Membranes, Volume 64 1063-5823/09 $35.00

DOI: 10.1016/S1063-5823(09)64006-0

ABBREVIATIONS

FERM, 4.1 protein, ezrin, radixin, moesin; GEF, guanine exchange factor; GPCR, G protein-coupled receptor; ICAM-1, intercellular adhesion molecule-1; LFA-1, lymphocyte function-associated antigen-1; PKC, protein kinase C; PLC, phospholipase C; Rac-1, Ras-related C3 botulinum toxin substrate-1; RhoA, Ras homologue A; VCAM-1, vascular cell adhesion molecule-1; VLA-4, very late antigen-4.

I. OVERVIEW

The arrest of rolling leukocytes at target vascular sites depends on rapid activation of their vascular integrins at endothelial contacts enriched with specific Ig superfamily (IgSF) ligands. Accumulating data suggest that these integrins acquire high affinity to these endothelial ligands in a process tightly regulated by a variety of cytoskeletal conformational changes in their α- and β-subunits. These alterations are controlled by inside-out signals which are induced primarily by chemokines displayed on the endothelium and transduced by G protein-coupled receptors (GPCRs) on the responding leukocytes. The GPCR signals can propagate within milliseconds at submicron ranges. Upon binding their ligands, GPCR-stimulated integrins also undergo outside-in conformational activation, which is accelerated by applied forces. The integrin activator, talin, a universal cytoskeletal adaptor in integrin-mediated focal adhesions, plays an instrumental role in this bidirectional activation process which is critical for leukocyte arrest on endothelial IgSF ligands. Multiple effectors in the immediate vicinity to integrin–talin complexes can determine the extent to which talin may translate a given chemokine signal into a fully productive integrin activation. Substantial heterogeneity exists in the pathways used by different cell types to translate a given combination of chemokines and integrin ligands into leukocyte arrest and subsequent adhesion strengthening on various target endothelia. This chapter highlights the main molecular players in the earliest events of chemokine signaling to lymphocyte integrins and proposes modalities used by rapid chemokine signals to trigger cytoskeletal rearrangements in integrin tails, leading to rapid integrin activation and leukocyte arrest on vascular endothelium.

II. INTRODUCTION

A. The Multistep Cascade of Integrin Activation on Rolling Leukocytes

Immune cells (leukocytes) and hematopoietic progenitor cells circulating the body must exit blood vessels near specific target sites of injury, infection, inflammation, differentiation, and proliferation (Butcher, 1991; Ley, Laudanna, Cybulsky, & Nourshargh, 2007; Springer, 1994). Recruitment of different subsets of leukocytes to these sites and to lymphoid organs is tightly regulated by sequential adhesive interactions between specific protein receptors on their surface and respective ligands on the blood vessel endothelial wall (Ley et al.; Luster, Alon, & von Andrian, 2005) (Fig. 1). Accumulated data from *in vivo* and *in vitro* studies suggest that the major players in this multi-step process are members of two adhesive families, selectins, and integrins, which are structurally and functionally adapted to operate under disruptive shear forces exerted on leukocytes at the vessel wall by the blood flow (Alon & Dustin, 2007). Selectins are the major adhesive receptors in the vasculature that mediate the initial tethering of flowing leukocytes to the vessel wall, and the propagation of these tethers into continuous rolling adhesions (McEver, 2002). They comprise of three member families, two inducibly expressed on endothelial surfaces or platelets, and one, L-selectin, expressed by most circulating leukocytes. The predominant role of selectins in leukocyte capture and reversible rolling adhesions has been traditionally attributed to their fast kinetics of association and dissociation, as well as to high tensile strength of individual bonds. This mechanical property resides in the ability of the ligand-binding domain of selectins to undergo extension upon occupancy by ligand, preferentially when experiencing applied forces (Springer, 2009). When present at sufficient density on the endothelial target, selectin-mediated interactions give rise to leukocyte rolling in the direction of flow, which allows the transiently attached immune cells to sample the endothelial lining, and brings them into proximity with activating chemoattractants, mostly small cytokines termed chemokines presented on the endothelial surfaces (Mackay, 2001). These chemotactic cytokines, displayed on the vessel wall in largely immobile states, bind specific GPCRs on recruited leukocytes, and rapidly switch the conformation of specific leukocyte integrins (Fig. 2), which upon encounter of cognate endothelial ligands, cause the immune cells to arrest on the blood vessel (Campbell & Butcher, 2000). Heterotrimeric Gi proteins are the main, if not the exclusive type of G proteins which upon coupling to chemoattractant/chemokine-occupied GPCRs trigger integrin activation (Kehrl, 2006; Thelen & Stein, 2008; Zarbock, Deem, Burcin, & Ley, 2007).

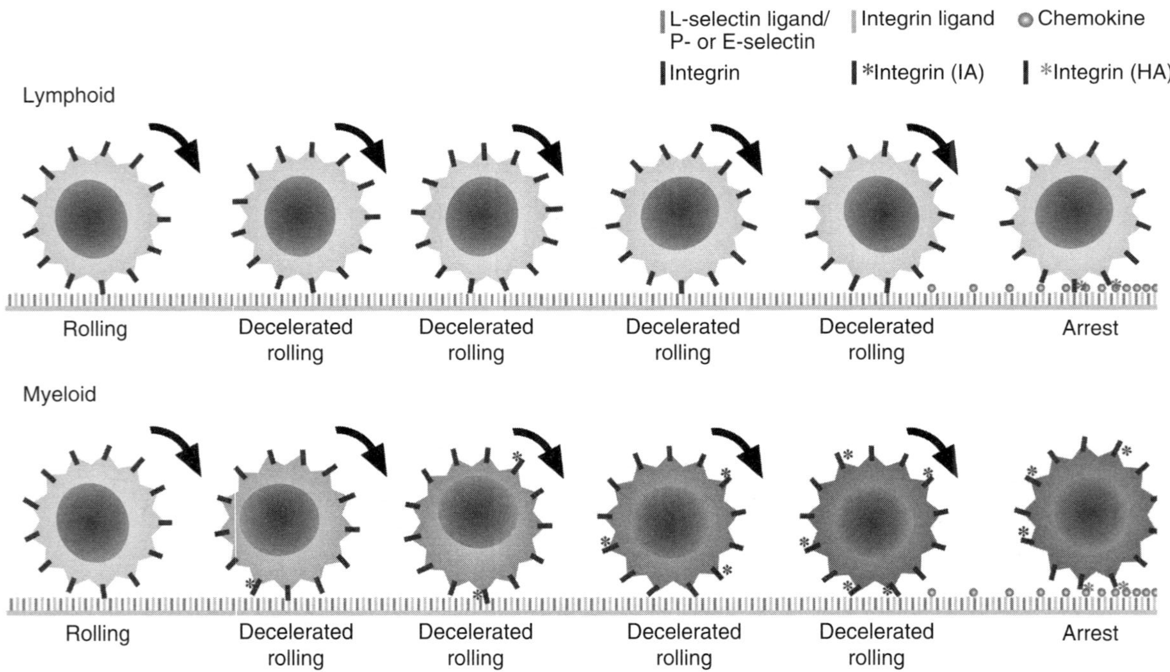

FIGURE 1 Postulated contribution of selectins to integrin activation in lymphoid and myeloid cells. Top: in lymphoid cells, selectin occupancy functions mainly for cell capture and to induce rolling, rather than signaling to integrin-activating effectors. In lymphoid cells, integrins do not undergo stepwise activation, though microvilli flattening during rolling can still contribute to GPCR and integrin exposure to counterligands (IA, intermediate affinity; HA, high affinity). Bottom: in myeloid cells, selectin occupancy of specific leukocyte glycoproteins can directly signal to PTKs and recruit secondary messengers such as Ca^{2+} which globally trigger intermediate-affinity LFA-1 or Mac-1 integrins, possibly in a stepwise manner. The increasing fraction of integrins acquiring an intermediate-affinity state slows down selectin-mediated rolling and thereby increases the probability of chemokine encounter and final integrin activation to a high-affinity states, which mediate rapid and firm arrest.

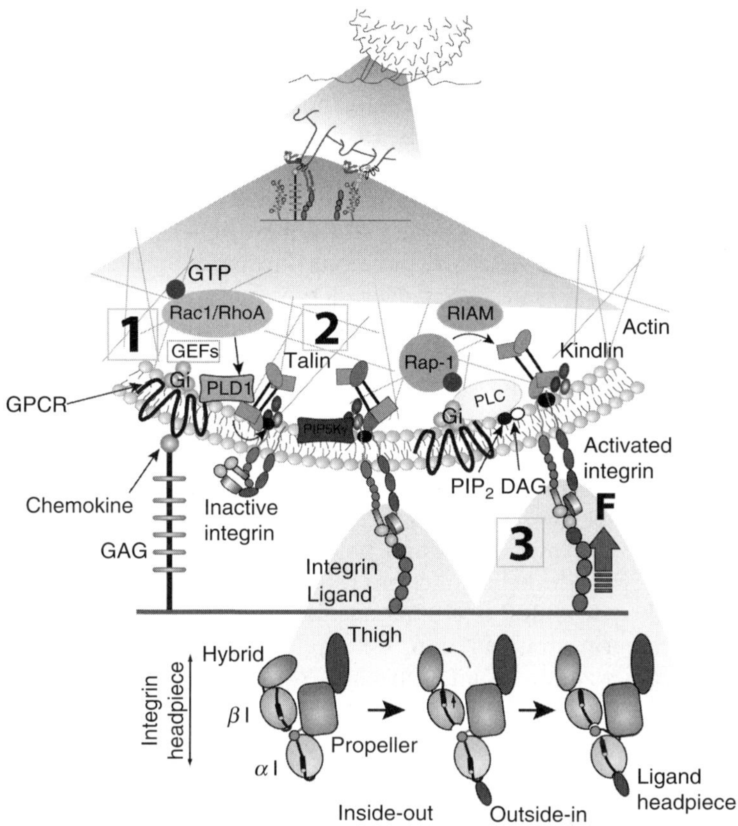

FIGURE 2 A postulated scheme for rapid chemokine signaling to lymphocyte LFA-1 under shear flow. (Top) A rolling leukocyte tethered to an integrin ligand must encounter juxtaposed chemokine at the site of integrin activation, possibly a single microvillus. A quaternary complex between integrin, ligand, chemokine, and G protein-coupled receptor (GPCR) must form within milliseconds (Laudanna & Alon, 2006) to locally activate the integrin–ligand complex via a bidirectional signaling event (bottom). Only a fully activated integrin can arrest the rolling leukocyte on the vessel wall while partially activated integrins can participate in rolling adhesions (see Fig. 1). Upon initial encounter, endothelial-bound chemokine transduces leukocyte GPCRs signals which convert the inactive (folded) integrin to its extended conformation (step 1). GTP-bound RhoA and Rac1 are involved in this step in the case of lymphocyte LFA-1. This critical chemokine-driven inside-out event primes the integrin to transiently bind endothelial ligands on the counter endothelial surface. The various I-domains undergo further conformational shifts upon extracellular ligand binding (step 2), resulting in further integrin activation (outside-in). This ligand-driven step is predicted to result in further separation of the integrin subunit tails, conditional on the presence of talin nearby the ligand-occupied integrin, and force transduction (step 3). Clustering of ligand-occupied integrins (not shown) can rapidly follow. Force transduction may lead to talin activation, recruitment of vinculin, and crosslinking of integrin–talin complexes to the cortical actin cytoskeleton, all within seconds after initial leukocyte arrest. The additional involvement of Rap1 and its effectors RIAM and RAPL (not shown) in GPCR-transduced inside-out activation is suggested for LFA-1 activation by chemokines, whereas PKC is necessary for full VLA-4 activation. For clarity, GPCR scaffolds such as filamins and β-arrestins are omitted.

The main integrin family members that recognize endothelial ligands and participate in leukocyte arrest on these ligands include the α_4-integrins, VLA-4 ($\alpha_4\beta_1$) and $\alpha_4\beta_7$, and the β_2-integrins, LFA-1 ($\alpha_L\beta_2$) and Mac-1 ($\alpha_M\beta_2$). These integrins are also known as "vascular integrins" (Kubes, 2002; Luster et al., 2005). A remarkable feature of integrins is that their affinity state and adhesive activity are regulated *in situ* by rapid extracellular signals, without an accompanying increase in their surface expression (Carman & Springer, 2003; Evans et al., 2009; Luo, Carman, & Springer, 2007). In most cell types studied to date, localized signals appear to be transmitted from the occupied GPCR and its Gi protein partner to the integrin target within a small adhesive zone, probably a single microvillus, which is simultaneously occupied by the integrin ligand and the activating chemokine (Fig. 2) (Alon & Ley, 2008). Recent evidence also suggests that lymphocyte integrins can undergo rapid activation and deactivation within submicron sized focal contacts on the ventral surface of arrested lymphocytes. This rapid turnover of integrin activation by chemokine signals allows lymphocytes and possibly other leukocytes to crawl toward and through endothelial junctions (Shulman et al., 2009). Thus, integrin-mediated leukocyte/endothelial contacts involve sequential and reversible changes in integrin conformation, cytoskeletal associations, IgSF ligand-induced oligomerization (Kim, Carman, Yang, Salas, & Springer, 2004), and lateral mobility (Laudanna & Alon, 2006).

While GPCR signals exert dramatic and reversible effects on integrin affinity and avidity to endothelial ligands, the intrinsic affinity of selectins to their glycoprotein ligands is generally not subject to *in situ* modulation by these signals. Instead, some endothelial-displayed chemokines may interfere with selectin–ligand recognition, possibly via an indirect steric effect of chemokine-clustered GPCRs (Grabovsky, Dwir, & Alon, 2002). Nevertheless, cytoskeletal associations of selectins and subsets of their glycoprotein ligands determine both the topography and mechanical properties of these counter-receptors (Dwir, Kansas, & Alon, 2001; Evans & Calderwood, 2007; Kansas, Ley, Munro, & Tedder, 1993; Killock et al., 2009; Setiadi & McEver, 2008; Setiadi, Sedgewick, Erlandsen, & McEver, 1998).

III. LEUKOCYTE INTEGRIN ACTIVATION AT ENDOTHELIAL CONTACTS

A. *Conformational Switches in VLA-4 and LFA-1 Induced by Chemokine Signals*

The ability of integrins to modulate their adhesive activities at endothelial contacts is much more robust than any of the selectin–ligand interactions. Indeed, selectin bonds fail to mediate leukocyte arrest even when some of the leukocyte endothelial contacts are stabilized by over 30 adhesive bonds/cell (Chen & Springer, 1999), way above the number of integrin–ligand bonds sufficient for leukocyte arrest (Shamri et al., 2005). Recent structural and biophysical studies predict that integrins can alternate between inactive bent conformers, and variable extended conformers with intermediate and high affinity to ligand, involving extensive allosteric changes to their headpiece (Fig. 2, bottom). Integrins bind their respective endothelial ligands under shear flow at much lower association rates than selectins (Springer, 1994). The majority of integrins on circulating leukocytes are kept in a low-affinity bent state, in which the ligand-binding headpiece is unavailable for ligand recognition and/or in a locked conformation that cannot respond to ligand-induced rearrangements (Yednock et al., 1995). In contrast, unfolded integrins (e.g., *in situ* chemokine-stimulated integrins) are predicted to extend 10–25 nm above the cell surface, and can readily bind immobilized ligand on a countersurface followed by extensive rearrangements of their I-domains (Luo et al., 2007). Interestingly, both the duration of contact and force application on the ligand-occupied integrins can affect the magnitude of integrin activation (Evans & Calderwood, 2007; Zhu et al., 2008). For example, the full acquisition of high-affinity VLA-4 state by VCAM-1 is favored not only on shear forces (Woolf et al., 2007), but in fact requires a critical duration of contact (Chen et al., 1999). Thus, lymphocyte VLA-4, in contrast to lymphocyte LFA-1, can mediate either rolling or firm arrest on its endothelial ligand VCAM-1 depending on the type of the chemokine, the chemokine density, as well as the density of VCAM-1 (D'Ambrosio et al., 2002; Feigelson et al., 2001). LFA-1 on neutrophils can, on the other hand, acquire an intermediate-affinity state during neutrophil rolling on E-selectin/ICAM-1-coated surfaces, and reverts to a high-affinity state that mediates firm adhesions immediately upon encounter of chemokine signals (Zarbock, Lowell, & Ley, 2007). Furthermore, studies on LFA-1 locked in an intermediate-affinity state, in which the integrin is extended but its headpiece is inactive, indicate that this LFA-1 conformer mediates leukocyte rolling even on high density ICAM-1 since it cannot arrest on this ligand regardless of force, ligand density and duration of contact (Salas, Shimaoka, Chen, Carman, & Springer, 2002; Salas et al., 2004). Thus, whether LFA-1, VLA-4, or other integrins can mediate rolling or firm arrest depends on both the magnitude and the kinetics of the signals they incorporate from the chemokine and integrin ligands, as well as on the cellular environment in which these integrins operate. Individual leukocyte integrins can assume a range of conformations and their range of conformational modulation by both inside-out and outside-in signals is very broad, possibly due to their divergent cytoskeletal associations (Chigaev, Waller, Zwartz, Buranda, & Sklar, 2007; Laudanna & Alon, 2006).

Different sets of cytoskeletal adaptors can determine whether a particular integrin–ligand pair will acquire high affinity and support firm adhesion (Legate & Fassler, 2009). These effectors not only determine the ability of integrins to undergo conformational activation but also the capacity of integrins to load external forces (Figs. 2 and 3). Cytoskeletal anchorage of integrins has been traditionally suggested to constrain integrin mobility and patching, and thereby to reduce integrin adhesiveness at intercellular contacts (Carman & Springer, 2003; Constantin et al., 2000; Kim et al., 2004). Nevertheless, recent evidence suggests that anchored integrins may undergo substantial rapid inside-out and outside-in activation before the less cytoskeletally constrained integrins undergo such activation (Alon et al., 2005; Cairo, Mirchev, & Golan, 2006). For example, the anchorage state of $\alpha_4\beta_1$ and the related integrin, $\alpha_4\beta_7$, indeed affect the ability of these α_4-integrins to mediate capture and undergo chemokine-stimulated adhesion strengthening on their respective ligands, VCAM-1 and MadCAM-1, selectively under shear stress conditions (Alon et al.; Manevich, Grabovsky, Feigelson, & Alon, 2007). It is also well recognized that ligand occupancy further anchors integrin to the cortical cytoskeleton (Cairo et al.). Notably, enhancement of integrin anchorage to the cytoskeleton does not always promote integrin activation and, in fact, may exert inhibitory outcomes on integrin affinity and avidity, if the integrin anchoring adaptor competes with the binding of a direct integrin activator, such as talin. For example, overexpression of the LFA-1 adaptor, filamin, results in reduced integrin adhesiveness, possibly due to steric hindrance of talin associations with LFA-1 (Takala et al., 2008).

B. Integrin Activation Requires Its Simultaneous Occupancy by Extracellular and Intracellular Ligands

Chemokine-stimulated signals induce conformational changes in only a fraction of integrins (Ley et al., 2007). Most integrins are inactive clasped heterodimers (Luo et al., 2007). It is well recognized that full conformational activation of integrins must be allosterically modulated in a bidirectional fashion (Hynes, 2002). Since integrin conformational transitions are very short lived, it is likely that a fully activated integrin must be simultaneously occupied by two ligands, an extracellular ligand and a cytoplasmic adaptor that can bind to and stabilize the integrin heterodimer in an activated (unclasped) high-affinity conformation. The key cytoplasmic regulator of integrin affinity is talin which can be regarded as a low-affinity ligand of most integrin β-subunit tails (Ratnikov, Partridge, & Ginsberg, 2005; Tadokoro et al., 2003; Wegener et al., 2008). Talin knock down in multiple

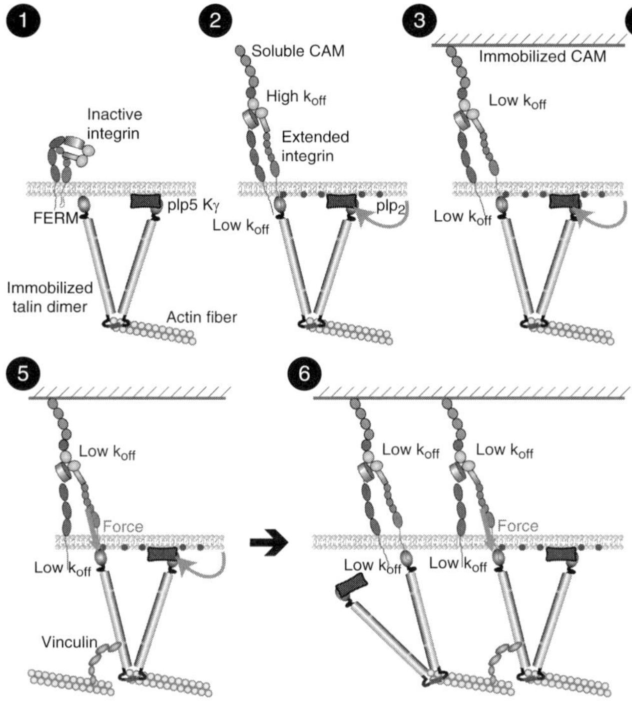

FIGURE 3 Bidirectional integrin activation requires simultaneous double occupancy of individual integrin molecules by extracellular and cytoplasmic ligands. Integrin conformation is allosterically modulated in a bidirectional manner by two sets of ligands, extracellular and cytoplasmic, that is, talin. The degree of activation is dictated primarily by unclasping of the integrin heterodimer, a process dependent on the binding of activated talin FERM to the integrin β-tail. (1) Inactive integrin. (2–4) Three distinct conformations of integrins responding within subseconds to initial Gi-triggered signaling. Talin FERM activation nearby the target integrin is a rate-limiting step in integrin activation (2, 3) and is controlled by an inside-out signal. Talin tethering to the plasma membrane by PIP2, generated by PIP5Kγ (purple) activates the talin FERM domain and enhances (possibly together with Kindlin-3) talin association with the target integrin. Only when the integrin is occupied by both its extracellular and cytoplasmic ligand, that is, talin, does maximal unclasping and acquisition of the highest affinity integrin–ligand bond take place (3). Full activation of integrins takes place when both the extracellular ligand and talin are linked to viscous scaffolds since immobilization of both the extracellular and cytoplasmic integrin ligands dramatically reduces their escape from the integrin (lowering the k_{off}) and thereby increases the probability of bidirectional activation. Note the separation of the integrin legs in the presence of immobilized ligand and shear force (4). Stable linkages allow the doubly occupied integrin to undergo extensive mechanical strengthening by low forces applied on the headpiece; these activate the headpiece I-domains and can separate the β- and α-subunits from each other, further stabilizing integrin unclasping (4). (5, 6) Talin activation by external forces and dimerization of adhesive integrin bonds. These forces stretch the talin rod domain and expose sites for vinculin and actin binding (5). Since integrin ligands are generally multivalent, rapid integrin microclustering can take place (6), even prior to recruitment of distant integrins by

cell types indeed results in nearly total loss of integrin activation (Lammermann et al., 2008; Petrich et al., 2007). Partial siRNA suppression of talin1 in primary human T-lymphocytes also significantly impairs both LFA-1 and VLA-4 activation by prototypic chemokines (Manevich et al., 2007; Shamri et al., 2005). I would therefore like to propose that only when both the extracellular ligand and talin co-occupy a given leukocyte integrin, can this integrin undergo full conformational activation necessary for high adhesiveness (Fig. 3). To maximize the probability of such double occupancy, both ligands must not only acquire high association rates to their target sites on the integrin's headpiece and tails, but once bound, both of these ligand–integrin interactions should maintain low effective k_{off}. Most soluble integrin ligands fail to bind physiologically activated integrins, probably because of intrinsically high k_{off} of their bonds (Luo et al.). Nevertheless, most of these extracellular integrin ligands interact with leukocyte integrins in their immobile states. Importantly, the immobilized state of the extracellular ligand can reduce its escape from the integrin once it is released, since reactants in viscous medium are more likely to recombine than to diffuse apart (Bell, 1978). Thus, immobilization does not only promote multimerization and "avidity regulation" of integrin–ligand contacts as traditionally proposed (van Kooyk & Figdor, 2000), but is also favorable for kinetic stabilization of individual bonds. Similar to cell surface extracellular integrin ligands, the cytoplasmic ligand talin is usually also immobile, since in its native dimeric state it is tethered to actin filaments (Critchley & Gingras, 2008). The association rate of talin with the target integrin tail is locally increased by multiple cofactors [e.g., PIP2; Martel et al., 2001 (Fig. 3) and possibly Kindlins; Larjava, Plow, & Wu, 2008] directly or indirectly modulated by chemokine signals. Once bound, these cytoskeletal linkages of talin as well as electrostatic associations with the plasma membrane can minimize its escape from the integrin tail it occupies and thereby maintain low k_{off} (Fig. 3). Thus, immobilization of both the extracellular and the cytoplasmic ligands greatly facilitate the probability that a given integrin gets co-occupied by both its extracellular and cytoplasmic ligands. Furthermore, if indeed both of these ligands are anchored, the double-occupied integrin is also much more prone to undergo extensive mechanical strengthening (Puklin-Faucher & Sheetz, 2009). Indeed, both the integrin headpiece and the integrin

lateral mobility. Soluble extracellular ligands usually fail to bind integrins unless these integrins are already unclasped by high local concentration of the talin FERM domain, a condition acquired mainly by activated platelet and neutrophil integrins. Talin cleavage generates highly active FERM which can also unclasp integrins. Nevertheless, without cytoskeleton immobilization via the talin rod, force transduction and full bidirectional activation are impossible. (See Color Insert.)

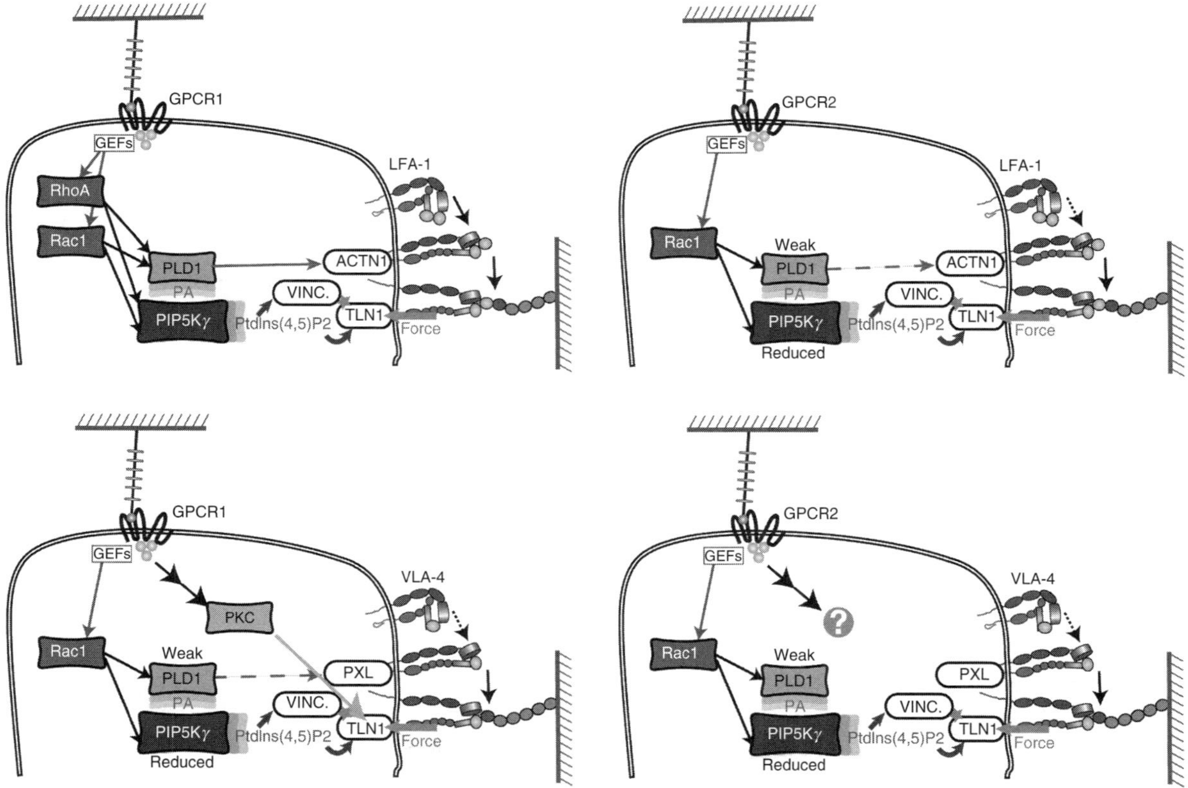

FIGURE 4 Modular Rho GTPase signals promote differential integrin affinity stimulation by distinct chemokine signals. Top left panel: three levels of LFA-1 affinity regulation are controlled by distinct Rho GTPases triggered by a high-affinity chemokine–GPCR1 interaction in T cells:

subunits appear to undergo faster opening and separation in the presence of applied forces, but do so only when integrins and their ligands are properly anchored (Alon & Dustin, 2007; Zhu et al., 2008). The ligand-occupied integrin can use talin–actin links to form a clutch to the cortical cytoskeleton with a mechanical resistance of 2 pN s (Jiang, Giannone, Critchley, Fukumoto, & Sheetz, 2003). Once binding the integrin beta subunit tail via its head FERM domain (Critchley & Gingras), talin can undergo local stretching, resulting in unmasking of vinculin-binding sites on its rod domain (del Rio et al., 2009) (Fig. 3). From the double-occupancy model proposed here (Fig. 3), it can be predicted that increasing the effective density of either the extracellular (external) ligand or the intracellular (cytoplasmic) ligand (i.e., talin) can result in similar activation outcome. Indeed, in the absence of chemokine signals, lymphocyte integrins can readily undergo outside-in activation by increasing densities of ICAM-1 or VCAM-1, especially in the presence of external forces (Feigelson & Alon, in preparation) and this ability is dramatically attenuated when talin expression is reduced (Manevich et al.; Shamri et al.).

In some physiological settings, such as platelet and leukocyte aggregation, inside-out activated integrins on these cells must bind soluble ligands with high affinity (Hynes, 2002). Fibrinogen is a soluble extracellular ligand for the platelet GpIIbβ_3 whose effective dissociation rate from the integrin can be minimized by oligomerization rather than by immobilization. The integrin–ligand complex in this case cannot undergo mechanical activation, since forces cannot be exerted on integrins occupied by soluble ligands.

(1) PLD1 activation enhances α-actinin (ACTN1) binding to and stabilization of LFA-1 in an extended intermediate-affinity state; (2) PLD1 produces PA, which stimulates the enzymatic activity of the talin1 (TLN1)-associated PIP5Kγ which produces local PIP2; and (3) PIP2 generation increases talin binding to integrin β-subunits, thereby promoting both inside-out and outside-in stabilization of the unclasped integrin heterodimer. Force transduction via talin and PIP2 can both activate vinculin (VINC) binding to talin and crosslinking of the complex. PIP5Kγ phosphorylation by Src and FAK (Ling et al., 2003) can enhance its binding to talin following leukocyte arrest. Top right panel: a suboptimal signal by a low-affinity chemokine–GPCR2 interaction fails to trigger RhoA and PLD1 but can still trigger Rac1 and PIP5Kγ at slightly reduced levels than in the first example. Consequently, the level of talin–integrin associations is low and LFA-1 conformational switch is not observed in the absence of extracellular ligand and applied forces. Bottom left panel: the high-affinity chemokine–GPCR1 pair assembles different GTPase and PKC effectors in the vicinity of VLA-4. Consequently, the level of RhoA activation proximal to VLA-4 is diminished and so is the amount of activated α-actinin and talin1 recruited at sites of VLA-4. Partial activation of PIP5Kγ and PKC control talin activation near ligand-occupied VLA-4. Bottom right panel: a suboptimal signal by a low-affinity chemokine–GPCR2 interaction fails to trigger PKC but can still trigger at low efficiency Rac1, PIP5Kγ, and yet unidentified effectors. Paxillin (PXL) associated selectively with the α_4-subunit tail may recruit these effectors to the vicinity of VLA-4 rather than to LFA-1. (See Color Insert.)

To bind fibrinogen with high affinity, GpIIbβ_3 may need to associate at high rates with soluble talin FERM fragments (e.g., calpain cleavage products; Yan, Calderwood, Yaspan, & Ginsberg, 2001). The extent of mechanical stabilization of this integrin–ligand complex is therefore predicted to be minimal since for mechanical stabilization both integrin ligands would need to be anchored to a viscous support. In analogy to platelet integrins, small subsets of the neutrophil and monocyte integrin Mac-1 and of chemokine activated LFA-1 can bind soluble ligands with high affinity (Constantin et al., 2000; Diamond & Springer, 1993). As with GpIIbβ_3 the low dissociation rates of these integrins from their cognate ligands could be controlled by ligand oligomerization rather than immobilization and the extent of mechanical activation of these integrin–ligand complexes would therefore be minimal unless ligands become anchored.

C. Integrin Activation by Endothelial Chemokines: A Local and Instantaneous Process

With the exception of subsets of effector/activated lymphocytes and monocytes, circulating leukocytes maintain their vascular integrins in generally nonadhesive states. This allows leukocytes to avoid nonspecific adhesion (clumping) to other circulating leukocytes and platelets, which constitutively express integrin ligands (del Pozo et al., 1998), as well as to vascular beds that express constitutively high levels of integrin ligands such as VCAM-1 and ICAM-1 (e.g., BM, lung capillaries; Galkina et al., 2005). Leukocytes arrest on endothelial integrin ligands after variable periods of selectin-mediated rolling (in the 1–100 s range), but only once their integrins have been properly activated *in situ* (Alon & Dustin, 2007; Laudanna & Alon, 2006). Whereas lymphocyte and monocyte arrest is generally triggered by a chemokine signal near or at the arrest site (Ley, 2003), neutrophils can accumulate integrin-activating signals as they roll along endothelial selectins until they encounter a strong integrin-activating chemoattractant signal and arrest on the proper endothelial IgSF ligand (Fig. 1) (Zarbock, Lowell, et al., 2007) (see Section V.A for more details).

Accumulating evidence over the past decade suggests that different integrins coexpressed on a given leukocyte can modulate their affinity and avidity via distinct mechanisms even when triggered by the same GPCR (Figs. 2 and 4). The activation of a given integrin can also vary considerably among different cell types (Franitza et al., 2004). Integrin activation by chemoattractants/chemokines is a transient process (Campbell, Hedrick, Zlotnik, Siani, & Thompson, 1998; Shimaoka et al., 2006) and must be fine tuned after arrest to allow the leukocyte to migrate from the site of arrest to the site of diapedesis (Shulman et al., 2009). Thus, immune cell integrins can rapidly adapt their adhesive behavior at specific endothelial sites within target tissues according to tissue- and context-restricted patterns of chemokine or chemoattractant expression at these sites. This diversity, the extremely rapid kinetics of integrin activation by chemokine signals, and the tight regulation of such adhesion by shear forces has required the introduction of multiple dynamic assays that simulate the various stages of leukocyte–endothelial contacts; such assays can be complemented by reductionist approaches that can dissect individual adhesive and signaling components, alone and in combination. *In vitro* studies based on flow chamber setups have indeed been useful in addressing these questions in a variety of genetically and biochemically manipulated populations of leukocytes (Alon & Feigelson, 2009; Campbell, Hedrick, et al.; Lawrence & Springer, 1991).

The first *in vitro* evidence for the crucial involvement of chemoattractants and their leukocyte GPCRs in triggering integrin adhesiveness to vascular endothelial ligands was provided by Lawrence and Springer (1991). Using a standard parallel plate flow chamber setup and real-time tracking, this group visualized neutrophils rolling on P-selectin reconstituted with ICAM-1 in a lipid bilayer, and showed that rapid β_2-integrin-mediated arrests on this endothelial like surface are stimulated *in situ* upon introduction of soluble fMLP to the rolling leukocytes (Lawrence & Springer). A similar role for a chemoattractants in lymphocyte arrest on vascular integrin ligands was demonstrated by Bargatze and Butcher (1993), *in vivo*. In a series of subsequent *in vitro* studies, both lymphocyte α_4- and β_2-integrins were shown to undergo rapid activation by immobilized chemokines and their cognate GPCRs (Grabovsky et al., 2000; Pachynski, Wu, Gunn, & Erle, 1998; Peled et al., 1999). The suggestions that both integrins undergo activation by chemokine signals at the original capture site came from an observation that a subsecond long encounter of a tethered T cell with immobilized CXCL12 and VCAM-1 was sufficient to initiate a stimulatory signal, without any prior adhesive interactions or encounter of chemokine signals upstream to the arrest site (Grabovsky et al.). The chemokine signal augmented by up to 50-fold the strength of the VLA-4–VCAM-1 interaction, yet only when the immobilized CXCL12 was encountered together with VCAM-1 (Grabovsky et al.). Likewise, LFA-1 priming by soluble chemokine signals encountered during the rolling phase upstream of the arrest site can transiently enhance LFA-1 affinity, but is insufficient to activate firm LFA-1 adhesiveness (Shamri et al., 2005). We have referred to this mode of chemokine signaling in lymphocytes as 2D-GPCR signaling, since the rapid integrin stimulatory GPCR signals are confined to a 2D interface (Cinamon et al., 2001).

Dissecting this transition in a flow chamber using mAb reporters for distinct LFA-1 conformations, we found that immobilized chemokines could trigger, within a fraction of a second, an extended LFA-1 conformation without inducing the full high-affinity integrin form (Shamri et al., 2005). Our studies also suggested that this subsecond chemokine signaling requires proximity of the chemokine to the integrin ligand at the adhesive contact, since only in the presence of the LFA-1 ligand can the extended LFA-1 undergo immediate stabilization and acquire the full high-affinity state critical for arrest and firm adhesiveness (Fig. 2). A powerful approach to elucidate the basis for this extremely efficient activation under shear flow has been to measure the kinetics of individual reversible tethers mediated by integrin/ligand interactions stimulated *in situ* by surface-bound chemokines at (transient) adhesive contacts (Alon, Hammer, & Springer, 1995). These transient tethers are the smallest units of adhesive interaction observable in shear flow. Using high speed camera measurements with temporal resolution of 4 ms, we found that in the absence of chemokines and selectins, T cells fail to form integrin-mediated adhesive bonds lasting more that 4 ms. Thus, *de novo* chemokine-stimulated integrin bonds must be triggered within this minimal time frame (V. Grabovsky, E. Manevich, & R. Alon, unpublished results). Although no selectins are required for immobilized chemokines to stimulate either α_4- or β_2-integrin adhesiveness, in the presence of endothelial selectins, the maximal shear rate levels permissive for VLA-4 or LFA-1 activation by these chemokines are significantly elevated (Grabovsky et al., 2000). Selectins are not only critical for initial cell capture on the endothelium but can increase both the closeness and the duration of contact between the otherwise sterically hindered GPCRs and their cognate chemokines on the countersurface.

Recent results suggest that this extremely efficient and rapid activation of integrin adhesiveness takes place more readily in the presence of a critical threshold of shear force (Woolf et al., 2007). Thus, immobilized chemokines were found to strongly activate lymphocyte integrin adhesiveness under shear force, while failing to stimulate similar integrin adhesiveness in extravascular shear-free environments (Woolf et al.). These results collectively suggest that rapid integrin activation by immobilized chemokines presented on the endothelium is a mechano-regulated process. Recent structural evidence supports this notion, showing that the α I-domain of LFA-1 (Fig. 2, bottom) changes its conformation and increases ligand-binding affinity under applied shear forces (Astrof, Salas, Shimaoka, Chen, & Springer, 2006). This dependence of chemokine-stimulated integrins on external forces for maximal activation can also explain why these integrins are readily activated by surface-immobilized chemokines. Such chemokines, but not their soluble counterparts, can localize nearby the target integrin cytoskeletal effectors which can confer ligand-occupied integrins with the ability to immediately respond to and undergo activation by external forces (Fig. 3).

IV. SIGNALING EVENTS IN RAPID INTEGRIN ACTIVATION BY GPCRs

A. *Preformed Signalosomes for Immobilized Chemokine Signaling to Integrins*

The complexity of the signaling mechanisms that orchestrate rapid integrin activation by *in situ* chemokine signals has begun to unfold (Laudanna & Alon, 2006). Several chemokine signaling networks assumed to initiate integrin activation by chemokines may involve an array of several key integrin cytoskeletal linkers, which regulate affinity modulation by both inside-out and outside-in signals (Alon & Dustin, 2007; Bolomini-Vittori et al., 2009; Laudanna & Alon; Ley et al., 2007) (Figs. 2–4). Since 2D chemokine signaling to integrin can take place during a fraction of a second, it was predicted that chemokine-occupied GPCRs and their associated Gi proteins must be able to organize this array within milliseconds (Laudanna & Alon). Integrin activation would therefore vary with the time frame of cell's encounter with the adhesive surface, the magnitude of the GPCR-transduced signal, which is related to the relative affinity between the chemokine and its cognate GPCR (D'Ambrosio et al., 2002), and the density of available chemokine. Some lateral diffusion of the activating GPCR may be also necessary to increase the association of chemokine within the adhesive contact. Thus, below a threshold of GPCR occupancy, chemokines can trigger only intermediate strength VLA-4–VCAM-1 interactions, which mediate rolling rather than firm arrests (Grabovsky et al., 2000). Interestingly, when a strong chemokine is present at high densities, but VCAM-1 is limiting, chemokine signals can trigger persistent rolling through intermediate strength VLA-4–VCAM-1 interactions. As none of the integrin-activating chemokines tested to date exhibit intrinsic adhesive activities on their own, these results suggest that *in situ* chemokine-stimulated VLA-4 is rapidly and reversibly stabilized in two major adhesive states that share a high association rate to immobilized VCAM-1, but differ in their dissociation rates; the first supports weak rolling adhesions, due to its fast off time, while the second supports immediate arrests, and requires both strong chemokine signals and a second signal from the integrin ligand to drive outside-in activation (Fig. 2).

B. Microvillar Compartments for GPCR Signalosomes

In light of the short time frames during which chemokine signals can be translated into integrin activation events, it is likely that preformed and previously compartmentalized protein networks exist in leukocytes encountering endothelial chemokines. These networks are most probably organized within microvilli (Fig. 2), the initial cell projections predicted to encounter endothelial presented chemokines (Laudanna & Alon, 2006). Indeed, prototypic GPCRs are constitutively clustered on lymphocyte microvilli (Singer et al., 2001). These projections are often enriched with particular integrin subsets ($\alpha_4\beta_7$, VLA-4) as well as with L-selectin and P-selectin ligands (Berlin et al., 1995; Moore et al., 1995). Prominent chemokines can also be presented on endothelial microvilli (Middleton et al., 1997). Given the importance of these projections for rapid leukocyte endothelial encounters under physiological shear flow conditions (von Andrian, Hasslen, Nelson, Erlandsen, & Butcher, 1995), integrin-/GPCR/Gi-effector assemblies within microvillar compartments may be particularly prone to chemokine activation at leukocyte endothelial contacts. Microvilli may also readily penetrate though the endothelial glycocalyx, a highly heterogeneous barrier which contains numerous glycosaminoglycan (GAG) sites for chemokine presentation (Johnson, Proudfoot, & Handel, 2005). Furthermore, the selectin-mediated rolling interactions may facilitate flattening of the leukocyte microvilli and thereby increase the accessibility of microvilli-associated GPCR assemblies and possibly of their target integrins to encounter their respective ligands on the endothelial surface (Fig. 2). Microvillar components may also actively regulate the formation of such GPCR/Gi-effector networks, since they are highly enriched in actin linker proteins (e.g., ERM family members; Bretscher, 1999), and large adaptors like supervillin (Chen et al., 2003) which may maintain the signaling GPCR assemblies in close proximity to their target integrins (Fig. 2). Interestingly, chemokine signals can dephosphorylate ERM proteins and thereby rapidly remodel rigidity of leukocyte microvilli (Brown et al., 2003).

Another interesting property of microvilli is that they can reduce the tension applied to individual GPCRs and ligand-occupied integrins and thereby allow gradual force loading by these bonds (Evans & Calderwood, 2007). Since integrin bonds rapidly dissociate in the presence of abruptly applied forces (de Chateau, Chen, Salas, & Springer, 2001), microvilli may enable such gradual force loading and thereby optimize integrin activation by both chemokine inside-out and integrin–ligand outside-in activation. Dispersal of shear forces on selectin bonds will also reduce the effective dissociation forces transduced through individual chemokine–GPCR bonds, and thereby facilitate a critical contact duration to enable Gi activation. This duration may be also positively controlled by the anchorage states of both the chemokine and its cognate GPCR. While chemokine-occupied GPCRs on their own cannot support adhesive interactions, it is still possible that endothelial anchored chemokines, retained via tail residues interacting at high affinity with specific endothelial GAGs (Johnson et al., 2005) may be superior to loosely immobilized chemokines in their ability to occupy leukocyte GPCRs, in analogy to the superior ability of immobilized integrin ligands to bind and activate immobilized integrins (Fig. 3). Chemokine–GPCR pairs with higher anchorage properties may also signal differently when experiencing shear forces (Vogel & Sheetz, 2006). Ligation of prototypic GPCRs such as CXCR4 brings a major fraction of these receptors to the detergent insoluble cytoskeleton (M. Sokolovsky & R. Alon, unpublished results). Thus, both chemokine and GPCR anchorage states may determine the capacity of a given chemokine–GPCR pair to rapidly trigger an integrin-activating Gi signal under shear flow. A major experimental challenge is thus to determine whether these and other subsets of chemokine receptors that are specialized in transmitting rapid integrin-activating signals under shear flow (see Section V.A) are not only topographically positioned nearby target integrins, but share specialized mechanical properties which facilitate their clustering and signaling. If so, microvillar GPCRs for endothelial chemokines may constitute a specialized subset of Gi-associated receptors adapted to trigger specific integrin-activating modules within specific integrin assemblies (Figs. 2 and 4).

Integrin-activating GPCRs could also operate within distinct plasma membrane platforms (Manes et al., 2001). Removal of cholesterol, a key stabilizer of lipid rafts, from the surface of lymphocytes retains spontaneous α_4-integrin function under flow, but completely abrogates CXCR4 and CCR7-triggering of both VLA-4 and of $\alpha_4\beta_7$-integrin avidity to their respective ligands. Identical treatment, however, retains, CXCR4, CCR7, or CXCR3 signaling to the LFA-1 integrin (Shamri et al., 2002) revealing high heterogeneity between lipid platforms of VLA-4 and LFA-1 activation by GPCRs. These results also suggest that individual GPCRs can be segregated in distinct membranal microdomains. As α_4-integrins are generally enriched on the tips and body of microvilli, whereas LFA-1 is localized to the base of microvilli (Abitorabi, Pachynski, Ferrando, Tidswell, & Erle, 1997), an attractive possibility is that GPCR–Gi complexes within or nearby lipid rafts are preferentially enriched within microvilli, whereas distinct subsets of GPCR–Gi complexes excluded from rafts are also excluded from microvilli. Proteomic analysis and differential immunogold labeling using EM microscopy will be required to confirm this interesting possibility.

C. GTPases as Key Effectors of GPCR Signaling to Leukocyte Integrins

Small GTP-binding proteins, GTPases, are critical for the ability of chemokine-triggered Gi signals to activate integrin adhesiveness (Bolomini-Vittori et al., 2009; Constantin et al., 2000; Shimonaka et al., 2003; Zhang et al., 2006). The RhoA GTPase was the first GTP-binding protein implicated in rapid chemokine-mediated activation of lymphocyte and neutrophil integrins (Laudanna, Campbell, & Butcher, 1996). As distinct regions of RhoA control either or both integrin affinity and clustering (Giagulli et al., 2004), this GTPase is ideally suited to bridge rapid chemokine signaling with both conformational integrin activation and the subsequent generation of high avidity and adhesion strengthening (Laudanna & Alon, 2006). Recent results implicate RhoA as well as Rac1, another major Rho family GTPase (Ridley, 2000), in chemokine-stimulated conformational switches of LFA-1, from a bent inactive to extended high-affinity state in T cells (Bolomini-Vittori et al.; Pasvolsky et al., 2008). Rac GTPases also control integrin clustering (Nolz et al., 2007), placing these GTPases as dual regulators of integrin avidity modulation by chemokine signals. A key Rac activator downstream of chemokine signals encountered by T cells is the Rac guanine exchange factor (GEF), DOCK2 (Fukui et al., 2001). Interestingly, deletion of this GEF in murine T cells does not impair triggering of high-affinity LFA-1 by chemokine signals and does not affect either rapid LFA-1 adhesiveness or subsequent adhesion strengthening (Shulman et al., 2006), whereas deletion of DOCK2 in B lymphocytes or in human T cells impairs subset of these LFA-1 properties (Garcia-Bernal et al., 2006; Nombela-Arrieta et al., 2004). Thus, distinct Rac GEFs may be involved in rapid Rac- and Rho-mediated LFA-1 activation by chemokines in different cellular environments. This raises the intriguing possibility that only small subsets of either RhoA or Rac1, which are constitutively associated with the plasma membrane, are involved in rapid integrin activation by chemokine signals. Other and probably larger pools of these GTPases do not participate in rapid integrin activation because they are cytosolic and may need to use cytosolic GEFs for the slower chemokine-stimulated actin remodeling events underlying lamellipodia formation and contractility (Ridley et al., 2003).

A third major integrin-activating GTPase in lymphocytes and probably in other leukocytes is Rap1 (Katagiri et al., 2004; Kinashi, 2005; Shimonaka et al., 2003). Rap1 controls both affinity triggering and clustering of LFA-1 at various cellular contacts including the immune synapse (Kinashi & Katagiri, 2004), yet the precise involvement of Rap1 signals in the earliest chemokine-triggered integrin activation events remains unclear. Activated Rap1 can bind multiple integrin regulatory effectors and recruit these effectors to the plasma membrane at sites of integrin occupancy by ligands. These effectors include the talin recruiting molecule, RIAM (Han et al., 2006; Watanabe et al., 2008) and the LFA-1-associated molecule, RAPL (Katagiri et al.). However, these effectors are more likely to be involved in post-arrest adhesion strengthening events necessary for leukocyte crawling away from the arrest site rather than in the earliest integrin activation by the most upstream adherent chemokine signals. Interestingly, transfection of primary lymphocytes with Rap1 dominant negative mutants or overexpression of the Rap1 GTPase-activating protein (GAP), SPA-1, attenuates rapid LFA-1 but not VLA-4-mediated adhesions induced by chemokines under shear flow (Ghandour, Cullere, Alvarez, Luscinskas, & Mayadas, 2007). However, in myeloid cells, Rap1 is critical for both β_2 and VLA-4 stimulation by chemokine signals (Bergmeier et al., 2007; Caron, Self, & Hall, 2000; Molteni et al., 2009), suggesting that this GTPase operates in distinct cell type-specific signalsomes. In myeloid cells, Rap1 activation may be triggered not only by chemokine signals but also by earlier selectin–ligand occupancy signals via a rise in intracellular Ca^{2+} and DAG, which can coactivate the key Rap1 GEF, CDG-I (Bergmeier et al.; Ghandour et al.).

D. Other Signaling Pathways Linking Rapid Chemokine Signals to Integrin Activation

Chemokines trigger a variety of additional machineries (discussed in excellent reviews by Kinashi, 2005; Ley et al., 2007) that can potentially regulate integrin valency rather than affinity. These pathways can either positively or negatively modulate integrin adhesiveness and control the high turnover of integrin-mediated focal contacts underlying leukocyte crawling and subsequent transendothelial migration (Shulman et al., 2009). Such chemokine-stimulated effectors include the atypical protein kinase C isoform, PKCζ, involved in β_2-integrin activation in multiple types of leukocytes (Giagulli et al., 2004; Laudanna, Mochly-Rosen, Liron, Constantin, & Butcher, 1998), the DAG-dependent PKCs, involved in VLA-4 activation in lymphocytes (Ghandour et al., 2007), and the Rac GEFs, VAV1–3 (Gakidis et al., 2004). Multiple lipid kinase phosphatidylinositol 3-OH kinases (PI(3)Ks) have been also implicated in rapid integrin activation by chemokines in both lymphocytes and monocytes (Constantin et al., 2000; Gerszten et al., 2001). In the context of LFA-1-mediated adhesion, general inhibition of these PI(3)Ks prevents chemokine-induced lymphocyte binding to diluted ICAM-1, but not chemokine triggering of LFA-1 affinity (Giagulli et al.) implicating these kinases in rapid clustering of LFA-1, downstream to initial affinity triggering of LFA-1. PI(3)K activity is not implicated, however, in chemokine stimulation of VLA-4 avidity in T cells (Grabovsky et al., 2000), consistent with the existence of specialized activation pathways for this integrin, distinct from machineries implicated in chemokine activation of LFA-1 and possibly other β_2-integrins.

E. Lipid Targets of Chemokine-Stimulated Rho GTPases

Talin associations with integrin tails can be regulated by at least two mechanisms. First, PI(4,5)P2 (PIP2) binding to the talin FERM domain directly activates this domain and reinforces its association with the NPXY motif on the β-integrin tail (Martel et al., 2001). In addition, phosphorylation of ser/thr residues on the talin rod or tail domain may also activate talin (Campbell & Ginsberg, 2004), perhaps by increasing its association with adaptors and actin.

PI(4,5)P2 is one of the key regulatory lipids generated and modified *in situ* in the plasma membrane (Itoh & Takenawa, 2002; Tall, Spector, Pentyala, Bitter, & Rebecchi, 2000). PI(4,5)P2 generation can be a critical limiting step in local talin recruitment to and activation of integrins (Nayal, Webb, & Horwitz, 2004). PI(4,5)P2 generation can be stimulated by RhoA and Rac GTPases in multiple types of cells (Anderson, Boronenkov, Doughman, Kunz, & Loijens, 1999; Matsui, Yonemura, & Tsukita, 1999), and a variety of type I PIP5 kinases have been implicated in PI(4,5)P2 generation in mammalian cells. The type I PIP5 kinases include a three member family (PI5KIα, β, and γ) that converts PtdIns(4)P into PtdIns(4,5) P2 (Toker, 1998). Recent studies implicate both PLD1 and PIP5Kγ in an integrated signaling complex that controls the transition of LFA-1 to high affinity in T cells (Bolomini-Vittori et al., 2009) (Fig. 4). PIP5Kγ is thought to be triggered by chemokine-stimulated RhoA, Rac1, and their downstream effector, PLD1 (Figs. 2 and 3). As PIP5Kγ is directly activated by PA, a product of PLD1, this membranal PLD isoform was proposed to link chemokine-stimulated RhoA and Rac1 to increased PIP5Kγ activity. Indeed, interfering with RhoA or Rac1 binding to PLD1 abolishes both the enzymatic activity of PLD1 in chemokine-stimulated lymphocytes and the triggering of intermediate- and high-affinity LFA-1 states (Bolomini-Vittori et al.). However, silencing the expression of PIP5Kγ but not of RhoA, Rac1 or PLD1 abrogates solely the chemokine-mediated induction of the high-affinity LFA-1 state (Bolomini-Vittori et al.). PLD1 may therefore coactivate LFA-1 independently of its PIP5Kγ-stimulating activity, possibly via its interaction with α-actinin, recently implicated in the induction of intermediate-affinity LFA-1 (Stanley et al., 2008).

Among various alternatively spliced PIP5Kγ isoforms, PIP5Kγ661 (Ling, Doughman, Firestone, Bunce, & Anderson, 2002) may be the predominant isoform implicated in rapid chemokine stimulation of leukocyte integrins. This isoform may be recruited to the plasma membrane via its interaction with the talin FERM (Di Paolo et al., 2002). Talin FERM occupied by PIP5Kγ can no longer bind to and activate integrins (Smith et al., 2005). Since talin is an antiparallel homodimer, the FERM domain of a talin dimer, associated with PIP5Kγ via its other FERM, may be readily available for binding to and activation of integrins (Fig. 3). Once a CAM–integrin–talin FERM complex is formed, the second talin FERM, if not already complexed with PIP5Kγ may recruit an additional copy of PIP5Kγ to the vicinity of the ligand-occupied integrin (Fig. 3). *In situ* generated PIP2 can then activate more talin FERM domains and coactivate the talin–actin linker, vinculin (Critchley, 2000). PIP2 can also recruit additional PLD molecules to the plasma membrane (Hammond et al., 1997), activate ERMs, and strengthen the overall attachments between the plasma membrane and the actin cytoskeleton (Raucher et al., 2000). Notably, local increases in PIP2 levels are rapidly offset by specific PLCs which can be activated either by chemokine-stimulated G proteins (e.g., PLCβ isoforms; Li et al., 2000) or by integrin occupancy (e.g., PLCγ1; Kanner, Grosmaire, Ledbetter, & Damle, 1993). Since these enzymes produce DAG and IP_3 (Rhee, 2001) which coactivate both Rap1 and a variety of PKC isoforms (Spitaler & Cantrell, 2004), PIP2 hydrolysis by PLC may trigger a spectrum of integrin activation machineries (Figs. 2 and 4). PIP2 hydrolysis products may, however, target different effectors in different cell types. Whereas in monocytes, VLA-4 activation by chemokines involves PLC-mediated Ca^{2+} mobilization and calmodulin activation (Hyduk et al., 2007), in T cells, VLA-4 activation by chemokines, is PLC and PKC dependent but independent of intracellular Ca^{2+} (Ghandour et al., 2007; Grabovsky et al., 2000).

V. MEMBRANAL PLATFORMS FOR INTEGRIN ACTIVATION BY CHEMOKINE SIGNALS

A. Specialization of Chemokine Signalosomes for Distinct Integrins

Our recent results in human T-lymphocytes indicate that weak chemokine signals fail to transduce LFA-1 extension in the absence of outside-in activation signals. These signals stimulate, however, rapid LFA-1 adhesiveness and do so independently of RhoA signaling (Pasvolsky et al., 2008) (Fig. 4). This suggests that distinct chemokine–GPCR pairs may trigger different sets of GTPases and their immediate integrin-activating effectors, depending on the relative strength, abundance and subcellular location of the signaling GPCR. It is still unclear whether one or more of the Rho/Rac GTPases also controls avidity modulation of VLA-4 and $\alpha_4\beta_7$-integrins by distinct chemokine–GPCR pairs. A most attractive possibility is that distinct integrins in distinct leukocyte subtypes may use specific signaling modules which differ in several Gi-triggered machineries but share the common downstream elements, possibly talin1 and Kindlin-3 (Moser et al., 2009; Shamri et al., 2005). Along these lines, recent results from our lab and from other groups suggest that interference with either Rho or Rap1 GTPase activities that inhibits LFA-1 activation by chemokines does not perturb VLA-4 activation by the same chemokines (Ghandour et al., 2007; Pasvolsky, and Alon, unpublished results). These results are consistent with the modern view of signal transduction networks as "modular dynamic systems" (Hartwell, Hopfield, Leibler, & Murray, 1999; Ravasz, Somera, Mongru, Oltvai, & Barabasi, 2002; Rives & Galitski, 2003). Gi-triggered integrin-activating effector systems may take part in submembranal modules, which are probably a subset of intracellular signaling networks triggered by chemokine receptors.

Each one of the Rho, Rac, and Rap GTPases responding to chemokine-triggered Gi signals can exist in either submembranal or cytosolic pools, and each pool has multiple GEFs with distinct locations in different compartments and intracellular compartments. It is therefore possible that certain pools of GTPase–GEF pairs, which preexist on or nearby the plasma membrane, participate in rapid integrin activation, while the nonmembranal GTPase–GEF complexes are activated by arrested leukocytes and promote postarrest spreading, polarization, crawling, and diapedesis.

Since one modular signaling system can preferentially activate a particular subset of integrins (Fig. 4), it is likely that distinct assemblies of GTPases reside in different membranal microdomains nearby their target integrins, and together with specific combinations of integrin adaptors, give rise to GPCR–integrin signalosomes.

Likely candidates to organize such GPCR–integrin signalosomes are filamins, large multifunctional adaptors which bind to and regulate the function of both GPCRs and integrins (Huang, Wu, Hujer, & Miller, 2006; Takala et al., 2008). Other potential scaffolds are β-arrestins, which not only regulate GPCR phosphorylation and internalization, but can also bind and possibly cluster some GPCRs even prior to chemokine binding (Luttrell & Lefkowitz, 2002). A recent study in myeloid leukocytes suggests that both β-arrestin-1 and -2 are required for rapid activation of integrins by 2D chemokine signals (Molteni et al., 2009). Notably, β-arrestin-2 participates in a subset of GPCR-triggered signaling events that activate Rap1 and promote adhesion strengthening following leukocyte arrest. It is likely that different types of leukocytes use distinct β-arrestin members for information transfer between particular GPCRs and integrins, on a subsecond time scale.

B. Specialization of Chemokine/GPCR Pairs in Stimulating Integrin Adhesiveness Under Shear Flow

Particular GPCRs, although abundantly expressed and functional in cell motility, can fail to stimulate integrin adhesiveness in specific cell types. For instance, CCR2 ligands known to stimulate integrin adhesion in eosinophils or monocytes, respectively (Gerszten et al., 1999; Weber, Alon, Moser, & Springer, 1996), fail to stimulate VLA-4 in human PBL subsets expressing this GPCR (Grabovsky et al., 2000). Weber et al. (2001) also demonstrated that a single chemokine, CCL5, transduces rapid integrin avidity stimulatory signals to monocytes and effector T cells exclusively through the CCL5 receptor, CCR1, even when another functional CCL5 receptor, CCR5, is coexpressed by the responding leukocytes. Similarly, the CCR5 ligands CCL3 and CCL4 fail to stimulate VLA-4 or LFA-1 integrins on PBL, although they can still trigger chemotaxis of these cells (Campbell, Bowman, et al., 1998; Grabovsky et al.). Thus, not all signaling chemokine–GPCR pairs, which mediate leukocyte spreading and motility, are capable of triggering integrin stimulation at adhesive contacts under shear flow (Huo et al., 2001; Ley, 2003; Weber, von Hundelshausen, Clark-Lewis, Weber, & Weber, 1999; Weber et al., 2001). As discussed in Sections III.B and IV.A, this may reflect possible differences in preformed Gi protein assemblies of GPCR subsets specializing in rapid integrin activation. GPCRs can associate not only with different heterotrimeric G proteins but also with different effectors (Amatruda, Gerard, Gerard, & Simon, 1993; Arai & Charo, 1996; Haribabu et al., 1999; Kuang, Wu, Jiang, & Wu, 1996; Loike et al., 1999; Xu et al., 2003). GPCRs and chemokines may assemble in distinct preformed or ligand-induced homo- or heterodimers with distinct signaling properties (Mellado et al., 2001). For example, CXCL4 can amplify CCL5-triggered arrest of monocytes (von Hundelshausen et al., 2005) possibly by ligating their cognate GPCRs. Each of these GPCRs appears to only weakly trigger integrin adhesiveness when occupied on its own, but upon heterodimerization, the GPCR assembly may acquire potent integrin stimulatory activities. Another example of specialization of GPCR subsets capable of stimulating integrin adhesiveness was recently provided in T cells. The nonsignaling CXCL12 receptor, CXCR7, was postulated to associate with a subset of CXCR4 involved in rapid integrin activation by their common ligand, CXCL12, while other pools of CXCR4, implicated in lymphocyte chemotaxis and ERK activation are not associated with this nonsignaling GPCR (Hartmann et al., 2008).

VI. PRIMING OF INTEGRINS TO CHEMOKINE SIGNALING IN ROLLING LEUKOCYTES

A. A Stepwise Model for Integrin Activation by Endothelial Selectin Ligands in Myeloid Cells

Leukocyte rolling on P- and E-selectins involves adhesive interactions that are several fold longer than leukocyte rolling mediated exclusively by L-selectin (Kunkel, Dunne, & Ley, 2000). Specific glycoprotein ligands for P- and E-selectins expressed by subsets of myeloid cells have been suggested to activate integrin subsets on slowly rolling neutrophils (Atarashi, Hirata, Matsumoto, Kanemitsu, & Miyasaka, 2005; Green, Pearson, Camphausen, Staunton, & Simon, 2004; Smith, Olson, & Ley, 2004). Accumulating data suggest that the tyrosine kinase Syk is activated by the P- and E-selectin ligand, PSGL-1 (Urzainqui et al., 2002) and that PSGL-1 engagement in neutrophils rolling on E-selectin switches inactive LFA-1 into an intermediate-affinity form (Zarbock, Lowell, et al., 2007). This LFA-1 subset participates in rolling adhesions on ICAM-1, which can slow down neutrophils rolling on E-selectin both *in vitro* and *in vivo* (Zarbock, Lowell, et al.). E-selectin engagements of both PSGL-1 and L-selectin can also trigger p38 MAPK and elevate intracellular free Ca^{2+} in a stepwise manner in the rolling neutrophils (Kunkel et al.), further suggesting that integrin activation is the result of a stepwise and global rise in multiple secondary messengers (Fig. 1) that activate β_2-integrin affinity and avidity, preferably under shear forces (Simon, Hu, Vestweber, & Smith, 2000). Shear-induced stretching in general, and particularly of microvilli (Sawada et al., 2001; Zwartz et al., 2004) may trigger both Rap1 and classical PKC isoforms implicated in the stimulation of integrin avidity (Beals, Edwards, Gottschalk, Kuijpers, & Staunton, 2001; Kucik, Dustin, Miller, & Brown, 1996). Notably, Syk-, MAPK DAG-, and Ca^{2+}-stimulated effectors triggered by endothelial selectin signals independently of chemokine signals have not been observed in T-lymphocytes, even in subsets that can slowly roll on E- or P-selectins (Z. Shulman, S. Feigelson, & F. Alon, unpublished results). Moreover, it is also unclear whether the neutrophil integrins activated by one or more of these secondary messengers and effectors are more prone to subsequent activation by chemokines, since slower rolling facilitates leukocyte encounter of endothelial chemokines and other endothelial cytokines (Hafezi-Moghadam, Thomas, Prorock, Huo, &

Ley, 2001; Jung, Norman, Scharffetter-Kochanek, Beaudet, & Ley, 1998). Notably, some selectin-triggered signals may antagonize rather than prime integrins for activation by chemokine signals. For instance, premature PKC activation by a selectin-mediated signal will reduce the ability of a GPCR to respond to a coencountered chemokine signal (Shamri et al., 2005).

B. Role of Serine/Threonine and Tyrosine Phosphorylation of Integrin Tails in Priming Integrins for Activation by Chemokine Signals

The ability of integrins to integrate *in situ* chemokine signals can be tightly regulated by their basal affinity states (Feigelson et al., 2001). For instance, VLA-4 in T cells exists in various intermediate-affinity states regulated by the Src kinase, Lck, possibly via its downstream substrates, PLCγ1, PKC, and VAV1 (Garcia-Bernal et al., 2005). Interestingly, higher affinity VLA-4 subsets undergo far more efficient *in situ* stimulation by a given chemokine signal, even though Lck is not required for GPCR activity (Feigelson et al.). Likewise, basal LFA-1 affinity states can dictate the ability of this integrin to undergo activation by chemokines and ICAM-1. For instance, phosphorylation of Ser1140 on the α_L-tail of LFA-1 occurs spontaneously on nearly 40% of the total LFA-1 expressed by resting T-lymphocytes (Fagerholm, Hilden, Nurmi, & Gahmberg, 2005). This posttranslational modification determines if LFA-1 can undergo both inside-out and outside-in conformational changes induced by chemokine and ligand signals (Fagerholm et al.). In contrast, tyrosine phosphorylation within the membrane proximal NPXY motif of VLA-4 may negatively regulate talin binding to this integrin (Legate & Fassler, 2009) and thereby reduce VLA-4 responsiveness to chemokine signals. Interestingly, high stoichiometry of serine phosphorylation is found also on ser 988 of the α_4-tail in resting lymphocytes (Han et al., 2001) but this modification inhibits paxillin binding to VLA-4 and thereby may suppress rather than prime the integrin for chemokine activation signals (Nishiya, Kiosses, Han, & Ginsberg, 2005). In addition to α-subunit ser/thr phosphorylation, β_2-ser/thr phosphorylation events (e.g., in the filamin-binding region of the β_2-tail) can determine the responsiveness of different β_2-integrins (e.g., LFA-1 and Mac-1 on neutrophils and monocytes) to *in situ* chemokine signals. For instance, leukocytes in which the LFA-1 β_2-threonine 758 phosphorylation site is negligible (Fagerholm et al.) may be more prone to chemokine stimulation, since this unphosphorylated β_2-tail binds less efficiently to filamin, a known suppressor of integrin activation (Takala et al., 2008).

So far, *de novo* ser/thr or tyrosine phosphorylation events have not been demonstrated to take place during leukocyte rolling or in chemokine-stimulated leukocytes. These events are hence more likely to be controlled prior to leukocyte capture and rolling on target endothelial beds. We therefore suggest that prior exposure of a leukocyte subset to specific cytokines, immunoreceptor ligands, and costimulatory signals may alter ser/thr and tyr phosphorylation signatures on different integrin tails, and thereby either prime or restrict these integrins from activation in leukocytes encountering *in situ* chemokine signals at target vessels.

VII. CONCLUSIONS

We have discussed multiple levels of affinity changes of integrins at leukocyte–endothelial contacts which are mediated by chemokine–GPCR signals to cytoskeletal integrin adaptors and by occupancy of these *in situ* activated integrins with either monovalent or oligovalent endothelial ligands. We have also discussed how particular selectin ligands occupied by endothelial selectins during rolling of myeloid leukocytes may prime particular integrins on these leukocytes and thereby facilitate encounter of and integrin activation by endothelial chemoattractants. We have also highlighted another level of integrin regulation controlled by covalent modifications of specific integrin tail residues which may recruit specific adaptors to the vicinity of these integrin subsets. We have postulated that these modifications can either positively or negatively modulate integrin responsiveness to *in situ* chemokine signals. Deep understanding of how these multiple levels of integrin activity differ between distinct integrins, cell types, and GPCRs is still missing. Molecular dissection of how cytoskeletal GPCR and integrin assemblies vary between different leukocytes and full elucidation of their *in situ* activation by distinct chemokine signals should introduce numerous new targets for manipulation of integrin adhesiveness and function in leukocyte–endothelial contacts. New generation therapies that can be designed to interfere with leukocyte trafficking as manifested in specific pathologies should combine general extracellular blockers of trafficking molecules with novel cell permeable compounds that can disrupt GPCR–integrin communications on pathogenic leukocyte subsets without interfering with normal immune surveillance.

Acknowledgments

I thank Dr. S. Schwarzbaum for editorial assistance and Dr. Sara W. Feigelson for fruitful discussions. I also thank Mrs. Channa Vega of the Weizmann Institute Graphics Department for assistance in scheme preparation. R. Alon is Incumbent of The Linda Jacobs Chair in Immune and Stem Cell Research. R.A. is supported by the Israel Science Foundation, the US–Israel Binational Science Foundation and by the FAMRI foundation, USA.

References

Abitorabi, M. A., Pachynski, R. K., Ferrando, R. E., Tidswell, M., & Erle, D. J. (1997). Presentation of integrins on leukocyte microvilli: A role for the extracellular domain in determining membrane localization. *Journal of Cell Biology, 139*, 563–571.

Alon, R., & Dustin, M. L. (2007). Force as a facilitator of integrin conformational changes during leukocyte arrest on blood vessels and antigen-presenting cells. *Immunity, 26*, 17–27.

Alon, R., & Feigelson, S. W. (2009). Chemokine signaling to lymphocyte integrins under shear flow. *Microcirculation, 16*, 3–16.

Alon, R., Feigelson, S. W., Rose, D. M., Schmitz, J., Manevich, E., Overby, D. R., et al. (2005). Integrin $\alpha_4\beta_1$(VLA-4)-dependent T cell tethering and adhesion strengthening under shear stress requires paxillin binding to the α_4 cytoplasmic domain. *Journal of Cell Biology, 171*, 1073–1084.

Alon, R., Hammer, D. A., & Springer, T. A. (1995). Lifetime of the P-selectin-carbohydrate bond and its response to tensile force in hydrodynamic flow. *Nature, 374*, 539–542.

Alon, R., & Ley, K. (2008). Cells on the run: Shear-regulated integrin activation in leukocyte rolling and arrest on endothelial cells. *Current Opinion in Cell Biology, 20*, 525–532.

Amatruda , T. T., III, Gerard, N. P., Gerard, C., & Simon, M. I. (1993). Specific interactions of chemoattractant factor receptors with G-proteins. *Journal of Biological Chemistry, 268*, 10139–10144.

Anderson, R. A., Boronenkov, I. V., Doughman, S. D., Kunz, J., & Loijens, J. C. (1999). Phosphatidylinositol phosphate kinases, a multifaceted family of signaling enzymes. *Journal of Biological Chemistry, 274*, 9907–9910.

Arai, H., & Charo, I. F. (1996). Differential regulation of G-protein-mediated signaling by chemokine receptors. *Journal of Biological Chemistry, 271*, 21814–21819.

Astrof, N. S., Salas, A., Shimaoka, M., Chen, J., & Springer, T. A. (2006). Importance of force linkage in mechanochemistry of adhesion receptors. *Biochemistry, 45*, 15020–15028.

Atarashi, K., Hirata, T., Matsumoto, M., Kanemitsu, N., & Miyasaka, M. (2005). Rolling of Th1 cells via P-selectin glycoprotein ligand-1 stimulates LFA-1-mediated cell binding to ICAM-1. *Journal of Immunology, 174*, 1424–1432.

Bargatze, R. F., & Butcher, E. C. (1993). Rapid G protein-regulated activation event involved in lymphocyte binding to high endothelial venules. *Journal of Experimental Medicine, 178*, 367–372.

Beals, C. R., Edwards, A. C., Gottschalk, R. J., Kuijpers, T. W., & Staunton, D. E. (2001). CD18 activation epitopes induced by leukocyte activation. *Journal of Immunology, 167*, 6113–6122.

Bell, G. (1978). Models for the specific adhesion of cells to cells. *Science, 200*, 618–627.

Bergmeier, W., Goerge, T., Wang, H., Crittenden, J. R., Baldwin, A. C. W., Cifuni, S. M., et al. (2007). Mice lacking the signaling molecule CalDAG-GEFI represent a model for leukocyte adhesion deficiency type III. *Journal of Clinical Investigation, 117*, 1699–1707.

Berlin, C., Bargatze, R. F., Campbell, J. J., von Andrian, U. H., Szabo, M. C., Hasslen, S. R., et al. (1995). α4 integrins mediate lymphocyte attachment and rolling under physiologic flow. *Cell, 80*, 413–422.

Bolomini-Vittori, M., Montresor, A., Giagulli, C., Staunton, D., Rossi, B., Martinello, M., et al. (2009). Regulation of conformer-specific activation of the integrin LFA-1 by a chemokine-triggered Rho signaling module. *Nature Immunology, 10*, 185–194.

Bretscher, A. (1999). Regulation of cortical structure by the ezrin–radixin–moesin protein family. *Current Opinion in Cell Biology, 11*, 109–116.

Brown, M. J., Nijhara, R., Hallam, J. A., Gignac, M., Yamada, K. M., Erlandsen, S. L., et al. (2003). Chemokine stimulation of human peripheral blood T lymphocytes induces rapid dephosphorylation of ERM proteins, which facilitates loss of microvilli and polarization. *Blood, 102*, 3890–3899.

Butcher, E. C. (1991). Leukocyte–endothelial cell recognition: Three (or more) steps to specificity and diversity. *Cell, 67*, 1033–1036.

Cairo, C. W., Mirchev, R., & Golan, D. E. (2006). Cytoskeletal regulation couples LFA-1 conformational changes to receptor lateral mobility and clustering. *Immunity, 25*, 297–308.

Campbell, J. J., Bowman, E. P., Murphy, K., Youngman, K. R., Siani, M. A., Thompson, D. A., et al. (1998a). 6-C-kine (SLC), a lymphocyte adhesion-triggering chemokine expressed by high endothelium, is an agonist for the MIP-3β receptor CCR7. *Journal of Cell Biology, 141*, 1053–1059.

Campbell, J. J., & Butcher, E. C. (2000). Chemokines in tissue-specific and microenvironment-specific lymphocyte homing. *Current Opinion in Immunology, 12*, 336–341.

Campbell, I. D., & Ginsberg, M. H. (2004). The talin–tail interaction places integrin activation on FERM ground. *Trends in Biochemical Sciences, 29*, 429–435.

Campbell, J. J., Hedrick, J., Zlotnik, A., Siani, M. A., & Thompson, D. A. (1998b). Chemokines and the arrest of lymphocytes rolling under flow conditions. *Science, 279*, 381–384.

Carman, C. V., & Springer, T. A. (2003). Integrin avidity regulation: Are changes in affinity and conformation underemphasized? *Current Opinion in Cell Biology, 15*, 547–556.

Caron, E., Self, A. J., & Hall, A. (2000). The GTPase Rap1 controls functional activation of macrophage integrin $\alpha M\beta 2$ by LPS and other inflammatory mediators. *Current Biology, 10*, 974–978.

Chen, C., Mobley, J. L., Dwir, O., Shimron, F., Grabovsky, V., Lobb, R. L., et al. (1999). High affinity VLA-4 subsets expressed on T cells are mandatory for spontaneous adhesion strengthening but not for rolling on VCAM-1 in shear flow. *Journal of Immunology, 162*, 1084–1095.

Chen, S., & Springer, T. A. (1999). An automatic braking system that stabilizes leukocyte rolling by an increase in selectin bond number with shear. *Journal of Cell Biology, 144*, 185–200.

Chen, Y., Takizawa, N., Crowley, J. L., Oh, S. W., Gatto, C. L., Kambara, T., et al. (2003). F-actin and myosin II binding domains in supervillin. *Journal of Biological Chemistry, 278*, 46094–46106.

Chigaev, A., Waller, A., Zwartz, G. J., Buranda, T., & Sklar, L. A. (2007). Regulation of cell adhesion by affinity and conformational unbending of alpha4beta1 integrin. *Journal of Immunology, 178*, 6828–6839.

Cinamon, G., Grabovsky, V., Winter, E., Franitza, S., Feigelson, S., Shamri, R., et al. (2001). Novel chemokine functions in lymphocyte migration through vascular endothelium under shear flow. *Journal of Leukocyte Biology, 69*, 860–866.

Constantin, G., Majeed, M., Giagulli, C., Piccio, L., Kim, J. Y., Butcher, E. C., et al. (2000). Chemokines trigger immediate $\beta 2$ integrin affinity and mobility changes: Differential regulation and roles in lymphocyte arrest under flow. *Immunity, 13*, 759–769.

Critchley, D. R. (2000). Focal adhesions—The cytoskeletal connection. *Current Opinion in Cell Biology, 12*, 133–139.

Critchley, D. R., & Gingras, A. R. (2008). Talin at a glance. *Journal of Cell Science, 121*, 1345–1347.

D'Ambrosio, D., Albanesi, C., Lang, R., Girolomoni, G., Sinigaglia, F., & Laudanna, C. (2002). Quantitative differences in chemokine receptor engagement generate diversity in integrin-dependent lymphocyte adhesion. *Journal of Immunology, 169*, 2303–2312.

de Chateau, M., Chen, S., Salas, A., & Springer, T. A. (2001). Kinetic and mechanical basis of rolling through an integrin and novel Ca(2+)-dependent rolling and Mg(2+)-dependent firm adhesion modalities for the alpha4beta7–MAdCAM-1 interaction. *Biochemistry, 40*, 13972–13979.

del Pozo, M. A., Nieto, M., Serrador, J. M., Sancho, D., Vicente-Manzanares, M., Martinez, C., et al. (1998). The two poles of the lymphocyte: Specialized cell compartments for migration and recruitment. *Cell Adhesion & Communication, 6*, 125–133.

del Rio, A., Perez-Jimenez, R., Liu, R., Roca-Cusachs, P., Fernandez, J. M., & Sheetz, M. P. (2009). Stretching single talin rod molecules activates vinculin binding. *Science, 323*, 638–641.

Di Paolo, G., Pellegrini, L., Letinic, K., Cestra, G., Zoncu, R., Voronov, S., et al. (2002). Recruitment and regulation of phosphatidylinositol phosphate kinase type 1 gamma by the FERM domain of talin. *Nature, 420*, 85–89.

Diamond, M. S., & Springer, T. A. (1993). A subpopulation of Mac-1 (CD11b/CD18) molecules mediates neutrophil adhesion to ICAM-1 and fibrinogen. *Journal of Cell Biology, 120*, 545–556.

Dwir, O., Kansas, G. S., & Alon, R. (2001). The cytoplasmic tail of L-selectin regulates leukocyte capture and rolling by controlling the mechanical stability of selectin:ligand tethers. *Journal of Cell Biology, 155*, 145–156.

Evans, E. A., & Calderwood, D. A. (2007). Forces and bond dynamics in cell adhesion. *Science, 316*, 1148–1153.

Evans, R., Patzak, I., Svensson, L., De Filippo, K., Jones, K., McDowall, A., et al. (2009). Integrins in immunity. *Journal of Cell Science, 122*, 215–225.

Fagerholm, S. C., Hilden, T. J., Nurmi, S. M., & Gahmberg, C. G. (2005). Specific integrin alpha and beta chain phosphorylations regulate LFA-1 activation through affinity-dependent and -independent mechanisms. *Journal of Cell Biology, 171*, 705–715.

Feigelson, S., Pasvolsky, R., Shulman, Z., Grabovsky, V., Ilani, T., Sagiv, A., et al. (In preparation). ICAM-1 is required for acquisition of high affinity LFA-1 critical for TCR-triggered lymphocyte stoppage on dendritic cells.

Feigelson, S. W., Grabovsky, V., Winter, E., Chen, L. L., Pepinsky, R. B., Yednock, T., et al. (2001). The src kinase p56Lck upregulates VLA-4 integrin affinity: Implications for rapid spontaneous and chemokine-triggered T cell adhesion to VCAM-1 and fibronectin. *Journal of Biological Chemistry, 276*, 13891–13901.

Franitza, S., Grabovsky, V., Wald, O., Weiss, I., Beider, K., Dagan, M., et al. (2004). Differential usage of VLA-4 and CXCR4 by CD3+CD56+ NKT cells and CD56+CD16+ NK cells regulates their interaction with endothelial cells. *European Journal of Immunology, 34*, 1333–1341.

Fukui, Y., Hashimoto, O., Sanui, T., Oono, T., Koga, H., Abe, M., et al. (2001). Haematopoietic cell-specific CDM family protein DOCK2 is essential for lymphocyte migration. *Nature, 412*, 826–831.

Gakidis, M. A., Cullere, X., Olson, T., Wilsbacher, J. L., Zhang, B., Moores, S. L., et al. (2004). Vav GEFs are required for beta2 integrin-dependent functions of neutrophils. *Journal of Cell Biology, 166*, 273–282.

Galkina, E., Thatte, J., Dabak, V., Williams, M. B., Ley, K., & Braciale, T. J. (2005). Preferential migration of effector CD8+ T cells into the interstitium of the normal lung. *Journal of Clinical Investigation, 115*, 3473–3483.

Garcia-Bernal, D., Sotillo-Mallo, E., Nombela-Arrieta, C., Samaniego, R., Fukui, Y., Stein, J. V., et al. (2006). DOCK2 is required for chemokine-promoted human T lymphocyte adhesion under shear stress mediated by the integrin alpha4beta1. *Journal of Immunology, 177*, 5215–5225.

Garcia-Bernal, D., Wright, N., Sotillo-Mallo, E., Nombela-Arrieta, C., Stein, J. V., Bustelo, X. R., et al. (2005). Vav1 and Rac control chemokine-promoted T lymphocyte adhesion mediated by the integrin $\alpha_4\beta_1$. *Molecular Biology of the Cell, 16*, 3223–3235.

Gerszten, R. E., Friedrich, E. B., Matsui, T., Hung, R. R., Li, L., Force, T., et al. (2001). Role of phosphoinositide 3-kinase in monocyte recruitment under flow conditions. *Journal of Biological Chemistry, 276*, 26846–26851.

Gerszten, R. E., Garcia-Zepeda, E. A., Lim, Y. C., Yoshida, M., Ding, H. A., Gimbrone, M. A. J., et al. (1999). MCP-1 and IL-8 trigger firm adhesion of monocytes to vascular endothelium under flow conditions. *Nature, 398*, 718–723.

Ghandour, H., Cullere, X., Alvarez, A., Luscinskas, F. W., & Mayadas, T. N. (2007). Essential role for Rap1 GTPase and its guanine exchange factor CalDAG-GEFI in LFA-1 but not VLA-4 integrin mediated human T-cell adhesion. *Blood, 110*, 3682–3690.

Giagulli, C., Scarpini, E., Ottoboni, L., Narumiya, S., Butcher, E. C., Constantin, G., et al. (2004). RhoA and zeta PKC control distinct modalities of LFA-1 activation by chemokines: Critical role of LFA-1 affinity triggering in lymphocyte in vivo homing. *Immunity, 20*, 25–35.

Grabovsky, V., Dwir, O., & Alon, R. (2002). Endothelial chemokines destabilize L-selectin-mediated lymphocyte rolling without inducing selectin shedding. *Journal of Biological Chemistry, 277*, 20640–20650.

Grabovsky, V., Feigelson, S., Chen, C., Bleijs, R., Peled, A., Cinamon, G., et al. (2000). Subsecond induction of α_4 integrin clustering by immobilized chemokines enhances leukocyte capture and rolling under flow prior to firm adhesion to endothelium. *Journal of Experimental Medicine, 192*, 495–505.

Green, C. E., Pearson, D. N., Camphausen, R. T., Staunton, D. E., & Simon, S. I. (2004). Shear-dependent capping of L-selectin and P-selectin glycoprotein ligand 1 by E-selectin signals activation of high-avidity beta2-integrin on neutrophils. *Journal of Immunology, 172*, 7780–7790.

Hafezi-Moghadam, A., Thomas, K. L., Prorock, A. J., Huo, Y., & Ley, K. (2001). L-selectin shedding regulates leukocyte recruitment. *Journal of Experimental Medicine, 193*, 863–872.

Hammond, S. M., Jenco, J. M., Nakashima, S., Cadwallader, K., Gu, Q., Cook, S., et al. (1997). Characterization of two alternately spliced forms of phospholipase D1. Activation of the purified enzymes by phosphatidylinositol 4,5-bisphosphate, ADP-ribosylation factor, and Rho family monomeric GTP-binding proteins and protein kinase C-alpha. *Journal of Biological Chemistry, 272*, 3860–3868.

Han, J., Lim, C. J., Watanabe, N., Soriani, A., Ratnikov, B., Calderwood, D. A., et al. (2006). Reconstructing and deconstructing agonist-induced activation of integrin alphaIIbbeta3. *Current Biology, 16*, 1796–1806.

Han, J., Liu, S., Rose, D. M., Schlaepfer, D. D., McDonald, H., & Ginsberg, M. H. (2001). Phosphorylation of the integrin alpha 4 cytoplasmic domain regulates paxillin binding. *Journal of Biological Chemistry, 276*, 40903–40909.

Haribabu, B., Zhelev, D. V., Pridgen, B. C., Richardson, R. M., Ali, H., & Snyderman, R. (1999). Chemoattractant receptors activate distinct pathways for chemotaxis and secretion. Role of G-protein usage. *Journal of Biological Chemistry, 274*, 37087–37092.

Hartmann, T. N., Grabovsky, V., Pasvolsky, R., Shulman, Z., Buss, E. C., Spiegel, A., et al. (2008). A crosstalk between intracellular CXCR7 and CXCR4 involved in rapid CXCL12-triggered integrin activation but not in chemokine-triggered motility of human T lymphocytes and CD34+ cells. *Journal of Leukocyte Biology, 84*, 1130–1140.

Hartwell, L. H., Hopfield, J. J., Leibler, S., & Murray, A. W. (1999). From molecular to modular cell biology. *Nature, 402*, C47–C52.

Huang, C., Wu, Z., Hujer, K. M., & Miller, R. T. (2006). Silencing of filamin A gene expression inhibits Ca2+-sensing receptor signaling. *FEBS Letters, 580*, 1795–1800.

Huo, Y., Weber, C., Forlow, S. B., Sperandio, M., Thatte, J., Mack, M., et al. (2001). The chemokine KC, but not monocyte chemoattractant protein-1, triggers monocyte arrest on early atherosclerotic endothelium. *Journal of Clinical Investigation, 108*, 1307–1314.

Hyduk, S. J., Chan, J. R., Duffy, S. T., Chen, M., Peterson, M. D., Waddell, T. K., et al. (2007). Phospholipase C, calcium, and calmodulin are critical for alpha4beta1 integrin affinity up-regulation and monocyte arrest triggered by chemoattractants. *Blood, 109*, 176–184.

Hynes, R. O. (2002). Integrins: Bidirectional, allosteric signaling machines. *Cell, 110*, 673–687.

Itoh, T., & Takenawa, T. (2002). Phosphoinositide-binding domains: Functional units for temporal and spatial regulation of intracellular signalling. *Cellular Signalling, 14*, 733–743.

Jiang, G., Giannone, G., Critchley, D. R., Fukumoto, E., & Sheetz, M. P. (2003). Two-piconewton slip bond between fibronectin and the cytoskeleton depends on talin. *Nature, 424*, 334–337.

Johnson, Z., Proudfoot, A. E., & Handel, T. M. (2005). Interaction of chemokines and glycosaminoglycans: A new twist in the regulation of chemokine function with opportunities for therapeutic intervention. *Cytokine & Growth Factor Reviews, 16*, 625–636.

Jung, U., Norman, K. E., Scharffetter-Kochanek, K., Beaudet, A. L., & Ley, K. (1998). Transit time of leukocytes rolling through venules controls cytokine-induced inflammatory cell recruitment *in vivo*. *Journal of Clinical Investigation, 102*, 1526–1533.

Kanner, S. B., Grosmaire, L. S., Ledbetter, J. A., & Damle, N. K. (1993). Beta 2-integrin LFA-1 signaling through phospholipase C-gamma 1 activation. *Proceedings of the National Academy of Sciences of the United States of America, 90*, 7099–7103.

Kansas, G. S., Ley, K., Munro, J. M., & Tedder, T. F. (1993). Regulation of leukocyte rolling and adhesion to high endothelial venules through the cytoplasmic domain of L-selectin. *Journal of Experimental Medicine, 177*, 833–838.

Katagiri, K., Ohnishi, N., Kabashima, K., Iyoda, T., Takeda, N., Shinkai, Y., et al. (2004). Crucial functions of the Rap1 effector molecule RAPL in lymphocyte and dendritic cell trafficking. *Nature Immunology, 5*, 1045–1051.

Kehrl, J. H. (2006). Chemoattractant receptor signaling and the control of lymphocyte migration. *Immunologic Research, 34*, 211–227.

Killock, D. J., Parsons, M., Zarrouk, M., Ameer-Beg, S. M., Ridley, A. J., Haskard, D. O., et al. (2009). *In vitro* and *in vivo* characterization of molecular interactions between calmodulin, ezrin/radixin/moesin, and L-selectin. *Journal of Biological Chemistry, 284*, 8833–8845.

Kim, M., Carman, C. V., Yang, W., Salas, A., & Springer, T. A. (2004). The primacy of affinity over clustering in regulation of adhesiveness of the integrin $\alpha_L\beta_2$. *Journal of Cell Biology, 167*, 1241–1253.

Kinashi, T. (2005). Intracellular signalling controlling integrin activation in lymphocytes. *Nature Reviews Immunology, 5*, 546–559.

Kinashi, T., & Katagiri, K. (2004). Regulation of lymphocyte adhesion and migration by the small GTPase Rap1 and its effector molecule, RAPL. *Immunology Letters, 93*, 1–5.

Kuang, Y., Wu, Y., Jiang, H., & Wu, D. (1996). Selective G protein coupling by C–C chemokine receptors. *Journal of Biological Chemistry, 271*, 3975–3978.

Kubes, P. (2002). Introduction: The complexities of leukocyte recruitment. *Seminars in Immunology, 14*, 65–72.

Kucik, D. F., Dustin, M. L., Miller, J. M., & Brown, E. J. (1996). Adhesion-activating phorbol ester increases the mobility of leukocyte integrin LFA-1 in cultured lymphocytes. *Journal of Clinical Investigation, 97*, 2139–2144.

Kunkel, E. J., Dunne, J. L., & Ley, K. (2000). Leukocyte arrest during cytokine-dependent inflammation *in vivo*. *Journal of Immunology, 164*, 3301–3308.

Lammermann, T., Bader, B. L., Monkley, S. J., Worbs, T., Wedlich-Soldner, R., Hirsch, K., et al. (2008). Rapid leukocyte migration by integrin-independent flowing and squeezing. *Nature, 453*, 51–55.

Larjava, H., Plow, E. F., & Wu, C. (2008). Kindlins: Essential regulators of integrin signalling and cell–matrix adhesion. *EMBO Reports, 9*, 1203–1208.

Laudanna, C., & Alon, R. (2006). Right on the spot. Chemokine triggering of integrin-mediated arrest of rolling leukocytes. *Thrombosis and Haemostasis, 95*, 5–11.

Laudanna, C., Campbell, J. J., & Butcher, E. C. (1996). Role of Rho in chemoattractant-activated leukocyte adhesion through integrins. *Science, 271*, 981–983.

Laudanna, C., Mochly-Rosen, D., Liron, T., Constantin, G., & Butcher, E. C. (1998). Evidence of zeta protein kinase C involvement in polymorphonuclear neutrophil integrin-dependent adhesion and chemotaxis. *Journal of Biological Chemistry, 273*, 30306–30315.

Lawrence, M. B., & Springer, T. A. (1991). Leukocytes roll on a selectin at physiologic flow rates: Distinction from and prerequisite for adhesion through integrins. *Cell, 65*, 859–873.

Legate, K. R., & Fassler, R. (2009). Mechanisms that regulate adaptor binding to beta-integrin cytoplasmic tails. *Journal of Cell Science, 122*, 187–198.

Ley, K. (2003). Arrest chemokines. *Microcirculation, 10*, 289–295.

Ley, K., Laudanna, C., Cybulsky, M. I., & Nourshargh, S. (2007). Getting to the site of inflammation: The leukocyte adhesion cascade updated. *Nature Reviews Immunology, 7*, 678–689.

Li, Z., Jiang, H., Xie, W., Zhang, Z., Smrcka, A. V., & Wu, D. (2000). Roles of PLC-β2 and -β3 and PI3Kgamma in chemoattractant-mediated signal transduction. *Science, 287*, 1046–1049.

Ling, K., Doughman, R. L., Firestone, A. J., Bunce, M. W., & Anderson, R. A. (2002). Type I gamma phosphatidylinositol phosphate kinase targets and regulates focal adhesions. *Nature, 420*, 89–93.

Ling, K., Doughman, R. L., Iyer, V. V., Firestone, A. J., Bairstow, S. F., Mosher, D. F., et al. (2003). Tyrosine phosphorylation of type I{gamma} phosphatidylinositol phosphate kinase by Src regulates an integrin–talin switch. *Journal of Cell Biology, 163*, 1339–1349.

Loike, J. D., Cao, L., Budhu, S., Marcantonio, E. E., El Khoury, J., Hoffman, S., et al. (1999). Differential regulation of β1 integrins by chemoattractants regulates neutrophil migration through fibrin. *Journal of Cell Biology, 144*, 1047–1056.

Luo, B. H., Carman, C. V., & Springer, T. A. (2007). Structural basis of integrin regulation and signaling. *Annual Review of Immunology, 25*, 619–647.

Luster, A. D., Alon, R., & von Andrian, U. H. (2005). Immune cell migration in inflammation: Present and future therapeutic targets. *Nature Immunology, 6*, 1182–1190.

Luttrell, L. M., & Lefkowitz, R. J. (2002). The role of beta-arrestins in the termination and transduction of G-protein-coupled receptor signals. *Journal of Cell Science, 115*, 455–465.

Mackay, C. R. (2001). Chemokines: Immunology's high impact factors. *Nature Immunology, 2*, 95–101.

Manes, S., Lacalle, R. A., Gomez-Mouton, C., del Real, G., Mira, E., & Martinez, A. C. (2001). Membrane raft microdomains in chemokine receptor function. *Seminars in Immunology, 13*, 147–157.

Manevich, E., Grabovsky, V., Feigelson, S. W., & Alon, R. (2007). Talin 1 and paxillin facilitate distinct steps in rapid VLA-4-mediated adhesion strengthening to vascular cell adhesion molecule 1. *Journal of Biological Chemistry, 282*, 25338–25348.

Martel, V., Racaud-Sultan, C., Dupe, S., Marie, C., Paulhe, F., Galmiche, A., et al. (2001). Conformation, localization, and integrin binding of talin depend on its interaction with phosphoinositides. *Journal of Biological Chemistry, 276*, 21217–21227.

Matsui, T., Yonemura, S., & Tsukita, S. (1999). Activation of ERM proteins *in vivo* by Rho involves phosphatidyl-inositol 4-phosphate 5-kinase and not ROCK kinases. *Current Biology, 9*, 1259–1262.

McEver, R. P. (2002). Selectins: Lectins that initiate cell adhesion under flow. *Current Opinion in Cell Biology, 14*, 581–586.

Mellado, M., Rodriguez-Frade, J. M., Vila-Coro, A. J., Fernandez, S., Martin de Ana, A., Jones, D. R., et al. (2001). Chemokine receptor homo- or heterodimerization activates distinct signaling pathways. *EMBO Journal, 20*, 2497–2507.

Middleton, J., Neil, S., Wintle, J., Clark-Lewis, I., Moore, H., Lam, C., et al. (1997). Transcytosis and surface presentation of IL-8 by venular endothelial cells. *Cell, 91*, 385–395.

Molteni, R., Crespo, C. L., Feigelson, S., Moser, C., Fabbri, M., Grabovsky, V., et al. (2009). {beta}-arrestin 2 is required for the induction and strengthening of integrin-mediated leukocyte adhesion during CXCR2-driven extravasation. *Blood, 114*, 1073–1082.

Moore, K. L., Patel, K. D., Bruehl, R. E., Fungang, L., Johnson, D. L., Lichenstein, H. S., et al. (1995). P-selectin glycoprotein ligand-1 mediates rolling of human neutrophils on P-selectin. *Journal of Cell Biology, 128*, 661–671.

Moser, M., Bauer, M., Schmid, S., Ruppert, R., Schmidt, S., Sixt, M., et al. (2009). Kindlin-3 is required for beta2 integrin-mediated leukocyte adhesion to endothelial cells. *Nature Medicine, 15*, 300–305.

Nayal, A., Webb, D. J., & Horwitz, A. F. (2004). Talin: An emerging focal point of adhesion dynamics. *Current Opinion in Cell Biology, 16*, 94–98.

Nishiya, N., Kiosses, W. B., Han, J., & Ginsberg, M. H. (2005). An alpha4 integrin–paxillin–Arf–GAP complex restricts Rac activation to the leading edge of migrating cells. *Nature Cell Biology, 7*, 343–352.

Nolz, J. C., Medeiros, R. B., Mitchell, J. S., Zhu, P., Freedman, B. D., Shimizu, Y., et al. (2007). WAVE2 regulates high-affinity integrin binding by recruiting vinculin and talin to the immunological synapse. *Molecular and Cellular Biology, 27*, 5986–6000.

Nombela-Arrieta, C., Lacalle, R. A., Montoya, M. C., Kunisaki, Y., Megias, D., Marques, M., et al. (2004). Differential requirements for DOCK2 and phosphoinositide-3-kinase gamma during T and B lymphocyte homing. *Immunity, 21*, 429–441.

Pachynski, R. K., Wu, S. W., Gunn, M. D., & Erle, D. J. (1998). Secondary lymphoid-tissue chemokine (SLC) stimulates integrin $\alpha4\beta7$-mediated adhesion of lymphocytes to mucosal addressin cell adhesion molecule-1 (MAdCAM-1) under flow. *Journal of Immunology, 161*, 952–956.

Pasvolsky, R., Grabovsky, V., Giagulli, C., Shulman, Z., Shamri, R., Feigelson, S. W., et al. (2008). RhoA is involved in LFA-1 extension triggered by CXCL12 but not in a novel outside-in LFA-1 activation facilitated by CXCL9. *Journal of Immunology, 180*, 2815–2823.

Peled, A., Grabovsky, V., Habler, L., Sandbank, J., Arenzana-Seisdedos, F., Petit, I., et al. (1999). The chemokine SDF-1 stimulates integrin-mediated arrest of $CD34^{+}$ cells on vascular endothelium under shear flow. *Journal of Clinical Investigation, 104*, 1199–1211.

Petrich, B. G., Marchese, P., Ruggeri, Z. M., Spiess, S., Weichert, R. A., Ye, F., et al. (2007). Talin is required for integrin-mediated platelet function in hemostasis and thrombosis. *Journal of Experimental Medicine, 204*, 3103–3111.

Puklin-Faucher, E., & Sheetz, M. P. (2009). The mechanical integrin cycle. *Journal of Cell Science, 122*, 179–186.

Ratnikov, B. I., Partridge, A. W., & Ginsberg, M. H. (2005). Integrin activation by talin. *Journal of Thrombosis and Haemostasis, 3*, 1783–1790.

Raucher, D., Stauffer, T., Chen, W., Shen, K., Guo, S., York, J. D., et al. (2000). Phosphatidylinositol 4,5-bisphosphate functions as a second messenger that regulates cytoskeleton-plasma membrane adhesion. *Cell, 100*, 221–228.

Ravasz, E., Somera, A. L., Mongru, D. A., Oltvai, Z. N., & Barabasi, A. L. (2002). Hierarchical organization of modularity in metabolic networks. *Science, 297*, 1551–1555.

Rhee, S. G. (2001). Regulation of phosphoinositide-specific phospholipase C. *Annual Review of Biochemistry, 70*, 281–312.

Ridley, A. (2000). Rho GTPases: Integrating integrin signaling. *Journal of Cell Biology, 150*, F107–F109.

Ridley, A. J., Schwartz, M. A., Burridge, K., Firtel, R. A., Ginsberg, M. H., Borisy, G., et al. (2003). Cell migration: Integrating signals from front to back. *Science, 302*, 1704–1709.

Rives, A. W., & Galitski, T. (2003). Modular organization of cellular networks. *Proceedings of the National Academy of Sciences of the United States of America, 100*, 1128–1133.

Salas, A., Shimaoka, M., Chen, S., Carman, C. V., & Springer, T. (2002). Transition from rolling to firm adhesion is regulated by the conformation of the I domain of the integrin lymphocyte function-associated antigen-1. *Journal of Biological Chemistry, 277*, 50255–50562.

Salas, A., Shimaoka, M., Kogan, A. N., Harwood, C., von Andrian, U. H., & Springer, T. A. (2004). Rolling adhesion through an extended conformation of integrin $\alpha_L\beta_2$ and relation to αI and βI-like domain interaction. *Immunity, 20*, 393–406.

Sawada, Y., Nakamura, K., Doi, K., Takeda, K., Tobiume, K., Saitoh, M., et al. (2001). Rap1 is involved in cell stretching modulation of p38 but not ERK or JNK MAP kinase. *Journal of Cell Science, 114*, 1221–1227.

Setiadi, H., & McEver, R. P. (2008). Clustering endothelial E-selectin in clathrin-coated pits and lipid rafts enhances leukocyte adhesion under flow. *Blood, 111*, 1989–1998.

Setiadi, H., Sedgewick, G., Erlandsen, S. L., & McEver, R. P. (1998). Interactions of the cytoplasmic domain of P-selectin with clathrin-coated pits enhance leukocyte adhesion under flow. *Journal of Cell Biology, 142*, 859–871.

Shamri, R., Grabovsky, V., Feigelson, S., Dwir, O., Van Kooyk, Y., & Alon, R. (2002). Chemokine-stimulation of lymphocyte α4 integrin avidity but not of LFA-1 avidity to endothelial ligands under shear flow requires cholesterol membrane rafts. *Journal of Biological Chemistry, 277*, 40027–40035.

Shamri, R., Grabovsky, V., Gauguet, J. M., Feigelson, S., Manevich, E., Kolanus, W., et al. (2005). Lymphocyte arrest requires instantaneous induction of an extended LFA-1 conformation mediated by endothelium-bound chemokines. *Nature Immunology, 6*, 497–506.

Shimaoka, M., Kim, M., Cohen, E. H., Yang, W., Astrof, N., Peer, D., et al. (2006). AL-57, a ligand-mimetic antibody to integrin LFA-1, reveals chemokine-induced affinity up-regulation in lymphocytes. *Proceedings of the National Academy of Sciences of the United States of America, 103*, 13991–13996.

Shimonaka, M., Katagiri, K., Nakayama, T., Fujita, N., Tsuruo, T., Yoshie, O., et al. (2003). Rap1 translates chemokine signals to integrin activation, cell polarization, and motility across vascular endothelium under flow. *Journal of Cell Biology, 161*, 417–427.

Shulman, Z., Pasvolsky, R., Woolf, E., Grabovsky, V., Feigelson, S. W., Erez, N., et al. (2006). DOCK2 regulates chemokine-triggered lateral lymphocyte motility but not transendothelial migration. *Blood, 108*, 2150–2158.

Shulman, Z., Shinder, V., Klein, E., Grabovsky, V., Yeger, O., Geron, E., et al. (2009). Lymphocyte crawling and transendothelial migration require chemokine triggering of high-affinity LFA-1 integrin. *Immunity, 30*, 384–396.

Simon, S. I., Hu, Y., Vestweber, D., & Smith, C. W. (2000). Neutrophil tethering on E-selectin activates β2 integrin binding to ICAM-1 through a mitogen-activated protein kinase signal transduction pathway. *Journal of Immunology, 164*, 4348–4358.

Singer, I. I., Scott, S., Kawka, D. W., Chin, J., Daugherty, B. L., DeMartino, J. A., et al. (2001). CCR5, CXCR4, and CD4 are clustered and closely apposed on microvilli of human macrophages and T cells. *Journal of Virology, 75*, 3779–3790.

Smith, A., Carrasco, Y. R., Stanley, P., Kieffer, N., Batista, F. D., & Hogg, N. (2005). A talin-dependent LFA-1 focal zone is formed by rapidly migrating T lymphocytes. *Journal of Cell Biology, 170*, 141–151.

Smith, M. L., Olson, T. S., & Ley, K. (2004). CXCR2- and E-selectin-induced neutrophil arrest during inflammation in vivo. *Journal of Experimental Medicine, 200*, 935–939.

Spitaler, M., & Cantrell, D. A. (2004). Protein kinase C and beyond. *Nature Immunology, 5*, 785–790.

Springer, T. A. (1994). Traffic signals for lymphocyte recirculation and leukocyte emigration: The multistep paradigm. *Cell, 76*, 301–314.

Springer, T. A. (2009). Structural basis for selectin mechanochemistry. *Proceedings of the National Academy of Sciences of the United States of America, 106*, 91–96.

Stanley, P., Smith, A., McDowall, A., Nicol, A., Zicha, D., & Hogg, N. (2008). Intermediate-affinity LFA-1 binds alpha-actinin-1 to control migration at the leading edge of the T cell. *EMBO Journal, 27*, 62–75.

Tadokoro, S., Shattil, S. J., Eto, K., Tai, V., Liddington, R. C., de Pereda, J. M., et al. (2003). Talin binding to integrin beta tails: A final common step in integrin activation. *Science, 302*, 103–106.

Takala, H., Nurminen, E., Nurmi, S. M., Aatonen, M., Strandin, T., Takatalo, M., et al. (2008). Beta2 integrin phosphorylation on Thr758 acts as a molecular switch to regulate 14-3-3 and filamin binding. *Blood, 112*, 1853–1862.

Tall, E. G., Spector, I., Pentyala, S. N., Bitter, I., & Rebecchi, M. J. (2000). Dynamics of phosphatidylinositol 4,5-bisphosphate in actin-rich structures. *Current Biology, 10*, 743–746.

Thelen, M., & Stein, J. V. (2008). How chemokines invite leukocytes to dance. *Nature Immunology, 9*, 953–959.

Toker, A. (1998). The synthesis and cellular roles of phosphatidylinositol 4,5-bisphosphate. *Current Opinion in Cell Biology, 10*, 254–261.

Urzainqui, A., Serrador, J. M., Viedma, F., Yanez-Mo, M., Rodriguez, A., Corbi, A. L., et al. (2002). ITAM-based interaction of ERM proteins with Syk mediates signaling by the leukocyte adhesion receptor PSGL-1. *Immunity, 17*, 401–412.

van Kooyk, Y., & Figdor, C. G. (2000). Avidity regulation of integrins: The driving force in leukocyte adhesion. *Current Opinion in Cell Biology, 12*, 542–547.

Vogel, V., & Sheetz, M. (2006). Local force and geometry sensing regulate cell functions. *Nature Reviews Molecular Cell Biology, 7*, 265–275.

von Andrian, U. H., Hasslen, S. R., Nelson, R. D., Erlandsen, S. L., & Butcher, E. C. (1995). A central role for microvillous receptor presentation in leukocyte adhesion under flow. *Cell, 82*, 989–999.

von Hundelshausen, P., Koenen, R. R., Sack, M., Mause, S. F., Adriaens, W., Proudfoot, A. E., et al. (2005). Heterophilic interactions of platelet factor 4 and RANTES promote monocyte arrest on endothelium. *Blood, 105*, 924–930.

Watanabe, N., Bodin, L., Pandey, M., Krause, M., Coughlin, S., Boussiotis, V. A., et al. (2008). Mechanisms and consequences of agonist-induced talin recruitment to platelet integrin alphaIIbbeta3. *Journal of Cell Biology, 181*, 1211–1222.

Weber, C., Alon, R., Moser, B., & Springer, T. A. (1996). Sequential regulation of alpha 4 beta 1 and alpha 5 beta 1 integrin avidity by CC chemokines in monocytes: Implications for transendothelial chemotaxis. *Journal of Cell Biology, 134*, 1063–1073.

Weber, K. S., von Hundelshausen, P., Clark-Lewis, I., Weber, P. C., & Weber, C. (1999). Differential immobilization and hierarchical involvement of chemokines in monocyte arrest and transmigration on inflamed endothelium in shear flow. *European Journal of Immunology, 29*, 700–712.

Weber, C., Weber, K. S., Klier, C., Gu, S., Wank, R., Horuk, R., et al. (2001). Specialized roles of the chemokine receptors CCR1 and CCR5 in the recruitment of monocytes and Th1-like/CD45RO^{+} T cells. *Blood, 97*, 1144–1146.

Wegener, K. L., Basran, J., Bagshaw, C. R., Campbell, I. D., Roberts, G. C., Critchley, D. R., et al. (2008). Structural basis for the interaction between the cytoplasmic domain of the hyaluronate receptor layilin and the talin F3 subdomain. *Journal of Molecular Biology, 382*, 112–126.

Woolf, E., Grigorova, I., Sagiv, A., Grabovsky, V., Feigelson, S. W., Shulman, Z., et al. (2007). Lymph node chemokines promote sustained T lymphocyte motility without triggering stable integrin adhesiveness in the absence of shear forces. *Nature Immunology, 8*, 1076–1085.

Xu, J., Wang, F., Van Keymeulen, A., Herzmark, P., Straight, A., Kelly, K., et al. (2003). Divergent signals and cytoskeletal assemblies regulate self-organizing polarity in neutrophils. *Cell, 114*, 201–214.

Yan, B., Calderwood, D. A., Yaspan, B., & Ginsberg, M. H. (2001). Calpain cleavage promotes talin binding to the β_3 integrin cytoplasmic domain. *Journal of Biological Chemistry, 276*, 28164–28170.

Yednock, T. A., Cannon, C., Vandevert, C., Goldbach, E. G., Shaw, G., Ellis, D. K., et al. (1995). $\alpha 4\beta 1$ integrin-dependent cell adhesion is regulated by a low affinity receptor pool that is conformationally responsive to ligand. *Journal of Biological Chemistry, 270*, 28740–28750.

Zarbock, A., Deem, T. L., Burcin, T. L., & Ley, K. (2007a). Galphai2 is required for chemokine-induced neutrophil arrest. *Blood, 110*, 3773–3779.

Zarbock, A., Lowell, C. A., & Ley, K. (2007b). Spleen tyrosine kinase Syk is necessary for E-selectin-induced alpha(L)beta(2) integrin-mediated rolling on intercellular adhesion molecule-1. *Immunity, 26*, 773–783.

Zhang, H., Schaff, U. Y., Green, C. E., Chen, H., Sarantos, M. R., Hu, Y., et al. (2006). Impaired integrin-dependent function in Wiskott–Aldrich syndrome protein-deficient murine and human neutrophils. *Immunity, 25*, 285–295.

Zhu, J., Luo, B. H., Xiao, T., Zhang, C., Nishida, N., & Springer, T. A. (2008). Structure of a complete integrin ectodomain in a physiologic resting state and activation and deactivation by applied forces. *Molecular Cell, 32*, 849–861.

Zwartz, G. J., Chigaev, A., Dwyer, D. C., Foutz, T. D., Edwards, B. S., & Sklar, L. A. (2004). Real-time analysis of very late antigen-4 affinity modulation by shear. *Journal of Biological Chemistry, 279*, 38277–38286.

CHAPTER 7

Biophysical Regulation of Selectin–Ligand Interactions Under Flow

Rodger P. McEver* **and Cheng Zhu**†

*Cardiovascular Biology Research Program, Oklahoma Medical Research Foundation, Oklahoma City, Oklahoma 73104, USA

†Coulter Department of Biomedical Engineering, Woodruff School of Mechanical Engineering, and Institute for Bioengineering and Biosciences, Georgia Institute of Technology, Atlanta, Georgia 30332, USA

I. OVERVIEW

Rolling adhesion on vascular surfaces is the first step in recruiting circulating leukocytes to secondary lymphoid organs or to sites of infection or injury. Rolling requires the rapid yet balanced formation and dissociation of adhesive bonds in the challenging environment of blood flow. This chapter explores how selectins interact through mechanically regulated kinetics with their ligands to meet these challenges. Remarkably, increasing force applied to adhesive bonds first prolongs their lifetimes (catch bonds) and then shortens their lifetimes (slip bonds). Catch bonds mediate the

Current Topics in Membranes, Volume 64 1063-5823/09 $35.00

DOI: 10.1016/S1063-5823(09)64007-2

counterintuitive phenomenon of flow-enhanced rolling adhesion. Force-regulated disruptions of receptor interdomain or intradomain interactions remote from the ligand-binding surface generate catch bonds. Adhesion receptor dimerization, clustering in membrane domains, and interactions with the cytoskeleton modulate the forces applied to bonds.

II. INTRODUCTION

Leukocytes use a multistep process in which they initially tether to and roll along the vessel wall, then decelerate and arrest, and finally emigrate into the underlying tissues (Ley, Laudanna, Cybulsky, & Nourshargh, 2007) (Fig. 1). For cells to tether, interactions between adhesion molecules must form rapidly. For cells to roll, these interactions must break rapidly. Rolling adhesion provides an important checkpoint for cells to encounter tissue-specific signals before committing to enter into a particular organ. It is the initial step in recruitment of naïve lymphocytes to secondary lymphoid organs and of myeloid leukocytes and effector lymphocytes to sites of inflammation. Interactions of selectins with their ligands mediate most tethering and rolling. Interactions of integrins with their ligands mediate arrest (firm adhesion) and migration but can also support rolling. This chapter focuses on the biophysical regulation of selectin–ligand interactions in the demanding hydrodynamic environment of flowing blood.

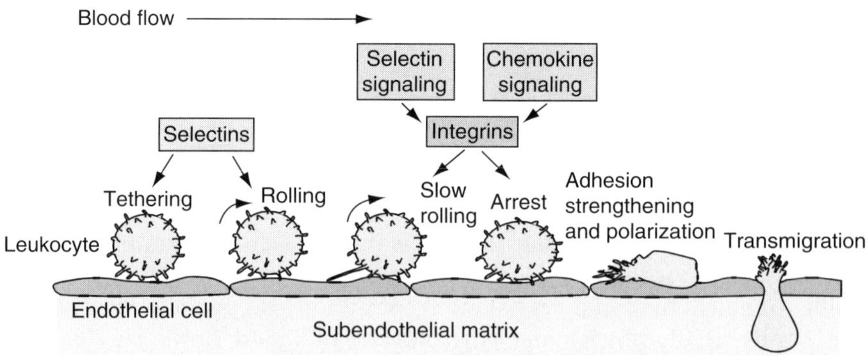

FIGURE 1 Multistep leukocyte adhesion cascade. Selectins initiate tethering and rolling of leukocytes. Depending on their activation state, integrins mediate slower rolling or cause the cells to arrest. Integrins also mediate spreading, crawling, and migration between or through endothelial cells into the underlying tissues.

III. SELECTINS

Each of the three selectins has an N-terminal carbohydrate-recognition domain characteristic of Ca^{2+}-dependent (C-type) lectins, followed by an epidermal growth factor (EGF)-like domain, a series of short consensus repeats like those in complement regulatory proteins, a transmembrane domain, and a short cytoplasmic tail (Fig. 2). L-selectin is constitutively expressed on most leukocytes (McEver, 2002; McEver, Moore, & Cummings, 1995; Vestweber & Blanks, 1999). E-selectin is constitutively expressed on endothelial cells of skin and bone marrow. In most other tissues, inflammatory cytokines such as TNF-α transiently induce its expression in endothelial cells of postcapillary venules through activation of NF-κB and other transcription factors. P-selectin is constitutively synthesized by megakaryocytes, where it is incorporated into platelets, and by endothelial cells, mostly in postcapillary venules. It is sorted to the membranes of α-granules of platelets and Weibel–Palade bodies of endothelial cells. Upon stimulation by secretagogues such as thrombin or histamine, the membranes of these organelles rapidly fuse with the plasma membrane and P-selectin redistributes to the cell surface.

IV. SELECTIN LIGANDS

Each selectin mediates adhesion in part through interactions of its N-terminal lectin domain with a sialyl Lewis x (sLe^x) capping structure ($NeuAc\alpha_{2-3}Gal\beta_{1-3}[Fuc\alpha_{1-3}]GlcNAc\beta_{1-R}$) on cell-surface glycoconjugates (McEver, 2002; McEver et al., 1995; Vestweber & Blanks, 1999). Crystal structures of sLe^x bound to the lectin domains of P- and E-selectin reveal multiple interactions between the fucose, a Ca^{2+} ion, and several amino acids, including those that coordinate the Ca^{2+} ion (Somers, Tang, Shaw, & Camphausen, 2000). This explains the Ca^{2+}-dependent binding to fucosylated glycans. Other lectin-domain residues make contacts with sialic acid and galactose. P- and L-selectin, but not E-selectin, bind in a Ca^{2+}-independent manner to sulfated glycans such as heparin, fucoidan, and glycosaminoglycans that lack sialic acid and fucose. Some sulfated proteoglycans on endothelial cells bind to L-selectin (Wang, Fuster, Sriramarao, & Esko, 2005), and soluble heparan sulfate proteoglycans like heparin might serve as endogenous inhibitors of P- and L-selectin interactions. However, cooperative sulfation, sialylation, and fucosylation of specific glycoproteins make them preferred ligands for P- and L-selectin.

P-selectin glycoprotein ligand-1 (PSGL-1) is a transmembrane, homodimeric mucin on leukocytes and some activated endothelial cells (McEver, 2002; McEver et al., 1995; Rivera-Nieves et al., 2006; Vestweber & Blanks, 1999)

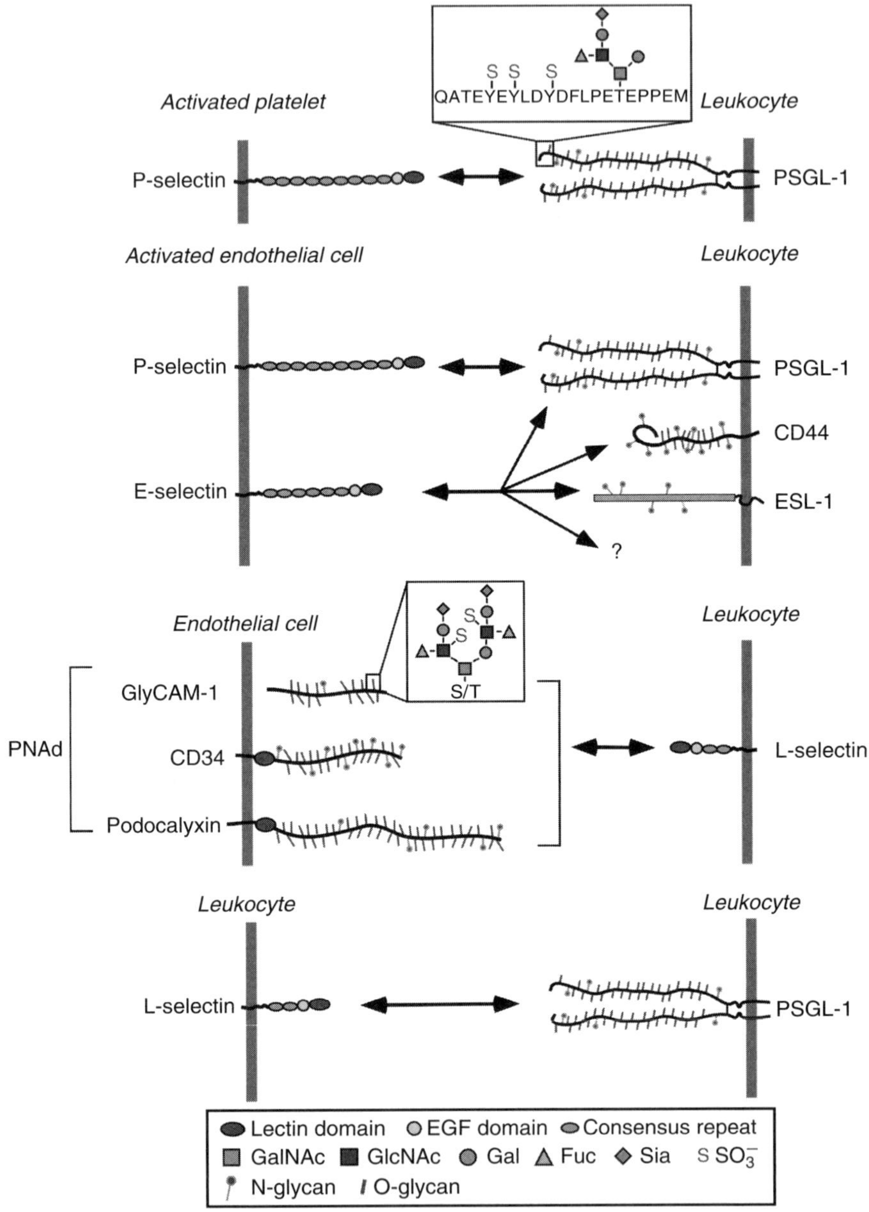

FIGURE 2 Selectins and their major glycoprotein ligands. The upper inset depicts the N-terminal glycosulfopeptide region of human PSGL-1 that binds to P-selectin (and L-selectin). The lower inset depicts an example of a sialylated, fucosylated, and sulfated *O*-glycan on murine PNAd mucins that binds to L-selectin.

(Fig. 2). The extracellular region of each subunit bears multiple *O*-glycans on serine and threonine residues that cause it to extend above the membrane. Antibody-blocking studies and genetic deletion of PSGL-1 indicate that PSGL-1 is the only physiologically relevant ligand for P- and L-selectin on leukocytes. This is due to stereospecific binding of P- and L-selectin to tyrosine sulfate residues, adjacent peptide determinants, and fucose, sialic acid, and galactose on a single core 2 *O*-glycan, all located at the extreme N-terminal region of PSGL-1 (Leppänen, White, Helin, McEver, & Cummings, 2000; Leppänen, Yago, Otto, McEver, & Cummings, 2003; Somers et al., 2000) (Fig. 2). These additional contacts explain why P- and L-selectin bind with higher affinity to PSGL-1 than to sLe^x alone.

L-selectin also binds to a group of mucins expressed on lymph node high endothelial venules and on some activated endothelial cells at sites of inflammation (Fig. 2). These mucins, collectively called peripheral node addressin (PNAd), include CD34, glycosylated cell adhesion molecule-1 (GlyCAM-1), and podocalyxin (Rosen, 2004). Unlike the sulfation of tyrosines on PSGL-1, PNAd mucins are sulfated at the C6 position of galactose and *N*-acetylglucosamine (GlcNAc) residues on multiple *O*-glycans and on some *N*-glycans. Targeted disruption of genes encoding specific glycosyltransferases and sulfotransferases reveals that a combination of *N*-glycans, branched core 2 *O*-glycans, and extended core 1 *O*-glycans capped with 6-sulfo-sLe^x (sLe^x modified with a sulfate ester attached to the C6 position of GlcNAc) confer optimal binding of PNAd mucins to L-selectin (Kawashima et al., 2005; Mitoma et al., 2007; Uchimura et al., 2005; Yeh et al., 2001) (Fig. 2). To date, there is no crystal structure of 6-sulfo-sLe^x bound to L-selectin to indicate where the sulfate docks to the lectin domain. The available evidence suggests that L-selectin binds to 6-sulfo-sLe^x but not to amino acids on PNAd, which contrasts with the cooperative binding to sLe^x, sulfated tyrosines, and other amino acids at the N-terminus of PSGL-1.

E-selectin was originally thought to mediate leukocyte rolling through interactions with multiple glycoproteins that bear sLe^x-capped glycans. However, gene knockout studies demonstrate that PSGL-1 and CD44 are the major glycoprotein ligands for E-selectin on murine leukocytes (Katayama, Hidalgo, Chang, Peired, & Frenette, 2005; Xia et al., 2002) (Fig. 2). Why these glycoproteins serve as preferential E-selectin ligands is not known. Because it has no affinity for sulfate, E-selectin binding is probably not limited to the N-terminus of PSGL-1. siRNA knockdown of transcripts for a glycoprotein termed E-selectin ligand-1 (ESL-1) suggests that it comprises the remaining E-selectin–ligand activity on murine leukocytes (Hidalgo, Peired, Wild, Vestweber, & Frenette, 2007). Unlike murine leukocytes, human neutrophils express sialylated glycosphingolipids with repeating Galβ_{1-4}GlcNAc units, some of which are modified with α_{1-3}-fucose (Nimrichter et al., 2008). When immobilized, these glycolipids support rolling of E-selectin-expressing cells. An inhibitor of glycolipid synthesis partially blocks rolling of human neutrophils on E-selectin, but this could be an indirect effect from altered cell deformability. Glycolipids are much shorter than most glycoproteins. Proteolytic digestion of neutrophil surfaces may allow glycolipids to interact with E-selectin under flow (Kobzdej, Leppänen, Ramachandran, Cummings, & McEver, 2002), but it remains unclear whether they do so within an intact glycocalyx.

V. KINETIC AND MECHANICAL PARAMETERS OF CELL TETHERING AND ROLLING UNDER FLOW

Cell adhesion is mediated through reversible interactions, or "bonds," between cell-surface receptors and their ligands, or counter-receptors, on other cell surfaces or in extracellular matrix. Here, an adhesive bond is defined as the sum of noncovalent interactions, for example, hydrogen bonds, electrostatic interactions, van der Waals forces, dipole–dipole interactions, between two macromolecules. As for other biochemical reactions, the equilibrium affinity of an adhesion receptor for its ligand is the ratio of the on-rate (k_{on}) to the off-rate (k_{off}). Proteins in solution may interact after they collide during diffusion in three-dimensional (3D) space. In contrast, adhesion receptors diffuse laterally on the cell membrane—a two-dimensional (2D) space—and bind ligands on another membrane or in extracellular matrix. A moving cell carries adhesion receptors to ligands on the surface of another cell or in matrix in both normal and tangential directions. Blood flow imposes additional transport and mechanical constraints on interacting molecules as one of them resides on a moving cell.

In steady laminar flow, the velocity of fluid (and cells freely flowing with it) increases with its distance from the vessel wall. The change in velocity per unit distance is the shear rate, usually expressed in units of s^{-1} (Fig. 3A). At a given shear rate, larger cells have higher velocities because they tend to extend further from the wall. A cell flowing near the vessel wall may be able to tether if its adhesion receptors contact ligands on the wall. Bond formation, however, involves two steps: transport, which brings two molecules into close proximity, and reaction, during which the molecules dock. Faster cell velocity produces more frequent collisions (Fig. 3B) but also shortens the contact time between adhesion molecules (Fig. 3C). Thus, the relative timescales for transport and docking affect the efficiency of tethering a flowing cell to the surface.

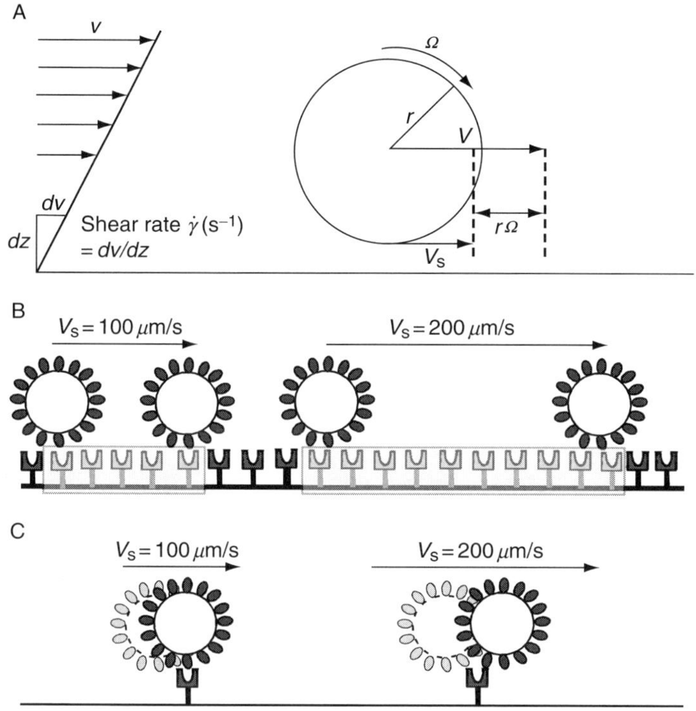

FIGURE 3 Parameters of tethering under flow. (A) The fluid velocity v of a Couette flow field bordered with a solid surface (x–y plane) is parallel to the surface and increases linearly with the distance away from the surface (z-direction). The shear rate $\dot{\gamma} = dv/dz$ is reciprocal to the slope of the velocity profile. Fluid mechanics theory predicts that the translational velocity V and angular velocity Ω of a sphere (or cell) of radius r freely moving above a surface in an otherwise Couette flow are proportional to $r\dot{\gamma}$ and r, respectively (Goldman, Cox, & Brenner, 1967). The sphere bottom has a sliding velocity $V_s \equiv V - r\,\Omega \propto r$ relative to the surface (Chang & Hammer, 1999) (adapted with permission from Yago et al., 2007). (B) Faster sliding velocity increases the number of surface ligands that an adhesion receptor on the flowing cell contacts per unit time. (C) Faster sliding velocity reduces the time that the receptor contacts the ligand before it moves away.

The product of shear rate and viscosity is shear stress, a measure of tangential force per unit area, usually expressed as dyne/cm^2 (Fig. 4A). A cell rolls by forming new adhesive bonds at the leading edge to replace bonds that dissociate at the trailing edge. Shear stress imposes a force F_s and a torque T_s to the rolling cell, which reach their maximal levels when the cell stops. F_s and T_s must be balanced by a tensile force F_t on the adhesive bonds at the trailing edge and a compressive force F_c on the cell bottom. F_t affects the off-rate of the bonds. A rolling cell stops when the adhesive bond

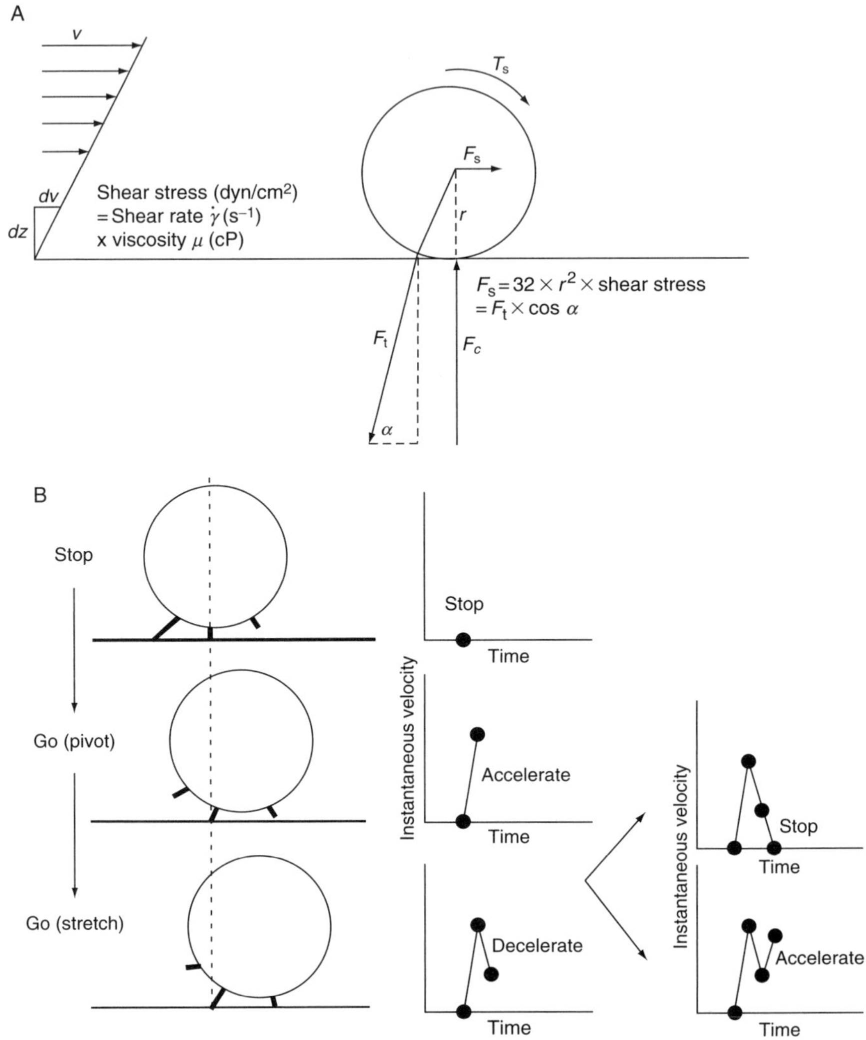

FIGURE 4 Parameters of rolling adhesion under flow. (A) The rolling motions of a sphere or cell of radius r are governed by the balance of the resultant force F_s and torque T_s exerted by the flowing fluid, the tether force F_t applied through the receptor–ligand bond, and the contact force F_c. The conversion of wall shear stress to F_t using the indicated variables is described in Yago et al. (2004). (B) A rolling sphere stops when the adhesive bond sustains the full load required to balance the maximum F_t and T_s. After the bond dissociates, the sphere accelerates as it pivots on a newly formed bond downstream and then decelerates as force develops in the bond. The sphere stops again if the new bond has sufficient strength to withstand the full load and lives long enough to survive loading, or it accelerates if the bond dissociates prematurely (both panels adapted with permission from Yago et al., 2004).

(or bond cluster) can withstand the force required to balance the maximal force and torque applied to the cell (Fig. 4B). After dissociation of this bond (or the last bond in a bond cluster), the cell accelerates as it pivots on a newly formed bond downstream and then decelerates as force develops in the bond. The cell stops again if the new bond persists long enough with sufficient strength to counter the maximal F_s and T_s. If the bond dissociates prematurely, the cell accelerates again before it can stop. The velocity of a rolling cell is primarily determined by the off-rate of adhesive bonds at the trailing edge of the cell. In turn, the mechanical regulation of these bonds, that is, how their off-rates respond to force, critically determines whether and if so, how cells roll under flow.

VI. FORCE-FREE KINETICS AND AFFINITY OF SELECTIN–LIGAND INTERACTIONS

Measurements of selectin–ligand interactions with surface plasmon resonance where selectin or ligand moves in 3D space confirm rapid on- and off-rates, as seems intuitively necessary for cell tethering and rolling. The measured k_{off} for P-selectin dissociating from PSGL-1 is $\sim 1.4\ s^{-1}$ (Klopocki et al., 2008; Mehta, Cummings, & McEver, 1998). Despite the rapid off-rate, the measured equilibrium affinity ($K_d \sim 0.3$–$1.5\ \mu M$) is unexpectedly high because the calculated k_{on} is remarkably fast: ~ 1–$4 \times 10^6\ M^{-1}\ s^{-1}$. The affinity of L-selectin for PSGL-1 is lower ($K_d \sim 47\ \mu M$) due to both a faster off-rate and a slower on-rate (Klopocki et al.). The k_{off} measured by surface plasmon resonance is at least $10\ s^{-1}$, the upper limit of resolution for the instrument. Measurements of 2D interactions of L-selectin with PSGL-1 by thermal fluctuations at high temporal resolution reveal the actual k_{off} to be $10.2\ s^{-1}$ and confirm that L-selectin binds to PSGL-1 with a higher off-rate but a slower on-rate than P-selectin (Chen, Evans, McEver, & Zhu, 2007). As determined by surface plasmon resonance, L-selectin binds with equivalent affinity ($K_d \sim 108\ \mu M$) to GlyCAM-1 and to 6-sulfo-sLe^x, suggesting that L-selectin binds only to glycan determinants on PNAd mucins (Klopocki et al.; Nicholson, Barclay, Singer, Rosen, & Van der Merwe, 1998). E-selectin binds to ESL-1 with a K_d of $62\ \mu M$, a measured k_{off} of $4.6\ s^{-1}$, and a calculated k_{on} of $7.4 \times 10^4\ M^{-1}\ s^{-1}$ (Wild, Huang, Schulze-Horsel, van Der Merwe, & Vestweber, 2001). Thus, all three selectins dissociate rapidly from their ligands, with the fastest off-rate observed for L-selectin. Compared to E-selectin, P- and L-selectin have higher on-rates, perhaps driven by electrostatic interactions with sulfate on their ligands, although this has not been experimentally tested. The fast on-rate of P-selectin for PSGL-1 is consistent with its dominant role in the initial capture, or tethering, of leukocytes to vascular surfaces (Smith, Sperandio, Galkina, & Ley, 2004; Xia et al., 2002). Measurements of 2D selectin–ligand interactions with a micropipette adhesion frequency assay are consistent with the 3D measurements (Huang et al., 2004; Long, Zhao, Huang, & Zhu, 2001).

VII. MECHANICAL REGULATION OF SELECTIN–LIGAND INTERACTIONS

Blood flow exerts force on the selectin–ligand bonds that anchor rolling cells, which affects their lifetimes by altering their off-rates. Indeed, the first demonstration that force affects adhesion receptor bond lifetimes was for interactions of P-selectin with PSGL-1 (Alon, Hammer, & Springer, 1995). Measuring force-dependent off-rates requires low densities of receptor and ligand to favor single-molecule interactions, sufficient temporal resolution to detect even short-lived bonds, and the application of a range of physiologically relevant forces. Initially, leukocytes perfused over low densities of an immobilized selectin or selectin ligand were observed by video microscopy (Alon et al., 1995; Alon, Chen, Puri, Finger, & Springer, 1997; Ramachandran et al., 1999). The lifetimes of transient leukocyte tethers, each assumed to represent a single selectin–ligand bond, were measured as a function of wall shear stress, which was used to estimate the force applied to the bond anchoring the cell. Bond lifetimes obey first-order dissociation kinetics, as required by single-state, single-molecular interactions. More recently, atomic force microscopy or a biomembrane force probe was used to apply tensile force to the bonds via a bent cantilever or a stretched red blood cell. These methods measure bond lifetimes over a range of constant forces (Lou et al., 2006; Marshall et al., 2003; Sarangapani et al., 2004) or rupture forces over a range of ramp rates (Evans, Leung, Hammer, & Simon, 2001).

Early theories suggested that force might shorten bond lifetimes, that is, accelerate dissociation, by lowering the energy barrier between the bound and free states (Bell, 1978). These are termed slip bonds. Conversely, force might prolong bond lifetimes, that is, decelerate dissociation, by deforming the molecules such that they lock more tightly. These are termed catch bonds (Dembo, Torney, Saxman, & Hammer, 1988). Initial studies detected only slip bonds between selectins and their ligands (Alon et al., 1995, 1997; Ramachandran et al., 1999). More recently, catch bonds have been demonstrated (Marshall et al., 2003; Sarangapani et al., 2004). Earlier reports failed to detect catch bonds because the forces studied were too high, because video frame speeds were too slow to measure the shortest bond lifetimes, or because rupture forces from constant-ramp experiments missed the catch bond regime (Evans, Leung, Heinrich, & Zhu, 2004). Thus, force exerts a biphasic effect on selectin–ligand interactions, first prolonging lifetimes (catch bonds) until a maximal value is reached and then shortening lifetimes (slip bonds) as force continues to increase (Fig. 5). In sharp contrast, selectin–antibody interactions exhibit only slip bonds in response to force, suggesting a specific role for catch bonds with physiological selectin ligands (Marshall et al.; Sarangapani et al.).

Two major models for catch bonds have been proposed. The first invokes allosteric change in the ligand-binding surface of the lectin domain (Springer, 2009; Waldron & Springer, 2009). The second invokes a sliding-rebinding mechanism (Lou & Zhu, 2007; Lou et al., 2006). The models are not mutually exclusive; indeed, the second is a form of allostery. Both models rely on distinct crystal structures of P-selectin–ligand complexes (Somers et al., 2000). In one, crystals of the lectin and EGF domains of P-selectin were soaked with sLe^x. In the other, the lectin and EGF domains of P-selectin were cocrystallized with an N-terminal glycosulfopeptide from PSGL-1. Figure 6A shows a ribbon overlay of the P-selectin structures without their bound ligands. A striking difference is the relative orientations of the lectin and EGF domains. In the P-selectin–sLe^x complex, the angle between the domains is more closed or bent. This bent conformation is also observed in crystal structures of all three selectins in the absence of ligand (Graves et al., 1994; Klopocki et al., 2008; Lou et al., 2006; Somers et al.). In the P-selectin–PSGL-1 complex, the angle between the lectin and EGF domains is more open or extended. The straightening of the interdomain hinge is associated with movement of several loops along one face of the lectin domain (Fig. 6A), including a loop near the Ca^{2+} coordination site that introduces

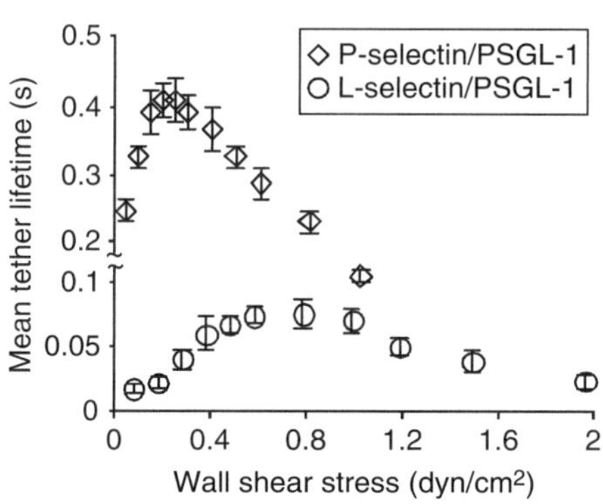

FIGURE 5 Force-dependent lifetimes of single bonds between PSGL-1 and P-selectin or L-selectin. Lifetimes of transient neutrophil tethers to low-density P-selectin or L-selectin at different wall shear stresses were measured by video microscopy (adapted with permission from Marshall et al., 2003; Sarangapani et al., 2004).

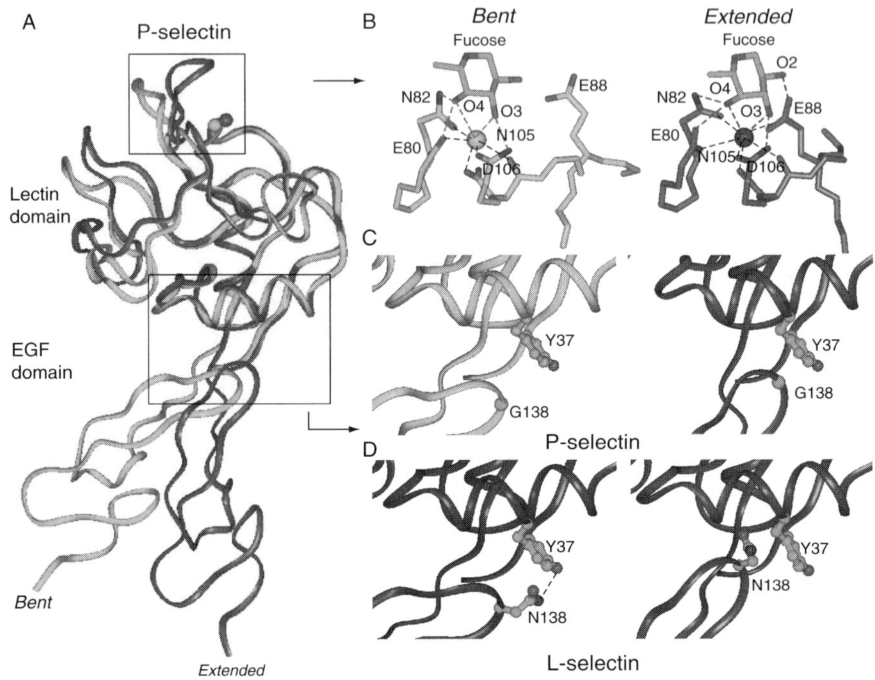

FIGURE 6 Bent and extended selectin structures. (A) Overlay of the lectin and EGF domains of P-selectin in its bent and extended forms. The magenta and blue spheres at the top represent the respective Ca^{2+} ion on each structure. (B) Binding of fucose (part of the sLe^x-binding determinant) to the bent and extended forms of P-selectin. The dashed black lines represent interactions of the Ca^{2+} ion with residues on P-selectin. The dashed red lines represent interactions of the fucose with the Ca^{2+} ion or residues on P-selectin. (C) Relative orientations of Y37 in the lectin domain and G138 in the EGF domain in the bent and extended forms of P-selectin. (D) Relative orientations of Y37 in the lectin domain and N138 in the EGF domain in the bent and extended forms of L-selectin. The black dashed line indicates a hydrogen bond [from Protein Data Bank (PDB) ID codes 1G1Q, 1G1R, 1G1S, and 3CFW; Klopocki et al., 2008; Somers et al., 2000]. The extended structure of L-selectin was derived by molecular modeling (Lou et al., 2006). (See Color Insert.)

a new contact for fucose (Fig. 6B) and optimizes docking to a sulfated tyrosine on PSGL-1. Both models assume equilibrium between the interdomain orientations. Force applied to bound ligand shifts the equilibrium to the more straight or extended orientation. In the allosteric model, force, by opening the interdomain hinge, propagates structural changes to the interface with ligand that decrease off-rate and prolong bond lifetime (Fig. 7A). The sliding-rebinding model neither requires nor excludes loop movement at the interface with ligand (Fig. 7B). In this model, force, by opening the

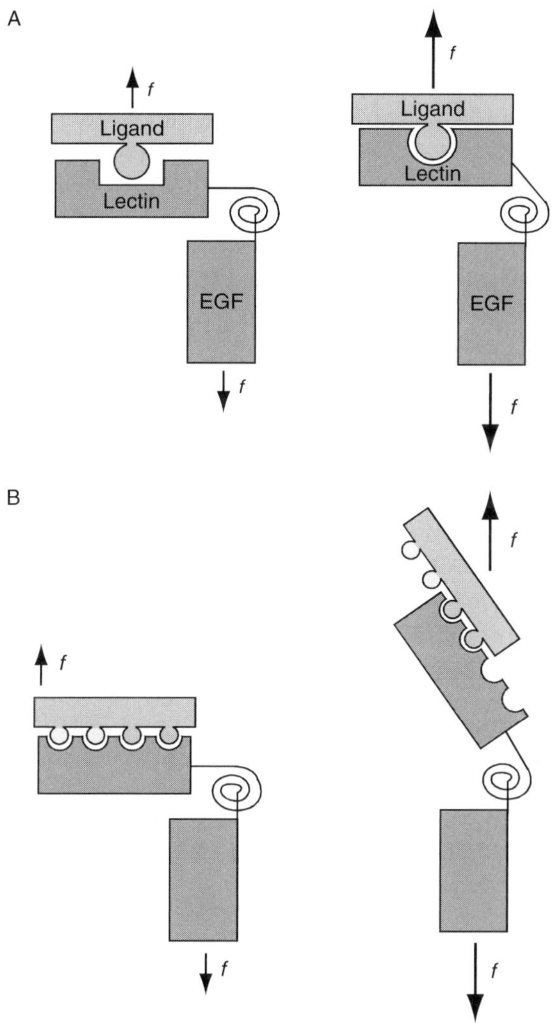

FIGURE 7 Models for catch bonds: (A) allosteric model and (B) sliding-rebinding model.

interdomain hinge, causes bound ligand to slide across the interface instead of pulling directly away. During sliding, existing interactions break but new interactions also form, decreasing off-rate and prolonging lifetime.

Two mutants of P-selectin increase the force-free affinity for PSGL-1, in part by reducing off-rate. One mutant introduces a glycan ''wedge'' between the lectin and EGF domains to force the interdomain angle open (Phan, Waldron, & Springer, 2006). The other substitutes a bulky residue

(A28H) to shift a loop in the lectin domain along the proposed allosteric pathway (Waldron & Springer, 2009). The effects of force on bond lifetimes were not examined. The force-free affinity data support the allosteric model, although the mutations might alter structure differently than does force.

Molecular dynamics (MD) simulations of force applied to the P-selectin–PSGL-1 complex support the sliding-rebinding model (Lou & Zhu, 2007). Transitions from catch to slip bonds occur at a higher force range for L-selectin than P-selectin (Sarangapani et al., 2004) (Fig. 5), which is due in part to different interdomain contacts. EGF-domain residue 138 is Gly in P-selectin but Asn in L-selectin. In the bent P-selectin structure, Gly138 is close to but does not interact with Tyr37 in the lectin domain (Fig. 6C). In the extended P-selectin structure, Tyr37 pivots about Gly138 without steric clashes. In the bent L-selectin structure, Asn138 forms a hydrogen bond with Tyr37 in the lectin domain (Lou et al., 2006) (Fig. 6D). Molecular modeling indicates that this hydrogen bond must break and steric clashes must be overcome for the interdomain angle to open. An N138G substitution in L-selectin increases hinge flexibility and produces more pronounced catch bonds at lower forces with longer lifetimes with both 6-sulfo-sLe^{x} and PSGL-1 (Lou et al.). Sliding-rebinding readily explains this effect because it can act on ligands with different structures as force is applied.

In summary, there is evidence supporting both allosteric and sliding-rebinding models. Further studies are required to determine whether either or both of these models, or perhaps another model, explain the mechanistic basis for catch bonds. Indeed, an L-selectin substitution in the lectin domain (A108H) predicted to create a contact with PSGL-1 peptide eliminates catch bonds with PSGL-1 but not with 6-sulfo-sLe^{x} (Klopocki et al., 2008), indicating that selectins might use different mechanisms for different ligands to generate catch bonds.

VIII. FLOW-ENHANCED ADHESION: THE SHEAR THRESHOLD PHENOMENON

Selectins require a minimal shear to support cell adhesion (Finger et al., 1996; Lawrence, Kansas, Kunkel, & Ley, 1997). This counterintuitive phenomenon is particularly striking for L-selectin. Below the shear threshold, few cells tether. As shear rises above the threshold, the tether rate increases, reaches a maximum at an optimal shear, and then declines as shear increases further (Fig. 8A). In parallel, cells roll more slowly and more regularly until an optimal shear is reached where rolling velocity is minimal; they then roll faster as shear increases further (Fig. 8B). Thus, flow-enhanced adhesion results from both increased tethering and slower rolling (Fig. 8C).

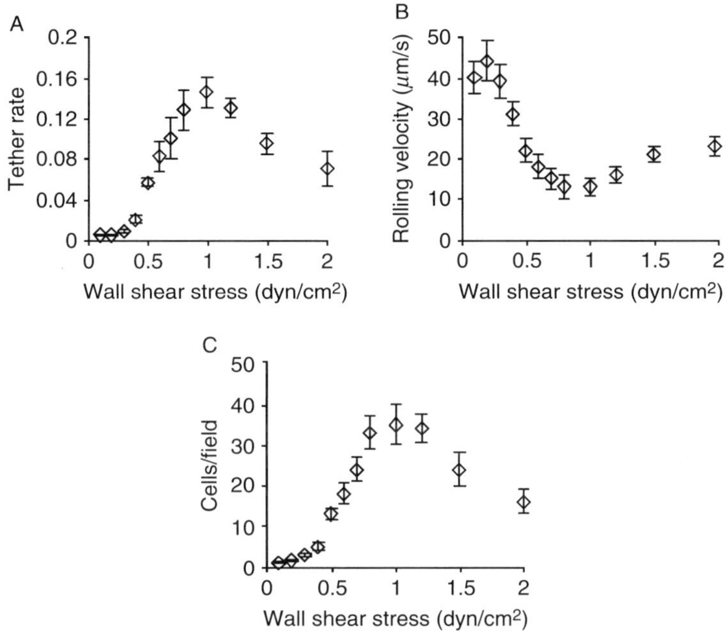

FIGURE 8 Flow-enhanced rolling adhesion. Human neutrophils were perfused over immobilized PSGL-1 at the indicated wall shear stress. The tethering rate (A), mean rolling velocity (B), and number of cells rolling per field (C) were measured (adapted with permission from Yago et al., 2004, 2007).

Transport governs flow-enhanced tethering through three mechanisms (Yago, Zarnitsyna, Klopocki, McEver, & Zhu, 2007). The first is the sliding velocity V_s of the cell bottom near the vascular surface (Fig. 3A), which is controlled by the product of the shear rate and the radius of the cell. The second is Brownian motion of the cell, which modulates the gap distance above and below the threshold where a selectin and its ligand can contact. The third is molecular diffusion, which allows a selectin and its ligand to orient their binding sites for docking. The N138G substitution in L-selectin increases its rotational diffusivity and augments tethering (Lou et al., 2006). As flow increases, transport increases encounters between L-selectin and its ligands; this favors productive interactions because the docking rate (a function of the k_{on}) is rapid (Yago et al.). Above the flow optimum, the tethering rate declines as the encounter times become shorter than the time for docking, and thus become limiting (Fig. 3C).

Force governs flow-enhanced rolling by eliciting catch bonds (Yago et al., 2004). As flow increases, rolling becomes slower and more regular as force lengthens the lifetimes of L-selectin bonds. Above the flow optimum, rolling becomes faster and less regular as higher forces shorten bond lifetimes (slip bonds). Adhesive dynamics simulations based on the experimental data confirm the importance of catch bonds for flow-enhanced rolling (Beste & Hammer, 2008). The N138G substitution in L-selectin reduces the shear threshold for rolling on both PSGL-1 and 6-sulfo-sLe^x by prolonging bond lifetimes at lower forces (Lou et al., 2006). The A108H substitution in L-selectin predicted to enhance binding to peptide but not glycan components of PSGL-1 increases the force-free affinity for PSGL-1 but not for 6-sulfo-sLe^x (Klopocki et al., 2008). Bond lifetimes with PSGL-1, but not with 6-sulfo-sLe^x, are longer at low forces and decrease monotonically as force increases. The A108H substitution eliminates the shear threshold for rolling on PSGL-1 but does not affect rolling on 6-sulfo-sLe^x, consistent with its selective prolongation of bond lifetimes with PSGL-1 at low forces (Klopocki et al.).

Flow-enhanced adhesion may have important biological functions. Circulating leukocytes do not aggregate even though they express both L-selectin and its ligand PSGL-1. Although colliding leukocytes might form bonds between L-selectin and PSGL-1, the force applied to these bonds is small because of the minor velocity differences between the fluid and circulating cells (Fig. 9). The bonds are therefore short-lived and the cells rapidly dissociate. Consistent with this hypothesis, microspheres bearing L-selectinN138G, but not L-selectin, form doublets and larger aggregates with neutrophils in a flow field (Lou et al., 2006). The same shear stress exerts larger forces on the bonds that anchor a cell to the stationary vessel wall, prolonging their lifetimes and enabling rolling (Fig. 9). Circulating platelets express high levels of P-selectin but do not mobilize it to the surface until they are activated (McEver, 2002; McEver et al., 1995; Vestweber & Blanks, 1999). PSGL-1 bonds with P-selectin have longer lifetimes than those with L-selectin even at low forces (Marshall et al., 2003; Sarangapani et al., 2004) (Fig. 5). The combination of high P-selectin densities and longer bond lifetimes initiate adhesion of circulating activated platelets to leukocytes. Subsequent signaling activates β_2-integrins that stabilize adhesion (Evangelista et al., 2007). Thus, circulating platelet–leukocyte aggregates are observed in disorders that increase platelet activation (Michelson, Barnard, Krueger, Valeri, & Furman, 2001). The shear threshold for leukocyte rolling through P-selectin is less evident than through L-selectin, because smaller forces elicit catch bonds and P-selectin bonds last longer than L-selectin bonds at all force levels. Similar principles likely explain the less evident shear threshold for rolling through E-selectin, although catch bonds between E-selectin and its ligands have not been reported. During rolling through P- or E-selectin, force distributed to multiple bonds might reduce the force on individual bonds to levels that produce catch bonds. Furthermore, bonds formed at the leading

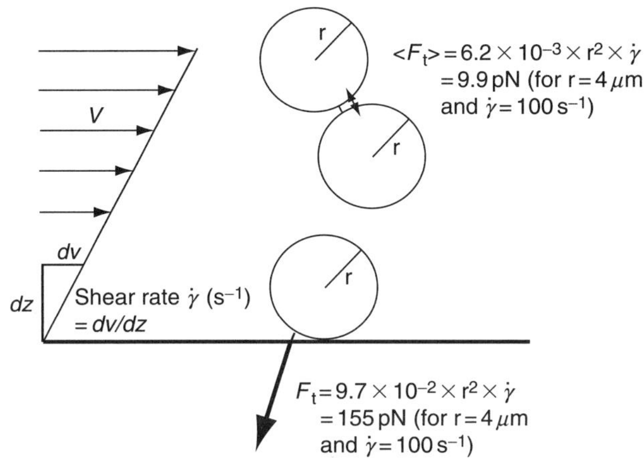

FIGURE 9 Schematic showing tensile forces (F_t) between cells in a flowing doublet as well as between a cell and the wall under a simple flow field. $\langle F_t \rangle$ indicates average force over a cycle of tumbling.

edge of the cell are gradually loaded by force as they move to the trailing edge. These bonds may experience force levels that prolong their survival until they reach the trailing edge to anchor the cell.

IX. CELLULAR FEATURES THAT MODULATE SELECTIN-MEDIATED LEUKOCYTE ROLLING

Microspheres bearing a selectin (or selectin ligand) roll on an immobilized selectin ligand (or selectin), establishing that the molecular features of selectins and their ligands are sufficient to support rolling (Brunk, Goetz, & Hammer, 1996; Greenberg, Brunk, & Hammer, 2000; Yago et al., 2002). However, cell activation regulates the densities of receptors and ligands on the plasma membrane (McEver, 2002; McEver et al., 1995; Vestweber & Blanks, 1999). Furthermore, the physical features of cells and the manner in which they present selectins and their ligands have major influences on rolling behavior. The surfaces of leukocytes are highly irregular because of numerous microvilli that extend ~1 μm from the cell body (Fig. 10). L-selectin and PSGL-1 are concentrated on the tips of microvilli; this enhances tethering by increasing the contacts with ligands on endothelial cells (Moore et al., 1995; Von Andrian, Hasslen, Nelson, Erlandsen, & Butcher, 1995; Yago et al.). Microvilli may penetrate the proteoglycan-rich glycocalyx of endothelial cells, estimated to extend ~0.5 μm above the cell

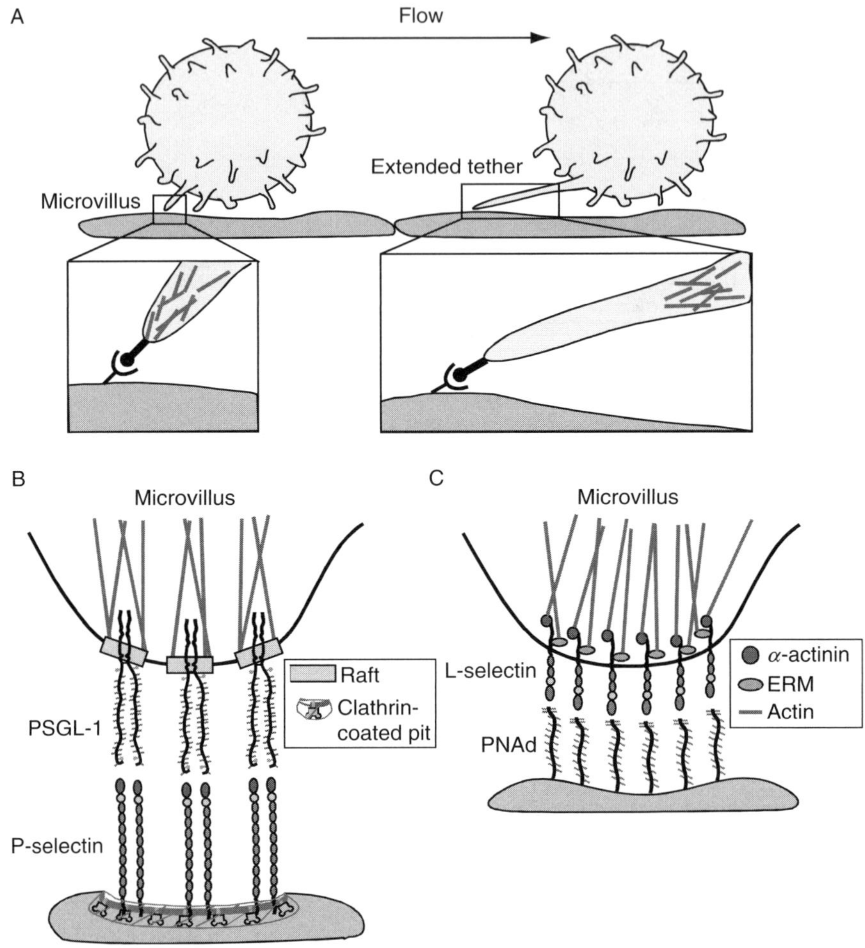

FIGURE 10 Effects of cell-surface organization on selectin–ligand interactions under flow. (A) Extension of a long membrane tether from a microvillus after disruption of cytoskeletal connections with the membrane. (B) P-selectin and PSGL-1 form dimers. P-selectin clusters in clathrin-coated pits through interactions of its cytoplasmic domain. PSGL-1 associates with lipid rafts and clusters in microvilli, perhaps indirectly through interactions of other raft components with the cytoskeleton. Note that the tip of a microvillus is actually larger than a clathrin-coated pit, and PSGL-1 molecules in different regions of the tip may interact with P-selectin molecules in two or more clustered pits. (C) L-selectin clusters in microvilli through direct interactions of its cytoplasmic domain with α-actinin and ERM proteins, which connect to actin filaments.

surface (Weinbaum, Tarbell, & Damiano, 2007), to favor rolling by bringing selectins and ligands into repeated contact. Cells are deformable. At higher wall shear stresses, the compressive force F_c acting on the cell bottom compresses the glycocalyx and enlarges the contact area so that more selectin–ligand bonds can form (Lei, Lawrence, & Dong, 1999). The tether force F_t also rapidly extrudes long membrane tethers at the trailing edge of the cell (Schmidtke & Diamond, 2000) (Fig. 10A). This process is highly dynamic, increasing at higher shears and decreasing at lower shears (Ramachandran, Williams, Yago, Schmidtke, & McEver, 2004). Tethers extend by stretching microvilli and by separating membrane around adhesion receptors from the cytoskeleton (Evans, Heinrich, Leung, & Kinoshita, 2005; Shao, Ting-Beall, & Hochmuth, 1998). Most tethers retract after selectin–ligand bonds dissociate (Ramachandran et al.; Schmidtke & Diamond, 2000), suggesting that the cytoskeleton maintains some interactions with the membrane. By altering the geometry of the structures that anchor cells under flow, long membrane tethers reduce force on adhesive bonds. This explains why leukocytes roll only slightly faster as wall shear stress increases to levels that generate slip bonds, whereas microspheres and fixed cells roll much faster until they detach (Yago et al.).

Both P-selectin and PSGL-1 are extended molecules with membrane-distal binding domains (Li et al., 1996; Ushiyama, Laue, Moore, Erickson, & McEver, 1993) (Fig. 10B). This architecture enhances tethering and rolling by increasing encounters between molecules (Huang et al., 2004; Patel, Nollert, & McEver, 1995; Yago et al., 2002). PSGL-1 forms noncovalent dimers through interactions of the transmembrane and cytoplasmic domains that are stabilized by a juxtamembrane disulfide bond (Epperson, Patel, McEver, & Cummings, 2000; Miner et al., 2008). P-selectin forms noncovalent dimers through interactions of the transmembrane domains (Barkalow, Barkalow, & Mayadas, 2000; Ushiyama et al.). Dimeric binding of P-selectin to PSGL-1 causes slower and more regular rolling in the catch bond regime because force is distributed between both subunits of dimeric bonds (Marshall et al., 2003; Ramachandran et al., 2001). Furthermore, the cell remains anchored when one pair of subunits dissociates, providing an opportunity for the pair to rebind. L-selectin may not dimerize (Sarangapani et al., 2004) (Fig. 10C), and no publication has reported whether E-selectin forms dimers.

Clustering of selectins and their ligands in membrane domains provides another mechanism to stabilize rolling by increasing bond number and reducing the force on individual bonds. The cytoplasmic domain links L-selectin to the cytoskeleton through a membrane-distal binding site for α-actinin (Pavalko et al., 1995) and a membrane-proximal site for ezrin/radixin/moesin (ERM) proteins (Ivetic, Deka, Ridley, & Ager, 2002) (Fig. 10C). In transfected cells, mutating the ERM-binding site shifts L-selectin from microvilli to the cell body; this reduces tethering but does not impair rolling (Ivetic et al., 2004). Deleting the α-actinin-binding site destabilizes rolling, and deleting both α-actinin- and ERM-binding sites virtually eliminates rolling (Dwir, Kansas,

& Alon, 2001). Thus, direct cytoskeletal anchorage of the cytoplasmic tail of L-selectin is essential to support tethering and/or rolling. The cytoplasmic domain of PSGL-1 also has a membrane-proximal binding site for ERM proteins (Serrador et al., 2002). However, contrary to an earlier report (Snapp, Heitzig, & Kansas, 2002), deleting the cytoplasmic domain of PSGL-1 does not impair rolling of transfected cells on P-selectin (Miner et al., 2008). Knock-in "ΔCD" mice that express a truncated form of PSGL-1 without the cytoplasmic domain have been generated (Miner et al.). Remarkably, ΔCD leukocytes still concentrate PSGL-1 in microvilli, exhibit at most minor defects in rolling on P-selectin, and extend and retract long membrane tethers at the trailing edge. Like WT PSGL-1, ΔCD PSGL-1 associates with lipid rafts (Miner et al.), which may anchor it indirectly to the cytoskeleton through interactions with raft-enriched proteins (Rossy, Schlicht, Engelhardt, & Niggli, 2009) (Fig. 10C). L-selectin, by contrast, is not associated with lipid rafts in leukocytes (Dwir et al., 2007) (Fig. 10C). The cytoplasmic domains of P- and E-selectin interact with clathrin-coated pits (Chuang et al., 1997; Kluger, Shiao, Bothwell, & Pober, 2002; Setiadi, Disdier, Green, Canfield, & McEver, 1995) (Fig. 10B). This is an important mechanism for endocytosis and eventual lysosomal degradation of P- and E-selectin, thus limiting inflammation by clearing the selectins from the cell surface. Clustering of P- and E-selectin in clathrin-coated pits supports slower, more regular rolling by forming bond clusters with PSGL-1 and other ligands (Setiadi & McEver, 2003, 2008; Setiadi, Sedgewick, Erlandsen, & McEver, 1998). Clustering of E-selectin in lipid rafts of endothelial cells further slows leukocyte rolling (Setiadi & McEver, 2008). Interactions of clathrin-coated pits and lipid rafts with the cytoskeleton may prevent force-induced extraction of P- and E-selectin from the membrane as leukocytes roll on endothelial cells. Endothelial cells also extrude long membrane tethers that may form selectin–ligand bonds with tethers extended at the trailing edges of rolling leukocytes (Girdhar & Shao, 2007).

X. CONCLUSIONS

The remarkable ability of circulating leukocytes to tether to and roll on vascular surfaces requires specialized, mechanically regulated kinetic properties of the interacting selectins and their ligands. Tethering requires fast on-rates that make transport the limiting factor at lower flow rates, providing the opportunity for flow to enhance tethering. Rolling requires a delicate balance between rapid formation and rapid breakage of adhesive bonds. Although both on-rates and off-rates contribute to this balance, the off-rates of bonds that anchor rolling cells at the trailing edge are the principal regulators of rolling velocities. These bonds are subjected to forces that must balance the forces imposed by shear stress and the vessel wall to which the leukocyte adheres. They must have sufficient tensile strength so that force does not accelerate dissociation so rapidly that the cells detach into the fluid stream. On the other hand, they must dissociate sufficiently quickly so that cells pivot about new bonds and continue to roll. Force-regulated transitions from catch bonds to slip bonds provide a mechanism to optimize rolling dynamics and explain the counterintuitive requirement for a shear threshold to support rolling. Catch bonds provide a mechanism to prevent inappropriate leukocyte aggregation in flowing blood, yet enable cells to roll on the vessel wall.

The structural features of selectins and their ligands are sufficient to support rolling in cell-free systems, but how cells organize these molecules on their surfaces is equally important. Common themes are self-association into dimers, clustering in membrane domains, and direct or indirect anchorage to the cytoskeleton. These presentations favor bond clusters that distribute force among individual bonds. Bond clusters, extension of microvilli, and pulling of long membrane tethers from the trailing edge cooperate to stabilize rolling velocities, particularly at higher wall shear stresses where slip bonds are dominant.

Although enormous progress has been made in the past 20 years, much remains to be learned about the molecular and cellular requirements for rolling cell adhesion. Further study of the mechanisms for catch bonds is required, with attention to the relative contributions of interactions within and between domains in regulating bond lifetime. We need to understand why force applied to selectin–ligand bonds promotes rolling rather than arrest. How partitioning of adhesion receptors into membrane domains regulates rolling is ripe for further investigation. Finally, the mechanisms that regulate rolling of leukocytes *in vitro* must be studied *in vivo* to clarify their biological consequences in health and disease.

Acknowledgments

The authors are supported by NIH grants HL34363, HL085607, HL090923, HL091020, AI44902, and AI077343. We thank Tadayuki Yago, Lijun Xia, and Wei Chen for assistance with figures.

References

Alon, R., Chen, S. Q., Puri, K. D., Finger, E. B., & Springer, T. A. (1997). The kinetics of L-selectin tethers and the mechanics of selectin-mediated rolling. *Journal of Cell Biology, 138*, 1169–1180.

Alon, R., Hammer, D. A., & Springer, T. A. (1995). Lifetime of the P-selectin: Carbohydrate bond and its response to tensile force in hydrodynamic flow. *Nature, 374*, 539–542.

Barkalow, F. J., Barkalow, K. L., & Mayadas, T. N. (2000). Dimerization of P-selectin in platelets and endothelial cells. *Blood, 96*, 3070–3077.

Bell, G. I. (1978). Models for the specific adhesion of cells to cells: A theoretical framework for adhesion mediated by reversible bonds between cell surface molecules. *Science, 200*, 618–627.

Beste, M. T., & Hammer, D. A. (2008). Selectin catch-slip kinetics encode shear threshold adhesive behavior of rolling leukocytes. *Proceedings of the National Academy of Sciences of the United States of America, 105*, 20716–20721.

Brunk, D. K., Goetz, D. J., & Hammer, D. A. (1996). Sialyl Lewis x/E-selectin-mediated rolling in a cell-free system. *Biophysical Journal, 71*, 2902–2907.

Chang, K.-C., & Hammer, D. A. (1999). The forward rate of binding of surface-tethered reactants: Effect of relative motion between two surfaces. *Biophysical Journal, 76*, 1280–1292.

Chen, W., Evans, E. A., McEver, R. P., & Zhu, C. (2007). Monitoring receptor–ligand interactions between surfaces by thermal fluctuations. *Biophysical Journal, 94*, 694–701.

Chuang, P. I., Young, B. A., Thiagarajan, R. R., Cornejo, C., Winn, R. K., & Harlan, J. M. (1997). Cytoplasmic domain of E-selectin contains a non-tyrosine endocytosis signal. *Journal of Biological Chemistry, 272*, 24813–24818.

Dembo, M., Torney, D. C., Saxman, K., & Hammer, D. (1988). The reaction-limited kinetics of membrane-to-surface adhesion and detachment. *Proceedings of the Royal Society of London B: Biological Science, 234*, 55–83.

Dwir, O., Grabovsky, V., Pasvolsky, R., Manevich, E., Shamri, R., Gutwein, P., et al. (2007). Membranal cholesterol is not required for L-selectin adhesiveness in primary lymphocytes but controls a chemokine-induced destabilization of L-selectin rolling adhesions. *Journal of Immunology, 179*, 1030–1038.

Dwir, O., Kansas, G. S., & Alon, R. (2001). Cytoplasmic anchorage of L-selectin controls leukocyte capture and rolling by increasing the mechanical stability of the selectin tether. *Journal of Cell Biology, 155*, 145–156.

Epperson, T. K., Patel, K. D., McEver, R. P., & Cummings, R. D. (2000). Noncovalent association of P-selectin glycoprotein ligand-1 and minimal determinants for binding to P-selectin. *Journal of Biological Chemistry, 275*, 7839–7853.

Evangelista, V., Pamuklar, Z., Piccoli, A., Manarini, S., Dell'elba, G., Pecce, R., et al. (2007). Src family kinases mediate neutrophil adhesion to adherent platelets. *Blood, 109*, 2461–2469.

Evans, E., Heinrich, V., Leung, A., & Kinoshita, K. (2005). Nano- to microscale dynamics of P-selectin detachment from leukocyte interfaces. I. Membrane separation from the cytoskeleton. *Biophysical Journal, 88*, 2288–2298.

Evans, E., Leung, A., Hammer, D., & Simon, S. (2001). Chemically distinct transition states govern rapid dissociation of single L-selectin bonds under force. *Proceedings of the National Academy of Sciences of the United States of America, 98*, 3784–3789.

Evans, E., Leung, A., Heinrich, V., & Zhu, C. (2004). Mechanical switching and coupling between two dissociation pathways in a P-selectin adhesion bond. *Proceedings of the National Academy of Sciences of the United States of America, 101*, 11281–11286.

Finger, E. B., Puri, K. D., Alon, R., Lawrence, M. B., Von Andrian, U. H., & Springer, T. A. (1996). Adhesion through L-selectin requires a threshold hydrodynamic shear. *Nature, 379*, 266–269.

Girdhar, G., & Shao, J. Y. (2007). Simultaneous tether extraction from endothelial cells and leukocytes: Observation, mechanics, and significance. *Biophysical Journal, 93*, 4041–4052.

Goldman, A. J., Cox, R. G., & Brenner, H. (1967). Slow viscous motion of a sphere parallel to a plane wall—Couette flow. *Chemical Engineering Science, 22*, 653–660.

Graves, B. J., Crowther, R. L., Chandran, C., Rumberger, J. M., Li, S., Huang, K.-S., et al. (1994). Insight into E-selectin/ligand interaction from the crystal structure and mutagenesis of the lec/EGF domains. *Nature, 367*, 532–538.

Greenberg, A. W., Brunk, D. K., & Hammer, D. A. (2000). Cell-free rolling mediated by L-selectin and sialyl Lewis x reveals the shear threshold effect. *Biophysical Journal, 79*, 2391–2402.

Hidalgo, A., Peired, A. J., Wild, M. K., Vestweber, D., & Frenette, P. S. (2007). Complete identification of E-selectin ligands on neutrophils reveals distinct functions of PSGL-1, ESL-1, and CD44. *Immunity, 26*, 477–489.

Huang, J., Chen, J., Chesla, S. E., Yago, T., Mehta, P., McEver, R. P., et al. (2004). Quantifying the effects of molecular orientation and length on two-dimensional receptor–ligand binding kinetics. *Journal of Biological Chemistry, 279*, 44915–44923.

Ivetic, A., Deka, J., Ridley, A., & Ager, A. (2002). The cytoplasmic tail of L-selectin interacts with members of the ezrin–radixin–moesin (ERM) family of proteins. *Journal of Biological Chemistry, 277*, 2321–2329.

Ivetic, A., Florey, O., Deka, J., Haskard, D. O., Ager, A., & Ridley, A. J. (2004). Mutagenesis of the ezrin–radixin–moesin binding domain of L-selectin tail affects shedding, microvillar positioning, and leukocyte tethering. *Journal of Biological Chemistry, 279*, 33263–33272.

Katayama, Y., Hidalgo, A., Chang, J., Peired, A., & Frenette, P. S. (2005). CD44 is a physiological E-selectin ligand on neutrophils. *Journal of Experimental Medicine, 201*, 1183–1189.

Kawashima, H., Petryniak, B., Hiraoka, N., Mitoma, J., Huckaby, V., Nakayama, J., et al. (2005). *N*-acetylglucosamine-6-*O*-sulfotransferases 1 and 2 cooperatively control lymphocyte homing through L-selectin ligand biosynthesis in high endothelial venules. *Nature Immunology, 6*, 1096–1104.

Klopocki, A. G., Yago, T., Mehta, P., Yang, J., Wu, T., Leppanen, A., et al. (2008). Replacing a lectin domain residue in L-selectin enhances binding to P-selectin glycoprotein ligand-1 but not to 6-sulfo-sialyl Lewis x. *Journal of Biological Chemistry, 283*, 11493–11500.

Kluger, M. S., Shiao, S. L., Bothwell, A. L. M., & Pober, J. S. (2002). Cutting edge: Internalization of transduced E-selectin by cultured human endothelial cells: Comparison of dermal microvascular and umbilical vein cells and identification of a phosphoserine-type di-leucine motif. *Journal of Immunology, 168*, 2091–2095.

Kobzdej, M. M. A., Leppänen, A., Ramachandran, V., Cummings, R. D., & McEver, R. P. (2002). Discordant expression of selectin ligands and sialyl Lewis x-related epitopes on murine myeloid cells. *Blood, 100*, 485–494.

Lawrence, M. B., Kansas, G. S., Kunkel, E. J., & Ley, K. (1997). Threshold levels of fluid shear promote leukocyte adhesion through selectins (CD62L, P, E). *Journal of Cell Biology, 136*, 717–727.

Lei, X., Lawrence, M. B., & Dong, C. (1999). Influence of cell deformation on leukocyte rolling adhesion in shear flow. *Journal of Biomechanical Engineering, 121*, 636–643.

Leppänen, A., White, S. P., Helin, J., McEver, R. P., & Cummings, R. D. (2000). Binding of glycosulfopeptides to P-selectin requires stereospecific contributions of individual tyrosine sulfate and sugar residues. *Journal of Biological Chemistry, 275*, 39569–39578.

Leppänen, A., Yago, T., Otto, V. I., McEver, R. P., & Cummings, R. D. (2003). Model glycosulfopeptides from P-selectin glycoprotein ligand-1 require tyrosine sulfation and a core 2-branched *O*-glycan to bind to L-selectin. *Journal of Biological Chemistry, 278*, 26391–26400.

Ley, K., Laudanna, C., Cybulsky, M. I., & Nourshargh, S. (2007). Getting to the site of inflammation: The leukocyte adhesion cascade updated. *Nature Reviews Immunology, 7*, 678–689.

Li, F., Erickson, H. P., James, J. A., Moore, K. L., Cummings, R. D., & McEver, R. P. (1996). Visualization of P-selectin glycoprotein ligand-1 as a highly extended molecule and mapping of protein epitopes for monoclonal antibodies. *Journal of Biological Chemistry, 271*, 6342–6348.

Long, M., Zhao, H., Huang, K. S., & Zhu, C. (2001). Kinetic measurements of cell surface E-selectin/carbohydrate ligand interactions. *Annals of Biomedical Engineering, 29*, 935–946.

Lou, J., Yago, T., Klopocki, A. G., Mehta, P., Chen, W., Zarnitsyna, V. I., et al. (2006). Flow-enhanced adhesion regulated by a selectin interdomain hinge. *Journal of Cell Biology, 174*, 1107–1117.

Lou, J., & Zhu, C. (2007). A structure-based sliding-rebinding mechanism for catch bonds. *Biophysical Journal, 92*, 1471–1485.

Marshall, B. T., Long, M., Piper, J. W., Yago, T., McEver, R. P., & Zhu, C. (2003). Direct observation of catch bonds involving cell-adhesion molecules. *Nature, 423*, 190–193.

McEver, R. P. (2002). Selectins: Lectins that initiate cell adhesion under flow. *Current Opinion in Cell Biology, 14*, 581–586.

McEver, R. P., Moore, K. L., & Cummings, R. D. (1995). Leukocyte trafficking mediated by selectin–carbohydrate interactions. *Journal of Biological Chemistry, 270*, 11025–11028.

Mehta, P., Cummings, R. D., & McEver, R. P. (1998). Affinity and kinetic analysis of P-selectin binding to P-selectin glycoprotein ligand-1. *Journal of Biological Chemistry, 273*, 32506–32513.

Michelson, A. D., Barnard, M. R., Krueger, L. A., Valeri, C. R., & Furman, M. I. (2001). Circulating monocyte-platelet aggregates are a more sensitive marker of in vivo platelet activation than platelet surface P-selectin: Studies in baboons, human coronary intervention, and human acute myocardial infarction. *Circulation, 104*, 1533–1537.

Miner, J. J., Xia, L., Yago, T., Kappelmayer, J., Liu, Z., Klopocki, A. G., et al. (2008). Separable requirements for cytoplasmic domain of PSGL-1 in leukocyte rolling and signaling under flow. *Blood, 112*, 2035–2045.

Mitoma, J., Bao, X., Petryanik, B., Schaerli, P., Gauguet, J. M., Yu, S. Y., et al. (2007). Critical functions of *N*-glycans in L-selectin-mediated lymphocyte homing and recruitment. *Nature Immunology, 8*, 409–418.

Moore, K. L., Patel, K. D., Bruehl, R. E., Fugang, L., Johnson, D. A., Lichenstein, H. S., et al. (1995). P-selectin glycoprotein ligand-1 mediates rolling of human neutrophils on P-selectin. *Journal of Cell Biology, 128*, 661–671.

Nicholson, M. W., Barclay, A. N., Singer, M. S., Rosen, S. D., & Van der Merwe, P. A. (1998). Affinity and kinetic analysis of L-selectin (CD62L) binding to glycosylation-dependent cell-adhesion molecule-1. *Journal of Biological Chemistry, 273*, 763–770.

Nimrichter, L., Burdick, M. M., Aoki, K., Laroy, W., Fierro, M. A., Hudson, S. A., et al. (2008). E-selectin receptors on human leukocytes. *Blood, 112*, 3744–3752.

Patel, K. D., Nollert, M. U., & McEver, R. P. (1995). P-selectin must extend a sufficient length from the plasma membrane to mediate rolling of neutrophils. *Journal of Cell Biology, 131*, 1893–1902.

Pavalko, F. M., Walker, D. M., Graham, L., Goheen, M., Doerschuk, C. M., & Kansas, G. S. (1995). The cytoplasmic domain of L-selectin interacts with cytoskeletal proteins via α-actinin: Receptor positioning in microvilli does not require interaction with α-actinin. *Journal of Cell Biology, 129*, 1155–1164.

Phan, U. T., Waldron, T. T., & Springer, T. A. (2006). Remodeling of the lectin-EGF-like domain interface in P- and L-selectin increases adhesiveness and shear resistance under hydrodynamic force. *Nature Immunology, 7*, 883–889.

Ramachandran, V., Nollert, M. U., Qiu, H., Liu, W. J., Cummings, R. D., Zhu, C., et al. (1999). Tyrosine replacement in P-selectin glycoprotein ligand-1 affects distinct kinetic and mechanical properties of bonds with P- and L-selectin. *Proceedings of the National Academy of Sciences of the United States of America, 96*, 13771–13776.

Ramachandran, V., Williams, M., Yago, T., Schmidtke, D. W., & McEver, R. P. (2004). Dynamic alterations of membrane tethers stabilize leukocyte rolling on P-selectin. *Proceedings of the National Academy of Sciences of the United States of America, 101*, 13519–13524.

Ramachandran, V., Yago, T., Epperson, T. K., Kobzdej, M. M. A., Nollert, M. U., Cummings, R. D., et al. (2001). Dimerization of a selectin and its ligand stabilizes cell rolling and enhances tether strength in shear flow. *Proceedings of the National Academy of Sciences of the United States of America, 98*, 10166–10171.

Rivera-Nieves, J., Burcin, T. L., Olson, T. S., Morris, M. A., McDuffie, M., Cominelli, F., et al. (2006). Critical role of endothelial P-selectin glycoprotein ligand 1 in chronic murine ileitis. *Journal of Experimental Medicine, 203*, 907–917.

Rosen, S. D. (2004). Ligands for L-selectin: Homing, inflammation, and beyond. *Annual Review of Immunology, 22*, 129–156.

Rossy, J., Schlicht, D., Engelhardt, B., & Niggli, V. (2009). Flotillins interact with PSGL-1 in neutrophils and, upon stimulation, rapidly organize into membrane domains subsequently accumulating in the uropod. *PLoS ONE, 4*, e5403.

Sarangapani, K. K., Yago, T., Klopocki, A. G., Lawrence, M. B., Fieger, C. B., Rosen, S. D., et al. (2004). Low force decelerates L-selectin dissociation from P-selectin glycoprotein ligand-1 and endoglycan. *Journal of Biological Chemistry, 279*, 2291–2298.

Schmidtke, D. W., & Diamond, S. L. (2000). Direct observation of membrane tethers formed during neutrophil attachment to platelets or P-selectin under physiological flow. *Journal of Cell Biology, 149*, 719–729.

Serrador, J. M., Urzainqui, A., Alonso-Lebrero, J. L., Cabrero, J. R., Montoya, M. C., Vicente-Manzanares, M., et al. (2002). A juxta-membrane amino acid sequence of P-selectin glycoprotein ligand-1 is involved in moesin binding and ezrin/radixin/moesin-directed targeting at the trailing edge of migrating lymphocytes. *European Journal of Immunology, 32*, 1560–1566.

Setiadi, H., Disdier, M., Green, S. A., Canfield, W. M., & McEver, R. P. (1995). Residues throughout the cytoplasmic domain affect the internalization efficiency of P-selectin. *Journal of Biological Chemistry, 270*, 26818–26826.

Setiadi, H., & McEver, R. P. (2003). Signal-dependent distribution of cell surface P-selectin in clathrin-coated pits affects leukocyte rolling under flow. *Journal of Cell Biology, 163*, 1385–1395.

Setiadi, H., & McEver, R. P. (2008). Clustering endothelial E-selectin in clathrin-coated pits and lipid rafts enhances leukocyte adhesion under flow. *Blood, 111*, 1989–1998.

Setiadi, H., Sedgewick, G., Erlandsen, S. L., & McEver, R. P. (1998). Interactions of the cytoplasmic domain of P-selectin with clathrin-coated pits enhance leukocyte adhesion under flow. *Journal of Cell Biology, 142*, 859–871.

Shao, J. Y., Ting-Beall, H. P., & Hochmuth, R. M. (1998). Static and dynamic lengths of neutrophil microvilli. *Proceedings of the National Academy of Sciences of the United States of America, 95*, 6797–6802.

Smith, M. L., Sperandio, M., Galkina, E. V., & Ley, K. (2004). Autoperfused mouse flow chamber reveals synergistic neutrophil accumulation through P-selectin and E-selectin. *Journal of Leukocyte Biology, 76*, 985–993.

Snapp, K. R., Heitzig, C. E., & Kansas, G. S. (2002). Attachment of the PSGL-1 cytoplasmic domain to the actin cytoskeleton is essential for leukocyte rolling on P-selectin. *Blood, 99*, 4494–4502.

Somers, W. S., Tang, J., Shaw, G. D., & Camphausen, R. T. (2000). Insights into the molecular basis of leukocyte tethering and rolling revealed by structures of P- and E-selectin bound to SLe(X) and PSGL-1. *Cell, 103*, 467–479.

Springer, T. A. (2009). Structural basis for selectin mechanochemistry. *Proceedings of the National Academy of Sciences of the United States of America, 106*, 91–96.

Uchimura, K., Gauguet, J. M., Singer, M. S., Tsay, D., Kannagi, R., Muramatsu, T., et al. (2005). A major class of L-selectin ligands is eliminated in mice deficient in two sulfotransferases expressed in high endothelial venules. *Nature Immunology, 6*, 1105–1113.

Ushiyama, S., Laue, T. M., Moore, K. L., Erickson, H. P., & McEver, R. P. (1993). Structural and functional characterization of monomeric soluble P-selectin and comparison with membrane P-selectin. *Journal of Biological Chemistry, 268*, 15229–15237.

Vestweber, D., & Blanks, J. E. (1999). Mechanisms that regulate the function of the selectins and their ligands. *Physiological Reviews, 79*, 181–213.

Von Andrian, U. H., Hasslen, S. R., Nelson, R. D., Erlandsen, S. L., & Butcher, E. C. (1995). A central role for microvillous receptor presentation in leukocyte adhesion under flow. *Cell, 82*, 989–999.

Waldron, T. T., & Springer, T. A. (2009). Transmission of allostery through the lectin domain in selectin-mediated cell adhesion. *Proceedings of the National Academy of Sciences of the United States of America, 106*, 85–90.

Wang, L., Fuster, M., Sriramarao, P., & Esko, J. D. (2005). Endothelial heparan sulfate deficiency impairs L-selectin- and chemokine-mediated neutrophil trafficking during inflammatory responses. *Nature Immunology, 6*, 902–910.

Weinbaum, S., Tarbell, J. M., & Damiano, E. R. (2007). The structure and function of the endothelial glycocalyx layer. *Annual Review of Biomedical Engineering, 9*, 121–167.

Wild, M. K., Huang, M. C., Schulze-Horsel, U., van Der Merwe, P. A., & Vestweber, D. (2001). Affinity, kinetics, and thermodynamics of E-selectin binding to E-selectin ligand-1. *Journal of Biological Chemistry, 276*, 31602–31612.

Xia, L., Sperandio, M., Yago, T., McDaniel, J. M., Cummings, R. D., Pearson-White, S., et al. (2002). P-selectin glycoprotein ligand-1-deficient mice have impaired leukocyte tethering to E-selectin under flow. *Journal of Clinical Investigation, 109*, 939–950.

Yago, T., Leppänen, A., Qiu, H., Marcus, W. D., Nollert, M. U., Zhu, C., et al. (2002). Distinct molecular and cellular contributions to stabilizing selectin-mediated rolling under flow. *Journal of Cell Biology, 158*, 787–799.

Yago, T., Wu, J., Wey, C. D., Klopocki, A. G., Zhu, C., & McEver, R. P. (2004). Catch bonds govern adhesion through L-selectin at threshold shear. *Journal of Cell Biology, 166*, 913–923.

Yago, T., Zarnitsyna, V. I., Klopocki, A. G., McEver, R. P., & Zhu, C. (2007). Transport governs flow-enhanced cell tethering through L-selectin at threshold shear. *Biophysical Journal, 92*, 330–342.

Yeh, J. C., Hiraoka, N., Petryniak, B., Nakayama, J., Ellies, L. G., Rabuka, D., et al. (2001). Novel sulfated lymphocyte homing receptors and their control by a Core1 extension β1, 3-*N*-acetylglucosaminyltransferase. *Cell, 105*, 957–969.

CHAPTER 8

Modeling Leukocyte Rolling

Maria K. Pospieszalska and Klaus Ley
Division of Inflammation Biology, La Jolla Institute for Allergy and Immunology, La Jolla, California 92037, USA

I. OVERVIEW

Computer modeling is a powerful tool giving detailed insights into *in vivo* biological processes and *in vitro* experiments. Typically, only some of the data generated by a model, and in a reduced form, can be observed or measured by an experimentalist. Leukocyte rolling, a behavior commonly observed in inflammation, is mediated by a continuous series of molecular bonds between the cell and the substrate that rapidly form and dissociate.

Current Topics in Membranes, Volume 64 1063-5823/09 $35.00

DOI: 10.1016/S1063-5823(09)64008-4

Molecular site densities and parameters such as molecular lengths and reaction rates have been experimentally measured or estimated from experiments. The leukocytes are viscoelastic bodies studded with viscoelastic microvilli. Some of their viscoelastic properties are accessible to experimental interrogation. This chapter discusses the need for modeling leukocyte rolling, summarizes the history of research in this field, and guides through the development process of a leukocyte rolling model including model parameters, model cellular, molecular, and environmental interaction rules, model algorithm, and model validation. Some rolling models use a direct approach, where key molecules, bonds, and cellular elements are tracked in time and space, whereas others use semianalytic, analytical, or agent-based modeling methods. We present published modeling approaches to leukocyte rolling, define open questions, and make suggestions for future work.

II. MOTIVATION FOR MODELING LEUKOCYTE ROLLING

In spite of tremendous progress in observing and measuring rolling leukocytes (Alon & Ley, 2008; Butcher, 1991; Hyduk & Cybulsky, 2009; Petri, Phillipson, & Kubes, 2008; Springer, 1995; Zarbock & Ley, 2009), the leukocyte rolling process cannot be fully understood without quantitative modeling. In most tissues, rolling is a necessary prerequisite for firm leukocyte adhesion (Lindbom, Xie, Raud, & Hedqvist, 1992), which in turn is required for transmigration, the process that allows the leukocyte to reach the tissue in which the insulting stimulus arose. This process involves hundreds of protein molecules on the cell surface (such as adhesion molecules and chemokine receptors), in the cytoskeleton (such as talin-1 and kindlin-3), in the cytosol (such as signaling molecules including kinases and phosphatases), and nonprotein players (such as lipids in the plasma membrane and sugars decorating glycoproteins and glycolipids). Therefore, a comprehensive model of leukocyte adhesion seems to be intractable because too many players and too many parameters exist. But rolling itself can be described using less than 50 parameters and involves, in the very simplest case, only two molecules, a selectin and a selectin ligand. Selectins are carbohydrate-binding molecules expressed on leukocytes and other vascular cells of all mammalians. Selectin ligands are scaffolding proteins that present carbohydrates to selectins. Their functionality is determined by a series of glycosyltransferases that determine the resulting sugar structures (Sperandio, Gleissner, & Ley, 2009).

Leukocyte rolling is a multiscale problem, where the smallest dimensions are atomic and the largest dimensions are cellular (Fig. 1). For example, the actual bond between a selectin and its ligand encompasses a gap of the order of less than 0.1 nm, whereas a typical leukocyte may be 7 μm in diameter,

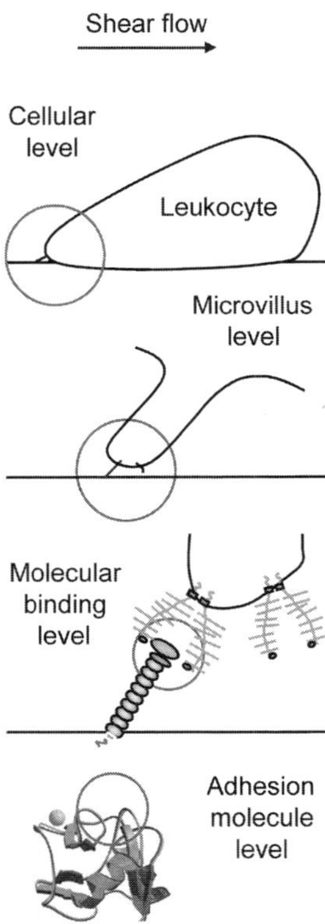

Typical scale	Typical characteristics	Measuring tools	Modeling approach
7 μm	Cortical tension Cytoplasmic viscosity Cytoplasmic relaxation time Deformation index	Light microscopy Micropipette aspiration	Direct Semianalytic Analytical Agent-based
200 nm	Microvillus spring constant Microvillus effective viscosity Microvillus length Tether effective viscosity Tether length	Micropipette aspiration Electron microscopy Atomic force microscopy Laser tweezers	Direct Semianalytic Analytical
50 nm	Bond formation rate Bond dissociation rate Transition-state spring constant Bound-state spring constant Unstressed bond length Bond length Conformational-state affinity	Atomic force microscopy Laser tweezers Sedimentation assay Rotary shadowing electron microscopy	Direct Semianalytic Analytical
5 nm	Geometrical constrains Van der Waals force Partial atomic charge Force constant	X-ray crystallography Nuclear magnetic resonance spectroscopy	Numerical (molecular dynamics) (not discussed)

FIGURE 1 Length Scales in Modeling Leukocyte Rolling.

a 70,000-fold range in scale. No single method can simultaneously measure events at both length scales, but direct models can incorporate this range. Next, the resolution of the optical microscope is of the order of 200 nm, but some of the molecular events determining rolling occur at 1/1000 of this length scale. Similarly, temporal resolution is unsatisfactory. Analog video yields 60 frames per second. Despite tremendous progress in the last few years, most digital cameras are slower yet. Although the fastest commonly available cameras (Kodak Motion Corder) can capture up to 10,000 frames per second, they can only record 2184 frames, limiting recording time to a quarter of a second. In addition, a lot of light is needed for these fast cameras, making them unsuitable for fluorescence work. Although other approaches exist to break these limits of spatial and temporal resolution, such as atomic force microscopy (AFM) (Marshall et al., 2003), not all aspects of rolling can be observed experimentally, and only the model can bring all the measurements from disparate experimental systems together and test whether we understand enough about the rolling process to make it work *in silico*.

Blood leukocytes include neutrophils, eosinophils, monocytes, basophils, natural killer cells, and many subsets of lymphocytes. Each cell type expresses specific sets of adhesion molecules and chemokine receptors. It is unreasonable to expect every experimental approach to be taken with every single cell type, but synthetic mathematical models can apply the available knowledge to these different cell types and take into account their different diameters, surface structures, molecular compositions, and viscoelastic properties. As mentioned above, the simplest rolling system can be modeled as a single type of selectin (L, E, or P) interacting with a single type of selectin ligand (typically P-selectin glycoprotein ligand-1, PSGL-1). However, real rolling events often involve two or three selectins and an unknown number of selectin ligands (this is because the true number of existing E-selectin ligands is unknown). Rolling velocity is modulated by integrins, a class of heterodimeric adhesion molecules (Hynes, 2002) that bind immunoglobulin-like adhesion molecules such as intercellular adhesion molecule-1 (ICAM-1) and vascular cell adhesion molecule-1 (VCAM-1). Both $\alpha_4\beta_1$-integrin (VLA-4) interactions with VCAM-1 (Alon et al., 1995; Berlin et al., 1995) and $\alpha_L\beta_2$-integrin (LFA-1) interactions with ICAM-1 participate in leukocyte rolling (Salas et al., 2004; Zarbock, Lowell, & Ley, 2007) by modulating rolling velocity. Again, it is unreasonable to expect that every possible combination of selectins and integrins will be tested experimentally, yet much of this work is possible *in silico*.

Cells including leukocytes are not just bags of cytosol, but have structured cytoskeleta, surface structures, and different viscoelastic properties at different length and time scales. As will be discussed throughout this chapter, this has a major, but currently underappreciated impact on leukocyte rolling. Although rigid, selectin-coated beads can roll (Brunk, Goetz, & Hammer, 1996; Park et al., 2002), they are not very good at it, that is, rolling becomes unstable at relatively modest wall shear rates (see Section IV.A.3 for the definition of the wall shear rate). To understand the exact cellular requirements that make leukocyte rolling and arrest such an efficient process requires a substantial modeling effort. Although under normal conditions almost no neutrophils reach a given tissue (i.e., the number of neutrophils exiting through the vein draining an organ is the same as the number entering through the artery), this extraction efficiency increases to an astounding 95% under conditions of acute inflammation (Kunkel, Dunne, & Ley, 2000). Rolling models should be able to help explain this phenomenon.

Rolling is not an intuitively obvious cellular behavior. Initially, rolling was thought to be caused by a nonspecific charge effect (Atherton & Born, 1972; Hubbe, 1981), and the first papers on the molecular mechanisms of rolling spent much of their introduction on dispelling this notion (Ley, Lundgren, Berger, & Arfors, 1989). Models of rolling clearly show that any stable rolling needs a set of molecular bonds characterized by high affinity, high bond formation rate, and high bond dissociation rate. Intuitively, rolling seems to be inherently unstable, and the first conceptual models assumed that a slight decrease in the propelling shear force would make the rolling cell stop and a slight increase would make it detach (House & Lipowsky, 1988). The first two rolling models (Hammer & Apte, 1992; Tözeren & Ley, 1992) showed that rolling can be stable over a range of wall shear rates.

Beyond leukocytes, certain metastasizing cancer cells show rolling behavior, especially those of colonic (Tözeren et al., 1995) and breast (Aigner et al., 1998) origin. These cancer cells express ligands for E-selectin (expressed on inflamed endothelium) and/or P-selectin (expressed on activated endothelium and platelets), both of which are different from the selectin ligands of leukocytes. Conceivably, modeling of rolling might, therefore, inform better strategies at limiting cancer metastasis.

A satisfactory model of leukocyte rolling is a source of detailed data, practically at any point in time (time resolution may be better than 10^{-7} s), that provide information about various aspects of the rolling process such as the cell displacement, cell translational and rotational velocities, number of bonds, bond force, number and duration of the cell resting periods, and more, depending on how detailed the model is. Typically, only some of the data generated by a model, and in a reduced form, can be observed or measured by an experimentalist. For example, if the model-generated data are filtered to 60 frames per second, a time resolution commonly used in experiments, the information about the process is not only reduced but can also be deformed, as illustrated by Ley et al. (2008) and Zhao, Chien, and Skalak (1995) for the cell translational velocity. The fact that the distribution of experimentally determined cell translational velocities depends on the frame rate at which observations are made was predicted by the stochastic model of Zhao et al. (1995) and verified experimentally.

A working model can be used to test the sensitivity of the rolling process to changes in the input parameters, to make predictions about the factors that are not observable/measurable as of today, or, if adopted, to test new ideas, as they emerge, such as the concept of catch–slip bonds (Marshall et al., 2003) or the concept of cell polarization during rolling (Green et al., 2006). Most models of leukocyte rolling can be simplified (by making the microvillus length equal to zero, eliminating the deformability of the "cell" and distributing the adhesion molecules randomly over the "cell" surface) to simulate rolling of hard spheres (Brunk et al., 1996).

III. HISTORY OF MODELING LEUKOCYTE ROLLING

The foundations for modeling leukocyte rolling were laid in the years 1915–1988. Jeffrey (1915) analyzed a sphere rotating about an axis perpendicular to the wall. Brenner (1961) and Maude (1961) analyzed a sphere translating perpendicular to the wall. Dean and O'Neill (1963) studied a sphere rotating about an axis parallel to the wall. O'Neill (1964) studied a sphere translating parallel to the wall. The physics of a sphere steadily translating and rotating in a linear flow field at a constant distance from a nearby wall was given in two papers (Goldman, Cox, & Brenner, 1967a,b). The translation of such a sphere is due to the fluid flow, while its rotation is caused by the fact that the fluid moves slower at the sphere surface facing the wall than on the other side, away from the wall. The sphere's translational and rotational velocities can be calculated from first principles with very few assumptions based on (Goldman et al., 1967). Both translational and rotational velocities depend on the shear rate, sphere radius, and sphere distance from the wall. The difference between the translational velocity and the rotational velocity (given as the product of the sphere radius and angular velocity with which the sphere rotates) is called the slip velocity. For a 7 μm (the average leukocyte diameter) sphere at a separation distance of 20 nm from the wall, exposed to a wall shear rate of 100 s^{-1}, the slip velocity is 79.5 $\mu m\ s^{-1}$.

Although a sphere under flow near a wall rotates, it does not roll. Rolling requires fast forming and dissociating bonds with the wall. Once the wall (in a flow chamber) or the endothelium (in a living blood vessel) and the rolling sphere surface are endowed with suitable pairs of adhesion molecules, rolling ensues. While rolling, such a sphere translates with an average velocity much lower than the translational velocity of a corresponding sphere lacking adhesion molecules. The translational velocity of a free-floating sphere near the substrate (the wall) is known as the hydrodynamic velocity (Lawrence & Springer, 1991) or critical velocity (Ley & Gaehtgens, 1991). This velocity separates all rolling from all free-flowing cells or spheres of the same type.

Interestingly, when in a model selectin molecules are added to the substrate, and selectin ligands are added to the sphere surface, immediate arrest (stop) of any sphere movement occurs. To achieve stable rolling, a repulsive force must be introduced to the model that keeps the sphere from running into the substrate. Theoretical estimates of the repulsive force have been made by Bell, Dembo, and Bongrand (1984) for cell–cell and cell–substrate interactions. Most authors consider the repulsive force as arising from negative charges and entanglement of the cell surface's and substrate surface's polymer–molecule glycocalyx layers (Bongrand & Bell, 1984; Bongrand, Capo, & Depieds, 1982).

Fundamental studies on receptor–ligand binding were conducted by Bell (1978), Dembo, Torney, Saxman, and Hammer (1988), and Evans (1985). Evans analytically determined the maximum tension induced in two membranes when their adhesive contact has been just formed and is about to spread, and the minimum tension required to separate the adherent membranes. His results show that the deviation between these two tensions can be very large and depends strongly on the surface density of cross-bridging sites (receptors). Bell proposed a formula for the receptor–ligand bond dissociation (called the Bell formula), formulas for the bond formation and dissociation rate constants (i.e., the unstressed rates), and a kinetic rate equation governing the bond formation in the contact area. Dembo et al. proposed a formula for the receptor–ligand bond dissociation (called the Dembo formula), and derived an exact formula for the critical tension required to overcome the membrane spreading after adhesive contact. Their analytical model predicted that catch bonds may occur. A characteristic of catch bonds is that their lifetimes are prolonged by applied force, in contrast to slip bonds (their lifetimes are shortened by force). The lifetimes of "ideal" or neutral bonds are independent of applied force.

Four basic types of leukocyte rolling models have emerged: direct, semianalytic, analytical, and agent-based. The direct models determine the number of receptor–ligand bonds by tracking the positions of the receptors and ligands during cell rolling and checking their association stage. Such models are very detailed but computationally intensive. The semianalytic models, instead of tracking the receptors and ligands, establish the bond density in the cell–substrate contact area from kinetic rate equations, such as the one proposed by Bell (1978). Those models run under simplified assumptions and, therefore, are less detailed and computationally less intensive compared with the direct models. However, they still capture the essentials of cell rolling. The analytical models describe the evolution of the rolling process through individual variables, the evolution of which can be described by mathematical equations. Agent-based models are multilevel, object-oriented models, which are typically based on available software toolkits, such as Swarm or Repast (Gilbert & Bankes, 2002), for their modeling and simulation framework.

The year 1992 marks the beginning of modeling leukocyte rolling. Hammer and Apte (1992) published their direct model of leukocyte rolling later named adhesive dynamics (AD). In the same year, Tözeren and Ley (1992) published their semianalytic model of steady-state leukocyte rolling. Detailed information about both models is given in Section V. The model of Tözeren and Ley was later refined by Krasik and Hammer (2004). The three-dimensional (3D) AD model, originally designed for one type of receptor–ligand bonds, was greatly developed over time by Hammer and his coworkers. They introduced a second type of receptor–ligand bonds corresponding to integrin bonds (Bhatia, King, & Hammer, 2003; Krasik, Caputo, & Hammer, 2008; Krasik, Yee, & Hammer, 2006), microvillus extension and tether formation (Caputo & Hammer, 2005), and catch–slip bonds with applications to P-selectin and L-selectin bonds (Beste & Hammer, 2008; Caputo, Lee, King, & Hammer, 2007). The same group used their AD model to establish state diagrams for adhesion (Bhatia et al., 2003; Chang, Tees, & Hammer, 2000), a convenient mapping that allows the prediction of different types of cell motion over a wide range of model parameters.

While the AD simulations progress by fixed time steps, a 3D, π-calculus based, direct event-tracking model of adhesion (ETMA) progresses by varying time steps driven by bond formation and dissociation events and tracking those events (Pospieszalska, Zarbock, Pickard, & Ley, 2009). ETMA provides high temporal resolution of bond events and is computationally efficient. More information on ETMA is given in Section V.B.

The 3D direct model of leukocyte rolling developed by Jadhav, Eggleton, and Konstantopoulos (2005) and Pawar, Jadhav, Eggleton, and Konstantopoulos (2008), and the 3D semianalytic model developed by Khismatullin and Truskey (2004, 2005) introduce whole-cell deformability to modeling of leukocyte rolling. In the first model, the leukocyte is assumed to be an elastic capsule. In the second model, the leukocyte is assumed to be a compound viscoelastic drop composed of a nucleus covered by a thick layer of cytoplasm (see Section V.D). Both models simulate the change in the leukocyte shape during short rolling in great detail. A similar 3D model of time-dependent cell shapes for leukocytes viewed as compound viscoelastic drops is presented in Jin et al. (2007).

The model of Zhao et al. (1995) is a classic example of an analytical model. From the theory of diffusion processes, this model obtains in analytical form the translational velocity distribution for a homogeneous population of leukocytes, and finds the distribution for the steady-state rolling case. Another example is the analytical model of Piper, Swerlick, and Zhu (1998) which, from kinetics of small systems, obtains the probability of a leukocyte having *n*-bonds in steady-state rolling. The model of Tang, Ley, and Hunt (2007) is the first agent-based model of leukocyte rolling. This model uses as a framework the Recursive Porous Agent Simulation Toolkit (Repast) (North, Collier, & Vos, 2006).

IV. DEVELOPMENT OF A LEUKOCYTE ROLLING MODEL

Detailed models are written in the form of computer programs. Starting with a set of initial parameters, and following the rules of cellular, molecular, and environmental physical and chemical interactions, a computer program simulates the process under investigation, instantaneously storing data about the status of the process for further analysis. Appropriate selection of the initial parameters and appropriate implementation of the interaction rules are key elements in designing a useful model. Having these two elements embedded into an algorithm, the model is ready to run. To validate the model, the results of the test runs are compared with the corresponding results obtained from experiments. Typically, the test runs are not fully successful, and the model needs further refinement. If the test and experimental results are quite close, then, most likely, the model parameters need a small adjustment. It is common that some model parameters are only approximately known from experiments, and pinpointing their exact values predicted by the model can spare experiments. If the test results clearly do not match the experimental results, one should not conclude that the model is wrong. The model just indicates that some important factors of the process, of which the modeler may not be aware, are not taken into account while building the model structure. In this case the assumptions of the model need to be redefined, which may imply major changes in the model design. Figure 2 shows a diagram illustrating the basic development process for a model simulating leukocyte rolling.

A. Model Parameters

The input parameters for models of leukocyte rolling include cellular parameters describing properties of the leukocyte, molecular parameters describing properties of the receptors and ligands in their unbound and bound states, environmental parameters describing fluid and substrate properties, and algorithmic parameters defining the resolution of simulations.

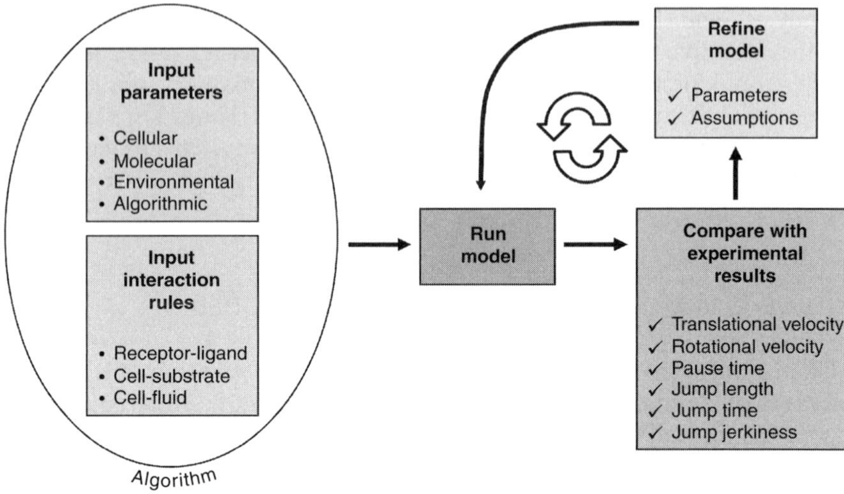

FIGURE 2 Basic development process for a model simulating leukocyte rolling. The model algorithm, based on input parameters and cellular, molecular, and environmental interaction rules (left boxes), is used to run simulations. The results of the test runs are compared with experimental data (bottom right box). The model may need refining rounds with more accurate parameters and/or under redefined assumptions (top right box).

The input parameters may vary, depending on the modeling approach. Table I lists the parameters for a basic model of neutrophil rolling on P-selectin. The transition-state spring constant (described in Section IV. A.2) is needed when the Dembo formula (Eq. 4) is used for the rate of bond dissociation, as in Hammer and Apte (1992). The reactive compliance (also described in Section IV.A.2) is used when the Bell formula (Eq. 7) is assumed for that rate, as in Bhatia et al. (2003). More advanced models of leukocyte rolling may need additional parameters. For modeling cell rolling with deformable microvilli, as in Caputo and Hammer (2005) and Pospieszalska and Ley (2009), the input data must include parameters describing the viscoelasticity of microvilli and their tethers. For modeling catch–slip behavior of specific bonds, as in Caputo et al. (2007) and Pawar et al. (2008), the input data must include parameters describing the dissociation pathways involved in the catch–slip process of those bonds. For modeling cell deformation during rolling, as in Jadhav et al. (2005) and Khismatullin and Truskey (2004), parameters describing the viscoelasticity of the cell are necessary to run simulations. Most likely, the set of possible input parameters for modeling leukocyte rolling will grow as our knowledge about leukocyte rolling develops.

TABLE I

Input Parameters for a Basic Model of Neutrophil Rolling on P-Selectin

	Value	References
Cellular parameters (symbol)		
Cell radius[b] (r_c)	3.8 μm	Schmid-Schönbein (1990)
Cell density[a] (ρ_c)	1.077 g cm^{-3}	Zipursky, Bow, Seshadri, and Brown (1976)
Microvillus length[c] (L^0_m)	200 nm	Finger, Bruehl, Bainton, and Springer (1996)
Microvillus radius[c]	50 nm	Finger, Bruehl, et al. (1996)
Microvillus density[c]	5 microvilli μm^{-2}	Bruehl et al. (1996)
Molecular parameters (symbol)		
PSGL-1 length[b]	50 nm	Li et al. (1996)
PSGL-1 site density[c] (n_L)	83 molecules μm^{-2}	Moore et al. (1991)
P-selectin length[b]	40 nm	Springer (1990)
P-selectin site density[b] (n_R)	150 molecules μm^{-2}	Moore et al. (1995)
Unstressed bond length[b] (λ)	70 nm	Patel et al. (1995)
Unstressed bond formation rate[c] (k^0_f)	1 s^{-1}	Mehta et al. (1998)
Unstressed bond dissociation rate[b] (k^0_d)	1 s^{-1}	Mehta et al. (1998)
Transition-state spring constant[d] (σ_{tr})	0.98 dyn cm^{-1}	Dembo (1994)
Bound-state spring constant[c] (σ)	1 dyn cm^{-1}	Dembo (1994)
Reactive compliance[c] (δ)	0.3 Å	Caputo and Hammer (2005)
Environmental parameters (symbol)		
Temperature[a] (T)	29 °K	
Suspending medium density[a] (ρ)	1.025 g cm^3	
Suspending medium viscosity[a] (μ)	0.01 g cm^{-1} s^{-1}	
Wall shear rate[b] (γ_w)	25–200 s^{-1}	
Glycocalyx effective thickness[c] (υ)	15 nm	Bell et al. (1984)
Glycocalyx repulsion constant[d] (ξ)	10 pN	Bell et al. (1984)
Algorithmic parameters (symbol)		
Time step[a] (Δt)	10^{-7}–10^{-5} s	

[a]Parameter known with a factor of 0.01 (i.e. with accuracy better that 1%).
[b]Parameter known with a factor of 0.1.
[c]Parameter known with a factor of 2.
[d]Parameter estimated with a factor of 10.

Different parameters are established with different levels of accuracy. For parameters in Table I, the approximate accuracy ranges are indicated by the subscripts following the parameter names. While the levels of accuracy are only rough estimates, they convey a useful overview about the absence of knowledge about many of them. Since some parameters are not known with great precision, parameter sensitivity tests are a staple of all modeling efforts. Perhaps surprisingly, some parameters make no real difference and can be changed over one or two orders of magnitude without influencing the modeling result, while others are critical to within 1% of their value. Ideally, all parameters used in a model should be scrutinized over an extended range, but this is only possible with computationally efficient models. For the experimentalist, the critical parameters are the most interesting to measure, because they predict the behavior of the system. As an example, the transition-state spring constant emerged as a critical parameter, where even small changes made big differences in the prediction (Hammer & Apte, 1992). There are essentially no measurements of the transition-state spring constant, but such measurements should certainly be made and are very much needed in modeling. Other critical parameters that remain as of today largely unexplored are those describing the cell–substrate glycocalyx interactions (see Section IV. A.3) and cell viscoelasticity. Equally important as parameters with large impact are those with small impact. As a case in point, temperature (between 25 and 37 °C) does not notably influence rolling and may be ignored (Pospieszalska & Ley, 2009).

1. Cellular Parameters

The cellular parameters are among those most critical for cell rolling, but for many of them quantitative estimates are lacking. The only two well-established cellular parameters are the cell radius and the cell density, as indicated in Table I. As a first approximation, a rolling cell was seen as a sphere (Tözeren & Ley, 1992). Next, it was realized that the relevant adhesion molecules are not evenly spaced, but rather clustered on a small fraction of the cell surface (Bruehl, Springer, & Bainton, 1996; Bruehl et al., 1997). The adhesion molecule-enriched portions turned out to be protruding structures that look like microvilli and ridges. The molecular composition of these structures is not well known, but bulk mechanical properties have been reported (Shao, Ting-Beall, & Hochmuth, 1998). Current models treat the surface protrusions as uniform microvilli, although experimental evidence (Bruehl et al., 1996; P. Sundd & K. Ley, 2009, unpublished data) suggests that such an approach may be inadequate. Some adhesion molecules are organized in lipid rafts within the cell membrane (Abbal et al., 2006), and the cytoskeletal attachment of adhesion molecules can change rapidly (Kim, Carman, & Springer, 2003; Kucik, Dustin, Miller, & Brown, 1996). Estimates and measurements of the relevant cellular parameters are needed before a complete model of leukocyte rolling can emerge.

The relatively well-defined molecular compositions of various cell components such as nucleus, organelles, cytosol, subcortical actin shell, stress fibers, and membranes have not been yet fully reflected in rolling models due to lack of relevant measurements. The cell nucleus is thought to be more rigid than the cytosol, and the cytosol is more rigid than the extracellular fluid. From this, it seems that using a three-fluid (nucleus, cytosol, and suspending medium) object as a model may be a reasonable first approach. That modeling idea was investigated by Khismatullin and Truskey (2004, 2005) based on estimates for eight different parameters specifically describing the fluids (see Section V.D). Their model's intermediate fluid represents the cytoplasm, which, in real life, contains not only the cytosol but also, for example, the mitochondria introducing their own material properties. The modeling of cell viscoelasticity becomes even more complicated in a case of neutrophils, where the nucleus is lobulated and may facilitate the cell deformation process. To model the change in the cell shape with such details requires additional cellular parameters to be established.

2. Molecular Parameters

The knowledge of molecular parameters for key receptor–ligand pairs involved in leukocyte rolling is not sufficient, except for the PSGL-1–P-selectin pair. The length of P-selectin has been measured by sedimentation methods (Patel, Nollert, & McEver, 1995), and the lengths of L- and E-selectin can be inferred based on their numbers of the consensus repeats (CRs). The length of PSGL-1 is also known (Li et al., 1996). PSGL-1 is a homodimer and P-selectin probably is mostly homodimeric. Much less structural information is available for the other selectin ligands, some of which, such as peripheral node addressin (PNAd), include many molecules (Rosen, 2004). Integrin molecules exist in multiple conformations. Their resting, low-affinity form protrudes not more than 10 nm from the plasma membrane, but their extended form reaches out 15–20 nm (Zhu et al., 2008). The extension of integrins allows them to participate in rolling, but the extended conformation does not *eo ipso* confer high-affinity binding. Integrin conformational changes and their impact on binding propensity and affinity are subject of active ongoing research (Chigaev et al., 2009; Luo, Carman, & Springer, 2007).

In addition to molecular length, the molecular site density, typically expressed as the number of molecules per μm^2, strongly influences rolling. Titrating selectin densities over just one order of magnitude can fundamentally change rolling behavior (Lawrence & Springer, 1991). *In vivo* site densities of the relevant molecules on vascular endothelium are virtually unknown, but may start as low as 20 molecules μm^{-2} for P-selectin (Hattori, Hamilton, Fugate, McEver, & Sims, 1989). Flow cytometry methods have established reasonable ranges for adhesion molecules on leukocytes. Some adhesion molecules are internalized upon ligand binding, causing their surface expression to change. Excellent information is available on the number of endothelial adhesion molecules per gram of tissue (Eppihimer, Russell, Anderson, Wolitzky, & Granger, 1997; Eppihimer, Wolitzky, Anderson, Labow, & Granger, 1996; Henninger et al., 1997), but it is not trivial to convert these into site densities, because the endothelial cell surface area per gram tissue is not well known. Also, much remains to be learned about organ- and vessel-specific site densities.

Modeling studies demonstrate that molecular parameters of biophysical nature, such as the unstressed bond formation and dissociation rates, bound-state and transition-state spring constants, and reactive compliance, are critical in shaping leukocyte rolling (Chang et al., 2000; Hammer & Apte, 1992). The translational velocity of a rolling leukocyte, as well as other variables in cell rolling, are very sensitive to changes in those parameters. Most of those parameters, with exception of the unstressed bond dissociation rate, are not well known, as indicated in Table I.

The unstressed reaction rates are molecular parameters of bond formation and dissociation kinetics. The reversible overall reaction between receptor R and ligand L resulting in the formation of bond RL can be written as

$$R + L \underset{k_{\mathrm{d}}}{\overset{k_{\mathrm{f}}}{\rightleftarrows}} RL, \tag{1}$$

where k_{f} is the bond formation rate also called the forward rate or the on-rate, and k_{d} is the bond dissociation rate also called the reverse rate or the off-rate. Rate k_{d} is typically given in a unit of s^{-1}. k_{f} represents the direct on-rate typically given in a unit of s^{-1}, or the 2D on-rate typically given in a unit of $\mu\mathrm{m}^2\ \mathrm{s}^{-1}$, or the 3D on-rate typically given in a unit of $\mathrm{M}^{-1}\ \mathrm{s}^{-1}$. The ratio $K = k_{\mathrm{f}}/k_{\mathrm{d}}$ represents the binding affinity. A parameter sensitivity analysis suggests that changes in k_{f} have a greater effect on leukocyte rolling than changes in k_{d} (Hammer & Apte, 1992). A common approach is to define rates k_{f} and k_{d} as functions of the separation distance between the receptor and ligand bases (see Eqs. 3, 4, 6, and 7, and Fig. 3). In the absence of force acting on the receptor–ligand bond, $k_{\mathrm{d}} = k_{\mathrm{d}}^0$ is the unstressed bond dissociation rate, and corresponding through (Eq. 1) $k_{\mathrm{f}} = k_{\mathrm{f}}^0$ is the unstressed bond formation rate. The unstressed rates k_{f}^0 and k_{d}^0 are also called equilibrium rates or rate constants, and serve as input parameters for leukocyte rolling models.

The unstressed off-rates were measured by Bell (1978), Mehta, Cummings, and McEver (1998), and Moore, Varki, and McEver (1991). The 3D unstressed on-rates come from experiments on multiple soluble receptors binding to multiple membrane-bound ligands (Mehta et al., 1998; Moore et al., 1991). The 2D unstressed on-rates come from experiments on multiple membrane-bound

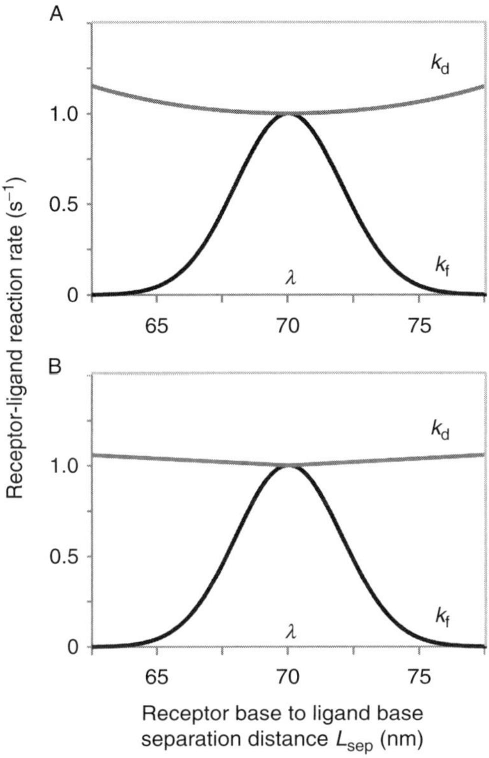

FIGURE 3 Receptor–ligand bond formation rate k_f (black curve) and dissociation rate k_d (gray curve) as functions of separation distance L_{sep} between the receptor and ligand bases for the Dembo approach (A) and Bell approach (B). λ is the unstressed bond length. The rates were calculated for $\lambda = 70$ nm, $k_f^0 = k_d^0 = 1\ s^{-1}$, $\sigma_{tr} = 0.98$ dyn cm^{-1}, $\sigma = 1$ dyn cm^{-1}, $\delta = 0.3$ Å, and $T = 290$ °K (Table I).

receptors binding to multiple membrane-bound ligands (Long, Zhao, Huang, & Zhu, 2001; Rinko, Lawrence, & Guilford, 2004). A rough estimate of the 2D unstressed on-rate can be also found from the 3D unstressed on-rate using a method described in Bell (1978). The direct unstressed on-rates should be measured in experiments on a single membrane-bound receptor binding to a single membrane-bound ligand with the distance between their bases equal to their unstressed (i.e., no force) bond length. Since such data have not been reported, the direct unstressed on-rates are estimated from the 3D unstressed on-rates (Mehta et al., 1998; Moore et al., 1991).

As suggested above, methods of estimating 2D rates involve experiments on receptor–ligand binding, where two surfaces are coated with multiple receptor and ligand molecules. Singling out the behavior of an individual receptor–ligand pair is a challenging task. The likelihood of reporting individual pairs can be increased by reducing the site densities of the receptors and ligands. A convincing indication that single consecutive bond events are observed is presented when the natural log of the number of bond formation events with a waiting time greater than *t*, and the natural log of the number of bond dissociation events with a waiting time greater than *t*, approximately follow decreasing linear functions (Chen, Evans, McEver, & Zhu, 2008; Marshall et al., 2003). The most promising method as of today for measuring the receptor–ligand reaction rates is by monitoring an abrupt decrease and resumption in thermal fluctuations of a biomembrane force probe (Chen et al., 2008). The other methods include laser trap experiments (Rinko et al., 2004), cell pause time distribution analysis (Smith, Berg, & Lawrence, 1999), and AFM experiments (Zhang, Wojcikiewicz, & Moy, 2002). Recently, new measurements for the 2D on-rates have become available (Chen et al., 2008; H.W. Guilford, 2009, unpublished data).

The receptor–ligand reaction rates can be defined as dependent on parameters describing the transition-state form and final bound-state form of the receptor–ligand molecule. Those parameters are the transition-state and bound-state spring constants (Dembo et al., 1988), or transition-state and bound-state resting lengths and bound-state spring constant (Dembo et al., 1988), or may include all of them as suggested by Zhu and McEver (2005). The transition state in a process of binding occurs after the entropic and enthalpic pain involved in bringing the two opposite binding domains into close proximity and before the gain from the stabilizing noncovalent bond interactions (a quote from Dembo, 1994). The transition state and the final bound state may have different stiffness characteristics (i.e., different spring constants), and/or different resting lengths (Dembo et al., 1988). In the models based on the Bell formula (Eq. 7) for the rate of bond dissociation, only the difference between the transition-state and bound-state resting lengths, called the reactive compliance, is of significance. Therefore, the reactive compliance can serve as an input parameter for those models. Rough estimates for the transition-state and bound-state parameters come from general studies of typical biological systems (Dembo, 1994; Dembo et al., 1988). Experimental estimates of the reactive compliance for some adhesion molecules were established from cell pause time distribution analysis (Smith et al., 1999) and AFM experiments (Zhang et al., 2002).

3. Environmental Parameters

In *in vivo* and *in vitro* experiments, the temperature, suspending medium density, and suspending medium viscosity can be measured with high accuracy. The wall shear rate characterizes the flow of the fluid near the wall (the substrate) under which a leukocyte rolls. The wall shear rate is defined as the derivative dv/dz, where $v = v(z)$ is the fluid velocity at distance z from the wall in the direction perpendicular to the fluid flow and the wall. The fluid velocity at the wall is $v(0) = 0$. The wall shear stress is defined as wall shear rate $v(z)$ multiplied by the viscosity of the fluid at distance z from the wall (Chien, Usami, & Skalak, 1984). The viscosity describes fluid's internal resistance to flow. In all existing models of leukocyte rolling, the wall shear rate and the fluid viscosity, and consequently the wall shear stress, are assumed to be constants. However, for leukocytes in blood, the wall shear stress changes as the viscosity changes systematically with distance from the wall (Long, Smith, Pries, Ley, & Damiano, 2004). Moreover, the rolling leukocyte itself disturbs the flow field in its vicinity (Pickard & Ley, 2009). These phenomena are currently not considered in rolling models. In flow chambers, wall shear stress can be derived with high accuracy from geometry and the pressure difference between inlet and outlet (see Section IV.D.7).

When studying leukocyte interactions with the vessel wall, the wall shear rate and wall shear stress must be estimated, because both rolling and adhesion are strongly dependent on these parameters. However, we have no true knowledge of the actual wall shear rate in microvessels because of the presence of an endothelial surface layer of glycocalyx (Vink & Duling, 1996), which does not allow significant plasma flow near the endothelial membrane (Damiano, Duling, Ley, & Skalak, 1996; Smith, Long, Damiano, & Ley, 2003). In venules with diameters between 15 and 50 μm, which are most relevant for leukocyte adhesion in inflammation, the shear rate at the interface between the endothelial surface layer and the free lumen is about $40V_{\text{mean}}/d$ (Long et al., 2004), where V_{mean} is the average blood flow velocity and d is the venule diameter. V_{mean} is about 60% of the commonly measured centerline velocity (Baker & Wayland, 1974). The above estimate for the shear rate is five times higher than the Newtonian (traditional) estimate reported in most papers. The Newtonian calculations do not account for the surface layer, and blood is not a Newtonian fluid. Therefore, the Newtonian estimate does not apply to blood flow in microvessels. More accurate determinations of wall shear rates can be achieved by microparticle velocimetry (Smith et al., 2003), which requires stroboscopic epifluorescence microscopy. The blood flow centerline velocity can be measured using a dual photodiode sensor system.

The compressible glycocalyx layers coating the leukocyte and *in vivo* substrate surfaces are composed of hydrophilic long-chain polymer molecules. As the two polymer-coated surfaces approach each other, the polymer layers overlap and some of the fluid is squeezed out of the contact area. A repulsive force, preventing the cell from colliding with the substrate, is thought to result, in part, from a combination of the osmotic tendency of fluid to return to its original location, and the steric compression of the polymer chains

(Bell et al., 1984). The electrostatic repulsion of glycocalyx layers adds to the total glycocalyx repulsion. Two parameters describing the glycocalyx buffer are the glycocalyx effective thickness and glycocalyx repulsion constant. The first one measures the cell–substrate separation distance at which the repulsive polymer layers are interpenetrated by $\sim$50%. The second one rises from the Flory–Krigbaum theory of steric stabilization (Bell et al., 1984; Bongrand & Bell, 1984; Napper, 1977) and measures the ease with which the polymer layers can be compressed. Theoretical estimates for the glycocalyx parameters come from work of Bell et al. (1984). *In vivo* estimates for the thickness of the endothelial glycocalyx layer range around 0.4 μm, based on dye exclusion (Vink & Duling, 1996) and near-wall micro-PIV (particle image velocimetry) (Smith et al., 2003). For leukocytes rolling *in vitro*, the substrate surface may lack a glycocalyx layer, as the glass plates of experimental flow chambers do, and the very thin leukocyte glycocalyx layer compresses directly against the substrate.

4. Algorithmic Parameters

The detailed models of leukocyte rolling progress by time steps. At the end of every time step the status of the rolling process is updated. Typically, the time step is predefined and fixed, as in AD (Hammer & Apte, 1992). In such models, the duration of the time step dictates the resolution of simulations. Using a shorter time step increases the resolution, but also the computational cost. An optimal time step is found when results are no longer different from those obtained at shorter time steps. The duration of the optimal time step varies, depending on the wall shear rate. In general, higher wall shear rates require shorter time steps. The resolution in ETMA (Pospieszalska et al., 2009) is improved by progressing by varying time steps which secure that the process updates occur at the key points for cell rolling, that is, when the bond formation and dissociation events occur (see details in Section V.B). There are models which concurrently use two different types of time steps (see an example in Section V.A).

B. Cell-Molecules-Environment Interaction Rules

The cell–molecules–environment interaction rules are model basic components which instruct the computer how cells and molecules behave under flow. The interaction rules can be quite different for different modeling approaches. As a first approach, the modeler may choose to include only the most influential interactions under assumed parameters and neglect the others. The interaction rules define receptor–ligand, cell–substrate, and cell–fluid interactions. Below, we list basic interaction rules in leukocyte rolling, used in direct and semianalytic models. Modelers can make a selection from the list, alternate a rule, or add a new one, depending on the assumptions and specifications of their models.

1. Receptor–Ligand Interaction Rules

A suitable receptor–ligand pair can form a bond which will later dissociate, as described by (Eq. 1). To simulate such a process, bond formation rate k_f and dissociation rate k_d need to be defined. A typical approach is to use the Dembo formula (Dembo et al., 1988) or Bell formula (Bell, 1978) for the off-rate, k_d. Dembo et al. (1988) require that affinity $K = k_f/k_d$ satisfies the following equation of Bell et al. (1984):

$$\frac{k_f}{k_d} = \left(\frac{k_f^0}{k_d^0}\right) \exp\left[\frac{-\sigma(L_{sep} - \lambda)^2}{2\kappa_B T}\right], \tag{2}$$

where T, κ_B, σ, λ, and L_{sep} are the absolute temperature, Boltzmann's constant, bound-state spring constant, unstressed bond length, and separation distance between the receptor and ligand bases, respectively. Therefore, the formula for on-rate, k_f, can be derived from k_d and K.

With the Dembo approach,

$$k_f = k_f^0 \exp\left[\frac{-\sigma_{ts}(L_{sep} - \lambda)^2}{2\kappa_B T}\right], \tag{3}$$

$$k_d = k_d^0 \exp\left[\frac{(\sigma - \sigma_{ts})(L_{sep} - \lambda)^2}{2\kappa_B T}\right], \tag{4}$$

and

$$\delta = 0, \tag{5}$$

where σ_{ts} is the transition-state spring constant and δ is the reactive compliance.

With the Bell approach,

$$k_f = k_f^0 \exp\left[\frac{\sigma|L_{sep} - \lambda|(\delta - 0.5|L_{sep} - \lambda|)}{\kappa_B T}\right], \tag{6}$$

$$k_d = k_d^0 \exp\left[\frac{\delta\sigma|L_{sep} - \lambda|}{\kappa_B T}\right], \tag{7}$$

and

$$\sigma = \sigma_{ts}. \tag{8}$$

Examples of reaction rate pairs, k_f and k_d, as functions of L_{sep} are given in Fig. 3A (the Dembo approach) and in Fig. 3B (the Bell approach). The models based on the Dembo approach assume that $\sigma > \sigma_{ts}$, that is, the molecular bonds are slip bonds. Caputo and Hammer (2005) use the Bell approach, but substitute k_f^0 with a two-value discrete function, with the larger value being valid when no bonds are present (see Section V.A). An on-rate as a function of slip velocity is introduced in Caputo et al. (2007) (see Section V.A). Some other formulas for the reaction rates, where the on-rate not necessarily depends on the off-rate, are summarized in Piper et al. (1998). For modeling P-selectin or L-selectin catch–slip bonds one can use the formula of Evans, Leung, Heinrich, and Zhu (2004) for the off-rate, as in Caputo et al. (2007).

The probability, P_f, that a receptor–ligand pair will form a bond in time interval Δt, and the probability, P_d, that a bond will dissociate in time interval Δt, are given by the following equations:

$$P_f = 1 - \exp(-k_f \Delta t), \tag{9}$$

$$P_d = 1 - \exp(-k_d \Delta t). \tag{10}$$

The time, t_f, to the next bond formation event, and the time, t_d, to the next bond dissociation event, can be calculated as follows:

$$t_f = \left(\frac{1}{k_f}\right) \ln\left(\frac{1}{a_1}\right), \tag{11}$$

$$t_d = \left(\frac{1}{k_d}\right) \ln\left(\frac{1}{a_2}\right), \tag{12}$$

where a_1 and a_2 are numbers selected randomly from the uniform distribution of values between 0 and 1.

If a model progresses by fixed time steps, then, based on Eqs. (9) and (10), a Monte Carlo method can be used to select all free (i.e., not bound) receptor–ligand pairs which will form bonds at the end of interval Δt, and all existing bonds which will dissociate at the end of that interval (Hammer & Apte, 1992; Pawar et al., 2008). A bond event actually occurs inside its time interval Δt, but with fixed time steps cannot be implemented until the interval ends. If a model allows for varying time steps, then a simulation can be conducted as follows. Each free receptor–ligand pair declares its time for bond formation based on (Eq. 11), and each existing bond declares its time for dissociation based on (Eq. 12). The fastest event is implemented at its time of occurrence, and the race for the next event starts again (Pospieszalska et al., 2009; see also Section V.B).

2. Cell–Substrate Interaction Rules

In *in vitro* experiments, gravitation brings the leukocyte to the vicinity of the substrate where the first receptor–ligand bond formation can occur. The gravitational force is caused by a small difference in density between the leukocyte and the suspending media. *In vivo*, a similar force may be provided by margination forces (Goldsmith & Spain, 1984). The gravitational force, F_{grav}, acting on the cell is calculated as (Hammer & Apte, 1992)

$$F_{grav} = \left(\frac{4}{3}\right)\pi(r_c + L_m^0)^3(\rho_c - \rho)g, \tag{13}$$

where r_c, L_m^0, ρ_c, ρ, and g are the leukocyte radius, microvillus length, leukocyte density, suspending medium density, and gravitational constant, respectively. The colloidal cell–substrate interactions involve the electrostatic force, F_{el}, which is typically repulsive, repulsive steric stabilization force, F_{ss} (because of the glycocalyx), and a weak attraction force called the van der Waals force, F_{vdw}. The overall colloidal force, F_C, acting on the leukocyte is given by the following formula (Hammer & Apte, 1992):

$$F_C = F_{el} + F_{ss} - F_{vdw} - F_{grav}. \tag{14}$$

The mechanical work, $\Gamma(h)$, needed to overcome forces F_{el} and F_{ss} to bring a unit area of cell membrane from an infinite separation from the planar surface to a separation distance of h is given by Bell et al. (1984) as shown below:

$$T(h) = \left(\frac{\xi}{h}\right)\exp\left(\frac{-h}{\upsilon}\right), \tag{15}$$

where ξ and υ are the repulsion constant and effective thickness of the glycocalyx, respectively. The total repulsive force, F_R, because of F_{el} and F_{ss}, is obtained by integrating the derivative $\mathrm{d}(-\Gamma)/\mathrm{d}h$ over the substrate area. If force F_{vdw} is ignored, then $F_C = F_R - F_{grav}$.

The bond density at time t, $n_B = n_B(t)$, in the cell–substrate contact area can be found by solving the kinetic equation of Bell (Bell, 1978):

$$\frac{\mathrm{d}n_B}{\mathrm{d}t} = k_f^0(n_R - n_B)(n_L - n_B) - k_d^0 n_B, \tag{16}$$

where n_R and n_L are, respectively, the site densities of receptors and ligands, and $n_B(0) = 0$. The bond force, F_b, exerted on the cell by a load-bearing bond (i.e., when $L_{sep} > \lambda$) can be calculated according to the Hookean spring model as follows:

$$F_b = \sigma(L_{sep} - \lambda). \tag{17}$$

The total bond force, F_B, acting on the cell is then calculated as

$$F_B = \Sigma F_b, \tag{18}$$

where the summation is over all the load-bearing bonds. The bond force components, F_{Bx}, F_{By}, and F_{Bz}, and the bond torques about the cell center, T_{Bx}, T_{By}, and T_{Bz}, imposed on the cell by F_B in all three Cartesian directions (x, y, and z, see Fig. 4A) can be calculated from first principles (Kuo, Hammer, & Lauffenburger, 1997).

It has been observed experimentally that the bond force causes microvillus deformation, as illustrated in Fig. 4C and D. Micropipette experiments of Shao et al. (1998) reveal that low forces (<34 pN) cause the microvillus to extend, while high forces (>61 pN) cause a thin membrane cylinder (a tether) to be formed at the tip of the microvillus. The microvillus deformation is

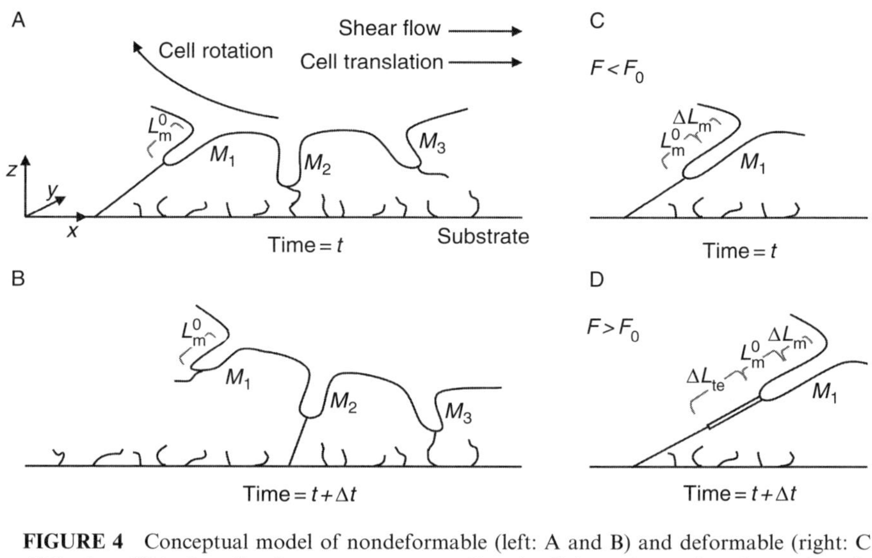

FIGURE 4 Conceptual model of nondeformable (left: A and B) and deformable (right: C and D) microvilli in the cell–substrate contact area. For simplicity, only one ligand per microvillus tip is shown. L^0_m is the microvillus length in the absence of force, and ΔL_m, F, and F_0 are microvillus extension, bond force, and threshold bond force, respectively. ΔL_{te} is the tether extension. (A, B) The microvillus length is assumed to be fixed, but the microvillus can pivot about its base. At time t microvillus M_1 has a load-bearing bond, microvillus M_2 has a bond bearing no load, and microvillus M_3 has no bond. At time $t + \Delta t$ the bond of M_1 is already broken, the bond of M_2 is load-bearing, and the newly formed bond of M_3 has no load yet. (C, D) Microvilli are allowed to extend and form tethers. If $F < F_0$ at time t, then microvillus M_1 will be longer than in (A). If $F > F_0$ at time $t + \Delta t$, then most likely the bond of microvillus M_1 will still exist (compare D with B), the microvillus will be longer than in (C), and a tether will develop.

included in the leukocyte rolling models of Caputo and Hammer (2005) and Pospieszalska and Ley (2009). For more details on microvilli and tethers, see Chapters 1 and 2 in this volume.

3. Cell–Fluid Interaction Rules

A rolling leukocyte, *in vivo* and *in vitro*, translates in the direction of the fluid flow and rotates in plane xz, as illustrated in Fig. 4A and B (other translations and rotations also can occur). If the total bond force becomes sufficiently large to balance the hydrodynamic forces of the fluid, the leukocyte stops rolling. The fluid force, F_S, called the shear force, acting on the immobilized cell in the x-direction, and the fluid torque, T_S, called the shear torque, acting on the immobilized cell in the y-direction, are given by the following equations:

$$F_S = 6\pi\mu\gamma_w r_c z_c C_1, \tag{19}$$

$$T_S = 4\pi\mu\gamma_w r_c^3 C_2, \tag{20}$$

where μ, γ_w, r_c, and z_c are the suspending medium viscosity, wall shear rate, cell radius, and distance of the cell center from the substrate. C_1 and C_2 are functions of z_c/r_c given by Goldman et al. (1967a,b).

For a leukocyte under flow, the vector $\mathbf{U} = (V_x, V_y, V_z, \Omega_x, \Omega_y, \Omega_z)$, where V_x, V_y, and V_z are the cell's translational velocities and Ω_x, Ω_y, and Ω_z are its angular velocities in all three Cartesian directions, can be calculated from the equation developed by Hammer and Apte (1992) as follows:

$$\mathbf{U} = M\mathbf{F}, \tag{21}$$

where $\mathbf{F}$ denotes a vector containing the three force components (due to F_C, F_B, and F_S given by Eqs. 14, 18, and 19, respectively) and three torque components (due to T_{Bx}, T_{By}, T_{Bz}, and T_S given by Eq. 20) listed below:

$$\mathbf{F} = (F_{Bx} + F_S, F_{By}, F_{Bz} + F_C, T_{Bx}, T_{By} + T_S, T_{Bz}). \tag{22}$$

Symbol M denotes the 6×6 mobility matrix concerning the motion of a sphere near a plane in Stokes flow, which is known from Brenner (1961), Goldman et al. (1967a,b), and Jeffrey (1915). The components of the matrix depend on μ, r_c, and z_c. Equations (21) and (22) arise from the assumption that the rolling leukocyte maintains a balance between the colloidal and fluid forces and fluid torques and corresponding bond forces and torques, by adjusting its velocities.

A modeler may chose to run simulations under drastically simplified assumptions. That may allow for deriving leukocyte rolling characteristics from a set of analytical equations that includes force and torque balance equations (Shao et al., 1998). It has been experimentally observed that rolling leukocytes deform under flow (Damiano, Westheider, Tözeren, & Ley, 1996).

The net effect is an increase in the cell–substrate contact area allowing for more molecular bonds to be formed. This is easy to observe but difficult to reproduce *in silico*. Indeed, cell deformation is just beginning to be modeled (Jadhav et al., 2005; Jin et al., 2007; Khismatullin & Truskey, 2004, 2005; Pawar et al., 2008) (see Section V.D).

C. Model Algorithm

A clearly defined procedure, called the algorithm, needs to be established to instruct the computer how to use the parameters and implement the interaction rules. Models simulating biological processes are often stochastic, as the direct leukocyte rolling models are, and the algorithm makes the decision about the next interaction implementation based on the instantaneous status of the process. In practice, those decision times must be discrete, implying that the simulation progresses by small time steps. There is no one universal formula for the algorithm structure. The algorithm design is a result of the modeler's skills, and architectural and artistic talents. Ultimate goals are to make the algorithm open-access structured (i.e., easy to be modified and allowing for its growth), and computationally efficient.

As an example of an algorithm, we present the algorithm of our direct ETMA (Pospieszalska et al., 2009) simulating leukocyte rolling on one class of receptors. ETMA progresses by varying time steps using Eqs. (11) and (12) for the waiting time to the next molecular event, as described in Section IV.B.1. To reduce the computational load, we neglect the cell's small instantaneous translations in the y-direction and rotations in plains other than plane xz (Fig. 4A). The cell velocities, translational $V_{\mathrm{tr}} = V_x$, rotational $V_{\mathrm{rot}} = \Omega_y r_{\mathrm{c}}$, and vertical $V_{\mathrm{ver}} = V_z$ (perpendicular to the substrate), are calculated from (Eq. 21).

Initialization. At the beginning of the simulation, a spherical cell with microvilli is positioned near the vertical balance point. The positions for microvillus base centers and receptor bases in the contact zone are selected randomly observing the microvillus radius and microvillus and receptor densities. Initial cell velocities, translational V_{tr1}, rotational V_{rot1}, and vertical V_{ver1}, are calculated. A competition for the first bond formation event reveals the first time step Δt_1 and points to a receptor–ligand pair that will realize the event. At the end of time step Δt_1 the following four steps are conducted:

- *Step 1.* The selected event is implemented.
- *Step 2.* Positions of the cell center, microvillus base centers, and receptor bases are updated from Δt_1 and V_{tr1}, V_{rot1}, and V_{ver1}. The new contact zone is organized including selecting microvillus base centers and receptor bases in the leading, new part of the zone.

- *Step 3*. New cell velocities, V_{tr2}, V_{rot2}, and V_{ver2}, are calculated from the updated positions and bond forces acting on the cell.
- *Step 4*. Competition for the next event determines the next time step Δt_2 and points to a free (i.e., not bound) receptor–ligand pair or to a bond that will realize the event.

Continuation. Repeating Steps 1–4 (using the most recently obtained time step instead of Δt_1, and the most recently calculated velocities instead of V_{tr1}, V_{rot1}, and V_{ver1}) a number of times completes the simulation. If the time to the next molecular event happens to be greater than $(\Delta t)_c$, a preferred time interval needed to track the cell motion, then the next time step is equal to $(\Delta t)_c$, and the simulation proceeds as described above with the exception of Step 1 (i.e., no event is implemented).

Typically, model algorithms are written in conventional programming languages such as Fortran, C++, Java, or other compiled languages. Such general purpose languages and their libraries are very well developed, kept growing with each new version being compatible with the previous one, and their technical support is easily available. A modeler needs to be fluent in one of those programming languages, but after achieving that, the freedom in modeling whatever is desired is practically unlimited. One may also choose agent-based modeling in the framework of one of the available library platforms such as Swarm, Ascape, or Repast (Gilbert & Bankes, 2002). The programmer still has to learn the general purpose language in which the platform operates. The advantage of those libraries is that they allow the programmer to built easily some specific simulation environments, collect data from simulations automatically, or build user-friendly interfaces. The disadvantage is that many processes, especially those very different from ones simulated by the platform developers, can be very difficult to model because of build-in platform assumptions. An example of an agent-based leukocyte rolling model is given in Section V.F.

D. Comparison with Experiment

Most leukocyte rolling experiments are not conducted with the modeler or any specific model in mind. Conversely, many modeling results are not accessible experimentally. So this represents an interesting intersection where small adjustments on either side can truly move the field forward. The modeler is often not aware of the experimental limitations dictated by physics, chemistry, and biology. The resolution of measurements in space and time is often severely limited. Equally as often, the experimentalist does not realize that certain processes such as cell shape changes may be easy to observe, but very difficult to simulate.

Leukocyte rolling is a complex process. The modeler relies on experimental measurements concerning model parameters, and on available information about the system behavior concerning model cellular, molecular, and environmental interaction rules. Both sources of information are established based on available experimental data. The degree of uncertainty in measurements typically ranges from a few percent to a factor of 2, in some cases to a factor of 10, and our knowledge about the rolling cell behavior is still far from being completed. With this said, it is surprising that any modeling result ever matches experimental data. Such a match does not prove that the model is valid. A reasonable model prediction may be a result of erroneous assumptions offsetting each other, or a result of excessive adjustments of parameters. The latter can happen when a parameter has a poor base in experimental data and the modeler allows the parameter to vary over a wide range of values.

One such parameter that is not well established experimentally is the transition-state spring constant, a concept derived from theoretical considerations with few supporting measurements (Heinrich, Wong, Halvorsen, & Evans, 2008). Based on Dembo et al. (1988), any value within a range of 0–2 dyn cm^{-1} may be considered for the transition-state spring constant (Hammer & Apte, 1992). However, even small deviations of an order of 0.02 dyn cm^{-1} significantly change modeling results (M.K. Pospieszalska, 2009, unpublished data). The same applies to the bond formation rate constant. The AD model uses different values at different occasions for the same adherent molecules explaining that the rate has not been experimentally measured. In King and Hammer (2001) $k_f^0 = 365\,s^{-1}$, while in Bhatia et al. (2003) $k_f^0 = 84\,s^{-1}$, for P-selectin/PSGL-1 or sLe^x molecules at a wall shear rate of 100 s^{-1}. Although such assumptions have no experimental support, they allow model cells or ligand-coated spheres to roll somewhat similarly to those in matching experiments.

A more interesting situation ensues when the model fails to match experimental data, even though it bases on sound measurements and interaction rules. These types of mismatches fall into four broad categories:

1. A parameter measurement was made, but it did not measure what it claimed to measure. For example, what may look like a single molecular bond may actually involve multiple molecules.
2. The parameter measurement was made at insufficient resolution or precision.
3. The model interprets a measurement or interaction incorrectly. That typically arises from the modeler's unfamiliarity with the conditions and precisions of the experiment being modeled.
4. The model neglects an important phenomenon or oversimplifies. A classic example of this is "Let's model the elephant as a sphere, which seems to be a good approximation."

To make progress in the field, it is necessary to address the above critical points rather than hide the discrepancies. A robust survey of parameter sensitivity can always help. If a certain parameter can be varied over a 10-fold range and the model cell still rolls in the same way, this means that this particular parameter is irrelevant, and experimental efforts into getting better estimates are wasted. However, if the model shows that a certain parameter is critical for cell rolling, going back to the bench and trying to get a better estimate is essential. Equally interesting is a situation where the modeler needs the value for a parameter that has never been measured. This is productive, because some of the requested parameters can be fairly easy to measure, yet no experimentalist had ever bothered to look at them.

In general, results of satisfactory models do not match experimental data over the full range of physiological parameters, and rolling models are no exception. The mismatch is a fertile starting point for more experiments and further improvement of the models. No biological process can be understood without modeling. Very few processes in biology have been modeled at a level that satisfies both the experimentalist and the modeler. Commonly, the experimentalists will describe their findings based on intuitive models with arrows and circles where molecular concentrations go up or down. Although this provides a first level of understanding, because even the arrows and circles constitute a conceptual model, the scrutiny of a quantitative model is much stricter and will teach much more. Since an expert in modeling cannot be an expert in experiments, and *vice versa*, a close cooperation of modelers and experimentalists is necessary to build successful models bridging between these two broad disciplines.

So what are the specific challenges of modeling leukocyte rolling the modeler and experimentalist face? One is the interface between molecular and cellular approaches. This so-called mesoscale area is notorious for uncertainty. At the molecular level, detailed molecular techniques (such as molecular dynamics) become computationally extremely expensive or even break down because of numerical instabilities. At the cellular level, keeping track of every molecule in the cell and their states is simply impossible. As a result, modelers chose between two-level approximations, such as modeling the microvillus deformation as of a spring and a dashpot (dynamics of both elements are known with mathematical precision), but this is obviously a simplification of the real process of microvillus elongation in rolling. There is no clear choice of an appropriate approximation satisfying both molecular- and cellular-level criteria. Another challenge is imposed by the physical limits of measurements. Light microscopy can see down to 200 nm, and electron microscopy is not compatible with live cells. The molecular reaction rates, bond forces and rates of their loading are notoriously hard to measure at the single molecule level. Even when the experimentalist succeeds, it can always be argued that the isolated molecule is not in its physiological environment and proper cellular context. While existing models of leukocyte rolling can match rolling behavior under certain conditions, they fail to do so in others. Distinguishing the fundamental oversights from a mere experimental error, or the failure of a model because of a design flaw from a real mismatch, is both challenging and interesting.

No single experiment can measure all the parameters needed for even the simplest of rolling models. Many molecular parameters come from biochemical and molecular biology experiments, morphological data from light and electron microscopy, and material property data from pulling and squeezing experiments using small micropipettes, atomic force microscopes, or laser tweezers. Below, we discuss experimental procedures which can result in observations and measurements useful for designing and validating rolling models. The experimental factors include:

1. Species
2. Cell types and primary cell isolation, cell lines
3. Substrates: endothelial cells and recombinant adhesion molecules
4. Suspending media: cell culture media and whole blood
5. Types of microscopy used
6. Data recorded
7. Types of flow chambers

1. Species

Mouse and human leukocytes are used in perhaps 90% of all published leukocyte rolling experiments. Of 1703 articles in PubMed (accessed 25 June 2009) on "leukocyte rolling," 641 mention "mouse" and 706 "human," but this is not an exact representation of the species actually used, because a mouse monoclonal antibody might have been used in a study of human leukocyte rolling. Human blood is readily available, and good separation methods exist for neutrophils, monocytes, and lymphocytes (Oh, Siano, & Diamond, 2008). Mice can be an advantageous model, because many transgenic and knockout mice are available, but the cell separation methods (Cotter, Norman, Hellewell, & Ridger, 2001) and the amount of blood (about 1 ml per mouse) are more limited. Since the mouse blood supply is so limited, bone marrow neutrophils and other leukocytes are often used (Zhang et al., 2006).

2. Cell Types and Primary Cell Isolation, Cell Lines

The large majority of leukocyte rolling experiments is conducted with primary blood leukocytes. Some cell lines like HL-60 (33 papers), Jurkat (17 papers), or U937 (10 papers) have been used, often as a control or for biochemical experiments that accompany the flow chamber work.

When using primary blood leukocytes, the types of cells and the methods of isolation must be considered. The most commonly investigated cell type in rolling studies is the neutrophil (738 papers), which is also the most abundant leukocyte in human blood, followed by lymphocytes (527 papers), monocytes (187 papers), and eosinophils (49 papers). Rolling of other cells such as basophils, mast cells, or dendritic cells is rarely investigated.

Each *in vitro* experiment starts with a cell isolation procedure. Most cell isolation procedures are based on density gradients of polymers such as Ficoll, Percoll, or Histopaque. All these methods lead to a small, but measurable activation of leukocytes (Forsyth & Levinsky, 1990; Glasser & Fiederlein, 1990; Kuijpers et al., 1991; Y. Kuwano & K. Ley, 2009, unpublished data). This is best documented for neutrophils, where commonly used activation markers indicate increased Mac-1 expression (reflects degranulation of secretory granules) and loss of cell surface L-selectin (reflects shedding by the ADAM-17 protease). Bead-based positive and negative selection methods exist for cell isolation, and the activation status of the resulting cells should be assessed. Cell activation affects rolling behavior and site densities of relevant adhesion molecules that the modeler needs to know.

Cell isolation can be circumvented by working in whole blood systems. Although such systems were developed early (Badimon, Turitto, Rosemark, Badimon, & Fuster, 1987), recent advances in microcapillaries have made it possible to develop whole mouse blood systems (Hafezi-Moghadam, Thomas, & Cornelssen, 2004; Smith, Sperandio, Galkina, & Ley, 2004) that can be made small enough for transillumination studies (Chesnutt et al., 2006) and have recently been adapted for human blood (Y. Kuwano & K. Ley, 2009, unpublished data). The advantages of whole blood are that the possible activation steps introduced by cell isolation are circumvented, wall shear stress can be tightly controlled, and the velocity profile near the wall is similar to the *in vivo* situation. Disadvantages include the short lifetime of whole blood systems (2–4 min) and the inability to distinguish cell types. The latter problem can be overcome by using transgenic mice that expresses green fluorescent protein (GFP) under the lysozyme M promoter (Faust, Varas, Kelly, Heck, & Graf, 2000; Zarbock et al., 2007), in which neutrophils are brightly fluorescent.

3. Substrates: Endothelial Cells and Recombinant Adhesion Molecules

Traditionally, rolling assays were conducted on cultured endothelial cells (Lawrence, McIntire, & Eskin, 1987). The endothelial cells provide a nearly physiologic substrate, where the adhesion molecules are expressed in their natural cellular context. However, the transcriptome of cultured endothelial cells overlaps only about 50% with that of endothelial cells *in situ* (Durr et al., 2004), the quality of the endothelial cell monolayers varies, cultured endothelial cells can become senescent, and they express many other molecules that influence leukocyte rolling. This means that the modeler cannot be sure that the behavior of the rolling cell is entirely due to a controlled set of proteins. Specificity controls are usually conducted with blocking monoclonal antibodies, but these controls only show that the molecule in question was necessary for rolling, yet other molecules may still contribute. This could be addressed by appropriate reconstitution experiments, but these are very difficult. Also, cultured endothelial cells do not express P-selectin, a commonly investigated and modeled rolling molecule.

An alternative approach is to use recombinant proteins on the surface of the flow chamber. Using this technique, the site density can be controlled tightly and measured accurately (Lawrence & Springer, 1991). Most recombinant adhesion molecules are Fc constructs, and there is concern that the Fc portions could engage Fc receptors on leukocytes. However, this concern is alleviated in whole blood, where the antibody concentration in plasma is sufficient to saturate Fc receptors.

4. Suspending Media: Cell Culture Media and Whole Blood

Cell culture media are typically rich in glucose and other nutrients and high in oxygen. Their viscosity is much lower than that of blood and about 30% lower than that of plasma. Their density is close to the density of water. All these factors must be taken into account when modeling. A consequence of the low density of cell culture media is that isolated leukocytes experience a gravitational force and will settle on the lower wall of the flow chamber. In some models, this gravitational force is used to bring the leukocyte into contact with the substrate (Hammer & Apte, 1992; Pospieszalska et al., 2009). In whole blood, leukocytes are practically neutrally buoyant, so there is no sedimentation, but they still get in contact with the substrate. Unlike the situation in cell culture media, they contact both the "floor" and the "ceiling" of the flow chamber. The exact nature of the forces that push leukocytes to the walls in whole blood is not fully understood, but seems to be related to the presence of red blood cells (Schmid-Schönbein, Usami, Skalak, & Chien, 1980) and red blood cell aggregation (Nobis, Pries, Cokelet, & Gaehtgens, 1985).

5. Types of Microscopy Used

With isolated leukocytes rolling on cultured endothelial cells, phase contrast microscopy is commonly used (Sircar et al., 2007). This can be combined with fluorescence microscopy, using live cell imaging probes that label specific proteins in the rolling cell or the endothelial substrate (Shaw et al., 2004). Such information can be very valuable for the modeler, because subcellular information including the localization of specific adhesion molecules becomes available and allows refining the model. In whole blood-perfused flow chambers, regular transillumination microscopy (Chesnutt et al., 2006) and stroboscopic epifluorescence microscopy (Smith et al., 2004) are used. With fluorescence, the nature of the rolling cells investigated can be identified by using GFP-transgenic mice. Transillumination microscopy shows subcellular structures like granules and the nucleus, which can be used to directly see the motion and deformation of the cell as it rolls. This can be used to measure the slip velocity (Tözeren & Ley, 1992). A recent development is the application of total internal reflection (TIRF) microscopy to flow chambers (Ley et al., 2008; S.I. Simon, unpublished data). TIRF provides valuable data to the modeler, because this new technique focuses on the structures near (within 200 nm of) the glass wall of the flow chamber and can be used to directly see the "footprints" of leukocyte microvilli as they engage the adhesive substrate.

6. Data Recorded

The parameters and variables measured are among the most important features of experiments. Any model, for a given wall shear rate and site density of substrate adhesion molecules, will predict the leukocyte translational velocity (rolling velocity), and most experiments report average rolling velocities. However, the exact way in which the average rolling velocity is calculated can make a difference. Averaging the velocities of all rolling cells in the field of view gives the volume average velocity. Averaging the velocities of all rolling cells crossing a line (across the flow chamber) perpendicular to the flow gives the flux average velocity (Ley & Gaehtgens, 1991). Since slow-rolling cells are overrepresented with the volume method, the volume average velocity is lower than the corresponding flux average velocity. To compare experimental data with modeling data, it is very useful to report the experimental data on a per-cell basis. Rolling leukocytes can be tracked with high resolution (Sperandio, Pickard, Unnikrishnan, Acton, & Ley, 2007), making it possible to report time–distance or time–velocity curves which can be directly compared with the model translational velocities filtered to match the time resolution of the experiment. Unfortunately, such data are rarely reported. Other experimental data that are necessary to be reported for modeling purposes are site densities of substrate adhesion molecules.

Any additional data are welcomed by modelers. The shape of rolling leukocytes can be assessed in the top view or in the side view. The latter requires a special flow chamber (Lei, Lawrence, & Dong, 1999). *In vivo* recordings by intravital microscopy also allow the assessment of leukocyte shape in the side view (Damiano, Westheider, et al., 1996; Pickard & Ley, 2009). To optimize the modeling of leukocyte rolling, an active communication between the modeler and the experimentalist is beneficial to ensure that all relevant data of the experiment are actually recorded. Each rolling model produces a rich output of data about various leukocyte rolling variables, some of which can be observed in rolling experiments. The measurable variables include leukocyte translational and rotational velocities, pause time, jump length (the distance between the cell detachment and reattachment), jump time, and jump jerkiness (the cell acceleration).

7. Types of Flow Chambers

By far the most commonly used flow chamber is the parallel-plate model (Lawrence, 2001). These dynamic *in vitro* assays helped to identify the shear threshold for selectin-mediated rolling (Finger, Puri, et al., 1996; Lawrence, Kansas, Ghosh, Kunkel, & Ley, 1997), the ability of α_4-integrins to initiate lymphocyte tethering and rolling in the absence of selectins (Berlin et al., 1995), and the role of chemokines in leukocyte arrest (Campbell et al., 1998; Shamri et al., 2005).

Parallel-plate flow chambers were described early (Toy & Bardawil, 1958), but were used for the investigation of leukocyte rolling only in the late 1980s (Lawrence et al., 1987). Commercially available parallel-plate flow chamber systems (http://www.glycotech.com) consist of two flat plates separated by a gasket, which forms a geometrically defined narrow channel between the two plates. Inflow and outflow ports are used to enable the perfusion of the cell suspension through the chamber. An additional port generates a negative pressure, which keeps both plates together. To control the flow of the cell suspension through the flow chamber, a high-precision perfusion pump is used. By knowing the geometry of the inner chamber, the flow rate of the perfusion pump, Q, and the fluid viscosity, μ, the wall shear stress, τ_w, can be calculated as follows:

$$\tau_w = \frac{3\mu Q}{2ab^2}, \tag{23}$$

where a is the chamber width and b is the channel half-height. The offline analysis of *in vitro* leukocyte rolling assays can be accomplished by counting the number of rolling cells per field of view and per time period. Leukocyte rolling velocity can be assessed by measuring the distance leukocytes have traveled in a certain time. Several automatic tracking systems have been developed to automatically quantify the number and the velocity of rolling cells *in vitro* (Mangan et al., 2005; Rao, Haskard, & Landis, 2002; Ray, Acton, & Ley, 2002).

Although flow chambers have been an important tool to study leukocyte rolling *in vitro*, there are several limitations such as the large dead volume of fluid within the flow system (300 μl in the Glycotech parallel-plate flow chamber), necessitating the isolation of large quantities of cells. This effectively prevented use of this type of compound flow chambers in the mouse system. The availability of genetically engineered mice that offer ideal conditions to study adhesive interactions between distinct pairs/groups of adhesion molecules stimulated the development of new flow systems, which can be conducted with small sample volumes. Smith et al. (2004) described an *ex vivo* autoperfused flow chamber in the mouse that consists of a rectangular glass capillary, of a cross section of 200 $\times$ 2000 μm, where adhesion molecules of choice can be immobilized. After immobilization the flow chamber is placed under an upright microscope and connected via polyethylene (PE) tubing to the jugular vein and the carotid artery of the mouse. Because of the pressure difference between the arterial and venous side, blood is continuously driven through the flow chamber. Additional tubing connected to the venous and the arterial side is used to continuously measure the pressure drop, ΔP, along the glass capillary. From this measurement in conjunction with the known geometry of the chamber, wall shear stress τ_w can be determined exactly and independent of any assumptions about blood viscosity, as follows:

$$\tau_\mathrm{w} = \frac{d_\mathrm{h}\Delta P}{4L}, \tag{24}$$

where L is the length of the capillary and d_h is the hydraulic diameter. The latter can be calculated using the following formula:

$$d_\mathrm{h} = \frac{4A}{p}, \tag{25}$$

where A is the chamber cross-sectional area and p is the wetted perimeter of the cross section (Smith et al., 2004). Adhesive interactions in the flow chamber between blood constituents (i.e., leukocytes) and immobilized proteins on the glass surface can then be observed and recorded (Smith et al., 2004). This flow chamber was further miniaturized to a cross section of 20 $\times$ 200 μm (Chesnutt et al., 2006). Because of the small volume in the system, the rectangular glass capillaries can also be used for *in vitro* rolling studies by simply connecting the glass capillary to a perfusion pump or a reservoir providing hydrostatic pressure. This provides superior optical conditions over round glass capillaries, which have also been used for leukocyte rolling studies (Nandi, Estess, & Siegelman, 2000). If appropriate condensers, objectives, and additional accessories are chosen, microflow chamber assays with cell suspensions can also be performed on the upright microscope using contrast enhancing techniques such as phase contrast, dark field, TIRF, or differential interference contrast (DIC) microscopy. Leukocyte rolling has also been studied in other types of flow chambers including radial and plate-and-cone flow chambers. Since these devices are rarely used and the data obtained have not been used for modeling efforts, they are not discussed here.

V. PUBLISHED MODELING APPROACHES

It has been 17 years since the first two leukocyte rolling models emerged in 1992, and many modeling papers have been published. Although all of them greatly contributed to the development of modeling in research, to stay within the scope of this chapter we had to make a selection. Our selection was based on desire to present at least one published model from each of the four basic categories, and also to illustrate how a developed model can be both a research tool and a source of new research ideas. The following models are discussed:

A. Adhesion dynamics (AD) model of Hammer and Apte (1992) as the first direct model
B. Event-tracking model of adhesion (ETMA) of Pospieszalska et al. (2009) as an efficient alternative to AD
C. Model of Tözeren and Ley (1992) as the fist semianalytic model
D. Model of Khismatullin and Truskey (2004) as the fist model of cell deformation in rolling
E. Model of Zhao et al. (1995) as an example of analytical models
F. *In silico* white blood cell (ISWBC) model of Tang et al. (2007) as the only agent-based model of rolling

A. Adhesion Dynamics Model by Hammer and Apte (1992)

The AD model, in its original version published in 1992, simulates rolling of a rigid sphere with rigid microvilli (oriented normal to the sphere surface) on a single class of receptors. The ligands are randomly distributed over the sphere surface. Molecular receptor–ligand interaction rates are modeled according to Dembo (Eqs. 3 and 4). The molecular bonds, forming perpendicularly to the substrate, are modeled as Hookean springs (Eq. 17). AD is a direct model progressing by fixed, predefined time steps. At the beginning of each time step all free receptor–ligand pairs are tested for the possibility of bond formation using (Eq. 9), and all existing bonds are tested for the possibility of dissociation using (Eq. 10). If selected, the molecular events are implemented at the end of the time step under consideration. At each time step, the sphere velocities are calculated from (Eq. 21), and the system is updated accordingly. The model yields translational velocity patterns consistent with the cell free motion, rolling, transient attachment, and firm adhesion observed experimentally. The authors have found one set of parameters for which the model results match the trend in translational velocity versus ligand density seen in experiments by Lawrence and Springer (1991) on human neutrophils rolling on P-selectin at a shear rate of 180 s^{-1}.

For the last 17 years, the AD model has been used by Hammer and his coworkers for various research studies, undergoing modifications and extensions. In Chang et al. (2000), the Dembo formula for the off-rate (Eq. 4) is substituted with the Bell formula (Eq. 7), while the on-rate is fixed at $k_f = 84\ s^{-1}$. The authors establish the state diagram for adhesion at a wall shear rate of 100 s^{-1} and a substrate receptor density of 3600 molecules μm^{-2}. The diagram shows boundaries between no adhesion, fast adhesion, transient adhesion, and firm adhesion regions as functions of unstressed off-rate versus reactive compliance.

The AD model in Bhatia et al. (2003) is extended to accommodate a second class of receptors, and simulates cell rolling through P-selectin–sLe^x and β_2-integrin–ICAM-1 bonds. In this version of AD, the Bell approach (Eqs. 6 and 7) is used to describe the molecular rates, with unstressed rates of $k_f^0 = 84\,s^{-1}$ and $k_d^0 = 2.4\,s^{-1}$ for selectin and $k_f^0 = 10{-}1000\,s^{-1}$ and $k_d^0 = 0.1\,s^{-1}$ for integrin bonds. The authors present the state diagrams for adhesion mediated by two receptors at wall shear rates of 100 and 1000 s^{-1}, for site densities of 40 molecules μm^{-2} for selectin and 1000 molecules μm^{-2} for ICAM-1. The diagrams show boundaries between rolling adhesion and firm adhesion as functions of β_2-integrin unstressed on-rate versus β_2-integrin density, or as functions of ICAM-1 density versus P-selectin density.

The AD model for a single class of receptors (Hammer & Apte, 1992) underwent major changes in Caputo and Hammer (2005). To reduce computational cost, the authors use two concurrent prefixed time steps, one for monitoring the cell motion and the other for monitoring the molecular events. They still use the Bell formulas for the molecular rates (Eqs. 6 and 7) with $k_d^0 = 0.9\,s^{-1}$, but substitute rate k_f^0 with a two-value discrete function, k_f^*, such that $k_f^* = 2$ or 3 s^{-1} when molecular bonds are present and $k_f^* = 116\,s^{-1}$ otherwise. This effectively allows the cell to recover from a "skip," during which it transiently loses contact with the substrate and rapidly gains speed. The wall shear rate is set at 180 s^{-1}, and the substrate receptor density is 124 molecules μm^{-2}. The new approach to the unstressed on-rate is based on Chang and Hammer (1999), and reasoning for it is discussed below. In this version of AD microvilli are deformable (a loaded microvillus can extend or form a tether at its tip), and ligands are clustered on the microvilli tips. The microvillus is modeled as a spring, and its tether as a dashpot, based on Shao et al. (1998). The authors find that the microvillus deformation stabilizes rolling on P-selectin, and note that their results do not match the experimental results for wall shear rates of 200 s^{-1} and higher.

The two-receptor AD model of Bhatia et al. (2003) is further developed in Krasik et al. (2006). The model is modified to account for three receptor–ligand pairs: PSGL-1/E-selectin, resting LFA-1/ICAM-1, and active LFA-1/ICAM-1. The unstressed reaction rates are $k_f^0 = 0.06\,\mu m^2\,s^{-1}$ and $k_d^0 = 2.6\,s^{-1}$ for the first pair, $k_f^0 = 0.3\,\mu m^2\,s^{-1}$ and $k_d^0 = 4\,s^{-1}$ for the second pair, and $k_f^0 = 115\,\mu m^2\,s^{-1}$ and $k_d^0 = 0.17\,s^{-1}$ for the third pair. It is assumed that selectin ligation can lead to integrin activation through MAPK (mitogen-activated protein kinase). AD is integrated with a subunit, which deterministically establishes the level of MAPK activation (i.e., the number of integrins on an individual microvillus that will be active in the next time step) from the number of existing PSGL-1–P-selectin bonds. The simulations are conducted at a wall shear rate of 100 s^{-1}, E-selectin site density of 3600 molecules μm^{-2}, and ICAM-1 site density of 210 molecules μm^{-2}. The authors present the state diagrams for adhesion mediated by the three types of receptor–ligand pairs, showing the boundaries between rolling adhesion and firm adhesion, as functions of LFA-1 site density versus PSGL-1 site density, for both resting and active integrin states. Modeling MAPK activation is further developed to be stochastic in Krasik et al. (2008).

The AD model, version of Caputo and Hammer (2005), is further modified in Caputo et al. (2007). Molecular binding is broken down into two steps. In the first step the receptor and ligand encounter each other by skimming the surface within binding distance, and in the second step they undergo the bond formation process (Bell, 1978). The authors introduce the concept of an overall unstressed on-rate for a cell–surface ligand, $(k_f^0)_L$, defined as follows:

$$(k_f^0)_L = k_0 P n_R, \tag{26}$$

where k_0, P, and n_R are the encounter rate, probability of bond formation, and density of the substrate receptors, respectively. Direct formulas for k_0 and P, which are functions of slip velocity, are derived. With the increasing slip velocity, encounter rate k_0 increases because the number of receptors encountered by the ligand increases. At the same time probability P decreases because the encounter duration decreases. Consequently, with increasing slip velocity, rate k_f^0 first increases (when P decreases slower than k_0 increases), and then decreases to zero (when P decreases faster that k_0 increases). The on-rate for the cell–surface ligand, $(k_f)_L$, is derived from Eqs. (2) and (26). For simulating catch–slip behavior of selectin bonds, the off-rate of Bell (Eq. 7) is substituted with the formula derived by Evans et al. (2004). The new off-rate is specific for bonds dissociating along two pathways according to the analytical model proposed by those authors. Other analytical models for catch–slip bonds exist (Barsegov & Thirumalai, 2005; Pereverzev, Prezhdo, Forero, Sokurenko, & Thomas, 2005; Thomas et al., 2006) and are waiting to be tested through modeling.

B. Event-Tracking Model of Adhesion by Pospieszalska et al. (2009)

ETMA is a newcomer to the field (Pospieszalska et al., 2009) and presents a computationally efficient alternative to AD. ETMA is a direct, 3D, stochastic π-calculus-driven model simulating leukocyte rolling on one class of receptors. The cell is modeled as a sphere with flexible microvilli (they can bend, and they can pivot at their base), which allows the microvillus to bend if the space between the cell body and the substrate cannot accommodate its full length. Unlike AD, ETMA tracks all free substrate receptors in the cell–substrate contact area and, therefore, imposes no restrictions concerning the angle between the bond and the substrate when the bond formation occurs, that is, the angle can differ from 90°. The model algorithm is described in Section IV.C. The physics of ETMA is almost the same as in AD (based on Eqs. 3, 4, 13, 15, and 17–21); however, the modeling method is quite different.

The basic idea of stochastic π-calculus (Milner, 1999) is to model the dynamic behavior of the system on a race condition basis. All the possible events at a given point in time compete for the next event by each declaring its time of occurrence (the time is selected according to the exponential distribution corresponding to the event's reaction rate from Eqs. 11 and 12), and the fastest event succeeds. The winning event is implemented and the race starts again. The continuity of exponential distributions ensures that the probability of having two simultaneous winner events is zero. ETMA progresses by varying time steps dictated by the time intervals between consecutive molecular events, and implements each event when it occurs. A time step may be shortened if it is longer than the preferred time interval needed to monitor the cell motion, as described in Section IV.C. Progressing by varying time steps gives each molecular event a chance to have an individual impact. The ability of tracking consecutive events independent of any predefined time resolution provides high temporal resolution of key events of the process which are bond formation and dissociation. ETMA is computationally efficient, minimizing the number of system updates for maximum accuracy of event implementation.

The precision of ETMA allowed for a comprehensive study of the subgroup of bonds which at some point in time become load bearing, providing their location, number, lifetime, history, and kinetics. ETMA produced a cumulative map of the load positions (the receptor base position in the contact area when the bond becomes load bearing) and corresponding break positions (the receptor base position when the bond breaks). Surprisingly, very few load-bearing bonds are needed to maintain rolling, and about 30% of bonds do not last long enough to become load bearing. The simulations for the above data were conducted for a wall shear rate of 50 s^{-1}, P-selectin site density of 150 molecules μm^{-2}, and $k_f^0 = k_d^0 = 1\,s^{-1}$. Our experience is similar to those of others. Although our model results match the experimental results of Lawrence and Springer (1991) and Yago et al. (2002) at low wall shear rates, they do not predict correctly the cell rolling for high shear rates.

In the second version of ETMA (Pospieszalska & Ley, 2009), we introduce fully deformable microvilli, modeling the microvillus in its extension state as a Kelvin–Voigt (a spring and a dashpot in parallel) viscoelastic material, while microvillus tether is modeled as a dashpot as in Shao et al. (1998). The microvillus extension, $\Delta L_{\mathrm{m}}(t)$, and its tether extension, $\Delta L_{\mathrm{te}}(t)$, at time t are calculated as follows:

$$\Delta L_{\mathrm{m}}(t) = \left(\frac{F}{\sigma_{\mathrm{m}}}\right)\left\{1 - \exp\left(\frac{-t}{\eta_{\mathrm{m}}/\sigma_{\mathrm{m}}}\right)\right\}, \tag{27}$$

$$\Delta L_{\mathrm{te}}(t) = \left[\frac{F - F_0}{\eta_{\mathrm{te}}}\right]t, \tag{28}$$

where σ_{m}, η_{m}, η_{te}, F, and F_0 are, respectively, the microvillus spring constant, microvillus effective viscosity, tether effective viscosity, microvillus bond force (assumed to be constant for the time step), and microvillus threshold force (the force at which a given microvillus stops extending and starts developing a tether). Conceptual models of nondeformable and deformable microvilli are compared in Fig. 4.

C. Model by Tözeren and Ley (1992)

The model of steady-state leukocyte rolling of Tözeren and Ley (1992) was the first semianalytic model of rolling. It considers a leukocyte-like sphere which has membrane folds and other unspecified surface projections. To accommodate the cell surface roughness, it is assumed that there is a fluid layer separating the cell and the substrate of a thickness of at least the length of the typical cell surface projections ($\sim 0.12 r_{\mathrm{c}}$, where r_{c} is the radius of the cell). The cell rolls with a uniform translational velocity V_x and slip velocity V_{s}, at a constant separation distance from the substrate of h_{c} (the Cartesian system is oriented as in Fig. 4A, with its origin located at the sphere's center and moving with the sphere). For numerical simplicity, the spherical surface in the cell–surface contact zone is replaced with a finite cylinder of radius r_{c}. Then the cell's cross sections parallel to plane xz are of the same shape within the contact zone. The bonds form at the trailing edge of the contact area perpendicular to the substrate having a length of λ (the unstressed bond length), and cannot withstand forces greater than 10^{-5} dyn (Bell, 1978). All bonds formed at time t look alike in rolling (i.e., their lengths and angular positions in reference to the substrate change in the same way) and can be lumped into one bond state in plane xz. Therefore, the case can be viewed as two-dimensional and the contact area as one-dimensional.

The length of the bond attached to the substrate at x is calculated from equations of kinematics (Beer & Johnston, 1984) as a function of V_x, V_s, and h_c. The bond force $\mathbf{F}_b = (F_{bx}(x), F_{bz}(x))$ is found from the bond length based on (Eq. 17). The bond density, $n_B = n_B(x)$, is found by solving the modified equation of Bell (1978), given below, assuming that the bond density at the trailing edge of the contact area is equal to zero:

$$-V_x \frac{dn_B}{dx} = k_f^*(n_R - n_B)(n_L - n_B) - k_d^* n_B, \tag{29}$$

where n_R and n_L are the receptor and ligand densities. The values for the rates in (Eq. 29) are $k_f^* = 1\,\mu\text{m}^2\,\text{s}^{-1}$ and $k_d^* = 0.5 \times 10^{-4}\,\text{s}^{-1}$ when the bond length is less than λ, and $k_f^* = 0$ and $k_d^* = 30\,\text{s}^{-1}$ otherwise. After bond lumping, density n_B can be viewed as a line density. The total bond force in the x-direction, F_{Bx}, acting on the cell is determined by integrating $F_{bx}(x)n_B(x)$ along the x-axis. Similarly, force F_{Bz} and then torque T_{By} imposed on the cell by $\mathbf{F}_B = (F_{Bx}, F_{Bz})$ are determined. The fluid force, F_F, and torque, T_F, acting on the cell are evaluated from Goldman et al. (1967a,b) and the assumed shear rate. The repulsive force, F_R, acting on the cell is evaluated using (Eq. 15). Since force F_{Bx} and torque T_{By} are assumed to balance those of the fluid, and force F_{Bz} is assumed to balance force F_R, the expression, $W = W(V_x, V_s, h_c)$, given below is optimized for V_x, V_s, and h_c to yield its minimum value:

$$W = (F_{Bx} + F_F)^2 + (F_{Bz} + F_R)^2 + (T_{By} + T_F)^2. \tag{30}$$

The obtained values of V_x, V_s, and h_c describe steady-state leukocyte rolling for the assumed shear rate. The model results are in the range of the experimental results of Lawrence and Springer (1991). The authors find that the bond length and the local cell stiffness near the bond play a critical role in enhancing leukocyte rolling.

D. Model by Khismatullin and Truskey (2004)

Khismatullin and Truskey (2004) were the first to take up the challenge of modeling whole-cell deformation in rolling. Their 3D semianalytic model simulates the leukocyte rolling in a parallel-plate flow chamber. The leukocyte is modeled as a compound viscoelastic drop composed of a viscoelastic nucleus surrounded by a thick layer of a viscoelastic cytoplasm. The cell membrane is assumed to have a cortical tension, a tension that pulls the cell into a spherical shape. The model uses six parameters specifically needed to simulate cell deformability, which are the volume fraction of the nucleus, nucleus viscosity, nucleus relaxation time, cytoplasm viscosity, cytoplasm relaxation time, and cortical tension, in addition to the suspending medium viscosity and density.

The membrane microvilli are modeled as springs, with spring constant σ_{m}, oriented normal to the cell surface. Also, the molecular bonds are modeled as springs with spring constant σ. In the model the bonds are assumed to be extensions of the microvillus filaments, aligned in the same direction as the filaments. That allows the microvillus and its bonds to be viewed as a system of N_{B} microvillus-bond springs in parallel, where N_{B} is the number of bonds present on the microvillus tip. It can be derived that the microvillus-bond spring constant, σ_{s}, is given by the following formula:

$$\sigma_{\mathrm{s}} = \frac{\sigma\sigma_{\mathrm{m}}}{\sigma + \sigma_{\mathrm{m}}}. \tag{31}$$

The molecular reaction rates are calculated from Eqs. (3) and (4) (the Dembo approach) and the bond force is calculated from (Eq. 17), with $L_{\mathrm{sep}} = L_{\mathrm{s}} - L_{\mathrm{m}}^{0}$, where L_{s} is the length of the microvillus-bond spring and L_{m}^{0} is the microvillus length in the absence of force. The unstressed reaction rates used in simulations are $k_{\mathrm{f}}^{0} = 100\,\mu\mathrm{m}^2\,\mathrm{s}^{-1}$ and $k_{\mathrm{d}}^{0} = 10\,\mathrm{s}^{-1}$, for monocytic cells. The rate k_{f}^{0} is higher than the measured value (Chesla, Selvaraj, & Zhu, 1998) to decrease the computation time. The density of bonds in the contact area is calculated from (Eq. 16), and the total bond force from (Eq. 18).

Tracking the cell deformation process is computationally very intensive. There are three different fluids (the nucleus, cytoplasm, and extracellular fluid) and two interfaces (the nucleus–cytoplasm interface, and the cytoplasm–extracellular fluid interface at the cell membrane) to consider. A 3D grid composed of cubes of size $\Delta x \Delta y \Delta z$ divides the volume of the cell into pieces, each to be processed. The cell shape is tracked with the volume-of-fluid method (VOF) (Guyeffier, Li, Nadim, Scardovelli, & Zaleski, 1999). The continuity and Navier–Stokes equations, describing the fluid motion inside and outside of the cell, are solved by Chorin's projection method (Chorin, 1967) on the established grid. The viscoelasticity of the cell nucleus and cytoplasm is captured by the Giesekus' model (Bird, Armstrong, & Hassager, 1987; Giesekus, 1982). The computer code is parallelized, and the IBM p690 server of the National Center for Supercomputing Applications (University of Illinois at Urbana-Champaign) is used to execute it. The longest cell rolling time shown in the figures is $t = 4$ ms.

At the beginning of a simulation a leukocyte is suspended in a low-viscosity fluid in the model parallel-plate flow chamber of height h_{ch}. The height of the chamber is small enough for the first molecular bonds to be

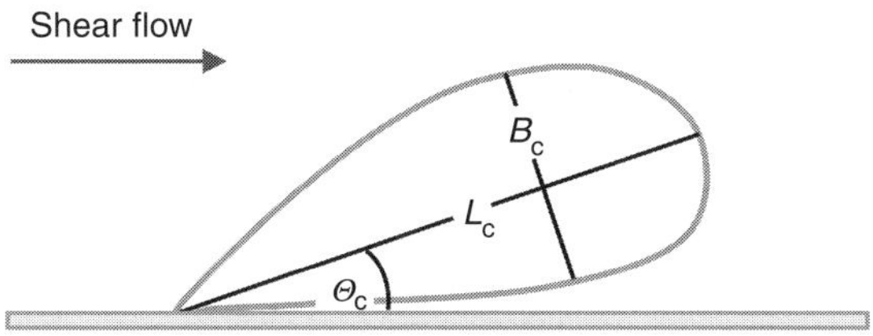

FIGURE 5 Illustration of the length L_c, breadth B_c, and inclination angle Θ_c for a deformed leukocyte.

formed. In the absence of flow, the leukocyte is at rest and has a spherical shape. After a sufficiently high shear flow is introduced, the cell starts to deform to a drop-like object of length L_c and breadth B_c, with inclination angle Θ_c between L_c and the lower plate of the chamber, as indicated in Fig. 5. The cell deformation index, D_c, is given by the following formula (Taylor, 1934):

$$D_c = \frac{L_c - B_c}{L_c + B_c}. \tag{32}$$

The deformation index takes a value from 0 to 1, with $D_c = 0$ and $\Theta_c = 90°$ for a spherical cell, and $D_c = 1$ and $\Theta_c = 0°$ for a flat cell of negligible thickness (a theoretical limit). A monocytic cell with a ligand surface density of 3000 molecules μm^{-2} barely deforms while rolling at a wall shear rate of 800 s^{-1} on receptors of site density of 1500 molecules μm^{-2}. The largest deformation, reached in $t = 3.4$ ms, is characterized by $D_c = 0.082$ and $\Theta_c = 44°$ in a chamber of height $h_{ch} = 80$ μm. The same cell rolling at a wall shear rate of 4000 s^{-1} reaches characteristics $D_c = 0.232$ and $\Theta_c = 37°$ at $t = 2.4$ ms. The authors find that at a constant wall shear rate, the cell–substrate contact area and shear force acting on the cell increase as the chamber height decreases. Depending on input parameters, the model can predict microvilli-like tethers (Khismatullin & Truskey, 2005).

E. Model by Zhao et al. (1995)

The analytical model of Zhao et al. (1995) describes in detail the translational velocity distribution for a homogeneous population of rolling leukocytes. The model is based on the assumption that at the microscopic level the leukocyte displacement is composed of random step-like jumps at random times. The lengths of the jumps are characterized by mean l_{mean} and standard deviation σ_l, and waiting times between consecutive jumps are characterized by mean t_{mean}. The time window of observations, Δt, which usually ranges from 1 to 20 s, is an input parameter. If Δt is large enough compared with t_{mean}, the cell translational velocity, V, defined below in accordance with the typical experimental method, can be viewed as an approximately continuous function of t:

$$V(t, x) = \frac{x(t + 0.5\Delta t) - x(t - 0.5\Delta t)}{\Delta t}, \tag{33}$$

where $x\,(t - 0.5\Delta t)$ and $x\,(t + 0.5\Delta t)$ are, respectively, the displacements of the leukocyte at times $t - 0.5\Delta t$ and $t + 0.5\Delta t$, and $t \geq 0.5\Delta t$. Since x is a realization of a random variable of displacement, $V(t, x)$ is a realization of a stochastic process which, under the model assumptions, can be viewed as a diffusion process. The probability density function (the distribution) of the process, $p(V, t)$, can be described by a system of the following three equations:

$$\frac{\partial p(V,t)}{\partial t} = -\frac{\partial J(V,t)}{\partial V}, \tag{34}$$

$$J(0, t) = 0, \tag{35}$$

$$\int_0^{V_{max}} p(V,t)\mathrm{d}V = 1. \tag{36}$$

Equation (34) is known as a Fokker–Planck equation. The integral limit V_{max} denotes the upper boundary for the rolling velocities. The function $J(V, t)$ represents the flux of probability flow defined as

$$J(V,t) = A(V)p(V,t) - (0.5)\frac{\partial[B(V)p(V,t)]}{\partial V}, \tag{37}$$

where A and B are the drift and diffusion coefficients, respectively. The authors derive the following formulas for A and B:

$$A(V) = \frac{l_{mean}/t_{mean} - V}{\Delta t}, \tag{38}$$

$$B(V) = 2Vl_{mean}\left[\frac{1 + (\sigma_l/l_{mean})^2}{(\Delta t)^2}\right]. \tag{39}$$

Solving (Eq. 34) for $p(V, t)$ with boundary conditions stated by Eqs. (35) and (36) yields a family of leukocyte translational velocity distributions, one distribution for each $t \geq 0.5\Delta t$, which describe with mathematical precision the transient evolution of the leukocyte rolling process. In steady-state rolling, $p(V, t)$ does not depend on t (i.e., the left-hand side of Eq. 34 is set to zero), and the family of distributions is practically composed of one member. The authors find that this one member is a γ-distribution, which gets more spread and more skewed to the right as Δt decreases, while the mean leukocyte translational velocity remains constant.

F. In Silico White Blood Cell Model by Tang et al. (2007)

ISWBC is the only published agent-based model of leukocyte rolling. The authors use Repast (North et al., 2006), a Java-based software developed at the University of Chicago, as a framework for creating and running their model. Software subunits are developed, verified, and plugged together to represent the components and mechanisms involved in leukocyte rolling. The leukocyte is represented by a flat, rectangular box-like object, with rounded edges. A uniform grid divides the leukocyte surface into 600 units, which does not directly correspond to actual cellular dimensions. The substrate is divided into the same size units, with 80 units corresponding to the contact area. Each leukocyte unit contains three objects, functioning as software components called agents, representing PSGL-1, VLA-4, and CXCR2 (chemokine receptor). Similarly, each substrate unit contains three agents representing P-selectin, VCAM-1, and GRO-α (chemokine). Each of these agents can map zero, one, or more than one adhesion molecules, depending on the local site density of the molecule. At the beginning of a simulation all VLA-4 molecules are in the low-affinity state. The model progresses by simulation cycles equivalent to 0.1 s of rolling *in vitro*. When there are no bonds, the leukocyte ratchets forward. During each ratchet event the leukocyte releases its trailing (back) row of units and adds a new row of units at the leading (front) side. If the leukocyte remains stationary for at least 100 simulation cycles, an extra row of units at the front and one extra column of units along one of its sides are added to mimic leukocyte spreading. If the stationary stage lasts for at least 200 simulation cycles, the leukocyte is considered adherent.

During each simulation cycle, each compatible pair of agents in overlapping (i.e., facing each other) units makes an attempt to establish N_B bonds, where $N_B = \min\{N_R, N_L\}$, and N_R and N_L are, respectively, the number of receptors and ligands represented by the two agents and available for bond formation. The PSGL-1/P-selectin molecules, and low- and high-affinity VLA-4/VCAM-1 molecules, form bonds according to probabilities based on Chigaev et al. (2001) and Mehta et al. (1998). The CXCR2/GRO-α molecules form bonds with a probability of 1. Each model bond maps to a single molecular bond. Only the bonds at the leukocyte trailing row of units are assumed to be loaded. The total bond force is an input parameter selected by comparing the model and corresponding experiment pause time characteristics. The bond force for each loaded bond is calculated by dividing the total bond force by the number of P-selectin and VCAM-1 bonds at the trailing row of leukocyte units. During each simulation cycle, each P-selectin and each VCAM-1 bond can break according to force-dependent dissociation probabilities established based on Park et al. (2002) and Zhang, Craig, Kirby, Humphries, and Moy (2004). Unlike other types of bonds, each GRO-α bond forms and breaks with a probability of 1 within the same simulation cycle, once per cycle. Such an event switches the affinity from low to high for one low-affinity VLA-4 molecule sharing the unit with the CXCR2 molecule, if there is any. The maximum percent of VLA-4 molecules on the leukocyte surface that can be induced into the high-affinity state when exposed to chemokines is assumed to be 12.5% (Diamond & Springer, 1993). The authors establish a set of model parameters at which the model simulations reasonably match six flow chamber experiments of Alon et al. (1995), Park et al. (2002) and Smith et al. (1999).

The authors use a large portion of their paper to present their contribution to the development of agent-based modeling in the Repast platform with application to leukocyte rolling. As the Repast is excellent for modeling involving agents located on a rectilinear grid, as illustrated above, it is less useful for modeling cell surface protrusions such as microvilli and ridges of them. The modeling limitations are reflected in a simplified leukocyte body structure, and, most likely, in the fact that the total bond force is one of input parameters rather than time-dependent output data.

VI. Future Directions

Although leukocyte rolling has been modeled for 17 years (Hammer & Apte, 1992; Tözeren & Ley, 1992), there is a long way to go before a complete model of leukocyte rolling may emerge. A shortcoming of the existing models is that they are still too basic to be able to explain leukocyte rolling at high wall shear rates (such as those commonly found in microvessels *in vivo*, of up to 2000 s^{-1}). Currently, neither the modelers nor the experimentalists know what critical parameter is missing or what important phenomenon is being overlooked. Another problem is that some of those models are computationally too intensive to carry out simulations for sufficiently long periods of cell rolling, as needed for comparison with experimental results and study purposes (see below). Leukocytes simulated by the moderate computational cost models with a reasonable set of parameters roll faster at high wall shear rates than their real counterparts and, as the wall shear rate increases, sooner become unstable (Pospieszalska & Ley, 2009). Since microvillus deformability (i.e., microvillus extension and tether formation) contributes to the stability of leukocyte rolling (Caputo & Hammer, 2005; Pospieszalska & Ley, 2009), modeling microvilli as extensible species improves the situation, but does not solve the problem.

During *in vivo* and *in vitro* rolling, most leukocytes, especially neutrophils, show large deformations of the whole cell body (Damiano, Westheider, et al., 1996), in addition to the deformation of microvilli. This phenomenon is thought to be one of the key factors shaping leukocyte rolling at high wall shear rates. Currently, the whole-cell deformation cannot be incorporated with sufficient detail into models of leukocyte rolling, because the material properties of the leukocyte numerous components are not known well enough, and the experimentally available measurements for the constitutive parameters are limited. In addition, as the first modeling attempts show, even implementation of a drastically simplified process of cell deformation into models of rolling hits a computational barrier. Detailed, analytical tracking of changes in the cell shape in response to the fluid shear is computationally very intensive. The models accounting for cell deformation (Khismatullin & Truskey, 2004; Pawar et al., 2008) simulate rolling for short periods of cell rolling time, usually no longer than 1 s.

While selectins fail to mediate leukocyte firm adhesion (arrest) even when over 30 of their bonds exist (Chen & Springer, 1999), integrins support adhesion at a lower number of bonds (Shamri et al., 2005). Leukocyte rolling *in vivo* and *in vitro* is followed by rapid adhesion upon exposure to chemokines, which induce a rapid affinity change of leukocyte integrins such as $\alpha_L\beta_2$ (LFA-1)- or $\alpha_4\beta_1$ (VLA-4)-integrins (Dustin, 2001; Hyduk & Cybulsky, 2001). This process is just beginning to be modeled (Bhatia et al., 2003; Krasik et al., 2006, 2008). In neutrophils, a second activation pathway starts with E-selectin binding, and results in activation of the Src family kinase Fgr, spleen tyrosine kinase Syk, and partial activation of LFA-1 (Zarbock et al., 2007, 2008). Since this process appears to result in a conformation of LFA-1 integrin that supports rolling rather than firm adhesion (Salas, Shimaoka, Chen, Carman, & Springer, 2002; Salas et al., 2004), incorporating participating bonds will be important in constructing more realistic models of rolling. Since the integrin affinity is thought to be over a 1000-fold higher in the high-affinity state than in the low-affinity state, by simulating changes in integrin affinity the model leukocyte should be able to arrest from rolling.

Acknowledgments

This work was supported by the NIH grant 2R01EB002185. We thank Michael B. Lawrence, Wladek Minor, and Hai-Bin Luo for helpful discussions and William H. Guilford, Scott I. Simon, Yoshihiro Kuwano, and Prithu Sundd for unpublished data.

References

Abbal, C., Lambelet, M., Bertaggia, D., Gerbex, C., Martinez, M., Arcaro, A., et al. (2006). Lipid raft adhesion receptors and Syk regulate selectin-dependent rolling under flow conditions. *Blood, 108*, 3352–3359.

Aigner, S., Ramos, C. L., Hafezi-Moghadam, A., Lawrence, M. B., Altevogt, P., & Ley, K. (1998). CD24 mediates rolling of breast carcinoma cells on P-selectin. *FASEB Journal, 12*, 1241–1251.

Alon, R., Kassner, P. D., Carr, M. W., Finger, E. B., Hemler, M. E., & Springer, T. A. (1995). The integrin VLA-4 supports tethering and rolling in flow on VCAM-1. *Journal of Cell Biology, 128*, 1243–1253.

Alon, R., & Ley, K. (2008). Cells on the run: Shear-regulated integrin activation in leukocyte rolling and arrest on endothelial cells. *Current Opinion in Cell Biology, 20*, 225–232.

Atherton, A., & Born, G. V. R. (1972). Quantitative investigations of the adhesiveness of circulating polymorphonuclear leukocytes to blood vessels. *Journal of Physiology (London), 222*, 447–474.

Badimon, L., Turitto, V., Rosemark, J. A., Badimon, J. J., & Fuster, V. (1987). Characterization of a tubular flow chamber for studying platelet interaction with biologic and prosthetic materials: Deposition of indium 111-labeled platelets on collagen, subendothelium, and expanded polytetrafluoroethylene. *Journal of Laboratory and Clinical Medicine, 110*, 706–718.

Baker, M., & Wayland, H. (1974). On-line volume flow rate and velocity profile measurement for blood in microvessels. *Microvascular Research, 7*, 131–143.

Barsegov, V., & Thirumalai, D. (2005). Dynamics of unbinding of cell adhesion molecules: Transition from catch to slip bonds. *Proceedings of the National Academy of Sciences of the United States of America, 102*, 1835–1839.

Beer, F. P., & Johnston, E. R. (Eds.), (1984). *Models for the specific adhesion of cells to cells.* New York, NY: McGraw-Hill.

Bell, G. I. (1978). Models for the specific adhesion of cells to cells. *Science, 200*, 618–627.

Bell, G. I., Dembo, M., & Bongrand, P. (1984). Cell adhesion. Competition between nonspecific repulsion and specific bonding. *Biophysical Journal, 45*, 1051–1064.

Berlin, C., Bargatze, R. F., Campbell, J. J., von Andrian, U. H., Szabo, M. C., Hasslen, S. R., et al. (1995). α_4 integrins mediate lymphocyte attachment and rolling under physiologic flow. *Cell, 80*, 413–422.

Beste, M. T., & Hammer, D. A. (2008). Selectin catch–slip kinetics encode shear threshold adhesive behavior of rolling leukocytes. *Proceedings of the National Academy of Sciences of the United States of America, 105*, 20716–20721.

Bhatia, S. K., King, M. R., & Hammer, D. A. (2003). The state diagram for cell adhesion mediated by two receptors. *Biophysical Journal, 84*, 2671–2690.

Bird, R. B., Armstrong, R. C., & Hassager, O. (Eds.), (1987). *Dynamics of polymeric liquids, Vol. 1: Fluid mechanics*. New York, NY: Wiley.

Bongrand, P., & Bell, G. L. (1984). Cell–cell adhesion: Parameters and possible mechanisms. In A. S. Perelson, C. DeLisi & F. Wiegel (Eds.), *Cell surface dynamics: Concepts and models* (pp. 459–493). New York, NY: Marcel Dekker.

Bongrand, P., Capo, C., & Depieds, R. (1982). Physics of cell adhesion. *Progress in Surface and Membrane Science, 12*, 217–285.

Brenner, H. (1961). The slow motion of a sphere through a viscous fluid towards a plane surface. *Chemical Engineering Science, 16*, 242–251.

Bruehl, R. E., Moore, K. L., Lorant, D. E., Borregaard, N., Zimmerman, G. A., McEver, R. P., et al. (1997). Leukocyte activation induces surface redistribution of P-selectin glycoprotein ligand-1. *Journal of Leukocyte Biology, 61*, 489–499.

Bruehl, R. E., Springer, T. A., & Bainton, D. F. (1996). Quantitation of L-selectin distribution on human leukocyte microvilli by immunogold labeling and electron microscopy. *Journal of Histochemistry and Cytochemistry, 44*, 835–844.

Brunk, D. K., Goetz, D. J., & Hammer, D. A. (1996). Sialyl Lewis(x)/E-selectin-mediated rolling in a cell-free system. *Biophysical Journal, 71*, 2902–2907.

Butcher, E. C. (1991). Leukocyte–endothelial cell recognition—Three (or more) steps to specificity and diversity. *Cell, 67*, 1033–1036.

Campbell, J. J., Hedrick, J., Zlotnik, A., Siani, M. A., Thompson, D. A., & Butcher, E. C. (1998). Chemokines and the arrest of lymphocytes rolling under flow conditions. *Science, 279*, 381–384.

Caputo, K. E., & Hammer, D. A. (2005). Effect of microvillus deformability on leukocyte adhesion explored using adhesive dynamics simulations. *Biophysical Journal, 89*, 187–200.

Caputo, K. E., Lee, D., King, M. R., & Hammer, D. A. (2007). Adhesive dynamics simulations of the shear threshold effect for leukocytes. *Biophysical Journal, 92*, 787–797.

Chang, K.-C., & Hammer, D. A. (1999). The forward rate of binding of surface-tethered reactants: Effect of relative motion between two surfaces. *Biophysical Journal, 76*, 1280–1292.

Chang, K. C., Tees, D. F. J., & Hammer, D. A. (2000). The state diagram for cell adhesion under flow: Leukocyte rolling and firm adhesion. *Proceedings of the National Academy of Sciences of the United States of America, 97*, 11262–11267.

Chen, W., Evans, E. A., McEver, R. P., & Zhu, C. (2008). Monitoring receptor–ligand interactions between surfaces by thermal fluctuations. *Biophysical Journal, 94*, 694–701.

Chen, S., & Springer, T. A. (1999). An automatic braking system that stabilizes leukocyte rolling by an increase in selectin bond number with shear. *Journal of Cell Biology, 144*, 185–200.

Chesla, S. E., Selvaraj, P., & Zhu, C. (1998). Measuring two-dimensional receptor–ligand binding kinetics by micropipette. *Biophysical Journal, 75*, 1553–1572.

Chesnutt, B. C., Smith, D. F., Raffler, N. A., Smith, M. L., White, E. J., & Ley, K. (2006). Induction of LFA-1-dependent neutrophil rolling on ICAM-1 by engagement of E-selectin. *Microcirculation, 13*, 99–109.

Chien, S., Usami, S., & Skalak, R. (1984). Blood flow in small tubes. In E. M. Renkin & C. C. Michel (Eds.), *Handbook of physiology. The cardiovascular system. Microcirculation* (pp. 217–249). Bethesda, MD: American Physiological Society.

Chigaev, A., Blenc, A. M., Braaten, J. V., Kumaraswamy, N., Kepley, C. L., Andrews, R. P., et al. (2001). Real time analysis of the affinity regulation of alpha 4-integrin. The physiologically activated receptor is intermediate in affinity between resting and Mn(2+) or antibody activation. *Journal of Biological Chemistry, 276*, 48670–48678.

Chigaev, A., Waller, A., Amit, O., Halip, L., Bologa, C. G., & Sklar, L. A. (2009). Real-time analysis of conformation-sensitive antibody binding provides new insights into integrin conformational regulation. *Journal of Biological Chemistry, 284*, 14337–14346.

Chorin, A. J. (1967). A numerical method for solving incompressible viscous flow problems. *Journal of Comparative Physiology, 2*, 12–26.

Cotter, M. J., Norman, K. E., Hellewell, P. G., & Ridger, V. C. (2001). A novel method for isolation of neutrophils from murine blood using negative immunomagnetic separation. *American Journal of Pathology, 159*, 473–481.

Damiano, E. R., Duling, B. R., Ley, K., & Skalak, T. C. (1996a). Axisymmetric pressure-driven flow of rigid pellets through a cylindrical tube lined with a deformable porous wall layer. *Journal of Fluid Mechanics, 314*, 163–189.

Damiano, E. R., Westheider, J., Tözeren, A., & Ley, K. (1996b). Variation in the velocity, deformation, and adhesion energy density of leukocytes rolling within venules. *Circulation Research, 79*, 1122–1130.

Dean, W. R., & O'Neill, M. E. (1963). A slow motion of viscous liquid caused by the rotation of a solid sphere. *Mathematika, 10*, 13–24.

Dembo, M. (1994). On peeling an adherent cell from a surface. *Lectures on Mathematics in the Life Sciences, 24*, 51–77.

Dembo, M., Torney, D. C., Saxman, K., & Hammer, D. A. (1988). The reaction-limited kinetics of membrane-to-surface adhesion and detachment. *Proceedings of the Royal Society of London B: Biological Science, 234*, 55–83.

Diamond, M. S., & Springer, T. A. (1993). A subpopulation of Mac-1 (CD11b/CD18) molecules mediates neutrophil adhesion to ICAM-1 and fibrinogen. *Journal of Cell Biology, 120*, 545–556.

Durr, E., Yu, J., Krasinska, K. M., Carver, L. A., Yates, J. R., Testa, J. E., et al. (2004). Direct proteomic mapping of the lung microvascular endothelial cell surface *in vivo* and in cell culture. *Nature Biotechnology, 22*, 985–992.

Dustin, M. L. (2001). *In vitro* flow models of leukocyte adhesion. In K. Ley (Ed.), *Physiology of inflammation* (pp. 242–262). New York, NY: Oxford University Press.

Eppihimer, M. J., Russell, J., Anderson, D. C., Wolitzky, B. A., & Granger, D. N. (1997). Endothelial cell adhesion molecule expression in gene-targeted mice. *American Journal of Physiology: Heart and Circulatory Physiology, 42*, H1903–H1908.

Eppihimer, M. J., Wolitzky, B., Anderson, D. C., Labow, M. A., & Granger, D. N. (1996). Heterogeneity of expression of E- and P-selectins *in vivo*. *Circulation Research, 79*, 560–569.

Evans, E. A. (1985). Detailed mechanics of membrane–membrane adhesion and separation. II. Discrete kinetically trapped molecular cross-bridges. *Biophysical Journal, 48*, 185–192.

Evans, E., Leung, A., Heinrich, V., & Zhu, C. (2004). Mechanical switching and coupling between two dissociation pathways in a P-selectin adhesion bond. *Proceedings of the National Academy of Sciences of the United States of America, 101*, 11281–11286.

Faust, N., Varas, F., Kelly, L. M., Heck, S., & Graf, T. (2000). Insertion of enhanced green fluorescent protein into the lysozyme gene creates mice with green fluorescent granulocytes and macrophages. *Blood, 96*, 719–726.

Finger, E. B., Bruehl, R. E., Bainton, D. F., & Springer, T. A. (1996a). A differential role for cell Shape in neutrophil tethering and rolling on endothelial selectins under flow. *Journal of Immunology, 157*, 5085–5096.

Finger, E. B., Puri, K. D., Alon, R., Lawrence, M. B., von Andrian, U. H., & Springer, T. A. (1996b). Adhesion through L-selectin requires a threshold hydrodynamic shear. *Nature, 379*, 266–269.

Forsyth, K. D., & Levinsky, R. J. (1990). Preparative procedures of cooling and re-warming increase leukocyte integrin expression and function on neutrophils. *Journal of Immunological Methods, 128*, 159–163.

Giesekus, H. (1982). A simple constitutive equation for polymeric fluids based on the concept of deformation-dependent tensorial mobility. *Journal of Non-Newtonian Fluid Mechanics, 11*, 69–109.

Gilbert, N., & Bankes, S. (2002). Platforms and methods for agent-based modeling. *Proceedings of the National Academy of Sciences of the United States of America, 99*, 7197–7198.

Glasser, L., & Fiederlein, R. L. (1990). The effect of various cell separation procedures on assays of neutrophil function: A critical appraisal. *American Journal of Clinical Pathology, 93*, 662–669.

Goldman, A. J., Cox, R. G., & Brenner, H. (1967a). Slow viscous motion of a sphere parallel to a plan wall. I. Motion through a quiescent fluid. *Chemical Engineering Science, 22*, 637–651.

Goldman, A. J., Cox, R. G., & Brenner, H. (1967b). Slow viscous motion of a sphere parallel to a plan wall. II. Couette flow. *Chemical Engineering Science, 22*, 653–660.

Goldsmith, H. L., & Spain, S. (1984). Margination of leukocytes in blood flow through small tubes. *Microvascular Research, 27*, 204–222.

Green, C. E., Schaff, U. Y., Sarantos, M. R., Lum, A. F., Staunton, D. E., & Simon, S. I. (2006). Dynamic shifts in LFA-1 affinity regulate neutrophil rolling, arrest, and transmigration on inflamed endothelium. *Blood, 107*, 2101–2111.

Guyeffier, D., Li, J., Nadim, A., Scardovelli, R., & Zaleski, S. (1999). Volume-of-fluid interface tracking and smoothed surface stress methods for three-dimensional flows. *Journal of Comparative Physiology, 152*, 423–456.

Hafezi-Moghadam, A., Thomas, K. L., & Cornelssen, C. (2004). A novel mouse-driven ex vivo flow chamber for the study of leukocyte and platelet function. *American Journal of Physiology: Cell Physiology, 286*, C876–C892.

Hammer, D. A., & Apte, S. M. (1992). Simulation of cell rolling and adhesion on surfaces in shear flow: General results and analysis of selectin-mediated neutrophil adhesion. *Biophysical Journal, 63*, 35–57.

Hattori, R., Hamilton, K. K., Fugate, R. D., McEver, R. P., & Sims, P. J. (1989). Stimulated secretion of endothelial von Willebrand factor is accompanied by rapid redistribution to the cell surface of the intracellular granule membrane protein GMP-140. *Journal of Biological Chemistry, 264*, 7768–7771.

Heinrich, V., Wong, W. P., Halvorsen, K., & Evans, E. (2008). Imaging biomolecular interactions by fast three-dimensional tracking of laser-confined carrier particles. *Langmuir, 24*, 1194–1203.

Henninger, D. D., Panes, J., Eppihimer, M., Russell, J., Gerritsen, M., Anderson, D. C., et al. (1997). Cytokine-induced VCAM-1 and ICAM-1 expression in different organs of the mouse. *Journal of Immunology, 158*, 1825–1832.

House, S. D., & Lipowsky, H. H. (1988). *In vivo* determination of the force of leukocyte–endothelium adhesion in the mesenteric microvasculature of the cat. *Circulation Research, 63*, 658–668.

Hubbe, M. A. (1981). Adhesion and detachment of biological cells *in vitro*. *Progress in Surface Science, 11*, 65–138.

Hyduk, S. J., & Cybulsky, M. I. (2001). *In vitro* flow models of leukocyte adhesion. In K. Ley (Ed.), *Physiology of inflammation* (pp. 263–284). New York, NY: Oxford University Press.

Hyduk, S. J., & Cybulsky, M. I. (2009). Role of alpha4beta1 integrins in chemokine-induced monocyte arrest under conditions of shear stress. *Microcirculation, 16*, 1–14.

Hynes, R. O. (2002). Integrins: Bidirectional, allosteric signaling machines. *Cell, 110*, 673–687.

Jadhav, S., Eggleton, C. D., & Konstantopoulos, K. (2005). A 3-D computational model predicts that cell deformation affects selectin-mediated leukocyte rolling. *Biophysical Journal, 88*, 96–104.

Jeffrey, G. B. (1915). On the steady rotation of a solid of revolution in a viscous fluid. *Proceedings of the London Mathematical Society, 14*, 327–338.

Jin, Q., Verdier, C., Singh, P., Aubry, N., Chotard-Ghodsnia, R, & Duperray, A (2007). Migration and deformation of leukocytes in pressure driven flows. *Mechanics Research Communications, 34*, 411–422.

Khismatullin, D. B., & Truskey, G. A. (2004). A 3D numerical study of the effect of channel height on leukocyte deformation and adhesion in parallel-plate flow chambers. *Microvascular Research, 68*, 188–202.

Khismatullin, D. B., & Truskey, G. A. (2005). Three-dimensional numerical simulation of receptor-mediated leukocyte adhesion to surfaces: Effects of cell deformability and viscoelasticity. *Physics of Fluids, 17*, 031505.1–031505.21.

Kim, M., Carman, C. V., & Springer, T. A. (2003). Bidirectional transmembrane signaling by cytoplasmic domain separation in integrins. *Science, 301*, 1720–1725.

King, M. R., & Hammer, D. A. (2001). Multiparticle adhesive dynamics. Interactions between stably rolling cells. *Biophysical Journal, 81*, 799–813.

Krasik, E. F., Caputo, K. E., & Hammer, D. A. (2008). Adhesive dynamics simulation of neutrophil arrest with stochastic activation. *Biophysical Journal, 95*, 1716–1728.

Krasik, E. F., & Hammer, D. A. (2004). A semianalytic model of leukocyte rolling. *Biophysical Journal, 87*, 2919–2930.

Krasik, E. F., Yee, K. L., & Hammer, D. A. (2006). Adhesive dynamics simulation of neutrophil arrest with deterministic activation. *Biophysical Journal, 91*, 1145–1155.

Kucik, D. F., Dustin, M. L., Miller, J. M., & Brown, E. J. (1996). Adhesion-activating phorbol ester Increases the mobility of leukocyte integrin LFA-1 in cultured lymphocytes. *Journal of Clinical Investigation, 97*, 2139–2144.

Kuijpers, T. W., Tool, A. T. J., van der Schoot, C. E., Ginsel, L. A., Onderwater, J. J. M., Roos, D., et al. (1991). Membrane surface antigen expression on neutrophils: A reappraisal of the use of surface markers for neutrophil activation. *Blood, 78*, 1105–1111.

Kunkel, E. J., Dunne, J. L., & Ley, K. (2000). Leukocyte arrest during cytokine-dependent inflammation *in vivo*. *Journal of Immunology, 164*, 3301–3308.

Kuo, C. S., Hammer, D. A., & Lauffenburger, D. A. (1997). Simulation of detachment of specifically bound particles from surfaces by shear flow. *Biophysical Journal, 73*, 517–531.

Lawrence, M. B. (2001). *In vitro* flow models of leukocyte adhesion. In K. Ley (Ed.), *Physiology of inflammation* (pp. 204–221). New York, NY: Oxford University Press.

Lawrence, M. B., Kansas, G. S., Ghosh, S., Kunkel, E. J., & Ley, K. (1997). Threshold levels of fluid shear promote leukocyte adhesion through selectins (CD62L, P, E). *Journal of Cell Biology, 136*, 717–727.

Lawrence, M. B., McIntire, L. V., & Eskin, S. G. (1987). Effect of flow on polymorphonuclear leukocyte/endothelial cell adhesion. *Blood, 70*, 1284–1290.

Lawrence, M. B., & Springer, T. A. (1991). Leukocytes roll on a selectin at physiologic flow rates: Distinction from and prerequisite for adhesion through integrins. *Cell, 65*, 859–873.

Lei, X., Lawrence, M. B., & Dong, C. (1999). Influence of cell deformation on leukocyte rolling adhesion in shear flow. *Journal of Biomechanical Engineering, 121*, 636–643.

Ley, K., & Gaehtgens, P. (1991). Endothelial, not hemodynamic differences are responsible for preferential leukocyte rolling in venules. *Circulation Research, 69*, 1034–1041.

Ley, K., Lundgren, E., Berger, E. M., & Arfors, K.-E. (1989). Shear-dependent inhibition of granulocyte adhesion to cultured endothelium by dextran sulfate. *Blood, 73*, 1324–1330.

Ley, K., Mestas, J., Pospieszalska, M. K., Sundd, P., Groisman, A., & Zarbock, A. (2008). Intravital microscopic investigation of leukocyte interactions with the blood vessel wall. *Methods in Enzymology, 445*, 255–279.

Li, F., Erickson, H. P., James, J. A., Moore, K. L., Cummings, R. D., & McEver, R. P. (1996). Visualization of P-selectin glycoprotein ligand-1 as a highly extended molecule and mapping of protein epitopes for monoclonal antibodies. *Biological Chemistry, 271*, 6342–6348.

Lindbom, L., Xie, X., Raud, J., & Hedqvist, P. (1992). Chemoattractant-induced firm adhesion of leukocytes to vascular endothelium *in vivo* is critically dependent on initial leukocyte rolling. *Acta Physiologica Scandinavica, 146*, 415–421.

Long, D. S., Smith, M. L., Pries, A. R., Ley, K., & Damiano, E. R. (2004). Microviscometry reveals reduced blood viscosity and altered shear rate and shear stress profiles in microvessels after hemodilution. *Proceedings of the National Academy of Sciences of the United States of America, 101*, 10060–10065.

Long, M., Zhao, H., Huang, K.-S., & Zhu, C. (2001). Kinetic measurements of cell surface E-selectin/carbohydrate ligand interactions. *Annals of Biomedical Engineering, 29*, 935–946.

Luo, B. H., Carman, C. V., & Springer, T. A. (2007). Structural basis of integrin regulation and signaling. *Annual Review of Immunology, 25*, 619–647.

Mangan, P. R., O'quinn, D., Harrington, L., Bonder, C. S., Kubes, P., Kucik, D. F., et al. (2005). Both Th1 and Th2 cells require p-selectin glycoprotein ligand-1 for optimal rolling on inflamed endothelium. *American Journal of Pathology, 167*, 1661–1675.

Marshall, B. T., Long, M., Piper, J. W., Yago, T., McEver, R. P., & Zhu, C. (2003). Direct observation of catch bonds involving cell-adhesion molecules. *Nature, 423*, 190–193.

Maude, A. D. (1961). End effects in a falling-sphere viscometer. *British Journal of Applied Physics, 12*, 293–295.

Mehta, P., Cummings, R. D., & McEver, R. P. (1998). Affinity and kinetic analysis of P-selectin binding to P-selectin glycoprotein ligand-1. *Journal of Biological Chemistry, 273*, 32506–32513.

Milner, R. (Ed.), (1999). *Communicating and mobile systems: The π-calculus*. Cambridge: Cambridge University Press.

Moore, K. L., Patel, K. D., Bruehl, R. E., Fugang, L., Johnson, D. A., Lichenstein, H. S., et al. (1995). P-selectin glycoprotein ligand-1 mediates rolling of human neutrophils on P-selectin. *Journal of Cell Biology, 128*, 661–671.

Moore, K. L., Varki, A., & McEver, R. P. (1991). GMP-140 binds to a glycoprotein receptor on human neutrophils: Evidence for a lectin-like interaction. *Journal of Cell Biology, 112*, 491–499.

Nandi, A., Estess, P., & Siegelman, M. H. (2000). Hyaluronan anchoring and regulation on the surface of vascular endothelial cells is mediated through the functionally active form of CD44. *Journal of Biological Chemistry, 275*, 14939–14948.

Napper, D. H. (1977). Steric stabilization. *Journal of Colloid and Interface Science, 58*, 390–407.

Nobis, U., Pries, A. R., Cokelet, G. R., & Gaehtgens, P. (1985). Radial distribution of white cells during blood flow in small tubes. *Microvascular Research, 29*, 295–304.

North, M. J., Collier, N. T., & Vos, J. R. (2006). Experiences creating three implementations of the repast agent modeling toolkit. *ACM Transactions on Modeling and Computer Simulation, 16*, 1–25.

Oh, H., Siano, B., & Diamond, S. (2008). Neutrophil isolation protocol. *Journal of Visualized Experiments, 745*, doi:10.3791/745.

O'Neill, M. E. (1964). A slow motion of viscous liquid caused by a slowly moving solid sphere. *Mathematika, 11*, 67–74.

Park, E. Y. H., Smith, M. J., Stropp, E. S., Snapp, K. R., DiVietro, J. A., Walker, W. F., et al. (2002). Comparison of PSGL-1 microbead and neutrophil rolling: Microvillus elongation stabilizes P-selectin bond clusters. *Biophysical Journal, 82*, 1835–1847.

Patel, K. D., Nollert, M. U., & McEver, R. P. (1995). P-selectin must extend a sufficient length from the plasma membrane to mediate rolling of neutrophils. *Journal of Cell Biology, 131*, 1893–1902.

Pawar, P., Jadhav, S., Eggleton, C. D., & Konstantopoulos, K. (2008). Roles of cell and microvillus deformation and receptor–ligand binding kinetics in cell rolling. *American Journal of Physiology: Heart and Circulatory Physiology, 295*, H1439–H1450.

Pereverzev, Y. V., Prezhdo, O. V., Forero, M., Sokurenko, E. V., & Thomas, W. E. (2005). The two-pathway model for the catch–slip transition in biological adhesion. *Biophysical Journal, 89*, 1446–1454.

Petri, B., Phillipson, M., & Kubes, P. (2008). The physiology of leukocyte recruitment: An *in vivo* perspective. *Journal of Immunology, 180*, 6439–6446.

Pickard, J. E., & Ley, K. (2009). Micro-PTV measurement of the fluid shear stress acting on adherent leukocytes *in vivo*. *Biophysical Journal, 96*, 4249–4259.

Piper, J. W., Swerlick, R. A., & Zhu, C. (1998). Determining force dependence of two-dimensional receptor–ligand binding affinity by centrifugation. *Biophysical Journal, 74*, 492–513.

Pospieszalska, M. K., & Ley, K. (2009). Dynamics of microvillus extension and tether formation in rolling leukocytes. *Cellular and Molecular Bioengineering, 2*, 207–217.

Pospieszalska, M. K., Zarbock, A., Pickard, J. E., & Ley, K. (2009). Event-tracking model of adhesion identifies load-bearing bonds in rolling leukocytes. *Microcirculation, 16*, 115–130.

Rao, R. M., Haskard, D. O., & Landis, R. C. (2002). Enhanced recruitment of Th2 and CLA-negative lymphocytes by the S128R polymorphism of E-selectin. *Journal of Immunology, 169*, 5860–5865.

Ray, N., Acton, S. T., & Ley, K. (2002). Tracking leukocytes *in vivo* with shape and size constrained active contours. *IEEE Transactions on Medical Imaging, 21*, 1222–1235.

Rinko, L. J., Lawrence, M. B., & Guilford, H. W. (2004). The molecular mechanics of P- and L-selectin lectin domains binding to PSGL-1. *Biophysical Journal, 86*, 544–554.

Rosen, S. D. (2004). Ligands for L-selectin: Homing, inflammation, and beyond. *Annual Review of Immunology, 22*, 129–156.

Salas, A., Shimaoka, M., Chen, S., Carman, C. V., & Springer, T. (2002). Transition from rolling to firm adhesion is regulated by the conformation of the I domain of the integrin lymphocyte function-associated antigen-1. *Journal of Biological Chemistry, 277*, 50255–50262.

Salas, A., Shimaoka, M., Kogan, A. N., Harwood, C., von Andrian, U. H., & Springer, T. A. (2004). Rolling adhesion through an extended conformation of integrin alphaLbeta2 and relation to alpha I and beta I-like domain interaction. *Immunity, 20*, 393–406.

Schmid-Schönbein, G. W. (1990). Leukocyte biophysics: An invited review. *Cell Biophysics, 17*, 107–135.

Schmid-Schönbein, G. W., Usami, S., Skalak, R., & Chien, S. (1980). The interaction of leukocytes and erythrocytes in capillary and postcapillary vessels. *Microvascular Research, 19*, 45–70.

Shamri, R., Grabovsky, V., Gauguet, J. M., Feigelson, S., Manevich, E., Kolanus, W., et al. (2005). Lymphocyte arrest on endothelial ICAM-1 requires a subsecond induction of an extended LFA-1 conformation by an endothelial-bound chemokine signal. *Nature Immunology, 6*, 497–506.

Shao, J.-Y., Ting-Beall, H. P., & Hochmuth, R. M. (1998). Static and dynamic lengths of neutrophil microvilli. *Proceedings of the National Academy of Sciences of the United States of America, 95*, 6797–6802.

Shaw, S. K., Ma, S., Kim, M. B., Rao, R. M., Hartman, C. U., Froio, R. M., et al. (2004). Coordinated redistribution of leukocyte LFA-1 and endothelial cell ICAM-1 accompany neutrophil transmigration. *Journal of Experimental Medicine, 200*, 1571–1580.

Sircar, M., Bradfield, P. F., Aurrand-Lions, M., Fish, R. J., Alcaide, P., Yang, L., et al. (2007). Neutrophil transmigration under shear flow conditions *in vitro* is junctional adhesion molecule-C independent. *Journal of Immunology, 178*, 5879–5887.

Smith, M. J., Berg, E. L., & Lawrence, M. B. (1999). A direct comparison of selectin-mediated transient, adhesive events using high temporal resolution. *Biophysical Journal, 77*, 3371–3383.

Smith, M. L., Long, D. S., Damiano, E. R., & Ley, K. (2003). Near-wall micro-PIV reveals a hydrodynamically relevant endothelial surface layer in venules *in vivo*. *Biophysical Journal, 85*, 637–645.

Smith, M. L., Sperandio, M., Galkina, E. V., & Ley, K. (2004). Autoperfused mouse flow chamber reveals synergistic neutrophil accumulation through P-selectin and E-selectin. *Journal of Leukocyte Biology, 76*, 985–993.

Sperandio, M., Gleissner, C. A., & Ley, K. (2009). Glycosylation in immune cell trafficking. *Immunological Reviews, 230*, 97–113.

Sperandio, M., Pickard, J., Unnikrishnan, S., Acton, S. T., & Ley, K. (2007). Analysis of leukocyte rolling *in vivo* and *in vitro*. *Methods in Enzymology, 416*, 346–371.

Springer, T. A. (1990). Adhesion receptors of the immune system. *Nature, 346*, 425–434.

Springer, T. A. (1995). Traffic signals on endothelium for lymphocyte recirculation and leukocyte emigration. *Annual Review of Physiology, 57*, 827–872.

Tang, J., Ley, K. F., & Hunt, A. (2007). Dynamics of in silico leukocyte rolling, activation, and adhesion. *BMC Systematic Biology, 1*, 14.

Taylor, G. I. (1934). The formation of emulsions in definable fields of flow. *Proceedings of the Royal Society of London A, 146*, 501–523.

Thomas, W., Forero, M., Yakovenko, O., Nilsson, L., Vicini, P., Sokurenko, E., et al. (2006). Catch-bond model derived from allostery explains force-activated bacterial adhesion. *Biophysical Journal, 90*, 753–764.

Toy, B. L., & Bardawil, W. A. (1958). A simple plastic perfusion chamber for continuous maintenance and cinematography of tissue cultures. *Experimental Cell Research, 14*, 97–103.

Tözeren, A., Kleinman, H. K., Grant, D. S., Morales, D., Mercurio, A. M., & Byers, S. W. (1995). E-selectin-mediated dynamic interactions of breast- and colon-cancer cells with endothelial-cell monolayers. *International Journal of Cancer, 60*, 426–431.

Tözeren, A., & Ley, K. (1992). How do selectins mediate leukocyte rolling in venules? *Biophysical Journal, 63*, 700–709.

Vink, H., & Duling, B. R. (1996). Identification of distinct luminal domains for macromolecules, erythrocytes, and leukocytes within mammalian capillaries. *Circulation Research, 79*, 581–589.

Yago, T., Leppänen, A., Qiu, H., Marcus, W. D., Nollert, M. U., Zhu, C., et al. (2002). Distinct molecular and cellular contributions to stabilizing selectin-mediated rolling under flow. *Journal of Cell Biology, 158*, 787–799.

Zarbock, A., Abram, C. L., Hundt, M., Altman, A., Lowell, C. A., & Ley, K. (2008). PSGL-1 engagement by E-selectin signals through Src kinase Fgr and ITAM adapters DAP12 and FcR gamma to induce slow leukocyte rolling. *Journal of Experimental Medicine, 205*, 2339–2347.

Zarbock, A., & Ley, K. (2009). Neutrophil adhesion and activation under flow. *Microcirculation, 16*, 31–42.

Zarbock, A., Lowell, C. A., & Ley, K. (2007). Spleen tyrosine kinase Syk is necessary for E-selectin-induced $\alpha_L\beta_2$ integrin mediated rolling on intercellular adhesion molecule-1. *Immunity, 26*, 773–783.

Zhang, X., Craig, S. E., Kirby, H., Humphries, M. J., & Moy, V. T. (2004). Molecular basis for the dynamic strength of the integrin $\alpha_4\beta_1$/VCAM-1 interaction. *Biophysical Journal, 87*, 3470–3478.

Zhang, H., Schaff, U. Y., Green, C. E., Chen, H., Sarantos, M. R., Hu, Y., et al. (2006). Impaired integrin-dependent function in Wiskott–Aldrich syndrome protein-deficient murine and human neutrophils. *Immunity, 25*, 285–295.

Zhang, X. H., Wojcikiewicz, E., & Moy, V. T. (2002). Force spectroscopy of the leukocyte function-associated antigen-1/intercellular adhesion molecule-1 interaction. *Biophysical Journal, 83*, 2270–2279.

Zhao, Y., Chien, S., & Skalak, R. (1995). A stochastic model of leukocyte rolling. *Biophysical Journal, 69*, 1309–1320.

Zhu, J., Luo, B. H., Xiao, T., Zhang, C., Nishida, N., & Springer, T. A. (2008). Structure of a complete integrin ectodomain in a physiologic resting state and activation and deactivation by applied forces. *Molecular Cell, 32*, 849–861.

Zhu, C., & McEver, R. P. (2005). Catch bonds: Physical models and biological functions. *Molecular & Cellular Biomechanics, 2*, 91–104.

Zipursky, A., Bow, E., Seshadri, R. S., & Brown, E. J. (1976). Leukocyte density and volume in normal subjects and in patients with acute lymphoblastic leukemia. *Blood, 48*, 361–371.

SECTION 3

ACTIVE ROLE OF ENDOTHELIAL CELLS

CHAPTER 9

Endothelial Adhesive Platforms Organize Receptors to Promote Leukocyte Extravasation

Olga Barreiro*,†

*Departamento de Biología Vascular e Inflamación, Centro Nacional de Investigaciones Cardiovasculares, 28029 Madrid, Spain

†Servicio de Inmunología, Hospital Universitario de la Princesa, Universidad Autónoma de Madrid, 28006 Madrid, Spain

I. OVERVIEW

Endothelium does not constitute merely a physical barrier but plays an active role during leukocyte transendothelial migration. Its contribution is based on the combined action of different adhesion receptors involved in subsequent steps of the process which are assisted by membrane organizers and cytoskeletal-associated components.

Current Topics in Membranes, Volume 64 1063-5823/09 $35.00

DOI: 10.1016/S1063-5823(09)64009-6

In this regard, endothelial adhesion receptors congregate in constitutive submicrometer-sized clusters termed endothelial adhesive platforms (EAPs), instead of being randomly distributed throughout the endothelial apical plasma membrane. The organizing units of such platforms are tetraspanin proteins, which interact among themselves and simultaneously with a broad variety of adhesion molecules, providing the adequate receptor avidity to overcome the threshold of adhesion required for leukocyte–endothelium interactions occurring under shear flow. Therefore, these adhesive platforms act as a supramolecular mechanism to spatiotemporally organize receptors with similar characteristics and functions at the plasma membrane to facilitate their efficient coordinated action during the process of extravasation, which requires rapid kinetics. Attempts to block tetraspanin function have unveiled their crucial role during extravasation, thereby emerging as potential therapeutic targets in a more general manner than existing therapies focused on the inhibition of a particular adhesion pathway mediated by a single pair of receptor/counter-receptor.

Upon leukocyte binding to inflamed endothelium, the signals emanating from the endothelial receptors bound to their leukocyte ligands trigger the coalescence of EAPs and the reorganization of the actin cytoskeleton into a three-dimensional microvilli-based docking structure to firmly attach the adherent leukocytes, prevented their detachment and promoting subsequent transmigration.

In summary, I present in this chapter the specific characteristics of tetraspanin-enriched microdomains (TEMs) in comparison with other types of membrane domains which make them well suited to organize endothelial receptors during extravasation, orchestrating the adhesion cascade on the endothelial side. In addition, we also discuss the technical details of the approaches employed to tackle the mechanisms by which adhesive platforms assemble and function.

II. INTRODUCTION

A. General Aspects of Plasma Membrane Organization

Lateral segregation of constituents of plasma membrane is required for the function of biological membranes (Jacobson, Mouritsen, & Anderson, 2007). Both lipids and proteins cluster into diverse specialized membrane domains which also contain signaling proteins associated in their inner leaflet. Moreover, the cortical cytoskeleton plays a role in the confinement of proteins, stabilizing these membrane microdomains (Goswami et al., 2008; Kusumi & Suzuki, 2005). In this regard, cholesterol- and sphingolipid-enriched rafts have been proposed as platforms for the sorting of glycosylphosphatidylinositol-anchored proteins and sites for the assembly of cytoplasmic signaling complexes (Lasserre et al., 2008; Simons & Toomre, 2000). Membrane microdomains primarily based on protein–protein interactions instead of lipid–protein interactions are also important for the organization of supramolecular structures at the plasma membrane (Douglass & Vale, 2005; Sieber et al., 2007).

Most knowledge on membrane microdomains was initially founded on biochemical analyses and model membranes, which are unable to reproduce the native conditions of biological plasma membranes. However, the development of analytical microscopy and spectroscopy techniques that "break" the diffraction limit and allow single-molecule analysis with high spatial and temporal resolution has shed light on the biophysical properties of distinct membrane domains (de Bakker et al., 2008; Eggeling et al., 2009; Greenfield et al., 2009; Larson, Gosse, Holowka, Baird, & Webb, 2005; Sharma et al., 2004; Suzuki, Fujiwara, Edidin, & Kusumi, 2007; Suzuki, Fujiwara, Sanematsu, et al., 2007). In this regard, the spatial distribution and steady-state dynamics of lipid rafts containing GPI-anchored proteins have been studied using fluorescence recovery after photobleaching (Kenworthy et al., 2004), fluorescence correlation spectroscopy (Lenne et al., 2006), single-particle tracking (Umemura et al., 2008), hetero- and homo-FRET (Förster resonance energy transfer) (Goswami et al., 2008; Sharma et al.), and etcetera. These studies revealed that GPI-anchored protein nanodomains are small (10–200 nm) and highly motile (Pike, 2006), characteristics that greatly increase the probability of protein encounters (Nicolau, Burrage, Parton, & Hancock, 2006). The biophysical characteristics of TEMs have been also analyzed recently (Barreiro et al., 2008; Espenel et al., 2008) and differences in diffusion properties and spatial organization have been found comparing TEMs with GPI-anchored protein-based lipid rafts. The nature of TEMs is extensively discussed in the following section and a brief definition of techniques is included in the technical appendix.

B. Characteristics of Tetraspanins and Tetraspanin-Enriched Microdomains

Tetraspanins are a family of small proteins (20–30 kDa protein core) abundantly expressed in cells. They share structural features since all have a small and a large extracellular loop, short N- and C-terminal tails and four transmembrane domains. The large extracellular loop (LEL) contains a variable region in its C-terminal domain with sites for specific protein–protein interactions. In addition, tetraspanins have membrane-proximal intracellular palmitoylation sites that are also critical for tetraspanin interactions. Due to all these characteristics, tetraspanins exhibit the ability to laterally associate among themselves and simultaneously with other transmembrane proteins (reviewed in Charrin et al., 2009; Yanez-Mo, Barreiro, Gordon-Alonso, Sala-Valdes, & Sanchez-Madrid, 2009). Apart from *cis*-interactions, tetraspanins may also bind ligands *in trans* (Pileri et al., 1998; Qi et al., 2006; Waterhouse, Ha, & Dveksler, 2002). However, as tetraspanins protrude around 5 nm from the plasma membrane, they do not typically exert the role of cell-surface receptors. There are 33 tetraspanins in mammals and at least a few different tetraspanins are expressed on nearly every cell, although the tetraspanin repertoire varies among different cell and tissue types. Based on their characteristics, tetraspanins have been proposed to form microdomains in the plasma membrane containing multiple transmembrane receptors, including integrins and other adhesion molecules (Barreiro et al., 2005; Berditchevski, 2001; Yanez-Mo et al., 1998), CD19–CD21 complex in B cells (Cherukuri et al., 2004; Levy & Shoham, 2005), peptide-major histocompatibility complex (Kropshofer et al., 2002), Fc receptors (Moseley, 2005), G protein-coupled receptors (Little, Hemler, & Stipp, 2004), and metalloproteinases (Lafleur, Xu, & Hemler, 2009; Xu, Sharma, & Hemler, 2009; Yanez-Mo et al., 2008). In addition, tetraspanins are not devoid of lipid interactions in that tetraspanins are highly palmitoylated proteins which are able to bind cholesterol and several gangliosides (Odintsova et al., 2006; Silvie et al., 2006). In fact, palmitoylation of tetraspanins and partner proteins has a crucial role during the assembly and maintenance of TEMs. Tetraspanins do not only associate with partners laterally at the plasma membrane, but also intracellularly with distinct cytoplasmic signaling proteins and cytoskeletal adaptors (reviewed in Charrin et al.). Furthermore, some members of the tetraspanin family are involved in trafficking and biosynthetic processing of associated receptors (Berditchevski & Odintsova, 2007).

The organization of these discrete and dynamic tetraspanin-enriched compartments accounts for the modulation of receptor function. Thus, tetraspanins exert a functional role in many fundamental physiological and pathological processes, such as egg–sperm fusion (Le Naour, Rubinstein, Jasmin, Prenant, & Boucheix, 2000; Miyado et al., 2000), antigen presentation (Unternaehrer, Chow, Pypaert, Inaba, & Mellman, 2007), pathogen infection (Gordon-Alonso et al., 2006; Jolly & Sattentau, 2007; Pileri et al., 1998; Silvie et al., 2003), angiogenesis (Takeda et al., 2007), renal function (Sachs et al., 2006), cell–cell adhesion (Barreiro et al., 2005; Chattopadhyay, Wang, Ashman, Brady-Kalnay, & Kreidberg, 2003), cell–matrix adhesion (Berditchevski, 2001; Sterk et al., 2000), cell migration and invasion (Hemler, 2003), and cancer (Zoller, 2009). The concept of TEMs emerged based on biochemical studies (Hemler), proteomics (Le Naour, Andre, Boucheix, & Rubinstein, 2006), structural biology (Min, Wang, Sun, & Kong, 2006), and analysis of fixed samples using conventional microscopy

(Nydegger, Khurana, Krementsov, Foti, & Thali, 2006). However, recent studies have bolstered their existence in living cells using fluorescence correlation spectroscopy (FCS), Förster resonance energy transfer-fluorescence lifetime imaging (FRET-FLIM) fluorescence lifetime imaging–Förster resonance energy transfer, fluorescence recovery after photobleaching (FRAP), total internal reflection fluorescence microscopy (TIRFM) for single-particle tracking and scanning electron microscopy combined with immunogold labeling (Barreiro et al., 2008; Espenel et al., 2008). These studies show that proteins embedded in tetraspanin microdomains exhibit different biophysical properties compared with typical markers of lipid rafts. Furthermore, tetraspanin mobility and partitioning seem to be dependent on palmitoylation and plasma membrane cholesterol (Espenel et al.). The analyses of individual molecules using either fluorescence correlation spectroscopy (Barreiro et al.) or total internal reflection fluorescence microscopy (Espenel et al.) have revealed the existence of two molecular subsets for each TEM component: one subset with faster and Brownian-like diffusion and another with slower and confined diffusion which takes part in platforms in permanent dynamic exchange with the rest of the membrane (Fig. 1 and Supplemental Video 1).

C. Tetraspanins Involved in Vascular Functions

Since many tetraspanin proteins are not sufficiently immunogenic to yield reliable antibodies, it has been difficult to accomplish a complete characterization of TEM components. Thus, studies are currently restricted to a few tetraspanin members. In the endothelial context, the most studied tetraspanins are CD9, CD151, CD81, and CD63.

CD9 is widely expressed in macrovasculature and microvasculature (Gutierrez-Lopez et al., 2003) as well as in lymphatics (Erovic, Neuchrist, Kandutsch, Woegerbauer, & Pammer, 2003). It is also expressed in a number of blood cell subsets (including platelets, neutrophils, monocytes, macrophages, eosinophils, basophils, and lymphoblasts) (Barreiro et al., 2005; Tohami, Drucker, Radnay, Shapira, & Lishner, 2004). CD9 has been implicated in the regulation of endothelial cell migration (Yanez-Mo et al., 1998), while its role in endothelium proliferation is controversial (Klein-Soyer, Azorsa, Cazenave, & Lanza, 2000; Ko et al., 2006). Participation of endothelial CD9 in leukocyte transendothelial migration has also been reported (Barreiro et al.). Monoclonal antibodies anti-CD9 have been reported to activate platelet integrin $\alpha_{IIb}\beta_3$ (Wu et al., 2001) and results from CD9 knockout mice displaying an increased $\alpha_{IIb}\beta_3$-activation indicate a role of the molecule as repressor of integrin $\alpha_{IIb}\beta_3$-activation (reviewed in Zhang, Kotha, Jennings, & Zhang, 2009). Moreover, this tetraspanin is also

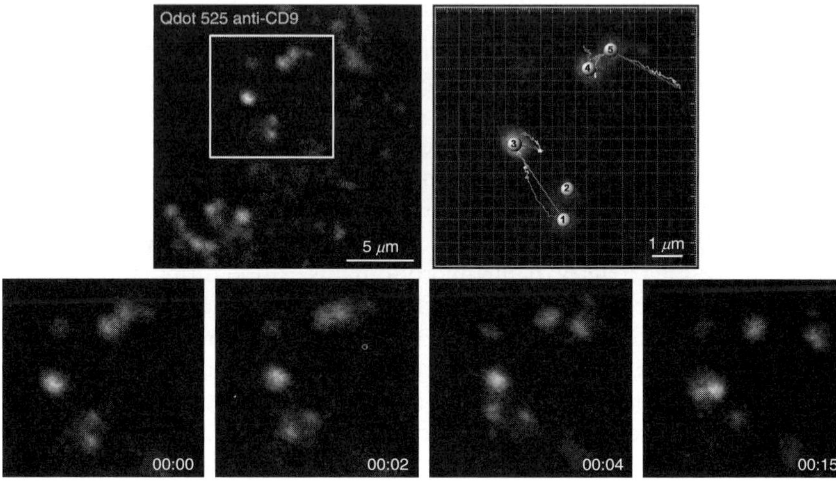

FIGURE 1 Study of tetraspanin CD9 dynamics in endothelial cells using single-particle tracking. Human primary endothelial cells from umbilical cord veins were stained with anti-CD9 antibodies coupled to Qdot 525 nm in a suboptimal amount in that only a small percentage of molecules were stained; thus, their mobility could be traced. From the full-field image, a region containing five particles has been magnified in the left panel. Particle 1 moved fast toward a bigger platform diffusing in a confined manner (number 3). Particle 2 and platform 3 remained confined during the whole acquisition time. Particles 4 and 5 codiffuse for a while and then, particle 5 moved further away while particle 4 continued with a confined movement that renders a productive displacement nearly null. Tracks are depicted in a pseudocolor scale, ranging from blue at the beginning to white at the end. Blue arrows indicate the productive displacement of each particle. The four frames below illustrate the molecular diffusion dynamics over time (s). The whole sequence is enclosed as Supplemental Video 1. (See Color Insert.)

important for the regulation of proliferation and migration of vascular smooth muscle cells (Kotha et al., 2009), and the overexpression of CD9 has been shown to ameliorate ventricular hypertrophy and myocardial infarction (reviewed in Zhang et al.).

CD151 is expressed in endothelial cells, smooth muscle cells, and hematopoietic cells such as platelets, megakaryocytes, erythrocytes, and activated T-lymphocytes (Geary, Cambareri, Sincock, Fitter, & Ashman, 2001). CD151 is an important regulator of vasculogenesis and angiogenesis, and also participates in maintaining cell–cell junctions (Takeda et al., 2007; Zheng & Liu, 2006). Moreover, endothelial CD151 promotes the collagenolytic activity and the association with TEMs of MT1-MMP (Yanez-Mo et al., 2008) and participates in leukocyte–endothelial interactions (Barreiro et al., 2005). Partial deletion of CD151 in patients produced severe defects in erythropoiesis (Karamatic Crew et al., 2004). CD151-null mice have impaired $\alpha_{IIb}\beta_3$-activation and delayed clot retraction (Lau et al., 2004).

Although CD81 is mostly involved in immune system functions (Levy & Shoham, 2005), it is also expressed in endothelium and participates in endothelial migration (Yanez-Mo et al., 1998) and leukocyte–endothelial interactions (Barreiro et al., 2005; Dijkstra et al., 2008; Rohlena et al., 2009).

CD63 is mostly an intracellular tetraspanin and is mainly localized to late endosomal and lysosomal compartments as well as in exosomes, actively participating in intracellular trafficking and exocytosis (Pols & Klumperman, 2009). It is expressed in endothelial cells, platelets, neutrophils, and basophils. In endothelial cells, CD63 is localized in Weibel–Palade bodies together with P-selectin and their expression at the plasma membrane is a hallmark of activation (Vischer & Wagner, 1993). CD63 also seems to be involved in monocyte and neutrophil interactions with endothelium (Toothill, Van Mourik, Niewenhuis, Metzelaar, & Pearson, 1990) and in the release of histamine in basophils (Kitani, Berenstein, Mergenhagen, Tempst, & Siraganian, 1991).

III. THE EMERGING CONCEPT OF ENDOTHELIAL ADHESIVE PLATFORMS

A. Endothelial Receptors Involved in Leukocyte Transendothelial Migration

To initiate the inflammatory response, circulating leukocytes in the bloodstream have to establish contact with the vascular wall and adhere to it, while withstanding shear forces (Alon & Ley, 2008). These initial contacts or tethering slows the leukocyte velocity and allows them to roll over the endothelial surface, favoring subsequent interactions. Tethering is largely mediated by selectins (P, E, and L) and their ligands. Selectins are type 1 transmembrane glycoproteins that bind to fucosylated and sialylated hydrocarbons present in their ligands in a Ca^{2+}-dependent manner. L-selectin is expressed by most leukocytes, whereas the E and P forms are expressed on endothelial cells activated by proinflammatory stimuli. In addition, to the interaction of leukocyte selectin (L-selectin) with endothelial selectins (P- and E-selectin), the P-selectin glycoprotein ligand-1 (PSGL-1) protein is a major ligand of the three selectins (reviewed in Ley & Kansas, 2004).

The following steps in the adhesion cascade are governed on the endothelial side by members of the immunoglobulin superfamily, most of them ligands of leukocyte integrins. First, the transition of rolling to leukocyte firm adhesion depends mostly on vascular cell adhesion molecule (VCAM)-1 and intercellular adhesion molecule (ICAM)-1, ligands for $\alpha_4\beta_1$ [very late antigen (VLA)-4] and $\alpha_L\beta_2$ [lymphocyte function-associated antigen (LFA)-1] leukocyte integrins, respectively (Elices et al., 1990; Marlin &

Springer, 1987). The binding of VCAM-1 and ICAM-1 with their ligands triggers the reorganization of the endothelial cortical actin cytoskeleton and generates a three-dimensional docking structure that surrounds the leukocyte and prevents the detachment of bound leukocytes under physiological flow conditions (Barreiro et al., 2002; Carman, Jun, Salas, & Springer, 2003). Once leukocytes arrest on the inflamed endothelium, they crawl searching for a suitable place to extravasate. ICAM-1 seems to play the central role during this process interacting with leukocyte $\alpha_M\beta_2$ (Mac-1) (Phillipson et al., 2006).

During leukocyte transendothelial migration, the endothelial junctions are partially disorganized to allow the passage of leukocytes but to avoid damage to the monolayer or substantial changes in permeability. Thus, the leukocyte membranes and the endothelium remain in close contact during diapedesis and, afterwards, the endothelial membranes reseal their links (Ley, Laudanna, Cybulsky, & Nourshargh, 2007). Once the leukocytes have reached an appropriate site for transmigration (preferably the intercellular junctions), they deploy exploratory pseudopods between adjacent endothelial cells. The pseudopods then transform into a lamella that moves across the open space on the monolayer. During this process, LFA-1 is the predominant integrin. This molecule is quickly relocalized to form a ring-shaped cluster at the contact interface between the leukocyte and endothelium, where it interacts with ICAM-1 (Shaw et al., 2004) and, in some other cell models, with junctional adhesion molecule (JAM)-A (Woodfin et al., 2007). Other proteins implicated in the transmigration process are ICAM-2, JAM-B, JAM-C, platelet endothelial cell adhesion molecule (PECAM)-1, endothelial cell-selective adhesion molecule (ESAM), or CD99 (Bixel et al., 2004; Bradfield et al., 2007; Lamagna et al., 2005; Wegmann et al., 2006; Woodfin et al., 2009). Many of these are able to interact both homophilically and heterophilically maintaining the interendothelial junctions or the leukocyte–endothelial interactions (Vestweber, 2007).

Interestingly, most of the endothelial molecules mentioned in this section (ICAM-1, ICAM-2, VCAM-1, CD44, PECAM-1, JAM-A, selectins) were found to interact to some extent with different tetraspanins (CD9, CD151, and CD81) (Barreiro et al., 2005, 2008; O. Barreiro et al., unpublished results), pointing to a joint organization of endothelial adhesion receptors in TEMs to ensure their functional interconnection.

B. Endothelial Adhesive Platforms as Adhesion Organizing Units

The first observation that entailed the study of specialized tetraspanin microdomains containing endothelial adhesion receptors refers to the ability of ICAM-1 and VCAM-1 to cocluster at endothelial docking structures around leukocytes that do not express the ligand for one of these endothelial receptors. Their coclustering occurs independently of actin cytoskeleton anchorage and it cannot be explained by the formation of ICAM-1/VCAM-1 heterodimers at the endothelial plasma membrane (Barreiro et al., 2008). The mechanistic insights of VCAM-1/ICAM-1 coclustering were unveiled using FLIM–FRET analysis. Direct interactions between ICAM-1 and the tetraspanin CD9, as well as between VCAM-1 and the tetraspanin CD151 were detected in primary human living endothelial cells with this analytical microscopy technique. In addition, both tetraspanins were found in direct association (Barreiro et al.). A complementary approach namely fluorescence cross-correlation spectroscopy demonstrated that these molecular species are able to codiffuse at the plasma membrane (O. Barreiro et al., unpublished results). Therefore, the specific interactions of endothelial adhesion receptors with different tetraspanins that simultaneously interact among themselves promote the formation of TEMs specially well suited for adhesion termed endothelial adhesive platforms (Barreiro et al.). A simplified view of EAPs is shown in Fig. 2. Once a leukocyte establishes contact with the endothelial monolayer, these preformed EAPs coalesce in bigger clusters found at the microvilli of the docking structures.

To elucidate the biophysical nature of EAPs, the diffusion of their components was determined using two different techniques; one that analyzes average apparent diffusion at the microscale (FRAP) and another that quantifies single-molecule diffusion at the nanoscale (FCS). Both techniques render similar results in that the diffusion coefficient of EAP components was below those described for typical raft markers (Barreiro et al., 2008; Lenne et al., 2006). In fact, the precise EAP diffusion coefficients obtained by FCS ranged from 0.05 to 0.3 $\mu m^2/s$ and might be divided into two different molecular subsets; one subset with faster and Brownian-like diffusion and another with slower and confined diffusion which is assigned to membrane platforms. Both molecular subsets seem to be in continuous dynamic exchange. Endothelial tetraspanins CD9 and CD151 showed a higher relative frequency of faster diffusion than adhesion receptors VCAM-1 and ICAM-1, result that concurs with the idea of tetraspanins being the active organizers of EAPs, while receptors could remain mainly embedded within the adhesive platforms (Barreiro et al.).

The spatial organization of EAPs was assessed by scanning electron microscopy combined with immunogold labeling of adhesion receptors. Gold-particle distribution was subjected to computational analysis which showed that EAPs are finite submicrometer-sized platforms distributed throughout the plasma membrane of inflamed endothelium (Barreiro et al., 2008). Thus, coalescence of EAPs promote the clustering of a wide range of endothelial receptors at the contact area with adherent leukocytes (probably applicable to all endothelial receptors found to interact with tetraspanins) which could be envisaged as a

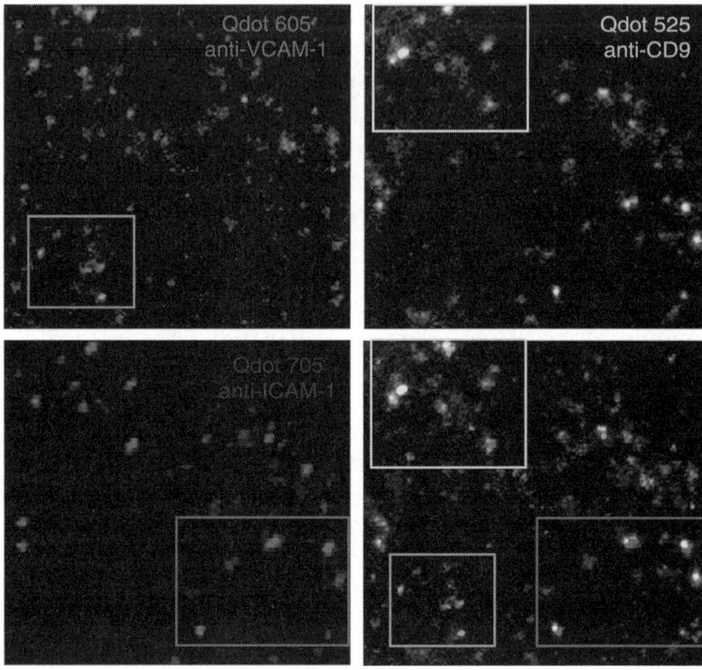

FIGURE 2 Visualizing endothelial adhesive platforms. Triple staining of CD9, VCAM-1, and ICAM-1 using antibodies coupled to Qdot 525, 605, and 705 nm, respectively, revealed the existence of small domains containing different mixtures of the three proteins. The red outline marks a region enriched in VCAM-1 platforms, the blue one depicts a region with several ICAM-1/CD9 platforms, and the green outline shows a region containing mixed microdomains with the three molecular species. The stoichiometry and size of the platforms could not be deducted from these images were only a very small amount of molecules are stained and there is not enough spatial resolution. (See Color Insert.)

mechanism to increase the availability of these receptors in the leukocyte surroundings, enhancing their adhesive properties and facilitating the transition to subsequent events during leukocyte extravasation (Fig. 3).

C. Functional Role of Endothelial Adhesive Platforms

In the bloodstream, leukocyte–endothelial cell interactions must be rapid and cooperative, to ensure leukocyte arrest under high shear stress conditions (Barreiro, de la Fuente, Mittelbrunn, & Sanchez-Madrid, 2007). The body of evidence to support the notion of EAPs as facilitators of leukocyte–endothelial cell interactions comes from functional analyses decreasing the

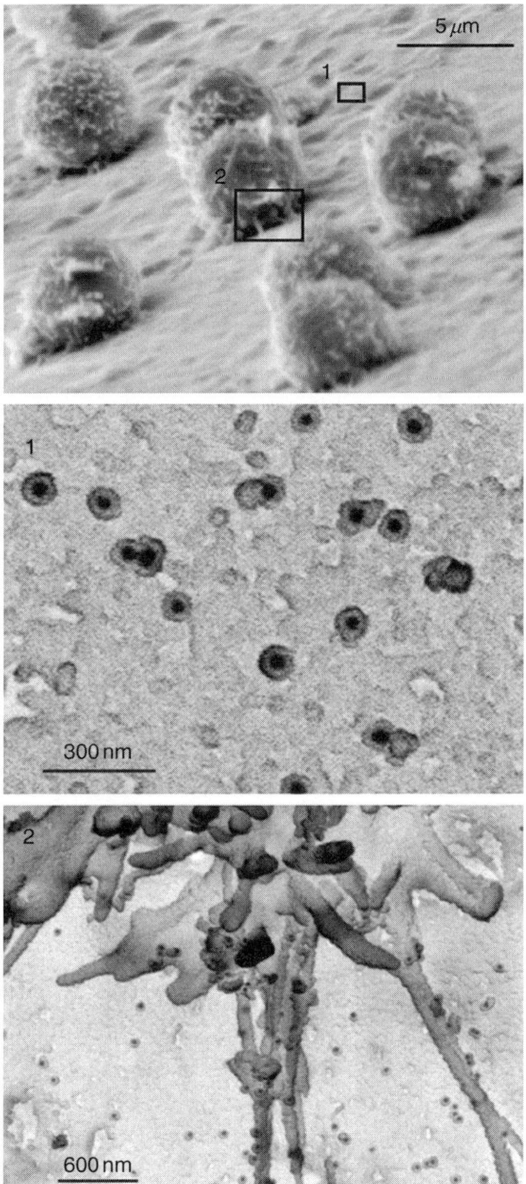

FIGURE 3 Endothelial adhesive platforms coalesce into docking structures upon leukocyte binding to endothelium. The scanning electron micrograph on the upper panel depicts the interaction of human peripheral blood lymphocytes with TNF-α-activated endothelium under physiological flow conditions ($\sim$2 dyn/cm^2). The middle panel numbered as 1 shows the

amount of tetraspanins in endothelial cells with small interference RNA (siRNA) or using tetraspanin large extracellular loops coupled to GST to compete with endogenous tetraspanins for binding with other tetraspanins and endothelial partners at the plasma membrane. The alteration of the appropriate amounts of endothelial tetraspanins or changes in the cell tetraspanin repertoire as well as the perturbation of EAP dynamics and receptor clustering by CD9–LEL–GST produced a clear impairment of VCAM-1 and ICAM-1 adhesiveness (Barreiro et al., 2005, 2008). In this sense, the insertion of endothelial receptors in EAPs seems to provide them with the required spatial distribution (avidity) to support leukocyte–endothelial interactions. This constitutes a new regulatory mechanism for endothelial ligands of integrins, which were believed to be regulated only at the transcriptional level (Collins et al., 1995). Other tetraspanins could be involved in the regulation of the molecular dynamics and organization of EAPs, as CD81 which also localizes at endothelial docking structures (Barreiro et al., 2005). In fact, CD81 has been described to regulate subsecond avidity of leukocyte integrin VLA-4 (Feigelson, Grabovsky, Shamri, Levy, & Alon, 2003).

Supramolecular organization in the nanometer range has been reported for other receptors, including LFA-1 on leukocytes (Cairo, Mirchev, & Golan, 2006; Cambi et al., 2006), and the C-type lectin dendritic cell-specific ICAM-3-grabbing nonintegrin (DC-SIGN) on dendritic cells (de Bakker et al., 2007; Neumann, Thompson, & Jacobson, 2008). The possible inclusion of these molecules in TEMs deserves further investigation.

Finally, molecular compartmentalization in TEMs is not only critical for adhesion but has been also demonstrated to be important for regulating metalloproteinase activity in endothelial cells, by either facilitating or impeding their accessibility to substrates (Yanez-Mo et al., 2008).

IV. CONCLUDING REMARKS AND THERAPEUTIC PERSPECTIVES

The biophysical characterization of EAPs is still incipient and, therefore, there are multiple remaining questions that deserve further investigation. Among them, it could be cited the study of the stoichiometry of the platforms and the discernment of the existence of different types of EAP based on

organization of endothelial adhesion receptors preassembled in nanoclusters on the apical membrane of the endothelium (endothelial adhesion platforms). When a leukocyte establishes contact with the endothelium, the endothelial adhesion receptors coalesce in the microvilli-based structure termed endothelial docking structure (2, lower panel), which keeps the leukocyte firmly adhered and prevents its detachment due to the shear forces that it has to withstand. The colloidal gold staining corresponds to ICAM-1.

molecular composition criteria versus the concept of common platforms for all kind of adhesion receptors interacting with tetraspanins. EAP composition may vary depending on vascular beds and physiological or pathological environmental conditions. New EAP components and novel specific interactions within these microdomains may be discovered. The dwell time of receptors and tetraspanins within EAPs, the half-life of these platforms, as well as the molecular mechanisms underlying EAP coalescence into larger structures remain to be analyzed. In addition, it could be possible that receptors embedded on EAPs also undergo conformational changes which could have an additive effect to the tetraspanin-based avidity regulation already demonstrated as responsible for the enhancement of receptor adhesive properties.

The ability of tetraspanins to regulate the binding properties of multiple endothelial adhesion receptors makes them potentially interesting anti-inflammatory targets (Dijkstra et al., 2008; Rohlena et al., 2009). The use of therapeutic blocking agents directed at reducing the adhesiveness of endothelial cells toward circulating leukocytes appears as a novel therapeutic strategy for the treatment of chronic inflammatory and autoimmune disorders in a more general manner than existing therapies focused on the inhibition of a single adhesion molecule (Hemler, 2008).

However, the administration of such reagents should be taken with caution and exhaustively tested in animal models and clinical trials, as tetraspanins, and particularly CD9, are very abundant not only in vasculature and lymphatics, but also it is very abundant in platelets and other hematopoietic subsets. As CD9 is involved in cell adhesion, proliferation and migration, platelet aggregation, and tumor metastasis, the administration of CD9 blockers would require a localized and controlled release in inflammatory foci. In this regard, an attractive alternative could be the use of ultrasound contrast agents (microbubbles) specifically targeted to inflammation by their coupling with antibodies against an inflammatory marker and also complexed with CD9-blocking agents which can be delivered at the inflammatory focus by destruction of microbubbles with high mechanical index insonation (Barreiro et al., 2009).

In conclusion, translational research targeting tetraspanins in vascular pathologies could provide useful preventive measures, diagnostic markers, and therapeutic agents for vascular diseases linked to inflammation such as atherosclerosis.

V. TECHNICAL APPENDIX

The concept of tetraspanin microdomains emerged based on techniques that can only provide a snapshot of the cells without any information on dynamics of tetraspanin complexes. The advent of new analytical microscopy and spectroscopy methods which render high spatial and temporal resolution has allowed us to tackle the study of membrane domains with sizes below the optical resolution limit imposed to conventional microscopy. Several of these techniques have been applied successfully to the study of TEMs and a brief description of them is enclosed in this section.

FRET. Förster resonance energy transfer microscopy is a powerful approach to visualize direct interactions in fixed and living cells. A target protein pair is tagged or labeled with spectrally overlapping donor–acceptor fluorescent proteins or fluorochromes. The energy transfer between fluorescent proteins and fluorochromes is directed from the higher-energy donor to the lower-energy acceptor when the protein pair is within about 10 nm, distance compatible with intramolecular and intermolecular direct interactions.

FLIM. Fluorescence lifetime imaging microscopy denotes a method to detect FRET by measuring the decrease of donor fluorescence lifetime (time mode) or the phase shift and demodulation of the fluorescence signal (frequency mode). This method is more reliable and accurate than other methods based on fluorescence intensity measurements.

FRAP. Fluorescence recovery after photobleaching is a technique to measure the average diffusion coefficient of a molecular population within a microscopic area on a living cell. This technique also yields information on the immobile fraction within the analyzed molecular population. Briefly, an area of a living cell labeled with a fluorochrome or fluorescent protein is bleached with a laser beam, and the diffusion of tagged molecules from the adjacent nonbleached area is monitored as fluorescence recovery into the bleached region.

FCS. Fluorescence correlation spectroscopy is an analytical technique based on the correlation of fluorescence fluctuations of a small number of molecules diffusing within a femtoliter volume on a living cell. This technique yields single-molecule diffusion measurements and also the stoichiometry of molecular complexes when the brightness of the diffusing particles is analyzed. Cross-correlation of two differentially fluorescently tagged molecular species is also possible, providing insights into the diffusion of multimolecular complexes over time.

TIRFM. Total internal reflection microscopy is a powerful technique that allows extremely thin optical sectioning with outstanding signal-to-noise ratio. This method is appropriate to track single fluorescent particles at the cell surface closest to the substratum.

Acknowledgment

I would like to thank Dr. F. Sánchez-Madrid for critical reading of the manuscript.

References

Alon, R., & Ley, K. (2008). Cells on the run: Shear-regulated integrin activation in leukocyte rolling and arrest on endothelial cells. *Current Opinion in Cell Biology, 20*(5), 525–532.

Barreiro, O., Aguilar, R. J., Tejera, E., Megías, D., de Torres-Alba, F., Evangelista, A., et al. (2009). Specific targeting of human inflamed endothelium and in situ vascular tissue transfection by the use of ultrasound contrast agents. *JACC: Cardiovascular Imaging, 2*(8), 997–1005.

Barreiro, O., de la Fuente, H., Mittelbrunn, M., & Sanchez-Madrid, F. (2007). Functional insights on the polarized redistribution of leukocyte integrins and their ligands during leukocyte migration and immune interactions. *Immunological Reviews, 218*, 147–164.

Barreiro, O., Yanez-Mo, M., Sala-Valdes, M., Gutierrez-Lopez, M. D., Ovalle, S., Higginbottom, A., et al. (2005). Endothelial tetraspanin microdomains regulate leukocyte firm adhesion during extravasation. *Blood, 105*(7), 2852–2861.

Barreiro, O., Yanez-Mo, M., Serrador, J. M., Montoya, M. C., Vicente-Manzanares, M., Tejedor, R., et al. (2002). Dynamic interaction of VCAM-1 and ICAM-1 with moesin and ezrin in a novel endothelial docking structure for adherent leukocytes. *Journal of Cell Biology, 157*(7), 1233–1245.

Barreiro, O., Zamai, M., Yanez-Mo, M., Tejera, E., Lopez-Romero, P., Monk, P. N., et al. (2008). Endothelial adhesion receptors are recruited to adherent leukocytes by inclusion in preformed tetraspanin nanoplatforms. *Journal of Cell Biology, 183*(3), 527–542.

Berditchevski, F. (2001). Complexes of tetraspanins with integrins: More than meets the eye. *Journal of Cell Science, 114*(Pt. 23), 4143–4151.

Berditchevski, F., & Odintsova, E. (2007). Tetraspanins as regulators of protein trafficking. *Traffic, 8*(2), 89–96.

Bixel, G., Kloep, S., Butz, S., Petri, B., Engelhardt, B., & Vestweber, D. (2004). Mouse CD99 participates in T-cell recruitment into inflamed skin. *Blood, 104*(10), 3205–3213.

Bradfield, P. F., Scheiermann, C., Nourshargh, S., Ody, C., Luscinskas, F. W., Rainger, G. E., et al. (2007). JAM-C regulates unidirectional monocyte transendothelial migration in inflammation. *Blood, 110*(7), 2545–2555.

Cairo, C. W., Mirchev, R., & Golan, D. E. (2006). Cytoskeletal regulation couples LFA-1 conformational changes to receptor lateral mobility and clustering. *Immunity, 25*(2), 297–308.

Cambi, A., Joosten, B., Koopman, M., de Lange, F., Beeren, I., Torensma, R., et al. (2006). Organization of the integrin LFA-1 in nanoclusters regulates its activity. *Molecular Biology of the Cell, 17*(10), 4270–4281.

Carman, C. V., Jun, C. D., Salas, A., & Springer, T. A. (2003). Endothelial cells proactively form microvilli-like membrane projections upon intercellular adhesion molecule 1 engagement of leukocyte LFA-1. *Journal of Immunology, 171*(11), 6135–6144.

Charrin, S., le Naour, F., Silvie, O., Milhiet, P. E., Boucheix, C., & Rubinstein, E. (2009). Lateral organization of membrane proteins: Tetraspanins spin their web. *Biochemical Journal, 420*(2), 133–154.

Chattopadhyay, N., Wang, Z., Ashman, L. K., Brady-Kalnay, S. M., & Kreidberg, J. (2003). $\alpha 3\beta 1$ integrin-CD151, a component of the cadherin–catenin complex, regulated PTPμ expression and cell–cell adhesion. *Journal of Cell Biology, 163*(6), 1351–1362.

Cherukuri, A., Shoham, T., Sohn, H. W., Levy, S., Brooks, S., Carter, R., et al. (2004). The tetraspanin CD81 is necessary for partitioning of coligated CD19/CD21-B cell antigen receptor complexes into signaling-active lipid rafts. *Journal of Immunology, 172*(1), 370–380.

Collins, T., Read, M. A., Neish, A. S., Whitley, M. Z., Thanos, D., & Maniatis, T. (1995). Transcriptional regulation of endothelial cell adhesion molecules: NF-kappa B and cytokine-inducible enhancers. *FASEB Journal, 9*(10), 899–909.

de Bakker, B. I., Bodnar, A., van Dijk, E. M., Vamosi, G., Damjanovich, S., Waldmann, T. A., et al. (2008). Nanometer-scale organization of the alpha subunits of the receptors for IL2 and IL15 in human T lymphoma cells. *Journal of Cell Science, 121*(Pt. 5), 627–633.

de Bakker, B. I., de Lange, F., Cambi, A., Korterik, J. P., van Dijk, E. M., van Hulst, N. F., et al. (2007). Nanoscale organization of the pathogen receptor DC-SIGN mapped by single-molecule high-resolution fluorescence microscopy. *ChemPhysChem, 8*(10), 1473–1480.

Dijkstra, S., Kooij, G., Verbeek, R., van der Pol, S. M., Amor, S., Geisert, E. E. Jr., et al. (2008). Targeting the tetraspanin CD81 blocks monocyte transmigration and ameliorates EAE. *Neurobiology of Disease, 31*(3), 413–421.

Douglass, A. D., & Vale, R. D. (2005). Single-molecule microscopy reveals plasma membrane microdomains created by protein–protein networks that exclude or trap signaling molecules in T cells. *Cell, 121*(6), 937–950.

Eggeling, C., Ringemann, C., Medda, R., Schwarzmann, G., Sandhoff, K., Polyakova, S., et al. (2009). Direct observation of the nanoscale dynamics of membrane lipids in a living cell. *Nature, 457*(7233), 1159–1162.

Elices, M. J., Osborn, L., Takada, Y., Crouse, C., Luhowskyj, S., Hemler, M. E., et al. (1990). VCAM-1 on activated endothelium interacts with the leukocyte integrin VLA-4 at a site distinct from the VLA-4/fibronectin binding site. *Cell, 60*(4), 577–584.

Erovic, B. M., Neuchrist, C., Kandutsch, S., Woegerbauer, M., & Pammer, J. (2003). CD9 expression on lymphatic vessels in head and neck mucosa. *Modern Pathology, 16*(10), 1028–1034.

Espenel, C., Margeat, E., Dosset, P., Arduise, C., Le Grimellec, C., Royer, C. A., et al. (2008). Single-molecule analysis of CD9 dynamics and partitioning reveals multiple modes of interaction in the tetraspanin web. *Journal of Cell Biology, 182*(4), 765–776.

Feigelson, S. W., Grabovsky, V., Shamri, R., Levy, S., & Alon, R. (2003). The CD81 tetraspanin facilitates instantaneous leukocyte VLA-4 adhesion strengthening to vascular cell adhesion molecule 1 (VCAM-1) under shear flow. *Journal of Biological Chemistry, 278*(51), 51203–51212.

Geary, S. M., Cambareri, A. C., Sincock, P. M., Fitter, S., & Ashman, L. K. (2001). Differential tissue expression of epitopes of the tetraspanin CD151 recognised by monoclonal antibodies. *Tissue Antigens, 58*(3), 141–153.

Gordon-Alonso, M., Yanez-Mo, M., Barreiro, O., Alvarez, S., Munoz-Fernandez, M. A., Valenzuela-Fernandez, A., et al. (2006). Tetraspanins CD9 and CD81 modulate HIV-1-induced membrane fusion. *Journal of Immunology, 177*(8), 5129–5137.

Goswami, D., Gowrishankar, K., Bilgrami, S., Ghosh, S., Raghupathy, R., Chadda, R., et al. (2008). Nanoclusters of GPI-anchored proteins are formed by cortical actin-driven activity. *Cell, 135*(6), 1085–1097.

Greenfield, D., McEvoy, A. L., Shroff, H., Crooks, G. E., Wingreen, N. S., Betzig, E., et al. (2009). Self-organization of the *Escherichia coli* chemotaxis network imaged with super-resolution light microscopy. *PLoS Biology, 7*(6), e1000137.

Gutierrez-Lopez, M. D., Ovalle, S., Yáñez-Mó, M., Sánchez-Sánchez, N., Rubinstein, E., Olmo, N., et al. (2003). A functionally relevant conformational epitope on the CD9 tetraspanin depends on the association with activated beta-1 integrin. *Journal of Biological Chemistry, 278*(1), 208–218.

Hemler, M. E. (2003). Tetraspanin proteins mediate cellular penetration, invasion, and fusion events and define a novel type of membrane microdomain. *Annual Review of Cell and Developmental Biology, 19*, 397–422.

Hemler, M. E. (2008). Targeting of tetraspanin proteins—Potential benefits and strategies. *Nature Reviews Drug Discovery, 7*(9), 747–758.

Jacobson, K., Mouritsen, O. G., & Anderson, R. G. (2007). Lipid rafts: At a crossroad between cell biology and physics. *Nature Cell Biology, 9*(1), 7–14.

Jolly, C., & Sattentau, Q. J. (2007). Human immunodeficiency virus type 1 assembly, budding, and cell–cell spread in T cells take place in tetraspanin-enriched plasma membrane domains. *Journal of Virology, 81*(15), 7873–7884.

Karamatic Crew, V., Burton, N., Kagan, A., Green, C. A., Levene, C., Flinter, F., et al. (2004). CD151, the first member of the tetraspanin (TM4) superfamily detected on erythrocytes, is essential for the correct assembly of human basement membranes in kidney and skin. *Blood, 104*(8), 2217–2223.

Kenworthy, A. K., Nichols, B. J., Remmert, C. L., Hendrix, G. M., Kumar, M., Zimmerberg, J., et al. (2004). Dynamics of putative raft-associated proteins at the cell surface. *Journal of Cell Biology, 165*(5), 735–746.

Kitani, S., Berenstein, E., Mergenhagen, S., Tempst, P., & Siraganian, R. P. (1991). A cell surface glycoprotein of rat basophilic leukemia cells close to the high affinity IgE receptor (Fc epsilon RI). Similarity to human melanoma differentiation antigen ME491. *Journal of Biological Chemistry, 266*(3), 1903–1909.

Klein-Soyer, C., Azorsa, D. O., Cazenave, J. P., & Lanza, F. (2000). CD9 participates in endothelial cell migration during *in vitro* wound repair. *Arteriosclerosis, Thrombosis, and Vascular Biology, 20*(2), 360–369.

Ko, E. M., Lee, I. Y., Cheon, I. S., Kim, J., Choi, J. S., Hwang, J. Y., et al. (2006). Monoclonal antibody to CD9 inhibits platelet-induced human endothelial cell proliferation. *Molecules and Cells, 22*(1), 70–77.

Kotha, J., Zhang, C., Longhurst, C. M., Lu, Y., Jacobs, J., Cheng, Y., et al. (2009). Functional relevance of tetraspanin CD9 in vascular smooth muscle cell injury phenotypes: A novel target for the prevention of neointimal hyperplasia. *Atherosclerosis, 203*(2), 377–386.

Kropshofer, H., Spindeldreher, S., Rohn, T. A., Platania, N., Grygar, C., Daniel, N., et al. (2002). Tetraspan microdomains distinct from lipid rafts enrich select peptide–MHC class II complexes. *Nature Immunology, 3*(1), 61–68.

Kusumi, A., & Suzuki, K. (2005). Toward understanding the dynamics of membrane-raft-based molecular interactions. *Biochimica et Biophysica Acta, 1746*(3), 234–251.

Lafleur, M. A., Xu, D., & Hemler, M. E. (2009). Tetraspanin proteins regulate membrane type-1 matrix metalloproteinase-dependent pericellular proteolysis. *Molecular Biology of the Cell, 20*(7), 2030–2040.

Lamagna, C., Meda, P., Mandicourt, G., Brown, J., Gilbert, R. J., Jones, E. Y., et al. (2005). Dual interaction of JAM-C with JAM-B and alpha(M)beta2 integrin: Function in junctional complexes and leukocyte adhesion. *Molecular Biology of the Cell, 16*(10), 4992–5003.

Larson, D. R., Gosse, J. A., Holowka, D. A., Baird, B. A., & Webb, W. W. (2005). Temporally resolved interactions between antigen-stimulated IgE receptors and Lyn kinase on living cells. *Journal of Cell Biology, 171*(3), 527–536.

Lasserre, R., Guo, X. J., Conchonaud, F., Hamon, Y., Hawchar, O., Bernard, A. M., et al. (2008). Raft nanodomains contribute to Akt/PKB plasma membrane recruitment and activation. *Nature Chemical Biology, 4*(9), 538–547.

Lau, L. M., Wee, J. L., Wright, M. D., Moseley, G. W., Hogarth, P. M., Ashman, L. K., et al. (2004). The tetraspanin superfamily member CD151 regulates outside-in integrin alphaIIb-beta3 signaling and platelet function. *Blood, 104*(8), 2368–2375.

Le Naour, F., Andre, M., Boucheix, C., & Rubinstein, E. (2006). Membrane microdomains and proteomics: Lessons from tetraspanin microdomains and comparison with lipid rafts. *Proteomics, 6*(24), 6447–6454.

Le Naour, F., Rubinstein, E., Jasmin, C., Prenant, M., & Boucheix, C. (2000). Severely reduced female fertility in CD9-deficient mice. *Science, 287*(5451), 319–321.
Lenne, P. F., Wawrezinieck, L., Conchonaud, F., Wurtz, O., Boned, A., Guo, X. J., et al. (2006). Dynamic molecular confinement in the plasma membrane by microdomains and the cytoskeleton meshwork. *EMBO Journal, 25*(14), 3245–3256.
Levy, S., & Shoham, T. (2005). The tetraspanin web modulates immune-signalling complexes. *Nature Reviews Immunology, 5*(2), 136–148.
Ley, K., & Kansas, G. S. (2004). Selectins in T-cell recruitment to non-lymphoid tissues and sites of inflammation. *Nature Reviews Immunology, 4*(5), 325–335.
Ley, K., Laudanna, C., Cybulsky, M. I., & Nourshargh, S. (2007). Getting to the site of inflammation: The leukocyte adhesion cascade updated. *Nature Reviews Immunology, 7*(9), 678–689.
Little, K. D., Hemler, M. E., & Stipp, C. S. (2004). Dynamic regulation of a GPCR–tetraspanin–G protein complex on intact cells: Central role of CD81 in facilitating GPR56–Galpha q/11 association. *Molecular Biology of the Cell, 15*(5), 2375–2387.
Marlin, S. D., & Springer, T. A. (1987). Purified intercellular adhesion molecule-1 (ICAM-1) is a ligand for lymphocyte function-associated antigen 1 (LFA-1). *Cell, 51*(5), 813–819.
Min, G., Wang, H., Sun, T. T., & Kong, X. P. (2006). Structural basis for tetraspanin functions as revealed by the cryo-EM structure of uroplakin complexes at 6-Å resolution. *Journal of Cell Biology, 173*(6), 975–983.
Miyado, K., Yamada, G., Yamada, S., Hasuwa, H., Nakamura, Y., Ryu, F., et al. (2000). Requirement of CD9 on the egg plasma membrane for fertilization. *Science, 287*(5451), 321–324.
Moseley, G. W. (2005). Tetraspanin-Fc receptor interactions. *Platelets, 16*(1), 3–12.
Neumann, A. K., Thompson, N. L., & Jacobson, K. (2008). Distribution and lateral mobility of DC-SIGN on immature dendritic cells—Implications for pathogen uptake. *Journal of Cell Science, 121*(Pt. 5), 634–643.
Nicolau , D. V., Jr., Burrage, K., Parton, R. G., & Hancock, J. F. (2006). Identifying optimal lipid raft characteristics required to promote nanoscale protein–protein interactions on the plasma membrane. *Molecular and Cellular Biology, 26*(1), 313–323.
Nydegger, S., Khurana, S., Krementsov, D. N., Foti, M., & Thali, M. (2006). Mapping of tetraspanin-enriched microdomains that can function as gateways for HIV-1. *Journal of Cell Biology, 173*(5), 795–807.
Odintsova, E., Butters, T. D., Monti, E., Sprong, H., van Meer, G., & Berditchevski, F. (2006). Gangliosides play an important role in the organization of CD82-enriched microdomains. *Biochemical Journal, 400*(2), 315–325.
Phillipson, M., Heit, B., Colarusso, P., Liu, L., Ballantyne, C. M., & Kubes, P. (2006). Intraluminal crawling of neutrophils to emigration sites: A molecularly distinct process from adhesion in the recruitment cascade. *Journal of Experimental Medicine, 203*(12), 2569–2575.
Pike, L. J. (2006). Rafts defined: A report on the keystone symposium on lipid rafts and cell function. *Journal of Lipid Research, 47*(7), 1597–1598.
Pileri, P., Uematsu, Y., Campagnoli, S., Galli, G., Falugi, F., Petracca, R., et al. (1998). Binding of hepatitis C virus to CD81. *Science, 282*(5390), 938–941.
Pols, M. S., & Klumperman, J. (2009). Trafficking and function of the tetraspanin CD63. *Experimental Cell Research, 315*(9), 1584–1592.
Qi, J. C., Wang, J., Mandadi, S., Tanaka, K., Roufogalis, B. D., Madigan, M. C., et al. (2006). Human and mouse mast cells use the tetraspanin CD9 as an alternate interleukin-16 receptor. *Blood, 107*(1), 135–142.

Rohlena, J., Volger, O. L., van Buul, J. D., Hekking, L. H., van Gils, J. M., Bonta, P. I., et al. (2009). Endothelial CD81 is a marker of early human atherosclerotic plaques and facilitates monocyte adhesion. *Cardiovascular Research, 81*(1), 187–196.

Sachs, N., Kreft, M., van den Bergh Weerman, M. A., Beynon, A. J., Peters, T. A., Weening, J. J., et al. (2006). Kidney failure in mice lacking the tetraspanin CD151. *Journal of Cell Biology, 175*(1), 33–39.

Sharma, P., Varma, R., Sarasij, R. C., Ira Gousset, K., Krishnamoorthy, G., Rao, M., et al. (2004). Nanoscale organization of multiple GPI-anchored proteins in living cell membranes. *Cell, 116*(4), 577–589.

Shaw, S. K., Ma, S., Kim, M. B., Rao, R. M., Hartman, C. U., Froio, R. M., et al. (2004). Coordinated redistribution of leukocyte LFA-1 and endothelial cell ICAM-1 accompany neutrophil transmigration. *Journal of Experimental Medicine, 200*(12), 1571–1580.

Sieber, J. J., Willig, K. I., Kutzner, C., Gerding-Reimers, C., Harke, B., Donnert, G., et al. (2007). Anatomy and dynamics of a supramolecular membrane protein cluster. *Science, 317*(5841), 1072–1076.

Silvie, O., Charrin, S., Billard, M., Franetich, J. F., Clark, K. L., van Gemert, G. J., et al. (2006). Cholesterol contributes to the organization of tetraspanin-enriched microdomains and to CD81-dependent infection by malaria sporozoites. *Journal of Cell Science, 119*(Pt. 10), 1992–2002.

Silvie, O., Rubinstein, E., Franetich, J. F., Prenant, M., Belnoue, E., Renia, L., et al. (2003). Hepatocyte CD81 is required for *Plasmodium falciparum* and *Plasmodium yoelii* sporozoite infectivity. *Nature Medicine, 9*(1), 93–96.

Simons, K., & Toomre, D. (2000). Lipid rafts and signal transduction. *Nature Reviews Molecular Cell Biology, 1*(1), 31–39.

Sterk, L. M., Geuijen, C. A., Oomen, L. C., Calafat, J., Janssen, H., & Sonnenberg, A. (2000). The tetraspan molecule CD151, a novel constituent of hemidesmosomes, associates with the integrin alpha6beta4 and may regulate the spatial organization of hemidesmosomes. *Journal of Cell Biology, 149*(4), 969–982.

Suzuki, K. G., Fujiwara, T. K., Edidin, M., & Kusumi, A. (2007a). Dynamic recruitment of phospholipase C gamma at transiently immobilized GPI-anchored receptor clusters induces IP3-Ca2+ signaling: Single-molecule tracking study 2. *Journal of Cell Biology, 177*(4), 731–742.

Suzuki, K. G., Fujiwara, T. K., Sanematsu, F., Iino, R., Edidin, M., & Kusumi, A. (2007b). GPI-anchored receptor clusters transiently recruit Lyn and G alpha for temporary cluster immobilization and Lyn activation: Single-molecule tracking study 1. *Journal of Cell Biology, 177*(4), 717–730.

Takeda, Y., Kazarov, A. R., Butterfield, C. E., Hopkins, B. D., Benjamin, L. E., Kaipainen, A., et al. (2007). Deletion of tetraspanin Cd151 results in decreased pathologic angiogenesis *in vivo* and *in vitro*. *Blood, 109*(4), 1524–1532.

Tohami, T., Drucker, L., Radnay, J., Shapira, H., & Lishner, M. (2004). Expression of tetraspanins in peripheral blood leukocytes: A comparison between normal and infectious conditions. *Tissue Antigens, 64*(3), 235–242.

Toothill, V. J., Van Mourik, J. A., Niewenhuis, H. K., Metzelaar, M. J., & Pearson, J. D. (1990). Characterization of the enhanced adhesion of neutrophil leukocytes to thrombin-stimulated endothelial cells. *Journal of Immunology, 145*(1), 283–291.

Umemura, Y. M., Vrljic, M., Nishimura, S. Y., Fujiwara, T. K., Suzuki, K. G., & Kusumi, A. (2008). Both MHC class II and its GPI-anchored form undergo hop diffusion as observed by single-molecule tracking. *Biophysical Journal, 95*(1), 435–450.

Unternaehrer, J. J., Chow, A., Pypaert, M., Inaba, K., & Mellman, I. (2007). The tetraspanin CD9 mediates lateral association of MHC class II molecules on the dendritic cell surface. *Proceedings of the National Academy of Sciences of the United States of America, 104*(1), 234–239.

Vestweber, D. (2007). Adhesion and signaling molecules controlling the transmigration of leukocytes through endothelium. *Immunological Reviews, 218*, 178–196.

Vischer, U. M., & Wagner, D. D. (1993). CD63 is a component of Weibel–Palade bodies of human endothelial cells. *Blood, 82*(4), 1184–1191.

Waterhouse, R., Ha, C., & Dveksler, G. S. (2002). Murine CD9 is the receptor for pregnancy-specific glycoprotein 17. *Journal of Experimental Medicine, 195*(2), 277–282.

Wegmann, F., Petri, B., Khandoga, A. G., Moser, C., Khandoga, A., Volkery, S., et al. (2006). ESAM supports neutrophil extravasation, activation of Rho, and VEGF-induced vascular permeability. *Journal of Experimental Medicine, 203*(7), 1671–1677.

Woodfin, A., Reichel, C. A., Khandoga, A., Corada, M., Voisin, M. B., Scheiermann, C., et al. (2007). JAM-A mediates neutrophil transmigration in a stimulus-specific manner in vivo: Evidence for sequential roles for JAM-A and PECAM-1 in neutrophil transmigration. *Blood, 110*(6), 1848–1856.

Woodfin, A., Voisin, M. B., Imhof, B. A., Dejana, E., Engelhardt, B., & Nourshargh, S. (2009). Endothelial cell activation leads to neutrophil transmigration as supported by the sequential roles of ICAM-2, JAM-A and PECAM-1. *Blood, 113*(24), 6246–6257.

Wu, H., Li, J., Peng, L., Liu, H., Wu, W., Zhou, Y., et al. (2001). Anti-human platelet tetraspanin (CD9) monoclonal antibodies induce platelet integrin alpha IIb beta 3 activation in a Fc receptor-independent fashion. *Chinese Medical Journal (England), 114*(1), 14–18.

Xu, D., Sharma, C., & Hemler, M. E. (2009). Tetraspanin12 regulates ADAM10-dependent cleavage of amyloid precursor protein. *FASEB Journal, [Epub ahead of print]*.

Yanez-Mo, M., Alfranca, A., Cabanas, C., Marazuela, M., Tejedor, R., Ursa, M. A., et al. (1998). Regulation of endothelial cell motility by complexes of tetraspan molecules CD81/TAPA-1 and CD151/PETA-3 with alpha3 beta1 integrin localized at endothelial lateral junctions. *Journal of Cell Biology, 141*(3), 791–804.

Yanez-Mo, M., Barreiro, O., Gonzalo, P., Batista, A., Megias, D., Genis, L., et al. (2008). MT1-MMP collagenolytic activity is regulated through association with tetraspanin CD151 in primary endothelial cells. *Blood, 112*(8), 3217–3226.

Yanez-Mo, M., Barreiro, O., Gordon-Alonso, M., Sala-Valdes, M., & Sanchez-Madrid, F. (2009). Tetraspanin-enriched microdomains: A functional unit in cell plasma membranes. *Trends in Cell Biology, 19*(9), 434–446.

Zhang, F., Kotha, J., Jennings, L. K., & Zhang, X. A. (2009). Tetraspanins and vascular functions. *Cardiovascular Research, 83*(1), 7–15.

Zheng, Z., & Liu, Z. (2006). CD151 gene delivery activates PI3K/Akt pathway and promotes neovascularization after myocardial infarction in rats. *Molecular Medicine, 12*(9–10), 214–220.

Zoller, M. (2009). Tetraspanins: Push and pull in suppressing and promoting metastasis. *Nature Reviews Cancer, 9*(1), 40–55.

CHAPTER 10

Transmigratory Cups and Invadosome-Like Protrusions: New Aspects of Diapedesis

Christopher V. Carman
Division of Molecular and Vascular Medicine, Department of Medicine,
Beth Israel Deaconess Medical Center, Center for Vascular Biology Research,
Harvard Medical School, Boston, Massachusetts 02115

I. OVERVIEW

Immune cell (i.e., blood leukocyte) functions are strongly coupled to their ability to traffic throughout the body as they conduct immune surveillance and respond to pathogens. Central aspects of this trafficking are the

Current Topics in Membranes, Volume 64 1063-5823/09 $35.00

DOI: 10.1016/S1063-5823(09)64010-2

continuous transitions from the tissue into the blood circulation and vice versa. The vascular endothelium represents the interface between these two tissue compartments, serving as both a barrier to leukocyte trafficking and a sentinel to instruct leukocyte adhesion, transmigration, and activation behaviors. The activation state and barrier properties of endothelium are also, in turn, strongly influenced by leukocytes. Thus, leukocyte–endothelial interactions represent a critical nexus for information exchange with strong implications for vascular and immune cell activities, particularly during immune and inflammatory responses. Basic mechanisms by which these interactions facilitate leukocyte trafficking and intercellular communication are, therefore, of central biomedical significance. Building on the "classic" three-step adhesion/activation cascade, recent studies have begun to reveal new and complex behaviors associated with leukocyte–endothelial interactions suggesting existence of highly dynamic and bidirectional proactive responses in both of these cell types. These ideas are illustrated by two very recently emerging activities/structures, namely, "transmigratory cups" formed by endothelium and "invadosome-like protrusions" (ILP) formed by leukocytes. In this chapter, the basic *in vitro* and *in vivo* observation of these structures, their molecular composition/regulation, and functional roles are summarized.

II. INTRODUCTION

A. An Overview of Leukocyte Trafficking

In order to fulfill their roles of immune surveillance and pathogen elimination, cells of the immune system (i.e., blood leukocytes, which include lymphocytes, monocytes, dendritic cells, and neutrophils) must continuously traffic throughout the body (von Andrian & Mackay, 2000). Leukocyte trafficking can be broken into two major phases—movement within the vascular and lymphatic circulation, and migration within tissues. The vascular and lymphatic circulatory systems are lined by monolayers of endothelial cells that grow on an abluminal layer of extracellular matrix (the basement membrane) and form organized intercellular junctional zones that include adherens junctions, tight junctions, and gap junctions (Baluk et al., 2007; Bazzoni & Dejana, 2004; Pepper & Skobe, 2003). In this way, the endothelium serves as the principal (selectively permeable) barrier between the circulation and the underlying tissues. Each phase change during trafficking (i.e., movement into (intravasation) or out of (extravasation) the circulation) therefore requires that leukocytes cross the endothelium (i.e., diapedese). It is recognized that this process is a critical and rate-limiting component of both constitutive and inflammation-specific trafficking and in this way is an important therapeutic target for immune-mediated and inflammatory disease (Ley, Laudanna, Cybulsky, & Nourshargh, 2007; von Andrian & Mackay, 2000).

The process of diapedesis occurs in diverse settings *in vivo* with a wide range of immediate purposes (von Andrian & Mackay, 2000). Blood leukocytes originate in the bone marrow, where they begin their life cycle by migrating into the bloodstream. T and B lymphocytes then enter and exit various lymphoid organs (including the thymus, lymph nodes, Peyer's patches, spleen, and tonsils) as part of their maturation processes and immune surveillance functions. Monocytes constitutively migrate from the circulation into the peripheral tissues, where they differentiate into antigen-presenting cells (APCs), including macrophages and dendritic cell-like APCs. These, in turn, traffic out of the tissue and into the lymphatic system through afferent lymphatic vessels that carry them to secondary lymphoid organs (SLO). In cases of infection, APCs bearing pathogen-derived antigen interact with and activate the expansion of antigen-specific lymphocytes. These lymphocytes then differentiate into effector or memory lymphocytes, which enter the vascular circulation and join innate immune cells (such as neutrophils) in migrating into the infected or inflamed peripheral tissues. Finally, several recent studies have begun to document so-called "reverse transmigration" whereby inflammatory leukocytes are thought to leave the peripheral tissues (during the resolution phase of inflammation) by reversing their migratory path and undergoing intravasation to reenter the vascular circulation (Huttenlocher & Poznansky, 2008).

The diapedesis associated with each of the above trafficking steps can occur in vastly different tissues and involve distinct endothelia (Aird, 2007a,b), leukocyte subtypes, and migration stimuli. Thus, while basic mechanisms and themes described herein likely have broad relevance, variation and additional mechanisms are also likely to exist.

B. Leukocyte Transmigration Across the Endothelium (i.e., Diapedesis)

1. The "Five-step" Cascade for Extravasation

Diapedesis, during extravasation, has been intensely investigated. The specialized postcapillary venules of SLO (i.e., high endothelial venules (HEV)) constitutively express adhesion molecules and chemokines that support entry of circulating naïve and memory lymphocytes (von Andrian & Mackay, 2000). Similarly, the endothelium of postcapillary venules in most inflamed peripheral tissues (or capillaries for inflamed lung and liver) upregulate adhesion molecules and chemokines that support diapedesis of effector/memory lymphocytes and innate immune cells (Springer, 1994). In both settings, extravasation begins with the accumulation of circulating leukocytes on the luminal surface of the endothelium through a well-characterized three-step adhesion and activation cascade (Butcher, 1991; Luscinskas et al., 1994; Springer, 1994), which is followed by two additional steps to complete the extravasation process (Fig. 1). Initially leukocytes undergo transient rolling-type interactions mediated by the selectin family of adhesion molecules (Step 1), which facilitate sensing of, and intracellular signaling responses to, chemokines presented on the glycocalyx of the endothelium (Step 2). This in turn triggers high-affinity interaction of lymphocyte integrin receptors (e.g., LFA-1, Mac1, and VLA-4) with their endothelial ligands (e.g., ICAM-1, ICAM-2, and VCAM-1) resulting in firm lymphocyte arrest (Step 3) (Carman & Springer, 2003; Luo, Carman, & Springer, 2007). Subsequently, lymphocytes undergo actin-dependent spreading, polarization, and integrin-dependent lateral migration on the luminal surface of the endothelium (Step 4). This activity seems to allow leukocytes to search out sites permissive for endothelial barrier penetration (Phillipson et al., 2006; Schenkel, Mamdouh, & Muller, 2004). Finally, the leukocyte must formally breach and transmigrate across the endothelium (Step 5), a process referred to specifically as "diapedesis." (Note, however, that this term often is employed more loosely to refer to the entire 5-step cascade.)

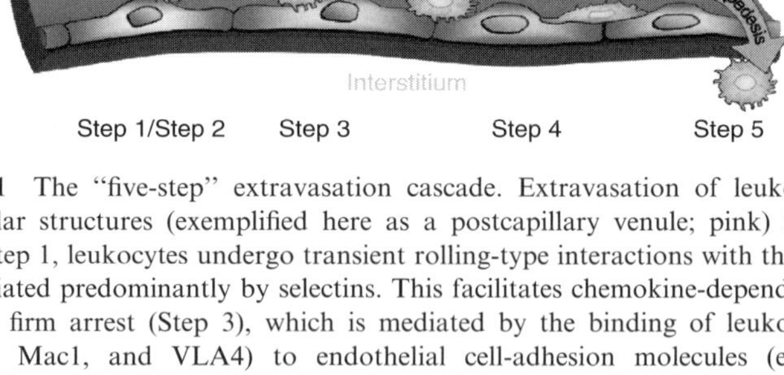

FIGURE 1 The "five-step" extravasation cascade. Extravasation of leukocytes (green) across vascular structures (exemplified here as a postcapillary venule; pink) is a multistep process. In Step 1, leukocytes undergo transient rolling-type interactions with the endothelium that are mediated predominantly by selectins. This facilitates chemokine-dependent activation (Step 2) and firm arrest (Step 3), which is mediated by the binding of leukocyte integrins (e.g., LFA1, Mac1, and VLA4) to endothelial cell-adhesion molecules (e.g., ICAM-1, ICAM-2, and VCAM-1). Subsequently, leukocytes spread, polarize, and migrate laterally over the surface of the endothelium, probing for a site to penetrate the endothelium (Step 4; see also Fig. 3). Finally, leukocytes cross the endothelial barrier (i.e., diapedese; Step 5), either para- or trans-cellularly, and enter the interstitium.

2. Mechanisms for Intravasation

The process of intravasation remains rather poorly investigated. Formally, the basic steps for intravasation should include interstitial migration toward the vessel (a process that may involve chemorepulsion, as well as chemoattraction (Huttenlocher & Poznansky, 2008)), transmigration across the basement membrane and endothelial barriers (i.e., diapedesis) and, ultimately, release of luminal leukocytes into circulation.

3. Two Routes for Crossing the Endothelium

Though multifaceted, perhaps the most fundamental role of the vascular endothelium is to serve as barrier, albeit a selectively permeable one. Thus, the question of precisely how this barrier is breached to accommodate leukocyte trafficking is a nontrivial one. Until recently, only one basic pathway, the "para-cellular" route, for diapedesis was widely recognized. Para-cellular diapedesis involves cooperative efforts on the part of both the leukocyte and the endothelium to locally disassemble the interendothelial junctions in order to form a para-cellular gap that will allow leukocyte transmigration (Burns, Smith, & Walker, 2003; Ley et al., 2007; Luscinskas, Ma, Nusrat, Parkos, & Shaw, 2002; Muller, 2001, 2003). In fact, however, there is a large body of literature (including some of the very first studies to directly investigate the route of diapedesis (Marchesi & Gowans, 1964; Williamson & Grisham, 1960, 1961) and nearly 50 subsequent studies recently reviewed in detail (Sage & Carman, 2009)) demonstrates the coexistence *in vivo* of the para-cellular route along with a second pathway termed that trans-cellular route, whereby leukocytes pass directly through individual endothelial cells via the formation of a trans-cellular pore. As a consequence of recent studies characterizing trans-cellular diapedesis *in vitro* for the first time, the mechanisms for this pathway (as discussed in Section IV) have begun to be elucidated and its physiologic relevance was more broadly appreciated (Carman & Springer, 2008). The relative roles, in distinct *in vivo* settings (see Section II.A), of the para- and trans-cellular pathways remain important open questions.

Whether para- or trans-cellular and in the setting of intravasation or extravasation, emerging evidence paint a picture of diapedesis as requiring intimate cell–cell contact and information exchange that is coupled to complex, dynamic, and proactive behaviors in both leukocytes and endothelial cells. These ideas are especially well illustrated by recent studies characterizing the endothelial "transmigratory cups" and leukocyte "ILP" described in the following section.

III. ENDOTHELIAL TRANSMIGRATORY CUPS

A. The Proactive Endothelium

Whereas, much of the early studies of leukocyte–endothelial interactions intuitively focused on the dynamics of the leukocytes, which are unquestionably active cells, it has become increasingly apparent that endothelial cells are highly dynamic and proactive partners in these interactions. In response to inflammatory cytokines endothelial cells mobilize preexisting internal pools and synthesize new amounts of cell surface adhesion molecules, cytokines, and chemokines in order to promote leukocyte adhesion and migration. Additionally, endothelial cells take up tissue-derived chemokines from their basolateral aspect, transcytose them luminally, and present them for leukocytes in association with the glycocalyx on the tips of microvilli (Middleton, Patterson, Gardner, Schmutz, & Ashton, 2002).

Upon direct binding of leukocytes, endothelial cells initiate intracellular signaling events that are coupled to a range of responses, which include active rearrangement of the actin cytoskeleton in ways that may influence leukocyte behaviors. For example, leukocyte engagement of the endothelial adhesion receptors ICAM-1, VCAM-1, PECAM-1, and E-selectin triggers intraendothelial calcium flux coupled to activation of the Rho GTPase and phosphorylation of cortactin, paxillin, p130cas, and focal-adhesion kinase and increased myosin contractility (Adamson, Etienne, Couraud, Calder, & Greenwood, 1999; Durieu-Trautmann, Chaverot, Cazaubon, Strosberg, & Couraud, 1994; Etienne et al., 1998; Etienne-Manneville et al., 2000; Huang et al., 1993; Muller, 2003; Pfau et al., 1995; Strey, Janning, Barth, & Gerke, 2002; Su, Chen, Huang, & Jen, 2000; Thompson, Randi, & Ridley, 2002; Wojciak-Stothard, Williams, & Ridley, 1999). Moreover, treatment of the endothelium with either chelators of intracellular calcium or inhibitors of actin (e.g., cytocholasin-D), Rho signaling (e.g., C3 transferase), or microtubules (e.g., colchicine) effectively inhibits leukocyte diapedesis (Adamson et al.; Durieu-Trautmann et al.; Etienne et al.; Etienne-Manneville et al.; Huang et al.; Muller; Pfau et al.; Strey et al.; Su et al.; Thompson et al.; Wojciak-Stothard et al.). Taken together, these and other studies have established a proactive role for endothelial signaling/cytoskeletal rearrangements in facilitating leukocyte transmigration. While opening of interendothelial junctions (to facilitate para-cellular gap opening) has been considered a major way in which this is achieved, the contribution of other mechanisms such as the formation of the transmigratory cups, discussed in the following section, has recently become evident.

B. *Transmigratory Cups:* In Vitro *Observations*

The formation of ICAM-1- VCAM-1- and actin-enriched upright microvilli-like projections that surround adherent and transmigrating leukocytes (termed "transmigratory-cups" or "docking structures" (Barreiro et al., 2002; Carman & Springer, 2003, 2004), represents a new mechanism by which endothelial cells proactively contribute to the adhesion and transmigration of leukocytes. Though at the time not fully elucidated as such, perhaps the first documentation of transmigratory cups/docking structures was in the context of imaging studies of monocyte adhesion to tumor necrosis factor-α (TNF-α)-activated human umbilical vein endothelial cells (HUVECs) *in vitro* (Wojciak-Stothard et al., 1999). In these studies, E-selectin, VCAM-1, and ICAM-1 were seen to form "clusters" around the edges of adherent monocytes that often exhibited a spike-like appearance (highly consisted with subsequently characterized structures, described in following section; see Fig. 2). These clusters were demonstrated to: (1) require intact actin (as demonstrated by cytochalasin D pretreatment); (2) be dependent on signaling through RhoA (as demonstrated with C3 transferase pretreatment or dominant negative Rho expression), but not Rac or Cdc42; and (3) be important for monocyte adhesion (Wojciak-Stothard et al.). Additionally, these clusters were found to colocalize with ezrin/radixin/moesin (ERM) proteins, cytoskeletal adaptor/regulatory proteins known to be important regulators of microvilli (Bretscher, Reczek, & Berryman, 1997).

Subsequent studies examining lymphocyte adhesion to HUVECs demonstrated the formation of three-dimensional "docking structures," apparently assembled from extended microvilli, that surrounded and partially embraced, the lymphocytes (Barreiro et al., 2002) (Figs. 2 and 3). From the *en face* perspective, these structures were remarkably similar to the "clusters" observed with monocytes (Wojciak-Stothard et al., 1999) (see Fig. 2). Also consistent with the previous studies was the demonstration that these structures: (1) were enriched in VCAM-1, ICAM-1, actin, and the ERM family members erzin and moesin; (2) were dependent on RhoA signaling; and (3) enhanced the adhesion between lymphocytes and endothelium (Barreiro et al.). This investigation went on to show that the docking structures were enriched in a range of additional cytoskeletal adaptor/regulatory proteins including α-actinin, vinculin, VASP, and paxillin (Barreiro et al.) (Table I). Moreover, phosphoinositides, the signaling molecules that, among others, are important regulators of ERM proteins and microvilli (Bretscher et al., 1997), were also found enriched in cups (Barreiro et al.).

Following this work, two largely complementary studies provided description of three-dimensional cell–cell interfaces, in which actin-, ICAM-1-, and VCAM-1-enriched microvilli-like projections were formed by endothelium

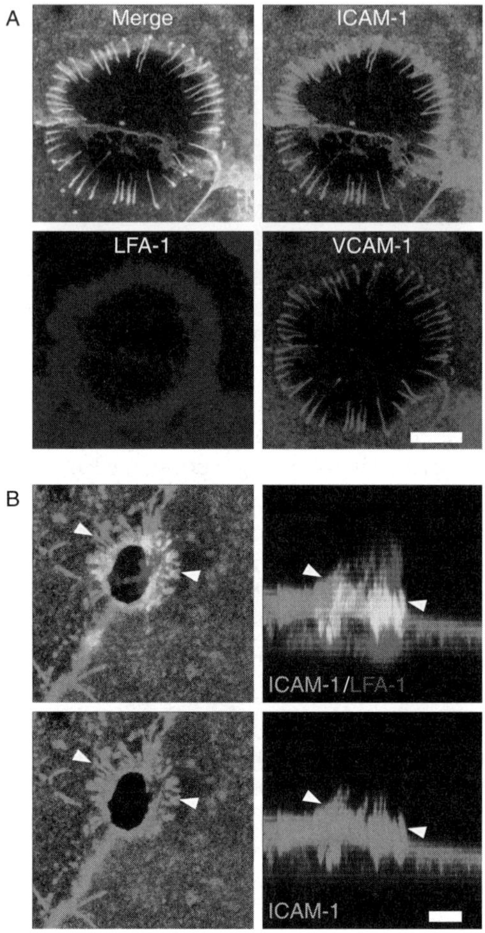

FIGURE 2 Fluorescence imaging of transmigratory cups. Panels depict fluorescence confocal micrographs of monocytes adherent to (A) and in the process of diapedesis across (B) a TNF-α-activated HUVEC monolayer *in vitro*. (A) Cells are stained for endothelial ICAM-1 (green) and VCAM-1 (red) and monocyte LFA-1 (blue). Panels depict a confocal z-stack projection viewed *en face* of a spread monocyte surrounded by a transmigratory cup. Individual microvilli-like projections emanating from the endothelium at the periphery of the monocyte extending up over its edge toward the center of the cell (For orientation, note the cross-sectional equivalents schematized in Fig. 3B and shown ultrastructurally in Fig. 4A). (B) Cells are stained for endothelial ICAM-1 (green) and LFA-1 on leukocyte (red). Left panels: An *en face* view, as in A, of a monocyte in the process of para-cellular diapedesis. Right panels: A three-dimensional projection of the panel shown on left, rotated 90° clearly showing the transmigratory cup microvilli embracing the migrating monocyte (arrows). Scale bars represent 5 μm. (See Color Insert.)

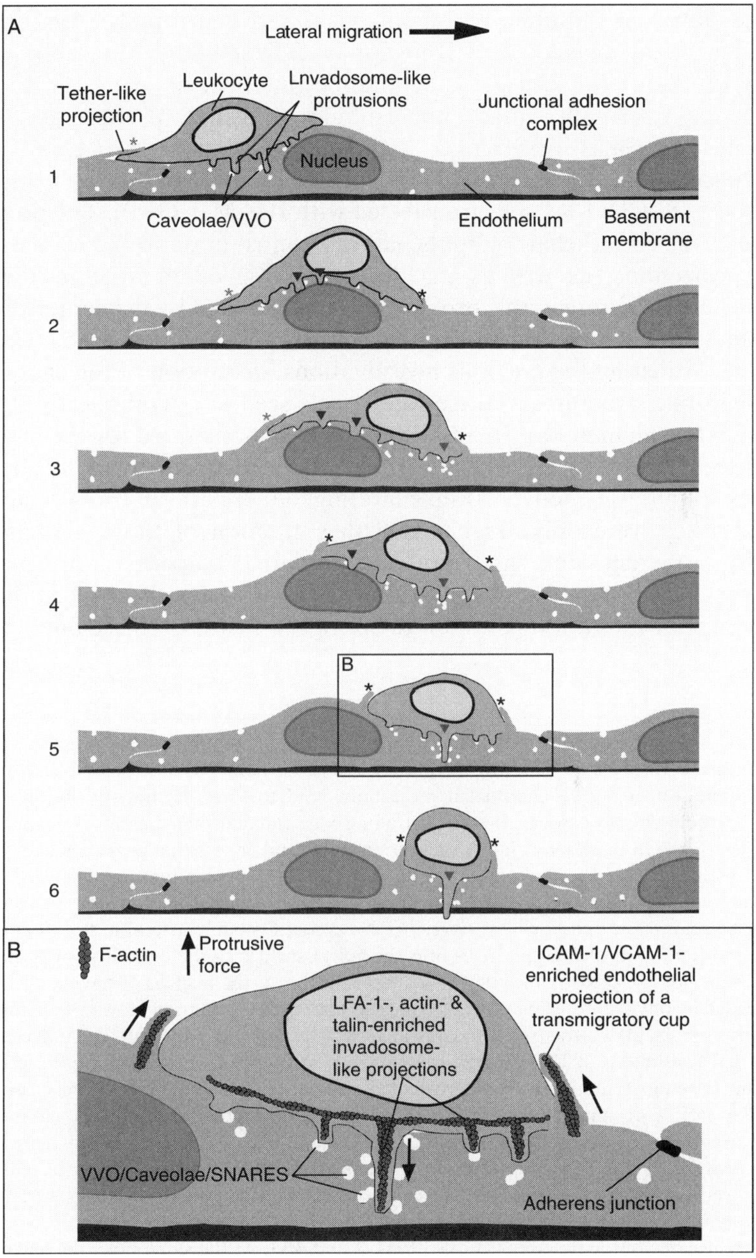

FIGURE 3 Transmigratory cup and invadosome-like protrusions in action. (A) Leukocyte (green) is depicted in the process of lateral migration over, and trans-cellular diapedesis through (see Fig. 1, Steps 4 and 5), an endothelial monolayer (pink). Numbered panels show successive time points that are intended to represent intervals of ~30–60 s. *Transmigratory cup dynamics*:

around adherent lymphocytes, monocytes, and neutrophils. Again, these were dependent on actin and Rho family GTPases and were additionally shown to require calcium signaling and (despite their undetectable presence within the microvilli) intact microtubules (Carman & Springer, 2004; Carman, Jun, Salas, & Springer, 2003). In addition to HUVECs, similar projections were demonstrated to form on epithelial cells expressing ICAM-1 (i.e., CHO-K1 cells transfected with ICAM-1-GFP). The proactive nature of the endothelial (or epithelial) cell contribution to the formation of these projections, as well as the dominant role of endothelial adhesion receptor cross-linking in this process, was underscored by the demonstration that similar structures formed around anti-ICAM-1-coated beads (Carman et al.). Contrasting the previous investigations, disruption of the endothelial microvilli-like structures in these studies was seen to only modestly effected adhesion (Carman et al.). In addition, it was demonstrated for the first time that these structures were, in fact, strongly associated with virtually all of the leukocytes that were actively transmigrating (either para- or trans-cellularly), whereas they were only associated with a fraction of those of that were apically adherent and had not yet initiated diapedesis (Carman & Springer). Moreover, disruption of these structures greatly reduced the efficiency of transmigration (Carman & Springer). Based on these findings and

---

ICAM-1-/VCAM-1-enriched endothelial projections that surround migrating leukocytes and form "transmigratory cups" are highlighted by asterisks. Panels 1–3 depict a phase of rapid lateral migration, in which endothelial projections tend to form asymmetrically around the leukocyte, preferentially associated with the trailing edge and take on a "tether-like" appearance (green asterisks). In panels 4–6, lateral migration slows and endothelial projections form more symmetrically around the leukocytes leading to "mature" transmigratory cups, which are thought to guide diapedesis and to provide a vertical traction substrate for protrusion of leukocytes against the endothelial surface. *ILP dynamics*: Dynamic insertion (to $\sim$0.2–1 μm in depth) and retraction of multiple ILPs into the apical surface of the endothelium is shown as lateral migration proceeds (panels 1–4). ILPs that form over the nucleus of the endothelial cell are impeded by physical resistance from the nuclear lamina and remain shallow (blue arrowheads in panels 2–4; see also ultrastructural equivalent in Fig. 4A). At a location of sufficiently low endothelial resistance, an ILP progressively extends several micrometers in depth, ultimately breaching the endothelium trans-cellularly (red arrowheads in panels 3–6; see also ultrastructural equivalent in Fig. 4C and D). Intracellular vesicular structures that are similar to caveolae and VVOs are enriched near, and fused to, endothelial invaginations ("podo-prints") that are formed by leukocyte ILPs (see also Fig. 4B). (B) A detailed view of the boxed region in panel 5 of A shows vertical endothelial microvillus-like projections (rich in actin, ICAM-1, VCAM-1, along with many other adhesion cytoskeletal and signaling molecules; see Table I) that surround the periphery of adherent leukocytes (see also Figs. 2 and 4A). At the same time, the leukocyte is driving actin-rich invadosomes into the endothelial cell surface (see also Fig. 4). Yellow circles in the endothelium depict VVO and caveolae that tend to accumulate at endothelial invaginations formed by ILPs (see also Fig. 4B). (See Color Insert.)

TABLE I

Molecules Involved in Endothelial Transmigratory Cups

Molecule type[a]	Molecule name[b]	Enriched in cups[c]	Functionally implicated in cup formation[d]	References
Adhesion	ALCAM	YES	ND	Cayrol et al. (2008)
	CD44	YES	ND	Barreiro et al. (2008)
	E-selectin	YES	ND	Wojciak-Stothard et al. (1999)
	ICAM-1	YES	YES	Barreiro et al. (2002, 2005, 2008), Carman and Springer (2004), Carman et al. (2003), Dittmar et al. (2008), Kanters et al. (2008), Millan et al. (2006), Rohlena et al. (2009), van Buul et al. (2007), Wojciak-Stothard et al. (1999)
	ICAM-2	YES	ND	Barreiro et al. (2008), Carman and Springer (2004)
	JAM-1	YES	ND	Barreiro et al. (2008)
	PECAM-1	YES	ND	Barreiro et al. (2008), Carman and Springer (2004)
	VCAM-1	YES	YES	Barreiro et al. (2002, 2005, 2008), Carman and Springer (2004), Carman et al. (2003), Dittmar et al. (2008), Rohlena et al. (2009), Wojciak-Stothard et al. (1999)
Tetraspanin	CD9	YES	YES	Barreiro et al. (2005, 2008)
	CD81	YES	YES	Rohlena et al. (2009)
	CD151	YES	YES	Barreiro et al. (2005, 2008)
Cytoskeletal	Actin	YES	YES	Barreiro et al. (2002), Carman and Springer (2004), Carman et al. (2003), Millan et al. (2006), van Buul et al. (2007), Wojciak-Stothard et al. (1999)
	Microtubules	NO	YES	Carman and Springer (2004), Carman et al. (2003)
	Vimentin	YES	YES	Nieminen et al. (2006)
Cyto-adaptor	α-Actinin	YES	ND	Barreiro et al. (2002)
	Ezrin	YES	YES	Barreiro et al. (2002), Wojciak-Stothard et al. (1999)
	Filamin B	YES	YES	Kanters et al. (2008)

(*continued*)

TABLE I (*continued*)

Molecule type[a]	Molecule name[b]	Enriched in cups[c]	Functionally implicated in cup formation[d]	References
	Moesin	YES	YES	Barreiro et al. (2002), Wojciak-Stothard et al. (1999)
	Paxillin	YES	ND	Barreiro et al. (2002)
	Talin	YES	ND	Barreiro et al. (2002)
	VASP	YES	ND	Barreiro et al. (2002)
	Vinculin	YES	ND	Barreiro et al. (2002)
Signaling	Calcium	ND	YES	Carman and Springer (2004), Carman et al. (2003)
	PI	YES	ND	Barreiro et al. (2002)
	RhoA	YES	YES	Barreiro et al. (2002), Wojciak-Stothard et al. (1999)
	RhoG	YES	YES	van Buul et al. (2007)
	ROCK	ND	YES	Barreiro et al. (2002)
	SGEF	YES	YES	van Buul et al. (2007)

[a]Type of molecule, related to functional classification. Cyto-adaptor, cytoskeletal adaptor/regulatory protein.

[b]Molecules that have been shown to have a relationship to transmigratory cups/docking structures.

[c]Demonstrated presence/enrichment in cup structures. "Yes" indicates positive data. For those molecules functionally implicated: "No" indicates negative data; "ND" indicates that localization has not yet been determined.

[d]Demonstrated functional role in the formation/stability of cup structures.

the previously noted similarly of this type of architecture to "phagocytic cups" (Barreiro et al., 2002), these structures were termed "transmigratory cups" (Carman & Springer) (Figs. 2 and 3).

Transmigratory cups/docking structures have now been broadly characterized in diverse *in vitro* settings that have included a range of leukocyte types, endothelial models (including HUVECs, as well as, human dermal, lung, brain microvascular endothelium, and lymphatic endothelium), and activation stimuli (Barreiro et al., 2005, 2008; Cayrol et al., 2008; Dittmar et al., 2008; Kanters et al., 2008; Millan et al., 2006; Nieminen et al., 2006; Riethmuller, Nasdala, & Vestweber, 2008; Rohlena et al., 2009; van Buul et al., 2007). Collectively, as discussed in Sections III.D and E, these studies have begun to uncover some of the basic mechanisms for cup formation, as well as their functional roles.

C. Transmigratory Cups: In Vivo Observations

Though currently limited, there is evidence to support transmigratory cup formation *in vivo*. It must be recognized, however, that the technical challenges of imaging the complex three-dimensional environment of the tissue microvasculature, as well as the inability to control the location, frequency, and timing of leukocyte–endothelial interactions, makes characterization of detailed subcellular structures (such as transmigratory cups) *in vivo* at least an order of magnitude more difficult than *in vitro* settings. Nonetheless, a variety of transmission electron microscopy (TEM) studies have provided *in vivo/in situ* cross-sectional views of microvilli-like endothelial projections embracing adherent and transmigrating leukocytes in a manner consistent with transmigratory cups (Faustmann & Dermietzel, 1985; Fujita, Puri, Yu, Travis, & Ferrans, 1991; Raine, Cannella, Duijvestijn, & Cross, 1990; Williamson & Grisham, 1961; Wolburg, Wolburg-Buchholz, & Engelhardt, 2005). In addition, in a recent technical tour de force, intravital fluorescence confocal microscopy, was used to define transmigratory cup-like structures, termed "domes," *in vivo* during neutrophil para- and trans-cellular diapedesis across postcapillary venules in mouse cremaster muscle (Phillipson, Kaur, Colarusso, Ballantyne, & Kubes, 2008). In this study parallel TEM analysis demonstrated that the ultrastructural equivalent of the "domes" defined by confocal microscopy were, in fact, endothelial projections remarkably similar in appearance to those observed in other *in vivo* TEM studies (Faustmann & Dermietzel; Fujita et al.; Raine et al.; Williamson & Grisham; Wolburg et al.) discussed earlier.

D. Molecular Composition and Regulation of Transmigratory Cups

1. Adhesion Molecules

Essentially all of the currently existing studies of transmigratory cups place endothelial cell adhesion molecules as critical markers and functional components of these structures. By far, the most broadly implicated are the immunoglobulin super family (IgSF) members ICAM-1, and VCAM-1 (Barreiro et al., 2002; Carman et al., 2003). These are ligands for leukocyte integrin adhesion receptors LFA-1 and Mac1 (ICAM-1) and VLA-4 (VCAM-1). Studies using leukocyte models that express either LFA-1 or VLA-4 (but no other receptors for major endothelial ligands) or beads coated with anti-ICAM-1 antibodies demonstrate that the engagement and clustering of either ICAM-1 or VCAM-1 alone is sufficient to form transmigratory cups/docking structures (Barreiro et al.; Carman et al.). Alternatively, anti-LFA-1 or anti-VLA-4 function blocking antibodies alone each partially blocked cup formation around physiologic leukocytes and had additive (but incomplete) effects when used in combination (Barreiro et al.; Carman et al.).

A growing list of additional adhesion molecules including IgSF members ALCAM (Cayrol et al., 2008), ICAM-2 (Barreiro et al., 2008; Carman & Springer, 2004), JAM-1 (Barreiro et al.), and PECAM-1 (Barreiro et al.; Carman & Springer), and the glycoproteins E-selectin (Wojciak-Stothard et al., 1999) and CD44 (Barreiro et al.) have been shown to be enriched in transmigratory cups (Table I). However, contrasting ICAM-1 and VCAM-1, direct demonstration of functional roles for any of these in the cup-formation process is currently lacking. Interestingly, though, studies of the ERM-interacting molecule ALCAM (a ligand for the lymphocyte costimulatory molecule CD6) have demonstrated that function blocking antibodies to ALCAM suppressed lymphocyte transmigration across the blood–brain barrier *in vitro* (Cayrol et al.).

2. Tetraspanins

Several studies have implicated tetraspanins in transmigratory cup formation apparently by enhancing the clustering and recruitment of important adhesion molecules (Barreiro et al., 2005, 2008; Rohlena et al., 2009) (Table I). Tetraspanins are proteins that posses four membrane-spanning segments and are largely implicated in forming/stabilizing lateral protein–protein associations within the plane of the plasma membrane (Hemler, 2005). In this way, they promote formation of multimeric protein complexes or domains sometimes referred to as "tetraspanin enriched microdomains" (Hemler). An initial study in activated HUVECs demonstrated that the tetraspanins CD9 and CD151 colocalized with ICAM-1 and VCAM-1 in transmigratory cups formed around T lymphobasts (Barreiro et al.). It was then further shown, that siRNA knockdown of CD9 and CD151 caused disorganization of ICAM-1 and VCAM-1 and destabilization of transmigratory cups, which correlated with decreased lymphocyte adhesion under shear flow conditions (Barreiro et al.). A subsequent study by the same group extended these finding demonstrating that tetraspanins may enhance cup formation and leukocyte adhesion explicitly through their ability to form preassembled microdomains (composed of CD9, CD151, ICAM-1, VCAM-1, PECAM-1, ICAM-2, CD44, JAM-1), which are poised for efficient incorporation into transmigratory cups (Barreiro et al.). Finally, the tetraspanin CD81 was also found to colocalize with VCAM-1 and ICAM-1 in cups (Rohlena et al.). Overexpression of CD81 in endothelium enhanced both the formation of these cups and the adhesion of monocytes, in an ICAM-1- and VCAM-1-dependent manner (Rohlena et al.).

3. Cytoskeletal Proteins

Transmigratory cups are strongly dependent on F-actin and are enriched in F-actin and a range of actin adaptor/regulator proteins (see Section III.D.4; Table I) (Barreiro et al., 2002; Carman & Springer, 2004; Carman et al., 2003; Millan et al., 2006; van Buul et al., 2007; Wojciak-Stothard et al., 1999). The intermediate filament vimentin was also found to localize to transmigratory cups and vimentin deficient endothelial were defective in cup formation and failed to support efficient diapedesis (Nieminen et al., 2006). Finally, whereas cups have been shown to be explicitly devoid of detectable microtubules, disruption of these structures, nonetheless, profoundly inhibits transmigratory cup formation (Barreiro et al.; Carman & Springer; Carman et al.). The finding that disruption of microfilaments, intermediate filaments, and microtubules each effect transmigratory cups is consistent with broad observations of strong interdependence between these cytoskeletal systems.

4. Cytoskeletal Adaptors/Regulatory Proteins

As discussed in Sections III.D.1 and III.D.3, transmigratory cups seem to be formed as a consequence of adhesion receptor signaling and coordination of cytoskeletal (especially actin-related; see Section III.D.3) responses. Critically, all of the identified adhesion molecules implicated in cup formation have relatively short cytoplasmic tails that lack intrinsic signaling domains. However, most of these have been demonstrated to interact with a variety of cytoskeletal adaptor proteins including ERM family members, α-actinin, paxillin, and filamin B. Thus, it is not surprising that these and other actin regulatory proteins (i.e., VASP, vinculin, talin) are enriched in transmigratory cups (Barreiro et al., 2002; Kanters et al., 2008; Wojciak-Stothard et al., 1999) (Table I). As mentioned in Section III.B, ERM proteins are perhaps the most interesting and intuitive adaptor proteins implicated in cup formation (Barreiro et al.; Wojciak-Stothard et al.), given their well-established roles as regulators of microvilli in general (Bretscher et al., 1997). However, direct characterization of functional roles played by ERM proteins is lacking, as is the case for most of the adaptors/regulators described earlier. However, studies, using, an siRNA-mediated knockdown approach, have recently provided direct evidence for a role of the actin cross-linking protein filamin B in ICAM-1 recruitment to cups as well as in leukocyte diapedesis (Kanters et al.).

5. Signaling Molecules

Several signaling molecules/activities have been demonstrated to have roles in transmigratory cup formation (Table I), yet clearly delineated pathways remain to be established. Perhaps, the most intuitive of these are the phosphoinositides $PI(4,5)P_2$, $PI(3,4)P_2$, and $PI(3,4,5)P_3$ (Barreiro et al.,

2002), key regulators of ERM proteins, and microvilli (Bretscher et al., 1997). However, the upstream signaling events causing enrichment of these lipid species in transmigratory cups remain unknown. Additionally, this calcium signaling is important in cup formation (Carman & Springer, 2004; Carman et al., 2003) fits well with previous studies demonstrating calcium flux as an important downstream consequence of ICAM-1 and VCAM-1 cross-linking in endothelium (Huang et al., 1993; Pfau et al., 1995; Su et al., 2000). However, the pathways downstream of calcium that ultimately effect cup formation remain to be identified.

Small Rho family GTPases have been implicated in transmigratory cup formation but consistent findings here also remain elusive. Early studies provided evidence for involvement of the small GTPase Rho A and the downstream kinase ROCK in cup formation (Barreiro et al., 2002; Wojciak-Stothard et al., 1999). These molecules are well known to be activated by ICAM-1 and VCAM-1 cross-linking in endothelial cells leading to formation of actin stress-fibers and contractility (Adamson et al., 1999; Durieu-Trautmann et al., 1994; Etienne et al., 1998; Etienne-Manneville et al., 2000; Huang et al., 1993; Muller, 2003; Pfau et al., 1995; Strey et al., 2002; Su et al., 2000; Thompson et al., 2002; Wojciak-Stothard et al., 1999). How RhoA/ROCK might couple to formation of microvilli-like structures is unclear. In another study, the broad-spectrum inhibitor *Clostridium difficile* toxin-B generally implicated small GTPases of the Rho, Rac, and CDC42 subfamilies in transmigratory cup formation, but specific inhibitors of RhoA or ROCK were found to have little effect on cups (Carman et al., 2003). In still further studies, the small GTPase RhoG was demonstrated to become activated downstream of ICAM-1 and shown to colocalized with ICAM-1, along with SGEF (SH3-containing guanine-nucleotide exchange factor, a RhoG-specific GEF) in transmigratory cups (van Buul et al., 2007). siRNA knockdown of either RhoG or sGEF significantly reduced cup formation without altered adhesion, but significantly disrupted transendothelial migration. This study also reported that the ROCK inhibitor Y27632 and RhoA knockdown partially inhibited cups, in a manner thought to be related to signaling through RhoG, through an as yet undefined mechanism (van Buul et al.).

E. Functional Roles for Transmigratory Cups

1. Adhesion/Docking

Based on the increased adhesive contact surface area, an intuitive role for transmigratory cups/docking structures is enhancing the strength of leukocyte adhesion to the endothelium. Via static and laminar shear flow adhesion assays, several studies have, indeed, provided support for such a function (Barreiro et al., 2002, 2005, 2008; Kanters et al., 2008; Rohlena et al., 2009; Wojciak-Stothard et al., 1999). However, several others were not able to observe significant contributions of cups to adhesive strength in similar systems (Carman & Springer, 2004; Carman et al., 2003; van Buul et al., 2007). Differences in these results likely lie in the intrinsic "baseline" cellular avidity of the specific models used (affected by factors such as the numbers and activation of state adhesion molecules, which will vary with different cell types and activation stimuli) and the details of the shear forces applied. In short, it seems likely that cups may enhance strength of adhesion in relatively low to medium strength avidity interactions, but may have diminished impact on inherently high-avidity interactions (e.g., highly activated leukocytes and endothelium).

2. Transmigration: Guidance and Traction

Transmigratory cups, as the name implies, are probably best established as serving roles of promoting transmigration across the endothelium, independently of adhesion *per se*. As mentioned in Section III.B, these structures are strongly associated with active transmigration events and studies in which cups have been perturbed show effective reduction of diapedesis, often without affecting the numbers of adherent leukocytes (Carman & Springer, 2004; Carman et al., 2003; Cayrol et al., 2008; Kanters et al., 2008; Millan et al., 2006; Nieminen et al., 2006; van Buul et al., 2007).

Detailed fixed and dynamic live-cell imaging studies of leukocytes at various stages of spreading/lateral migration (extravasation Step 4; Section II.B) and diapedesis suggest that the basis the transmigratory cup's contribution to transmigration efficiency is most likely related to the provision of guidance and traction (Carman & Springer, 2004). At initial stages of lateral migration endothelial microvilli-like structures were found associated asymmetrically with the trailing edge of migrating leukocytes and seemed to function more like "tethers" than cups (Carman & Springer) (Fig. 3). As leukocytes slowed their lateral migration, seemingly as a consequence of the "tether-like" projections, more symmetric cup-like arrangements ensued with an appearance of "corralling" the leukocytes and encouraging repolarization of protrusive forces against the surface of the endothelium rather that parallel to it (Carman & Springer) (Figs. 2 and 3). Importantly, the vertical ICAM-1/VCAM-1-enriched projections of the "mature" transmigratory cup, which are aligned perpendicular to the plane of the endothelium and parallel to the direction of diapedesis, may provide a physical basis (i.e., traction scaffold) for oriented migration (Carman & Springer). The demonstration that at sites opposing the endothelial projections, leukocyte integrins formed linear clusters reminiscent of fibrillar adhesions, on which cells are known exert forces during migration (Dzamba, Bultmann, Akiyama, & Peters, 1994), is consistent with this idea (Carman & Springer).

3. Barrier Maintenance

Transmigratory cups may serve a role in the maintenance of endothelial barrier properties during diapedesis. As mentioned earlier, arguably the most important role of the endothelium is to provide a barrier between circulation and tissue. Excessive disruption of this barrier during pathologic inflammatory responses (as opposed to the normal healthy and protective ones that routinely protect us from pathogens and heal wounds) leads to inappropriate fluid accumulation in the tissue (edema) and unregulated leukocyte diapedesis, effects that carry significant secondary pathologic consequences. Thus, the endothelium must balance its role as a sentinel/conduit for leukocyte trafficking with the critical need to maintain its barrier function. Transmigratory cups have been suggested to provide a mechanism for achieving this balance (Phillipson et al., 2008). *In vivo* imaging of neutrophil diapedesis in an inflammatory model demonstrated that modest local permeability increases during diapedesis correlated with formation of *extreme* transmigratory cups termed "domes" that seem to envelop migrating neutrophils before releasing them abluminally (Phillipson et al.). In this way, cups were interpreted as providing a kind of "air-lock" system to limit fluid permeability, while accommodating leukocyte extravasation (Phillipson et al.).

4. Intercellular Communication

Though largely speculative, the possibility exists that the large intimate cell–cell surface area provided by the transmigratory cup plays an explicit (though likely interrelated with other functions) role in enhancing intercellular communication. For example, it is established that synthesized and transcytosed chemokines are expressed preferentially at the tips of microvilli (Middleton et al., 2002). It is tempting to speculate that the microvilli-like structures that make up transmigratory cups may provide a particularly effective way to deliver chemokines signals to adherent leukocytes. In addition, growing evidence supports roles for an ability of endothelial cells to "instruct" innate and adaptive immune cell activation and differentiation behaviors (Nourshargh & Marelli-Berg, 2005). This may even occur through a process of antigen presentation by endothelial cells to effector lymphocytes (Choi, Enis, Koh, Shiao, & Pober, 2004). In this regard, it is intriguing to note the enrichment in transmigratory cups of ALCAM (Cayrol et al., 2008), a ligand the lymphocyte costimulatory molecules CD6 that is known to function in stabilizing immunological synapses during antigen recognition (Zimmerman et al., 2006). It is also interesting to consider how the large and intimate transmigratory cup interface may provide a basis for containment, and, therefore more efficient exchange, of secreted molecules, such as cytokines.

5. Pathogenesis/Subversion

Studies suggest that transmigratory cups may play roles in pathogenesis or indeed, become, subverted by invading pathogens. For example, one study revealed through genetic screens that CD81, a tetraspanin known to facilitate cup formation (Section III.D.2), is significantly upregulated in endothelial cells of atherosclerotic plaques (Rohlena et al., 2009). *In vitro* studies demonstrated that CD81 overexpression enhanced the recruitment of ICAM-1 and VCAM-1 into transmigratory cups and facilitated adhesion of monocytes. These studies suggest a specific molecular mechanism by which aberrantly effective/stable transmigratory cups may contribute to atherosclerosis (Rohlena et al.). Another study demonstrated that infection of primary human lymphocytes with the measles virus lead to concomitant upregulation of LFA-1 and VLA-4 adhesiveness and perturbation of migration capacity (Dittmar et al., 2008). The net result of these aberrations was enhanced engagement of ICAM-1 and VCAM-1 on brain microvascular endothelial cells and formation of sustained transmigratory cups, which provided intimate cell–cell contacts that facilitated virus transfer to the endothelium and ultimate release on their abluminal side. These results were interpreted as providing a possible mechanism by which the measles virus gains entry to the central nervous system (Dittmar et al.).

Neisseria meningitides bacteria have apparently subverted the endothelial cup responses directly in order to cross the blood–brain barrier. These bacteria have adapted cell surface molecules (type IV pili) to effectively bind endothelial cell surface ICAM-1 and CD44. Such binding was shown to drive actin- and ERM-dependent formation of microvilli-like structures that surrounded individual bacterial cells (in effect, "mini-transmigratory cups"), which ultimately facilitated bacterial transcytosis across the blood–brain barrier (Eugene et al., 2002; Nassif, Bourdoulous, Eugene, & Couraud, 2002). The idea that *N. meningitides*, in fact, subverts the same cup machinery used by leukocytes is underscored by the finding that bacteria present on the endothelial surface can prevent leukocyte transmigratory cup formation and diapedesis through sequestration of requisite endothelial ezrin (Doulet et al., 2006). In this way, *N. meningitides* bacteria may concomitantly gain access to the CNS and prevent trafficking of those cells that would eliminate them. With respect to this later point, it is not difficult to imagine that targeting cup formation pharmaceutically might be able to reduce unwanted leukocyte trafficking associated with inflammatory pathology.

IV. LEUKOCYTE INVADOSOME-LIKE PROTRUSIONS

A. "Classical" Invadosomes

Of interest to leukocyte–endothelial interactions, and diapedesis in particular, are the leukocyte protrusive activities used to facilitate adhesion, locomotion, and crossing of the endothelial barrier. Well-characterize protrusive structures formed by migratory cells in general include filopodia, lamellipodia, and related pseudopodia, each of which have been ascribed various roles in mediating locomotion and adhesion (Ridley et al., 2003). Recently, these structures have been joined by two novel protrusive organelles termed podosomes ("foot-protrusions") and invadopodia (Buccione, Orth, & McNiven, 2004; Linder & Aepfelbacher, 2003; Yamaguchi, Wyckoff, & Condeelis, 2005). Given, their apparent structural and functional interrelation, these have been collectively subtended under the term "invadosomes." As we describe in Section IV.B, structures related to invadosomes are emerging as important new elements of diapedesis. In the current section, we will first describe basic features of "classical" invadosomes, as determined in wide ranging cell types including leukocytes, osteoclasts, fibroblasts, endothelial cells, and various transformed cells (Buccione et al.; Gimona, Buccione, Courtneidge, & Linder, 2008; Linder, 2009; Linder & Aepfelbacher; Yamaguchi et al.). Importantly, despite the fact that the original descriptions of podosomes was in the context of physiologic cell–cell interactions (Section IV.B), essentially all of the contemporary ideas about invadosomes, including the fundamental definition of what constitutes "classical" invadosomes, derive from cells adhering to rigid glass and plastic substrates (for podosomes) or models of the extracellular matrix (for invadopodia) (Linder). Thus, the relationship of such structures to the invadosomes-like protrusions as described in Section IV.B must be considered cautiously.

Classical invadosomes are micron-scale cylindrical protrusions ($\sim$500 nm in depth and diameter) formed explicitly and uniquely on the *ventral* aspect of adherent cells. These have the defining architecture of a peripheral ring of $\beta 2$ or $\beta 3$ integrin, talin and vinculin, and a central core of F-actin (Buccione et al., 2004; Linder & Aepfelbacher, 2003). Both podosomes and invadopodia share a strong dependence on src kinase and the cytoskeletal regulator Wiskott–Aldrich Syndrome (WAS) protein (WASP) or, in nonhematopoietic cells, N-WASP (Calle, Chou, Thrasher, & Jones et al., 2004; Linder & Aepfelbacher). Integrin ligation is thought to elicit src-dependent signals that activate the Rho-family GTPase Cdc42, which in turn triggers WASp- and Arp2/3-mediated actin polymerization, thereby driving invadosome protrusion (Linder & Aepfelbacher). Leukocytes from immunodeficient WAS patients (that lack functional WASp expression), or from mice in which WASp or WIP (WASp interacting protein) has been knocked out, fail to form podosomes (Linder, Nelson, Weiss, & Aepfelbacher, 1999). The src substrate cortactin in nonhematopoietic cells has also been shown to be essential for podosome formation (Webb, Eves, & Mak, 2006). It is anticipated that in leukocytes the hematopoietic homologue of cortactin, HS1, plays this role. Another important, and defining feature shared by classical podosomes and invadopodia is their WASP/cortactin-dependent secretion of matrix-degrading metalloproteases (Gimona et al., 2008; Linder, 2008).

Distinction between podosomes and invadopodia are seen in their size (with significantly greater protrusion depths observed in invadopodia), dynamics, and apparent functions (Buccione et al., 2004; Yamaguchi et al., 2005). Podosomes have been described in a wide range of migratory cell types, most notably leukocytes, where they form dynamic clustered arrays or "rosettes" (Buccione et al.; Evans, Correia, Krasavina, Watson, & Matsudaira, 2003; Linder & Aepfelbacher, 2003). However, their function is unclear though generally linked to adhesion and migration. Invadopodia, In contrast, which have only been described in transformed cell types, have a more defined function in burrowing across matrix barriers, thereby conferring invasiveness to metastatic tumor cells (Buccione et al.; Yamaguchi et al.).

B. *ILP During Leukocyte-Endothelial Interactions* In Vitro

1. Defining ILP Structure and Dynamics

Interestingly, among of the earliest *in vitro* characterizations of podosomes was a study (performed by the same group credited with conducting the first detailed molecular characterization of podosomes (Tarone, Cirillo, Giancotti, Comoglio, & Marchisio, 1985)) that demonstrated leukocytes (natural killer cells) formed podosome-like adhesions and actin puncta when bound to endothelium (Allavena et al., 1991). These, authors predicted that such structures might function in endothelial penetration during diapedesis (Allavena et al.). However, until recently this idea was abandoned as most of the invadosome field focused instead on models that employ artificial substrates (Section IV.A).

Recent live-cell imaging studies of endothelial cells expressing fluorescent membrane markers, led to the unexpected observation of fluorescent ring-like structures (~500 nm in diameter) that formed dynamically (half-lives ~20 s) on the surface of endothelial cells as lymphocytes and monocytes migrated over and through them (Carman et al., 2007; Hidalgo & Frenette, 2007). Subsequent confocal imaging studies revealed that the "rings," in fact, represented invaginations in the surface of the endothelium (i.e., termed "podo-prints"), which were formed as a result of transient lymphocyte protrusions (Carman et al.). Studies using live-cell total internal reflection fluorescence (TIRF) microscopy demonstrate the formation of similarly dynamic lymphocyte-induced endothelial invaginations (Millan et al., 2006).

High-resolution immunofluorescence studies of these protrusions revealed features related to podosomes including an F-actin-rich core and peripheral enrichment in LFA-1, talin, and vinculin (Carman et al., 2007). The podosome-like nature of these structures was further supported by their demonstrated dependence on WASp and src family kinases (Carman et al.). Furthermore, TEM analysis revealed a continuum of protrusion depths ranging from ~100 to ~2000 nm suggesting both podosome- and invadosome-like features. Some of the more extended projections could be seen spanning nearly the entire endothelial-cell depth, placing the apical and basal membranes in close apposition (Carman et al.; Gerard, van der Kammen, Janssen, Ellenbroek, & Collard, 2009). Similar protrusions were also observed in TEM studies of neutrophil diapedesis (Cinamon, Shinder, Shamri, & Alon, 2004). Although the protrusive structures described here share many features of classical invadosomes, other properties are unique or remain undefined (e.g., metalloprotease secretion). In addition, the endothelium may be viewed as an *atypical* (albeit much more physiologic) substrate compared to the vast majority of studies characterizing invadosomes, which use plastic, glass, or model extracellular matrices as substrates. Thus, we refer to the structures formed on endothelium as ILPs to explicitly denote the current uncertainty in their precise relationship to the invadosomes that have been defined in other settings (Section IV.A).

2. A Migratory Pathfinding Model for ILPs

As discussed in Section II.B.1, following the accumulation of leukocytes on the luminal surface of the endothelium, they undergo integrin-dependent lateral migration, a critical step that seems to allow leukocytes to search out sites for endothelial barrier penetration (Phillipson et al., 2006; Schenkel et al., 2004). Lossinsky & Shivers (2004) have proposed that diapedesis occurs through the "path of least resistance," an idea that was recently supported by *in vivo* studies (Wang et al., 2006). However, precisely how leukocytes identify and exploit such paths remains incompletely understood. This has been particularly unclear for the trans-cellular mode of diapedesis. For para-cellular diapedesis, interendothelial cell junctions provide defined and intuitive sites for crossing the endothelium and mechanistic models for this process have been proposed (Muller, 2003). In contrast, for trans-cellular diapedesis neither an intuitive locus nor an obvious mechanism for trans-cellular pore formation exists.

The dynamic ILP behavior discussed in Section IV.B.1) provides a mechanism for leukocytes to both discover and exploit trans-cellular pathways for diapedesis. As a consequence of the dynamic extension of leukocyte ILPs, endothelial cell surface invaginations (i.e., podo-prints) are formed, which displace and distort the cytoplasm, cytoskeleton, and other organelles (Carman et al., 2007). Though shallow invadosomes and podo-prints are frequently observed to encounter the endothelial nucleus, trans-cellular pore formation never occurs at these sites (see Fig. 3), presumably because the underlying lamina provides too great a resistance for deeper penetration. This finding illustrates that leukocytes do not know *a priori* that they cannot diapedesis directly through the nucleus, but must discover a route for transmigration through trial and error. Even in nonnuclear areas, the vast majority of invadasomes inserted into the endothelium are shallow and short lived (i.e., retracted after ~10–30 s) and not coupled to trans-cellular pore formation (Carman et al.) (Figs. 3 and 4). This suggests that endothelial locations permissive for deeper protrusion (and trans-cellular diapedesis) are limited by the sum of local resistances provided by the plasma membrane and other intraendothelial organelles. The ILP-mediated *probing* or *palpation* during lateral migration provides leukocytes with a stochastic mechanism to efficiently identify locations of relatively low total endothelial resistance (it should be recognized that endothelial surface chemokines and adhesion molecules play critical roles in this process at minimum by driving lateral movement in general, but also potentially by providing gradients that encourage movement toward certain regions on the endothelium). Furthermore, the ability of ILPs to progressively extend (in invadopodia fashion) at such locations (Carman et al.; Gerard et al., 2009) indicates their ability to exploit the path of least (or sufficiently low) resistance in order to drive trans-cellular pore formation (Fig. 4). In deed, functional perturbation of ILPs (via WASp or src inactivation) strongly inhibits trans-cellular diapedesis (Carman et al.; Gerard et al.).

3. ILP as Mechanotransducers?

An implicit component of the model presented in previous section is that ILPs can somehow sense mechanical resistance and, in response, alter their signaling and dynamics. Although this has not been widely investigated, several recent studies demonstrate that podosomes can serve as "dynamic mechanosensors" (Alexander et al., 2008; Collin et al., 2006, 2008; Enderling et al., 2008). Using polyacrylamide collagen-coated substrates of defined rigidity, Collin et al. demonstrated that the lifespan and density of fibroblast podosomes depended on substrate flexibility and that such mechanosensory properties were mediated in part through myosin II motor proteins. Similar findings were made with tumor cell invadopodia on matrices of varied

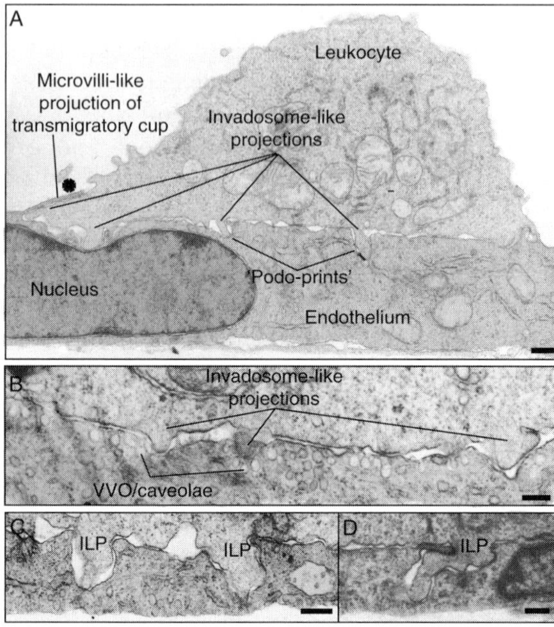

FIGURE 4 Ultrastructure of invadosome-like protrusions and transmigratory cup. Shown are transmission electron micrographs of primary effector lymphocytes (green) migrating on activated microvascular endothelial cells (pink) as described (Carman et al., 2007). (A) Formation of typical shallow ILPs on the endothelial-cell surface. Note one ILP (left) protruding over the endothelial nucleus and being "frustrated" by the relatively rigid nuclear lamina. Also, note the presence of a microvilli-like endothelial projection of a transmigratory cup that was captured in this cross-section (asterisk, left). (B) Several shallow ILP are protruding into the endothelium where local enrichment of plasma-membrane fusion of vesicles, VVOs, and/or caveolae are evident (yellow). (C, D) Extended ILP protruding deeply into the endothelium and in one case (C, ILP on right) placing the apical and basal membranes in close apposition as though poised for trans-cellular pore formation than those shown in (A). (D) An ILP that has nearly crossed an individual endothelial cell. Scale bars = 300 nm. (See Color Insert.)

densities (Alexander et al.; Enderling et al.). These observations support the plausibility of a model in which mechanical probing of the endothelial surface by ILPs transduces inside-out and outside-in signals that regulate ILP activity and ultimately influence the efficiency of trans-cellular diapedesis.

4. Endothelial Responses to ILP Probing

Though the leukocyte clearly plays a dominant role in the ILP probing/pathfinding model, studies suggest that responses somehow triggered in the endothelium may also contribute to trans-cellular pore formation. *In vitro* studies show enrichment of the caveolar marker caveolin-1, plasmalemma vesicle associated protein-1 (PV-1), various vesicles and vesiculo-vacuolar organelles (VVOs), and fusogenic proteins (namely, the SNAREs VAMP2 and VAMP3) in the endothelium at sites of ILP protrusion and trans-cellular pore formation (Carman et al., 2007; Migliorisi, Folkes, Pawlowski, & Cramer, 1987; Millan et al., 2006) (Figs. 3 and 4A). Similar observations have been made *in vivo* (Carman et al.; De Bruyn, Cho, & Michelson, 1989; Greenwood, Howes, & Lightman, 1994; Olah & Glick, 1985), leading one research group to speculate that "the fusing of the vesicles may form a gradual transendothelial channel in which the pseudopod of the lymphocyte penetrates and by this mechanism the lymphocyte may cross the endothelium" (Olah & Glick). Vesicle fusion may also facilitate local delivery of adhesion receptors (Mamdouh, Chen, Pierini, Maxfield, & Muller, 2003; Millan et al.) and chemokines (Middleton et al., 2002) that might modify ILP activity and the efficiency of trans-cellular migration. The functional significance of vesicle-fusion activity was suggested by studies in which siRNA-mediated knockdown of caveolin-1 and pharmacologic perturbation of the NSF/SNAP/SNARE membrane-fusogenic machinery in endothelium significantly reduced the efficiency of trans-cellular diapedesis (Carman et al.; Millan et al.). However, much more detailed characterization is still required to understand mechanisms by which ILPs "trigger" such responses, and of the consequences of this triggering activity.

C. *ILP Probing* In Vivo

Critically, the ILP probing model in its basic form has been previously suggested and supported by a large number of diverse *in vivo* observations (Table II). Indeed, the first description of a podosome (and use of the term "podosome") was made in the context of trans-cellular migration of lymphocytes and eosinophils across bone-marrow endothelium *in vivo* (Wolosewick, 1984). These authors predicted that the protrusive force supplied by such actin-enriched structures might drive trans-cellular pore formation. Moreover, although the structures visualized have been variously referred to as "microvillus-, filopodium-, and finger-like protrusions," "processes", "pseudopodia," and "probing pseudopods," an extensive accumulation of *in vivo* studies (most of which are associated specifically trans-cellular diapedesis events) have also demonstrated the endothelium-directed protrusion of structures with clearly invadosome-like morphology emanating from lymphocytes, monocytes, neutrophils, eosinophils, and acute myeloid leukemia tumor cells (Astrom, Webster, & Arnason, 1968; Azzali, Arcari, & Caldara, 2008; Bamforth, Lightman, & Greenwood, 1997;

TABLE II

In Vivo and *In Vitro* Settings for ILPs During Leukocyte–Endothelial Interactions

Setting[a]	Tissue[b]	Leukocyte[c]	Stimulus[d]	Method[e]	Reference
Intravasation from BM	BM	Ly, Gran, Ret	Untreated	Serial TEM[f]	De Bruyn et al. (1971)
	BM	Ly, Gran, Ret, Me	Untreated	SEM, TEM	Becker and De Bruyn (1976)
	BM	Ly, Eo	Untreated	TEM, IF	Wolosewick (1984)
Intravasation from thymus	Thymus	Ly	Untreated	SEM, TEM	Ushiki (1986)
Extravasation in SLO (HEV)	LN (c,m,p,pa)	Ly	Untreated	Serial TEM	Farr and De Bruyn (1975)
	LN	Ly	Untreated	TEM	Anderson and Anderson (1976)
	Tonsil	Ly	Untreated	TEM	Holibka (1991)
	PP	Ly	Irritant	Serial TEM	Azzali et al. (2008)
Lymphatics	Lymphatic sinusoid	Ly	Untreated	TEM	Olah and Glick (1985)
Peripheral inflammation	Mesentary	Neu, Eo, Mono	Mechanical trauma	TEM	Marchesi and Florey (1960))
	Pancreas	Leukocytes	Ischemia	TEM	Williamson and Grisham (1961)
	Li, Lu, Sp, Ki, He	Ly	IL-2	TEM	Fujita et al. (1991)
	Skin	Neu, Eo	fMLP	Serial TEM	Feng et al. (1998)
BBB	BBB	Ly	EAN	TEM	Astrom et al. (1968)
	BBB	Ly	Post-op	TEM	Matthews and Kruger (1973)
	BBB	Ly	Degeneration	TEM	Barron et al. (1974)
	BBB	Ne	α-Bungaro toxin	SEM, TEM	Faustmann and Dermietzel (1985)
	BBB	Leukocytes	EAE	SEM, TEM	Lossinsky et al. (1989)
	BBB	Ly	EAE	TEM	Raine et al. (1990)

	BBB	Leukocytes	EAE	SEM, TEM, HVEM	Lossinsky et al. (1991)
	BBB	Ly	EAE	Serial TEM	Wolburg et al. (2005)
BRB	BRB	Ly	EAU	Serial TEM	Greenwood et al. (1994)
	BRB	Mac, Gran, Ly	IL-1β	TEM	Bamforth et al. (1997)
In vitro	BBB	Ly	TNF-α	TEM	Wong, Prameya, and Dorovini-Zis (1999)
	HUVEC	Ne	TNF-α	TEM, LM	Cinamon et al. (2004)
	HDMVEC	Ly	TNF-α	IF	Millan et al. (2006)
	HUVEC	Ly	TNF-α	IF, TEM	Millan et al. (2006)
	Lymphatic	Ly	TNF-α	IF	Millan et al. (2006)
	HUVEC	Ly	TNF-α	IF	Carman et al. (2007)
	HDMVEC	Ly	TNF-α	IF, TEM	Carman et al. (2007)
	HLMVEC	Ly	TNF-α	IF, TEM	Carman et al. (2007)
	Lymphatic	Ly	TNF-α	IF	Carman et al. (2007)
	HDMVEC	Ly	TNF-α	IF	Shulman et al. (2009)
	HDMVEC	Ly	TNF-α	IF	Gerard et al. (2009)
	HUVEC	Ne	TNF-α	TEM	Furie, Naprstek, and Silverstein (1987)

[a]BBB, blood–brain barrier; BM, bone marrow; BRB, blood–retinal barrier; SLO, secondary lymphoid organs.

[b]a, axillary; ALPA, absorbing lymphatic peripheral apparatus; ALV, absorbing lymphatic vessel; c, cervical; p, popliteal; pa, para-aortic; m, mesenteric; si, small intestine; TAAL, tumor-associated absorbing lymphatic.

[c]Eo, Eosinophil; Gran, granulocyte; Ly, lymphocyte; Leukocyte, unspecified leukocyte; Mac, macrophage; Me, megakaryocyte; Mono, monocyte; Ne, Neutrophil; Ret, reticulocyte.

[d]C5a, complement component 5a; EAE, experimental autoimmune encephalomyelitis; EAN, experimental autoimmune neuritis; EAU, experimental autoimmune uveoretinitis; fMLP, formyl-met-leu-phe; IL, interleukin; LTB4, leukotriene B4; MIP-2, macrophage inflammatory protein 2; NAP, neutrophil activating peptide.

[e]TEM, transmission electron microscopy; SEM, scanning EM; IF, immunofluorescence microscopy.

Barron, Means, Feng, & Harris, 1974; Becker & De Bruyn, 1976; De Bruyn, Michelson, & Thomas, 1971; De Bruyn et al., 1989; Farr & De Bruyn, 1975; Faustmann & Dermietzel, 1985; Feng, Nagy, Pyne, Dvorak, & Dvorak, 1998; Fujita et al., 1991; Greenwood et al., 1994; Lossinsky & Shivers, 2004; Lossinsky, Badmajew, Robson, Moretz, & Wisniewski, 1989; Lossinsky et al., 1991; Marchesi & Florey, 1960; Matthews & Kruger, 1973; Olah & Glick, 1985; Raine et al., 1990; Williamson & Grisham, 1960; Wolburg et al., 2005; Wolosewick, 1984) (Table II). The actions of these protrusions have often been interpreted as "probing" the endothelium and potentially driving trans-cellular pore formation (Astrom et al.; Bamforth et al.; Farr & De Bruyn; Greenwood et al.; Lossinsky & Shivers; Lossinsky et al.; Marchesi & Florey; Olah & Glick; Raine et al.; Wolburg et al.; Wolosewick). Thus, the ability and tendency of leukocytes to extend ILPs into the endothelial surface during diapedesis seems to be broadly relevant.

D. Additional Functions for Leukocyte ILPs

1. ILP Pathfinding: Para-cellular Diapedesis?

Though the initial studies clearly demonstrate a dominant role for ILPs in trans-cellular diapedesis (Carman et al., 2007; Gerard et al., 2009), the possibility remains that ILPs may have a role in para-cellular migration, as well. Interendothelial junctions *in vitro* are known to be relatively weak compared to their *in vivo* counterparts. Thus, it seems plausible that ILPs could be functionally important *in vivo*, but dispensable in many *in vitro* setting. Interestingly, very recent studies have characterized lymphocyte structures termed "invasive filopodia" that are highly analogous to ILPs (in terms of dynamics, molecular composition, and morphology), that were similarly ascribed a role in probing for sites for diapedesis, though not explicitly by the trans-cellular route (Shulman et al., 2009).

2. ILP Pathfinding: Crossing the Basement Membrane?

In both intravasation and extravasation, the endothelial barrier is only one of the key obstacles to be overcome. The basement membrane also provides a formidable and often rate-limiting barrier. Precisely, how the basement membrane is crossed by leukocytes remains poorly defined. Some studies have suggested that proteolytic degradation of the basement membrane enables the passage of leukocytes. At least for metastatic tumor cells during their intravasation across lymphatic endothelial layers, this idea is well supported (Gimona et al., 2008). It seems conceivable that if proteases are enriched in leukocyte ILPs as they are in classical invadosomes (Chavrier, 2009), this could facilitate basement-membrane degradation and crossing.

However, the requirement for basement-membrane degradation during diapedesis remains somewhat controversial, and several nonproteolytic modes of basement-membrane transmigration have been hypothesized (Rowe & Weiss, 2008). In this vein, it is interesting to note that neutrophils extravasating *in vivo* have been observed to transmigrate preferentially at preexisting sites of relatively attenuated basement membrane (Wang et al., 2006). By analogy with their roles in probing for attenuated regions of endothelium, ILPs might act similarly in locating regions of basement membrane that are permissive for migration.

3. ILP Pathfinding: Negotiating Nonendothelial Cell Barriers

ILP probing may not be limited to trans-diapedesis across endothelium. The trans-cellular mode of cell migration might also be used to cross nonendothelial barriers. For example, it has been demonstrated *in vivo* that neutrophils can migrate trans-cellularly through the pericytes that underlie the vascular endothelium (Feng et al., 1998). In addition, extensive migration of leukocytes across epithelial cell layers (e.g., the mucosal epithelia of the intestine, airway, and urinary tract) occurs *in vivo* (Zen & Parkos, 2003). Routes and mechanisms for this migration remain either only partially understood or uninvestigated (Zen & Parkos). One recent study demonstrated that para-cellular diapedesis of neutrophils occurs in an *in vitro* model system (Porter, Falzon, & Hall, 2008). In an alternative (albeit less physiologic) system, we recently found that lymphocytes could readily form trans-cellular pores across epithelial cells (i.e., CHO-K1 cells expressing GFP-tagged ICAM1) by means of probing ILPs (Carman et al., 2007).

4. ILPs in Biochemical Sensing

In essence, the pathfinding model discussed earlier posits that ILPs are sensory organelles and probes, as well as invasive structures. The observations made thus far focus on a putative role for ILPs in sensing the biomechanical properties on cellular, and possibly matrix substrates (i.e., a mechanotransduction role). It is interesting, however, to consider whether such a "probing" function might also include biochemical sensing of the local environment.

The surface of all cells is modified by a gel-like polysaccharide coating that is termed the glycocalyx, which is made up of diverse proteoglycans and glycoproteins (Reitsma, Slaaf, Vink, van Zandvoort, & oude Egbrink, 2007). On a typical cell the glycocalyx is at least 45 nm thick, but it can be substantially thicker (up to ~500 nm) in some cases, such as on the lumen-facing surface of endothelial cells (Reitsma et al.; Weinbaum, Tarbell, & Damiano, 2007). The glycocalyx can serve as a reservoir of noncovalently immobilized chemokines and other chemoattractants (Middleton et al., 2002;

Proudfoot et al., 2003). In addition, the glycocalyx provides a formidable energy barrier (by means of both steric and electrostatic repulsion) to close membrane–membrane encounter between cells. Thus, relatively small cell-surface adhesion and signaling molecules would seem to be effectively shielded (Bell, 1978; Bell, Dembo, & Bongrand, 1984; Reitsma et al.; Springer, 1990; Weinbaum et al.). Protrusive forces provided by ILPs appear to provide sufficient energy to overcome this barrier, driving close membrane–membrane apposition and thereby promoting molecular interactions that might otherwise be highly inefficient or impossible. Indeed, our recent studies suggest that recognition of endothelial MHC-antigen complexes by T cell receptors on lymphocytes seem to be dependent on ILPs (C. V.Carman, unpublished observations). Thus, ILPs might facilitate a kind of informational scanning of local membrane surfaces, literally allowing cells to get a "deeper" understanding of their local environment.

V. SUMMARY AND PERSPECTIVE

The recent emergence of transmigratory cups and ILP provide new insights to the process of leukocyte diapedesis. In particular, these findings underscore the highly dynamic and intimate "conversation" that takes place between leukocytes and endothelial cells in order to orchestrate transmigration and information exchange at this critical locus. Studies described in Section III.E.5 demonstrate that aberrant regulation transmigratory cup activity can play direct roles in pathologic processes. Moreover, lymphocytes from patients with WAS, an immunodeficiency associated with profound leukocyte trafficking defects, have been shown to exhibit correlated perturbations in ILP activity and trans-cellular diapedesis (Sections IV.B.1 and II.B.2). It may be anticipated that aberrant upregulation of ILP activity may conversely drive excessive/inappropriate leukocyte trafficking (akin to invadopodia-conferred metastatic properties in tumors cells) in certain inflammatory or autoimmune disease setting. Thus, the detailed cellular and molecular investigation of these processes may well yield therapeutic targets for immune-related pathologies. It is clear that while the current results of such inquiry, described herein, have offered some basic direction, a great deal more exists to be learned about these processes. Given, that the vast majority of what has been learned so far, derives from *in vitro* models, it will be critically important to continue to invest in and develop new approaches for characterizing the cell biological details of leukocyte–endothelial interactions *in vivo*.

Acknowledgement

Electron micrographs (Fig. 4) were provided through collaboration with Tracey Sciuto and Ann Dvorak. CVC was supported by a grant from the Arthritis Foundation.

References

Adamson, P., Etienne, S., Couraud, P. O., Calder, V., & Greenwood, J. (1999). Lymphocyte migration through brain endothelial cell monolayers involves signaling through endothelial ICAM-1 via a rho-dependent pathway. *Journal of Immunology, 162*(5), 2964–2973.

Aird, W. C. (2007a). Phenotypic heterogeneity of the endothelium: II Representative vascular beds. *Circulation Research, 100*(2), 174–190.

Aird, W. C. (2007b). Phenotypic heterogeneity of the endothelium: I. Structure, function, and mechanisms. *Circulation Research, 100*(2), 158–173.

Alexander, N. R., Branch, K. M., Parekh, A., Clark, E. S., Iwueke, I. C., Guelcher, S. A., et al. (2008). Extracellular matrix rigidity promotes invadopodia activity. *Current Biology, 18*(17), 1295–1299.

Allavena, P., Paganin, C., Martin-Padura, I., Peri, G., Gaboli, M., Dejana, E., et al. (1991). Molecules and structures involved in the adhesion of natural killer cells to vascular endothelium. *The Journal of Experimental Medicine, 173*(2), 439–448.

Anderson, A. O., & Anderson, N. D. (1976). Lymphocyte emigration from high endothelial venules in rat lymph nodes. *Immunology, 31*(5), 731–748.

Astrom, K. E., Webster, H. D., & Arnason, B. G. (1968). The initial lesion in experimental allergic neuritis. A phase and electron microscopic study. *The Journal of Experimental Medicine, 128*(3), 469–495.

Azzali, G., Arcari, M. L., & Caldara, G. F. (2008). The "mode" of lymphocyte extravasation through HEV of Peyer's patches and its role in normal homing and inflammation. *Microvascular Research, 75*(2), 227–237.

Baluk, P., Fuxe, J., Hashizume, H., Romano, T., Lashnits, E., Butz, S., et al. (2007). Functionally specialized junctions between endothelial cells of lymphatic vessels. *The Journal of Experimental Medicine, 204*(10), 2349–2362.

Bamforth, S. D., Lightman, S. L., & Greenwood, J. (1997). Ultrastructural analysis of interleukin-1 beta-induced leukocyte recruitment to the rat retina. *Investigative Ophthalmology and Visual Science, 38*(1), 25–35.

Barreiro, O., Yanez-Mo, M., Serrador, J. M., Montoya, M. C., Vicente-Manzanares, M., Tejedor, R., et al. (2002). Dynamic interaction of VCAM-1 and ICAM-1 with moesin and ezrin in a novel endothelial docking structure for adherent leukocytes. *The Journal of Cell Biology, 157*(7), 1233–1245.

Barreiro, O., Yanez-Mo, M., Sala-Valdes, M., Gutierrez-Lopez, M. D., Ovalle, S., Higginbottom, A., et al. (2005). Endothelial tetraspanin microdomains regulate leukocyte firm adhesion during extravasation. *Blood, 105*(7), 2852–2861.

Barreiro, O., Zamai, M., Yanez-Mo, M., Tejera, E., Lopez-Romero, P., Monk, P. N., et al. (2008). Endothelial adhesion receptors are recruited to adherent leukocytes by inclusion in preformed tetraspanin nanoplatforms. *The Journal of Cell Biology, 183*(3), 527–542.

Barron, K. D., Means, E. D., Feng, T., & Harris, H. (1974). Ultrastructure of retrograde degeneration in thalamus of rat. 2. Changes in vascular elements and transvascular migration of leukocytes. *Experimental and Molecular Pathology, 20*(3), 344–362.

Bazzoni, G., & Dejana, E. (2004). Endothelial cell-to-cell junctions: Molecular organization and role in vascular homeostasis. *Physiological Reviews, 84*(3), 869–901.

Becker, R. P., & De Bruyn, P. P. (1976). The transmural passage of blood cells into myeloid sinusoids and the entry of platelets into the sinusoidal circulation; a scanning electron microscopic investigation. *The American Journal of Anatomy, 145*(2), 183–205.

Bell, G. I. (1978). Models for the specific adhesion of cells to cells. *Science, 200*(4342), 618–627.

Bell, G. I., Dembo, M., & Bongrand, P. (1984). Cell adhesion. Competition between nonspecific repulsion and specific bonding. *Biophysical Journal, 45*(6), 1051–1064.

Bretscher, A., Reczek, D., & Berryman, M. (1997). Ezrin: A protein requiring conformational activation to link microfilaments to the plasma membrane in the assembly of cell surface structures. *J. Cell Sci., 110*(Pt. 24), 3011–3018.

Buccione, R., Orth, J. D., & McNiven, M. A. (2004). Foot and mouth: Podosomes, invadopodia and circular dorsal ruffles. *Nature Reviews. Molecular Cell Biology, 5*, 647–657.

Burns, A. R., Smith, C. W., & Walker, D. C. (2003). Unique structural features that influence neutrophil emigration into the lung. *Physiological Reviews, 83*(2), 309–336.

Butcher, E. C. (1991). Leukocyte-endothelial cell recognition: Three (or more) steps to specificity and diversity. *Cell, 67*, 1033–1036.

Calle, Y., Chou, H. C., Thrasher, A. J., & Jones, G. E. (2004). Wiskott-Aldrich syndrome protein and the cytoskeletal dynamics of dendritic cells. *Journal of Pathology, 204*, 460–469.

Carman, C. V., & Springer, T. A. (2003). Integrin avidity regulation: Are changes in affinity and conformation underemphasized? *Current Opinion in Cell Biology, 15*(5), 547–556.

Carman, C. V., & Springer, T. A. (2004). A transmigratory cup in leukocyte diapedesis both through individual vascular endothelial cells and between them. *The Journal of Cell Biology, 167*(2), 377–388.

Carman, C. V., & Springer, T. A. (2008). Trans-cellular migration: Cell–cell contacts get intimate. *Current Opinion in Cell Biology, 20*(5), 533–540.

Carman, C. V., Jun, C. D., Salas, A., & Springer, T. A. (2003). Endothelial cells proactively form microvilli-like membrane projections upon intercellular adhesion molecule 1 engagement of leukocyte LFA-1. *Journal of Immunology, 171*(11), 6135–6144.

Carman, C. V., Sage, P. T., Sciuto, T. E., de la Fuente, M. A., Geha, R. S., Ochs, H. D., et al. (2007). Transcellular diapedesis is initiated by invasive podosomes. *Immunity, 26*(6), 784–797.

Cayrol, R., Wosik, K., Berard, J. L., Dodelet-Devillers, A., Ifergan, I., Kebir, H., et al. (2008). Activated leukocyte cell adhesion molecule promotes leukocyte trafficking into the central nervous system. *Nature Immunology, 9*(2), 137–145.

Chavrier, P. (2009). ECM degradation and invasion. *Journal of Cell Science, 122*, 3015–3024.

Choi, J., Enis, D. R., Koh, K. P., Shiao, S. L., & Pober, J. S. (2004). T lymphocyte-endothelial cell interactions. *Annual Review of Immunology, 22*, 683–709.

Cinamon, G., Shinder, V., Shamri, R., & Alon, R. (2004). Chemoattractant signals and beta 2 integrin occupancy at apical endothelial contacts combine with shear stress signals to promote transendothelial neutrophil migration. *Journal of Immunology, 173*(12), 7282–7291.

Collin, O., Tracqui, P., Stephanou, A., Usson, Y., Clement-Lacroix, J., & Planus, E. (2006). Spatiotemporal dynamics of actin-rich adhesion microdomains: Influence of substrate flexibility. *Journal of Cell Science, 119*(Pt. 9), 1914–1925.

Collin, O., Na, S., Chowdhury, F., Hong, M., Shin, M. E., Wang, F., et al. (2008). Self-organized podosomes are dynamic mechanosensors. *Current Biology, 18*(17), 1288–1294.

De Bruyn, P. P., Michelson, S., & Thomas, T. B. (1971). The migration of blood cells of the bone marrow through the sinusoidal wall. *Journal of Morphology, 133*(4), 417–437.

De Bruyn, P. P., Cho, Y., & Michelson, S. (1989). Endothelial attachment and plasmalemmal apposition in the transcellular movement of intravascular leukemic cells entering the myeloid parenchyma. *The American Journal of Anatomy, 186*(2), 115–126.

Dittmar, S., Harms, H., Runkler, N., Maisner, A., Kim, K. S., & Schneider-Schaulies, J. (2008). Measles virus-induced block of transendothelial migration of T lymphocytes and infection-mediated virus spread across endothelial cell barriers. *Journal of Virology, 82*(22), 11273–11282.

Doulet, N., Donnadieu, E., Laran-Chich, M. P., Niedergang, F., Nassif, X., Couraud, P. O., et al. (2006). *Neisseria meningitidis* infection of human endothelial cells interferes with leukocyte transmigration by preventing the formation of endothelial docking structures. *Journal of Cell Biology, 173*(4), 627–637.

Durieu-Trautmann, O., Chaverot, N., Cazaubon, S., Strosberg, A. D., & Couraud, P. O. (1994). Intercellular adhesion molecule 1 activation induces tyrosine phosphorylation of the cytoskeleton-associated protein cortactin in brain microvessel endothelial cells. *Journal of Biological Chemistry, 269*(17), 12536–12540.

Dzamba, B. J., Bultmann, H., Akiyama, S. K., & Peters, D. M. (1994). Substrate-specific binding of the amino terminus of fibronectin to an integrin complex in focal adhesions. *Journal of Biological Chemistry, 269*, 19646–19652.

Enderling, H., Alexander, N. R., Clark, E. S., Branch, K. M., Estrada, L., Crooke, C., et al. (2008). Dependence of invadopodia function on collagen fiber spacing and cross-linking: Computational modeling and experimental evidence. *Biophysical Journal, 95*(5), 2203–2218.

Etienne, S., Adamson, P., Greenwood, J., Strosberg, A. D., Cazaubon, S., & Couraud, P. O. (1998). ICAM-1 signaling pathways associated with Rho activation in microvascular brain endothelial cells. *Journal of Immunology, 161*(10), 5755–5761.

Etienne-Manneville, S., Manneville, J. B., Adamson, P., Wilbourn, B., Greenwood, J., & Couraud, P. O. (2000). ICAM-1-coupled cytoskeletal rearrangements and transendothelial lymphocyte migration involve intracellular calcium signaling in brain endothelial cell lines. *Journal of Immunology, 165*(6), 3375–3383.

Eugene, E., Hoffmann, I., Pujol, C., Couraud, P. O., Bourdoulous, S., & Nassif, X. (2002). Microvilli-like structures are associated with the internalization of virulent capsulated *Neisseria meningitidis* into vascular endothelial cells. *Journal of Cell Science, 115*(6), 1231–1241.

Evans, J. G., Correia, I., Krasavina, O., Watson, N., & Matsudaira, P. (2003). Macrophage podosomes assemble at the leading lamella by growth and fragmentation. *Journal of Cell Biology, 161*, 697–705.

Farr, A. G., & De Bruyn, P. P. (1975). The mode of lymphocyte migration through postcapillary venule endothelium in lymph node. *American Journal of Anatomy, 143*(1), 59–92.

Faustmann, P. M., & Dermietzel, R. (1985). Extravasation of polymorphonuclear leukocytes from the cerebral microvasculature. Inflammatory response induced by alpha-bungarotoxin. *Cell Tissue Research, 242*(2), 399–407.

Feng, D., Nagy, J. A., Pyne, K., Dvorak, H. F., & Dvorak, A. M. (1998). Neutrophils emigrate from venules by a transendothelial cell pathway in response to FMLP. *Journal of Experimental Medicine, 187*(6), 903–915.

Fujita, S., Puri, R. K., Yu, Z. X., Travis, W. D., & Ferrans, V. J. (1991). An ultrastructural study of *in vivo* interactions between lymphocytes and endothelial cells in the pathogenesis of the vascular leak syndrome induced by interleukin-2. *Cancer, 68*(10), 2169–2174.

Furie, M. B., Naprstek, B. L., & Silverstein, S. C. (1987). Migration of neutrophils across monolayers of cultured microvascular endothelial cells. An *in vitro* model of leucocyte extravasation. *Journal of Cell Science, 88*(Pt. 2), 161–175.

Gerard, A., van der Kammen, R. A., Janssen, H., Ellenbroek, S. I., & Collard, J. G. (2009). The Rac activator Tiam1 controls efficient T-cell trafficking and route of transendothelial migration. *Blood, 113*(24), 6138–6147.

Gimona, M., Buccione, R., Courtneidge, S. A., & Linder, S. (2008). Assembly and biological role of podosomes and invadopodia. *Current Opinion in Cell Biology, 20*(2), 235–241.

Greenwood, J., Howes, R., & Lightman, S. (1994). The blood-retinal barrier in experimental autoimmune uveoretinitis. Leukocyte interactions and functional damage. *Laboratory Investigation, 70*(1), 39–52.

Hemler, M. E. (2005). Tetraspanin functions and associated microdomains. *Nature Reviews. Molecular Cell Biology, 6*(10), 801–811.
Hidalgo, A., & Frenette, P. S. (2007). Leukocyte podosomes sense their way through the endothelium. *Immunity, 26*(6), 753–755.
Holibka, V. (1991). Lymphocyte migration through the wall of postcapillary venules in the fetal development of human palatine tonsil. *Acta Universitatis Palackianae Olomucensis Facultatis Medicae, 130*, 71–86.
Huang, A. J., Manning, J. E., Bandak, T. M., Ratau, M. C., Hanser, K. R., & Silverstein, S. C. (1993). Endothelial cell cytosolic free calcium regulates neutrophil migration across monolayters of endothelial cells. *Journal of Cell Biology, 120*, 1371–1380.
Huttenlocher, A., & Poznansky, M. C. (2008). Reverse leukocyte migration can be attractive or repulsive. *Trends in Cell Biology, 18*(6), 298–306.
Kanters, E., van Rijssel, J., Hensbergen, P. J., Hondius, D., Mul, F. P., Deelder, A. M., et al. (2008). Filamin B mediates ICAM-1-driven leukocyte transendothelial migration. *Journal of Biological Chemistry, 283*(46), 31830–31839.
Ley, K., Laudanna, C., Cybulsky, M. I., & Nourshargh, S. (2007). Getting to the site of inflammation: The leukocyte adhesion cascade updated. *Nature Reviews of Immunology, 7*(9), 678–689.
Linder, S. (2009). Podosomes and invadopodia at a glance. *Journal of Cell Science, 122*, 3009–30013.
Linder, S., & Aepfelbacher, M. (2003). Podosomes: Adhesion hot-spots of invasive cells. *Trends in Cell Biology, 13*, 376–385.
Linder, S., Nelson, D., Weiss, M., & Aepfelbacher, M. (1999). Wiskott-Aldrich syndrome protein regulates podosomes in primary human macrophages. *Proceedings of the National Academy of Sciences of the United States of America, 96*(17), 9648–9653.
Lossinsky, A. S., & Shivers, R. R. (2004). Structural pathways for macromolecular and cellular transport across the blood-brain barrier during inflammatory conditions. *Review of Histology and Histopathology, 19*(2), 535–564.
Lossinsky, A. S., Badmajew, V., Robson, J. A., Moretz, R. C., & Wisniewski, H. M. (1989). Sites of egress of inflammatory cells and horseradish peroxidase transport across the blood-brain barrier in a murine model of chronic relapsing experimental allergic encephalomyelitis. *Acta Neuropathologica, 78*(4), 359–371.
Lossinsky, A. S., Pluta, R., Song, M. J., Badmajew, V., Moretz, R. C., & Wisniewski, H. M. (1991). Mechanisms of inflammatory cell attachment in chronic relapsing experimental allergic encephalomyelitis: A scanning and high-voltage electron microscopic study of the injured mouse blood-brain barrier. *Microvascular Research, 41*(3), 299–310.
Luo, B. H., Carman, C. V., & Springer, T. A. (2007). Structural basis of integrin regulation and signaling. *Annual Review of Immunology, 25*, 619–647.
Luscinskas, F. W., Kansas, G. S., Ding, H., Pizcueta, P., Schleiffenbaum, B. E., Tedder, T. F., et al. (1994). Monocyte rolling, arrest and spreading on IL-4-activated vascular endothelium under flow is mediated via sequential action of L-selectin, beta 1-integrins, and beta 2-integrins. *Journal of Cell Biology, 125*(6), 1417–1427.
Luscinskas, F. W., Ma, S., Nusrat, A., Parkos, C. A., & Shaw, S. K. (2002). Leukocyte transendothelial migration: A junctional affair. *Seminars in Immunology, 14*(2), 105–113.
Mamdouh, Z., Chen, X., Pierini, L. M., Maxfield, F. R., & Muller, W. A. (2003). Targeted recycling of PECAM from endothelial surface-connected compartments during diapedesis. *Nature, 421*(6924), 748–753.
Marchesi, V. T., & Florey, H. W. (1960). Electron micrographic observations on the emigration of leucocytes. *Quarterly Journal of Experimental Physiology and Cognate Medical Sciences, 45*, 343–348.

Marchesi, V. T., & Gowans, J. L. (1964). The migration of lymphocytes through the endothelium of venules in lymph nodes: An electron microscope study. *Proceedings of the Royal Society of London. Series B Biological Sciences, 159*, 283–290.

Matthews, M. A., & Kruger, L. (1973). Electron microscopy of non-neuronal cellular changes accompanying neural degeneration in thalamic nuclei of the rabbit. I. Reactive hematogenous and perivascular elements within the basal lamina. *Journal of Comparative Neurology, 148*(3), 285–312.

Middleton, J., Patterson, A. M., Gardner, L., Schmutz, C., & Ashton, B. A. (2002). Leukocyte extravasation: Chemokine transport and presentation by the endothelium. *Blood, 100*(12), 3853–3860.

Migliorisi, G., Folkes, E., Pawlowski, N., & Cramer, E. B. (1987). *In vitro* studies of human monocyte migration across endothelium in response to leukotriene B4 and f-Met-Leu-Phe. *American Journal of Pathology, 127*(1), 157–167.

Millan, J., Hewlett, L., Glyn, M., Toomre, D., Clark, P., & Ridley, A. J. (2006). Lymphocyte transcellular migration occurs through recruitment of endothelial ICAM-1 to caveola- and F-actin-rich domains. *Nature Cell Biology, 8*(2), 113–123.

Muller, W. A. (2001). Migration of leukocytes across endothelial junctions: Some concepts and controversies. *Microcirculation, 8*(3), 181–193.

Muller, W. A. (2003). Leukocyte-endothelial-cell interactions in leukocyte transmigration and the inflammatory response. *Trends in Immunology, 24*(6), 327–334.

Nassif, X., Bourdoulous, S., Eugene, E., & Couraud, P. O. (2002). How do extracellular pathogens cross the blood-brain barrier? *Trends in Microbiology, 10*(5), 227–232.

Nieminen, M., Henttinen, T., Merinen, M., Marttila-Ichihara, F., Eriksson, J. E., & Jalkanen, S. (2006). Vimentin function in lymphocyte adhesion and transcellular migration. *Nature Cell Biology, 8*(2), 156–162.

Nourshargh, S., & Marelli-Berg, F. M. (2005). Transmigration through venular walls: A key regulator of leukocyte phenotype and function. *Trends in Immunology, 26*(3), 157–165.

Olah, I., & Glick, B. (1985). Lymphocyte migration through the lymphatic sinuses of the chicken's lymph node. *Poultry Science, 64*(1), 159–168.

Pepper, M. S., & Skobe, M. (2003). Lymphatic endothelium: Morphological, molecular and functional properties. *Journal of Cell Biology, 163*(2), 209–213.

Pfau, S., Leitenberg, D., Rinder, H., Smith, B. R., Pardi, R., & Bender, J. R. (1995). Lymphocyte adhesion-dependent calcium signaling in human endothelial cells. *Journal of Cell Biology, 128*, 969–978.

Phillipson, M., Heit, B., Colarusso, P., Liu, L., Ballantyne, C. M., & Kubes, P. (2006). Intraluminal crawling of neutrophils to emigration sites: A molecularly distinct process from adhesion in the recruitment cascade. *Journal of Experimental Medicine, 203*(12), 2569–2575.

Phillipson, M., Kaur, J., Colarusso, P., Ballantyne, C. M., & Kubes, P. (2008). Endothelial domes encapsulate adherent neutrophils and minimize increases in vascular permeability in paracellular and transcellular emigration. *PLoS ONE, 3*(2), e1649.

Porter, J. C., Falzon, M., & Hall, A. (2008). Polarized localization of epithelial CXCL11 in chronic obstructive pulmonary disease and mechanisms of T cell egression. *Journal of Immunology, 180*(3), 1866–1877.

Proudfoot, A. E., Handel, T. M., Johnson, Z., Lau, E. K., LiWang, P., Clark-Lewis, I., et al. (2003). Glycosaminoglycan binding and oligomerization are essential for the *in vivo* activity of certain chemokines. *Proceedings of the National Academy of Sciences of the United States of America, 100*(4), 1885–1890.

Raine, C. S., Cannella, B., Duijvestijn, A. M., & Cross, A. H. (1990). Homing to central nervous system vasculature by antigen-specific lymphocytes. II. Lymphocyte/endothelial cell adhesion during the initial stages of autoimmune demyelination. *Laboratory Investigation, 63*(4), 476–489.

Reitsma, S., Slaaf, D. W., Vink, H., van Zandvoort, M. A., & oude Egbrink, M. G. (2007). The endothelial glycocalyx: Composition, functions, and visualization. *Pflugers Archiv, 454*(3), 345–359.

Ridley, A. J., Schwartz, M. A., Burridge, K., Firtel, R. A., Ginsberg, M. H., Borisy, G., et al. (2003). Cell migration: Integrating signals from front to back. *Science, 302*, 1704–1709.

Riethmuller, C., Nasdala, I., & Vestweber, D. (2008). Nano-surgery at the leukocyte-endothelial docking site. *Pflugers Archiv, 456*(1), 71–81.

Rohlena, J., Volger, O. L., van Buul, J. D., Hekking, L. H., van Gils, J. M., Bonta, P. I., et al. (2009). Endothelial CD81 is a marker of early human atherosclerotic plaques and facilitates monocyte adhesion. *Cardiovascular Research, 81*(1), 187–196.

Rowe, R. G., & Weiss, S. J. (2008). Breaching the basement membrane: Who, when and how? *Trends in Cell Biology, 18*, 560–574.

Sage, P. T., & Carman, C. V. (2009). Settings and mechanisms for trans-cellular diapedesis. *Frontiers in Bioscience, 14*, 5066–5083.

Schenkel, A. R., Mamdouh, Z., & Muller, W. A. (2004). Locomotion of monocytes on endothelium is a critical step during extravasation. *Nature Immunology, 5*(4), 393–400.

Shulman, Z., Shinder, V., Klein, E., Grabovsky, V., Yeger, O., Geron, E., et al. (2009). Lymphocyte crawling and transendothelial migration require chemokine triggering of high-affinity LFA-1 integrin. *Immunity, 30*(3), 384–396.

Springer, T. A. (1990). Adhesion receptors of the immune system. *Nature, 346*(6283), 425–434.

Springer, T. A. (1994). Traffic signals for lymphocyte recirculation and leukocyte emigration: The multistep paradigm. *Cell, 76*(2), 301–314.

Strey, A., Janning, A., Barth, H., & Gerke, V. (2002). Endothelial Rho signaling is required for monocyte transendothelial migration. *FEBS Letters, 517*(1–3), 261–266.

Su, W. H., Chen, H. I., Huang, J. P., & Jen, C. J. (2000). Endothelial [Ca(2+)](i) signaling during transmigration of polymorphonuclear leukocytes. *Blood, 96*(12), 3816–3822.

Tarone, G., Cirillo, D., Giancotti, F. G., Comoglio, P. M., & Marchisio, P. C. (1985). Rous sarcoma virus-transformed fibroblasts adhere primarily at discrete protrusions of the ventral membrane called podosomes. *Experimental Cell Research, 159*(1), 141–157.

Thompson, P. W., Randi, A. M., & Ridley, A. J. (2002). Intercellular adhesion molecule (ICAM)-1, but not ICAM-2, activates RhoA and stimulates c-fos and rhoA transcription in endothelial cells. *Journal of Immunology, 169*(2), 1007–1013.

Ushiki, T. (1986). A scanning electron-microscopic study of the rat thymus with special reference to cell types and migration of lymphocytes into the general circulation. *Cell Tissue Research, 244*(2), 285–298.

van Buul, J. D., Allingham, M. J., Samson, T., Meller, J., Boulter, E., Garcia-Mata, R., et al. (2007). RhoG regulates endothelial apical cup assembly downstream from ICAM1 engagement and is involved in leukocyte trans-endothelial migration. *Journal of Cell Biology, 178*(7), 1279–1293.

von Andrian, U. H., & Mackay, C. R. (2000). T-cell function and migration. Two sides of the same coin. *New England Journal of Medicine, 343*(14), 1020–1034.

Wang, S., Voisin, M. B., Larbi, K. Y., Dangerfield, J., Scheiermann, C., Tran, M., et al. (2006). Venular basement membranes contain specific matrix protein low expression regions that act as exit points for emigrating neutrophils. *Journal of Experimental Medicine, 203*(6), 1519–1532.

Webb, B. A., Eves, R., & Mak, A. S. (2006). Cortactin regulates podosome formation: Roles of the protein interaction domains. *Experimental Cell Research, 312*(6), 760–769.

Weinbaum, S., Tarbell, J. M., & Damiano, E. R. (2007). The structure and function of the endothelial glycocalyx layer. *Annual Review of Biomedical Engineering, 9*, 121–167.

Williamson, J. R., & Grisham, J. W. (1960). Leucocytic emigration from inflamed capillaries. *Nature, 188*, 1203.

Williamson, J. R., & Grisham, J. W. (1961). Electron microscopy of leukocytic margination and emigration in acute inflammation in dog pancreas. *Americal Journal of Pathology, 39*, 239–256.

Wojciak-Stothard, B., Williams, L., & Ridley, A. J. (1999). Monocyte adhesion and spreading on human endothelial cells is dependent on Rho-regulated receptor clustering. *Journal of Cell Biology, 145*(6), 1293–1307.

Wolburg, H., Wolburg-Buchholz, K., & Engelhardt, B. (2005). Diapedesis of mononuclear cells across cerebral venules during experimental autoimmune encephalomyelitis leaves tight junctions intact. *Acta Neuropathologica, 109*(2), 181–190.

Wolosewick, J. J. (1984). Distribution of actin in migrating leukocytes *in vivo*. *Cell Tissue Research, 236*(3), 517–525.

Wong, D., Prameya, R., & Dorovini-Zis, K. (1999). *In vitro* adhesion and migration of T lymphocytes across monolayers of human brain microvessel endothelial cells: regulation by ICAM-1, VCAM-1, E-selectin and PECAM-1. *Journal of Neuropathology and Experimental Neurology, 58*(2), 138–152.

Yamaguchi, H., Wyckoff, J., & Condeelis, J. (2005). Cell migration in tumors. *Current Opinion in Cell Biology, 17*(5), 559–564.

Zen, K., & Parkos, C. A. (2003). Leukocyte-epithelial interactions. *Current Opinion in Cell Biology, 15*(5), 557–564.

Zimmerman, A. W., Joosten, B., Torensma, R., Parnes, J. R., van Leeuwen, F. N., & Figdor, C. G. (2006). Long-term engagement of CD6 and ALCAM is essential for T-cell proliferation induced by dendritic cells. *Blood, 107*(8), 3212–3220.

CHAPTER 11

How Endothelial Cells Regulate Transendothelial Migration of Leukocytes: Molecules and Mechanisms

William A. Muller

Department of Pathology, Northwestern University Feinberg School of Medicine, Chicago, Illinois 60611, USA

I. OVERVIEW

A great deal of progress has been made recently in understanding the molecules and mechanisms that regulate transendothelial migration (TEM) of leukocytes, or diapedesis, a critical step in the inflammatory response. This chapter focuses on the active role of the endothelial cell in this process as it occurs at endothelial cell borders. It discusses some of the many molecules that have been reported to play a role in transendothelial migration and asks

Current Topics in Membranes, Volume 64 1063-5823/09 $35.00

DOI: 10.1016/S1063-5823(09)64011-4

why so many molecules seem to be involved. The concept is emerging that diapedesis itself can be dissected into sequential molecularly dissectible steps controlled by specific molecule(s) at the endothelial cell border. Several mechanisms have been shown to play a critical role transendothelial migration including signals derived from clustering of apically disposed intercellular adhesion molecule-1 (ICAM-1) and vascular cell adhesion molecule-1 (VCAM-1), disruption or loosening of adherens junctions, and targeted recycling of platelet/endothelial cell adhesion molecule (PECAM) and other molecules from the recently described lateral border recycling compartment (LBRC). A hypothesis that integrates the various known mechanisms is proposed.

II. INTRODUCTION

The inflammatory response is the body's stereotyped reaction to tissue damage of any kind. It involves rapidly and transiently delivering preformed soluble elements in the blood to the site of injury followed by a more prolonged delivery of leukocytes. Since leukocytes cannot swim, they are recruited locally at the site of inflammation in a series of adhesive steps that allow them to attach to the vessel wall, locomote along the wall to the endothelial borders, traverse the endothelium and the subendothelial basement membrane and migrate through the interstitial tissue (Ley, Laudanna, Cybulsky, & Nourshargh, 2007; Muller, 2003). Transendothelial migration or diapedesis is arguably the point-of-no-return in the inflammatory response. The preceding steps of leukocyte rolling, activation, adhesion, and locomotion are all reversible, and most leukocytes that attach to the post-capillary venule at the site of inflammation re-enter the circulation. However, once the leukocyte commits to diapedesis, it does not go back—at least not as the same cell type (Muller, 2007). The regulation of the leukocyte recruitment steps of capture, rolling, activation, and adhesion have been well studied and reviewed. While less is known about diapedesis, there has been a significant increase in our knowledge of the molecules and mechanisms controlling transendothelial migration relatively recently. Most TEM takes place at endothelial borders. Recently, there has been a flurry of interest in TEM through the endothelial cell body (transcellular migration). While this clearly can occur *in vitro* (Carman & Springer, 2008) and *in vivo* (Feng, Nagy, Pyne, Dvorak, & Dvorak, 1998), this chapter focuses on transendothelial migration at cell borders. I shall focus on some general types of molecules and mechanisms that have been implicated in TEM and try to show how these observations may be related.

III. ENDOTHELIAL MOLECULES REGULATING TRANSMIGRATION

A number of endothelial cell molecules have been implicated in transmigration due to the fact that genetic deletion or antibody blockade of these molecules impairs diapedesis. These molecules all are enriched at or restricted to the endothelial cell borders. In addition to adhesive functions, these molecules have signaling functions that contribute to their role in TEM. ICAM-1 and VCAM-1 will be discussed in this section, as well. Although they are not involved in the diapedesis step *per se*, they seem to be involved in events directly preceding diapedesis and are recruited to the endothelial cell border during transmigration. The interactions are summarized in Table I.

A. Luminal Surface Molecules

ICAM-1 is involved in firm adhesion of leukocytes to the apical surface of endothelial cells through interactions with leukocyte CD11a/CD18 and/or CD11b/CD18. Dimers of ICAM-1 on the endothelial surface (i.e., in *cis*) are the preferential ligands for CD11/18 (Miller et al., 1995; Reilly et al., 1995). Once adherent, ICAM-1 becomes enriched under the leukocyte as it migrates to the endothelial cell border and continues to surround it during transmigration (Shaw et al., 2004). The actin cytoskeleton is involved in this process;

TABLE I

Endothelial Cell Molecules Participating in Leukocyte Transmigration (in order of appearance in the text)

Endothelial molecule	Leukocyte ligand	Endothelial ligand
ICAM-1	CD11a/CD18, CD11b/CD18	N/A
VCAM-1	CD49d/CD29	N/A
JAM-A	CD11a/CD18	JAM-A
JAM-B	Not described on leukocytes	JAM-B, JAM-C
JAM-C	CD11b/CD18	JAM-C, JAM-B
ESAM	None known	ESAM
PECAM-1	PECAM-1	PECAM-1
CD99	CD99	CD99
CD99L2	CD99L2	CD99L2
VE-cadherin	None known	VE-cadherin

The main endothelial molecules described in the text and their ligands on endothelial cells and leukocytes are shown. Refer to text for functions and abbreviations.

specifically the src-dependent phosphorylation of the actin-binding molecule cortactin is required for ICAM-1 clustering (Yang et al., 2005; Yang, Kowalski, Zhan, Thomas, & Luscinskas, 2006).

VCAM-1 is involved in the firm adhesion of monocytes and lymphocytes bearing CD49d/CD29. VCAM-1 clustering has been observed in the steps leading up to diapedesis. Both ICAM-1 and VCAM-1 are enriched over actin-rich "docking structures" that form prior to TEM (Barreiro et al., 2002; Carman & Springer, 2004). Engagement of VCAM-1 activates intracellular calcium release and the small GTPase Rac-1. This in turn activates the endothelial NADPH oxidase Nox2 (Cook-Mills et al., 2004). Reactions downstream of this have effects on the adherens junctions (see below).

ICAM-2, another CD11a/CD18 ligand, is constitutively expressed on endothelial cells, where it is concentrated at cell borders, but retains considerable surface expression. Antibodies against ICAM-2 do not seem to have a major effect on TEM *in vitro* and compared to ICAM-1, ICAM-2 seems to play a lesser role (Reiss, Hoch, Deutsch, & Engelhardt, 1998). However, in some inflammatory models *in vivo*, blocking antibodies or genetic deletion of ICAM-2 inhibit transmigration of neutrophils (Huang et al., 2006; Woodfin et al., 2009).

B. Junctionally Enriched Molecules

JAM-A is concentrated at endothelial cell borders. While it normally engages in homophilic adhesion, during inflammation it can bind to CD11a/CD18 on the leukocyte (Ostermann, Weber, Zernecke, Schroder, & Weber, 2002). Blocking JAM-A on human endothelial cells using a polyclonal antibody *in vitro* has been shown to reduce TEM (Ostermann et al.); however, other investigators using polyclonal or monoclonal antibodies have seen no effect (Liu et al., 2000; Schenkel, Mamdouh, & Muller, 2004; Shaw et al., 2004). On the other hand, *in vivo* studies show decreased inflammation (Martin-Padura et al., 1998) and transendothelial migration (Woodfin et al., 2009) when JAM-A is blocked. *JAM-C* is likewise concentrated at endothelial cell borders. In can engage in homophilic adhesion with JAM-C, or heterophilic adhesion with JAM-B or CD11b/CD18. The latter interaction is implicated in transendothelial migration *in vitro* (Johnson-Leger, Aurrand-Lions, Beltraminelli, Fasel, & Imhof, 2002) and *in vivo* (Chavakis et al., 2004). For an extensive review of the roles of JAM family members in the inflammatory response, the reader is referred to a recent review (Weber, Fraemohs, & Dejana, 2007).

Endothelial cell-selective adhesion molecule (*ESAM*) is molecularly related to the JAMs, but has a long cytoplasmic domain. As its name implies, its distribution is mostly limited to endothelial junctions, but it is expressed on activated platelets (Nasdala et al., 2002). It binds homophilically, and a ligand on leukocytes has not been described. ESAM-deficient mice have no defect in lymphocyte extravasation, but had a transient decrease in neutrophil emigration (marked decrease at 2 h that had recovered by 4 h) (Wegmann et al., 2006).

PECAM-1 (CD31) is an Ig superfamily member concentrated at the borders of endothelial cells as well as expressed diffusely on platelets and leukocytes. Homophilic interaction of leukocyte PECAM with endothelial PECAM is required for transendothelial migration (Mamdouh, Chen, Pierini, Maxfield, & Muller, 2003; Muller, Weigl, Deng, & Phillips, 1993). Blockade with mAb against the amino-terminal homophilic interaction domain, soluble PECAM-Fc chimeras, and genetic deletion of PECAM inhibit transendothelial migration *in vitro* and *in vivo* (reviewed in Muller, 2007). When PECAM is transfected into cells that normally lack it, expression of PECAM imparts on them the ability to support TEM (Dasgupta, Dufour, Mamdouh, & Muller, 2009). This gain of function has not been demonstrated with other adhesion molecules. When PECAM–PECAM interactions are blocked, leukocytes are arrested tightly adherent to the apical surface of the cell (Liao et al., 1995) and actively migrate along the junctions as if searching for a place to transmigrate (Schenkel, Mamdouh, et al., 2004). *In vivo*, at sites of inflammation leukocytes are able to get to the postcapillary venules at the site of inflammation, but are unable to transmigrate efficiently. They are seen in vastly increased numbers apparently adherent to the endothelial cell luminal surface (Bogen, Pak, Garifallou, Deng, & Muller, 1994; Schenkel, Chew, & Muller, 2004), reminiscent of the block to TEM seen *in vitro* (Liao et al.; Muller et al.).

This phenotype is seen with human cells and in all mouse strains examined except for C57BL/6 (Schenkel, Chew, et al., 2004; Seidman, Chew, Schenkel, & Muller, 2009). Interestingly, this mouse strain seems to be unique in that genetic deletion of PECAM or administration of blocking antibody or mouse PECAM-Fc to these mice has no effect in a variety of inflammatory models (Duncan et al., 1999; Schenkel, Chew, et al., 2004; Schenkel, Chew, Chlipala, Harbord, & Muller, 2006). Even the closely related C57BL/10 strain responds to anti-PECAM therapy (Seidman et al.). The ability to circumvent the need for PECAM in the thioglycollate peritonitis model of inflammation has been linked to a small locus at the proximal end of chromosome 2 (Seidman et al.). Therefore, earlier studies carried out in C57BL/6 mice that found no role or only a minor role for PECAM in inflammation need to be re-evaluated. See Muller (2007) for a detailed discussion of the role of PECAM in various *in vivo* models.

There is a role for leukocyte PECAM in traversing the basal lamina (Wakelin et al., 1996). C57BL/6 mice in which PECAM has been knocked out (Duncan et al., 1999) or blocked with antibody (Woodfin et al., 2009) are defective in their ability to migrate across this extravascular barrier.

CD99 is a relatively unique molecule unrelated to any other molecule in the human genome except the closely related paralog CD99-like 2 (CD99L2), which may have arisen from a common ancestral gene (Suh et al., 2003). The gene encoding CD99 is in the pseudoautosomal region of the human X chromosome (Smith, Goodfellow, & Goodfellow, 1993). In mice, the region of the genome syntenic to the pseudoautosomal region of the human X chromosome is on chromosome 4 (Park et al., 2005), and this is where mouse CD99 is encoded. Similar to PECAM, homophilic interaction between CD99 at the endothelial cell border and CD99 on monocytes (Schenkel, Mamdouh, Chen, Liebman, & Muller, 2002) and neutrophils (Lou, Alcaide, Luscinskas, & Muller, 2007) is required for transmigration. However, CD99 regulates a later step in transmigration than PECAM. Leukocytes in which PECAM has been blocked can still be prevented from transmigrating if anti-CD99 is added after the anti-PECAM block has been removed. Conversely, when CD99 interaction is first blocked, leukocytes can no longer be inhibited from transmigrating by anti-PECAM antibody when the anti-CD99 block is removed (Schenkel et al.). In support of this, confocal images of leukocytes blocked in the act of transmigration by anti-CD99 show their leading edge under the endothelial cytoplasm, their cell body lodged at the border between endothelial cells, and the trailing uropod on the apical surface (Lou et al.; Schenkel et al.). As long as the block continues, they migrate along the junctions over the surface of the endothelium in this manner, unable to finish transmigration (Lou et al.). There is indirect evidence that CD99 in fact cannot function unless PECAM acts first (Dasgupta et al., 2009). Blocking antibodies against mouse CD99 inhibit inflammation in several animal models. Migration of T-lymphocytes into skin (Bixel et al., 2004) and neutrophils and monocytes into the peritoneal cavity (Dufour, Deroche, Bae, & Muller, 2008) are blocked by interfering with CD99 function.

CD99L2 is a molecule ancestrally related to CD99. It is encoded by a gene on the X chromosome, as is CD99, but unlike CD99, the gene encoding CD99L2 is not in the pseudoautosomal region (Park et al., 2005). CD99L2 expression in mice seems similar to that of CD99. That is, it is expressed on vascular endothelium of all tissues examined (Bixel et al., 2007; Schenkel, Dufour, Chew, Sorg, & Muller, 2007) and is expressed at the borders of endothelial cells. It is expressed to varying degrees on all circulating blood cells. Only polyclonal antibodies against murine CD99L2 have been tested *in vivo*. They block neutrophil and monocyte influx in the thioglycollate peritonitis model (Bixel et al.,; Schenkel et al.). It is tempting to speculate that the incomplete blockade of inflammation seen when interfering with either CD99 or CD99L2 is due to partial redundancy of the function of these molecules.

VE-cadherin is the major adhesion molecule of the endothelial adherens junction. It negatively regulates transmigration. Antibodies against VE-cadherin enhance early migration into a site of inflammation *in vivo* (Gotsch et al., 1997). *In vitro* studies show that VE-cadherin is transiently removed from the site of transmigration at the cell junction (Allport, Muller, & Luscinskas, 2000; Shaw, Bamba, Perkins, & Luscinskas, 2001). Mutation of the cytoplasmic tail of VE-cadherin so that it cannot interact with p120 or β-catenin or overexpression of β-catenin prevents clearance of VE-cadherin from the cell border and blocks transmigration (Alcaide et al., 2008; Allingham, van Buul, & Burridge, 2007).

C. Why So Many Molecules?

Other endothelial molecules that have been shown to play a role in TEM by virtue of the inhibition of TEM by blocking antibodies include poliovirus receptor (CD155) (Reymond et al., 2004), MUC18 (CD146) (Bardin et al., 2009), activated leukocyte cell adhesion molecule (ALCAM/CD166) (Masedunskas et al., 2006), integrin-associated protein (IAP/CD47) (Stefanidakis, Newton, Lee, Parkos, & Luscinskas, 2008), and nepmucin/CLM-9 (Jin et al., 2008). It seems each month brings a new report of an endothelial cell or leukocyte molecule that is implicated in diapedesis. When added to the well-characterized molecules discussed in the previous section, this raises the question of why so many molecules are required for TEM. Is this just an artifact of clogging up the junction with antibody or turning the cell junctions into immune complexes? This is unlikely, as most of these studies used control antibody, Fab or $F(ab')_2$ fragments, soluble recombinant adhesion molecules, siRNA knockdown, or genetic deletion to buttress their claims.

The process of diapedesis itself can be further dissected into a series of molecularly defined steps controlled by specific molecules acting in sequence. Sequential blocking experiments demonstrated that PECAM regulates a step in diapedesis that is "upstream" of the step regulated by CD99 (Schenkel et al., 2002). As mentioned above, images of the blocked cells showed that blocking PECAM arrests leukocytes before the start of transmigration, blocking CD99 arrests leukocytes during the process of transmigration. Sequential blockade analysis has not been performed with other pairs of molecules, but images of leukocytes blocked by antibodies *in vivo* in C57BL/6 mice show that ICAM-2 arrests neutrophils on the apical surface of the endothelium, anti-JAM-A arrests them at the cell junctions, and anti-PECAM arrests them between the endothelial cell and basal lamina (Woodfin et al., 2009). This begs the questions of whether each molecule controls its own defined step in the sequence, whether multiple molecules control each step, and how many steps are there? Until sequential blockade studies can be performed with each of these molecules, this question will remain unanswered. The answer is likely to be different for different leukocyte types, vascular beds, and inflammatory stimuli, as well as the time after the initiation of the stimulus. However, it seems unlikely that there is a separate unique step in diapedesis controlled by each molecule reported to be important for transmigration.

What if most of the endothelial molecules reported to control transmigration were part of a large multimolecular "transmigration complex," or a series of multimolecular transmigration complexes (one for each successive step in diapedesis) that combined to make a platform to support transmigration analogous to the way that multiple transcription factors and coactivators combine to make DNA accessible to transcription? Loss of or interference with any one of the molecules in that case could make the complex less efficient at supporting diapedesis and could account for the published results.

IV. MECHANISMS REGULATING TRANSMIGRATION

A. Clustering Surface ICAM-1 and VCAM-1

The adhesion step immediately upstream of diapedesis is an obvious prerequisite for diapedesis, and there is reason to think that some of the events that occur during this step signal the events that regulate transmigration. Clustering of ICAM-1 and VCAM-1 on the endothelial cell has been observed as the leukocyte approaches the endothelial cell border (Barreiro et al., 2002; Carman & Springer, 2004) The initial leukocyte-faciliated clustering of ICAM-1 requires src-dependent phosphorylation of the actin-binding protein cortactin, which is also associated with actin filament remodeling that takes place during transmigration (Yang et al., 2006). On the other hand, ICAM-1 engagement or clustering induces src-dependent phosphorylation of cortactin (Durieu-Trautmann, Chaverot, Cazaubon, Strosberg, & Couraud, 1994). Rather than being results that are at odds, these observations may belie a self-amplification cycle. The initial recruitment of ICAM-1 and VCAM-1 may be due to adhesion to their leukocyte ligands. This clustering induces phosphorylation of cortactin, which leads to the actin polymerization and the recruitment of more ICAM-1 to the site of leukocyte adhesion, which induces more cortactin phosphorylation. Clustering of ICAM-1 and VCAM-1 stimulates signaling in the endothelial cells that promote diapedesis in ways that are discussed later.

Clustering of ICAM-1 and VCAM-1 may occur in three dimensions. "Docking structures" or "transmigratory cups" are terms used to describe finger-like projections of endothelial apical surface membrane reported to surround the lower portion of adherent leukocytes. The membrane is enriched in ICAM-1 and VCAM-1 and overlies cytoplasm enriched in f-actin and actin-binding proteins. Sanchez-Madrid and coworkers first used the term "docking structures" to describe these finger-like projections that engaged polyclonally activated lymphocytes and lymphoblasts adherent to cytokine-activated HUVEC (Barreiro et al., 2002). Most of their observations were made with a cell line (4M7) that expressed VLA-4 but not LFA-1. These cells could adhere strongly, but not transmigrate and thus there was plenty of time to strengthen VLA-4/VCAM-1 interactions and recruit more VCAM-1, actin, and ERM proteins to the site of adhesion. However, similar extensions bearing ICAM-1 were seen around adherent LFA-1-expressing lymphoblasts.

Subsequently, Carman, Jun, Salas, and Springer (2003) demonstrated similar projections of ICAM-1 that seemed to rise up off the endothelial surface and surround at least the lower part and sides of leukocytes engaging cytokine-activated endothelial cells or ICAM-1-transfected CHO cells. Disruption of the cytoskeleton abolished these structures, but had no effect on leukocyte adhesion. The authors commented that this might belie a role in transmigration. Interestingly, however, Barreiro et al. (2002) had found that the docking structures rapidly vanished as lymphocytes began to migrate through the monolayers. The Carman-Springer team went on to show that these projections were associated with transmigrating neutrophils, monocytes, and lymphocytes, at least under their experimental conditions that involved apical application of a chemokine or leukocyte activator and interaction with activated endothelium. They referred to these structures, associated with both paracellular and transcellular migration, as "transmigratory cups" (Carman & Springer, 2004). Subsequently, most transmigratory cups have been reported in association with transcellular migration.

The experimental conditions used for these studies involved monolayers of TNFα-activated human umbilical vein endothelial cells grown on glass coverslips. Transmigration experiments were performed in the presence or absence of fluid shear, but this did not seem to make a difference (Carman & Springer, 2004; Carman et al., 2007). In all cases, the investigators reported that a ring of enriched ICAM-1 (and VCAM-1) fluorescence was seen on the apical surface of the endothelial cell at the point of contact with the adherent or transmigrating leukocyte.

However, not everyone who reports rings of ICAM-1 enrichment around transmigrating leukocytes has seen docking structures or transmigratory cups. For example, Ridley's group, using a similar system (but without application of apical SDF-1) showed distinct ICAM-1 enrichment around transmigrating lymphoblasts, but no docking structures (Millan et al., 2006). Luscinskas' group also demonstrated local enrichment of ICAM-1 around transmigrating neutrophils undergoing transmigration (Shaw et al., 2004) and commented that they did not see such actin-rich microvilli.

At sites of documented ongoing transcellular migration *in vivo* (Feng et al., 1998), cup-like structures were not visualized during migration. Instead, the PMN appeared to be migrating through a fenestra in the flat endothelial cell. A similar EM image was shown for a neutrophil migrating through an ICAM-1-enriched zone of endothelial cell cytoplasm *in vitro* (Yang et al., 2005). A true apical cup has been demonstrated (van Buul, Allingham, et al., 2007) by scanning electron microscopy. This was surrounding a differentiated HL60 cell adherent to a TNFα-activated HUVEC for 30 min. The cell adhered but did not transmigrate.

What do these docking structures represent and why are they not universally seen? One possibility is that they represent a response of the endothelial cell to leukocytes that are either highly activated or tightly adherent. The structures were seen under conditions where the leukocytes were adherent but could not transmigrate, allowing time for recruitment of additional ICAM-1 and/or VCAM-1 molecules (Barreiro et al., 2002; van Buul, Allingham, et al., 2007) or under which the leukocytes were additionally activated by the exogenous application of PAF or chemokines on the apical surface of the endothelial cells (Carman & Springer, 2004; Carman et al., 2003). One could easily imagine that under these conditions, enhanced leukocyte integrin activation could result in greater recruitment of counter-receptors from the endothelial surface. The scanning electron micrograph (van Buul, Allingham, et al.) is reminiscent of ligand-mediated phagocytosis in macrophages, and endothelial cells are known to be phagocytic under certain conditions. In contrast, under conditions where the transmigrating neutrophils (Shaw et al., 2004) or lymphoblasts (Millan et al., 2006) were activated by interactions with the cytokine-activated endothelium without additional apical chemokine provided, ICAM-1 enrichment was not accompanied by formation of transmigratory cups.

B. Loosening the Junctions

Several lines of evidence show that loosening the endothelial cell junctions is important for efficient transmigration. Clustering of ICAM-1 and VCAM-1 on endothelial cells transmits a number of signals into the endothelial cell (reviewed in van Buul, Kanters, & Hordijk, 2007), some of which appear to be relevant to diapedesis. Crosslinking VCAM-1 (Lorenzon et al., 1998) and

ICAM-1 (van Buul, Kanters, et al.) on the endothelial cell stimulates an increase in cytosolic-free calcium ions, which has long been known to be a requirement for diapedesis (Huang et al., 1993). The increase in cytosolic-free calcium ion has been shown to activate myosin light chain kinase (MLCK), leading to actin–myosin fiber contraction. This is believed to help endothelial cells separate (Hixenbaugh et al., 1997).

Stimulation of ICAM-1 leads to phosphorylation of VE-cadherin, which is a prerequisite for adherens junction disassembly (Turowski et al., 2008). In HUVEC, the kinases Src and Pyk2 phosphorylate VE-cadherin on the p120 and β-catenin-binding sites, tyrosine residues 658 and 731, respectively (Allingham et al., 2007). This inhibits the binding of p120 and β-catenin to VE-cadherin. Since the interaction of these proteins with VE-cadherin is critical for retaining VE-cadherin at the adherens junction, this destabilizes the junctions.

Crosslinking VCAM-1 also activates Rac1 (van Wetering et al., 2002) and stimulates an increase in reactive oxygen species in endothelial cells (Cook-Mills et al., 2004) that leads to loosening of adherens junctions. In other systems Rac1 activation leads to phosphorylation of VE-cadherin on serine 665, which signals its clathrin-dependent internalization (Gavard & Gutkind, 2006). The net result is "loosening" of junctional structures.

Under resting conditions the vascular endothelial protein tyrosine phosphatase (VE-PTP) associates with VE-cadherin via plakoglobin (γ-catenin), maintaining VE-cadherin in a hypophosphorylated state at the junction. Interaction of leukocytes with cytokine-activated endothelial cells triggers rapid dissociation of VE-PTP from VE-cadherin, allowing it to be phosphorylated on tyrosine, increasing junctional permeability and facilitating transendothelial migration (Nottebaum et al., 2008). A role for another VE-cadherin accessory molecule, p120 catenin, has been demonstrated recently (Alcaide et al., 2008). Overexpression of p120 prevented VE-cadherin phosphorylation and the formation of "gaps" in VE-cadherin staining along the endothelial junction during engagement of leukocytes. (These gaps were not spaces between cells, but disruption of the staining pattern of VE-cadherin.) This was associated with a significant decrease in transmigration. Interestingly, the authors did not find evidence for VE-cadherin internalization during gap formation (Alcaide et al.).

In a similar manner, clustering of ICAM-1 activates RhoA, which activates Rho kinase (ROCK) (reviewed in Cernuda-Morollon & Ridley, 2006). This in turn phosphorylates and inactivates PP1c, the major phosphatase inactivating MLCK. The end result is potentiation of actin–myosin contraction. It is important to point out that, although intercellular gaps that are visible in the light microscope can be produced on endothelial cells cultured on glass coverslips, *in vivo* the gaps produced between endothelial cells by even the strongest inducers of vascular permeability (e.g., histamine and serotonin) are in the order of hundreds of angstroms (Majno & Palade, 1961), and these are resealed by the time most leukocytes are recruited. This does not mean that these gaps are not important, but it means that leukocytes must still crawl through closely adherent endothelial cells; they do not fall into holes between endothelial cells.

C. The Lateral Border Recycling Compartment

Even under steady-state conditions, there is a considerable amount of membrane movement taking place at the endothelial cell borders. Membrane is internalized into and recycled from an interconnected reticulum of tubulovesicular structures that resides just beneath the plasma membrane of the endothelial cell borders (Mamdouh et al., 2003). This compartment, which we have come to call the lateral border recycling compartment (Mamdouh, Kreitzer, & Muller, 2008) is distinct from caveolae, typical recycling endosomes, and vesiculovacuolar organelles (Mamdouh et al., 2003). By electron microscopy most of the components appear to be 50-nm vesicles. About 30% of the cell's PECAM resides in this compartment and recycles with a half time of about 10 min (Mamdouh et al., 2003). This compartment also contains CD99 and JAM-A, but not VE-cadherin (unpublished data). In high endothelial venule endothelium, the Ig superfamily molecule nepmucin (CLM-9), which promotes lymphocyte TEM is in the LBRC (Jin et al., 2008).

The purpose of the constitutive recycling is not known. However, when a leukocyte transmigrates, membrane from the LBRC is redirected. It is targeted to and exteriorized at the cell border at the position where the leukocyte is transmigrating (Mamdouh et al., 2003, 2008). Blocking homophilic PECAM–PECAM interactions between leukocyte and endothelial cell blocks targeted recycling from the LBRC and blocks transmigration. Moreover, there is accumulating evidence that targeted recycling from the LBRC is an essential step in TEM: LBRC membrane is trafficked to the site of transmigration by kinesin molecular motors along microtubules (Mamdouh et al., 2008). Disrupting or bundling microtubules, or inhibiting the motor domain of kinesin blocks targeted recycling and blocks TEM. Lymphocytes and activated lymphoblasts transmigrate in a manner that cannot be blocked by anti-PECAM antibodies (Bird, Spragg, Ager, & Matthews, 1993; Muller, 2001). Nevertheless, transmigration of lymphoblasts can be efficiently blocked by disrupting targeted recycling of the LBRC (Mamdouh et al., 2008). A tyrosine → phenylalanine mutation on the cytoplasmic tail of PECAM blocks the ability of PECAM to support TEM. It turns out that this mutation interferes with the ability of PECAM to enter and leave the

LBRC and markedly diminishes its ability to participate in targeted recycling (Dasgupta et al., 2009). In confluent endothelial cell monolayers a minority of PECAM is phosphorylated; however, essentially all of the phosphorylated PECAM resides in the LBRC (Dasgupta & Muller, 2008).

Targeted recycling of LBRC membrane during TEM potentially solves many of the "problems" inherent in the process. Rather than having to "unzip" high-density homophilic adhesions of VE-cadherin, PECAM, JAM-A, CD99, etc., these molecules (and other structural components of the junction) may be pushed aside by membrane from the LBRC. This then presents unligated molecules that the leukocyte must interact with (e.g., PECAM, JAM-A, CD99, nepmucin) on its path across the endothelial cell while removing structural barriers to transmigration (e.g., the adherens junction complex of VE-cadherin and associated catenins). Hypothetically, once the leukocyte has moved across the junction, the LBRC may be pulled back into the cell, allowing the other components to diffuse back into place, re-establishing the endothelial junction without having to reform all of the complex three-dimensional interactions.

D. A Unifying Concept of Transmigration?

ICAM-1 and VCAM-1 signaling, cytosolic-free calcium flux, RhoA and Rac1 activation, VE-cadherin removal from the junction, MLCK activation, and targeted recycling of the LBRC have all been shown to be necessary for efficient transmigration. How are these diverse phenomena related? Are they sequential links in a chain, or events occurring in parallel with all required for TEM to occur? Considering that many second messenger signaling systems interact with each other and feedback loops exist, this may be a question of semantics. However, the following undoubtedly oversimplified scheme seems to be consistent with all of the published data and at least provides a testable hypothesis (see Fig. 1).

LFA-1 preferentially binds to ICAM-1 dimers (Miller et al., 1995; Reilly et al., 1995), which initiates clustering of ICAM-1. This stimulates phosphorylation of cortactin, enhancing the further actin-induced clustering of ICAM-1. This self-enhancing cycle leads to the enrichment of ICAM-1 around tightly adherent leukocytes. ICAM-1 multimerization leads to increases in cytosolic-free calcium and activation of RhoA (Fig. 1A).

In the meantime, if the leukocytes express VLA-4 and the endothelial cells are expressing VCAM-1, clustering of VCAM-1 also stimulates an increase in cytosolic-free calcium, activation of Rac-1, and production of reactive oxygen species in endothelial cells (Cook-Mills, 2002; van Wetering et al., 2003). The latter activates PKCα (Abdala-Valencia & Cook-Mills, 2006). The net result is loosening of endothelial cell junctions (Fig. 1A and B).

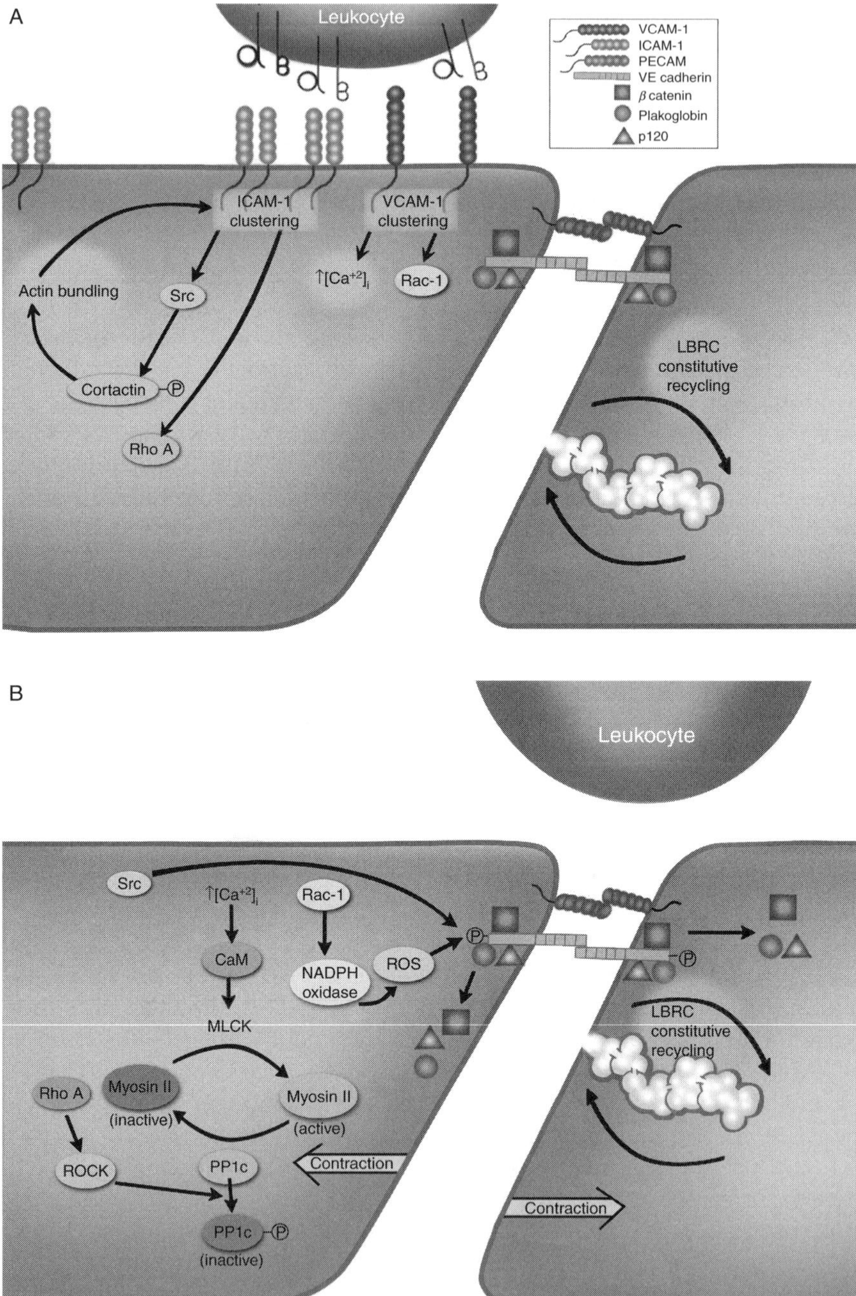
A
Leukocyte
VCAM-1
ICAM-1
PECAM
VE cadherin
β catenin
Plakoglobin
p120
ICAM-1 clustering
VCAM-1 clustering
Actin bundling
Src
↑[Ca+2]i
Rac-1
Cortactin
Rho A
LBRC constitutive recycling
B
Leukocyte
Src
↑[Ca+2]i
Rac-1
CaM
NADPH oxidase
ROS
MLCK
Rho A
Myosin II (inactive)
Myosin II (active)
ROCK
PP1c
PP1c (inactive)
Contraction
LBRC constitutive recycling
Contraction

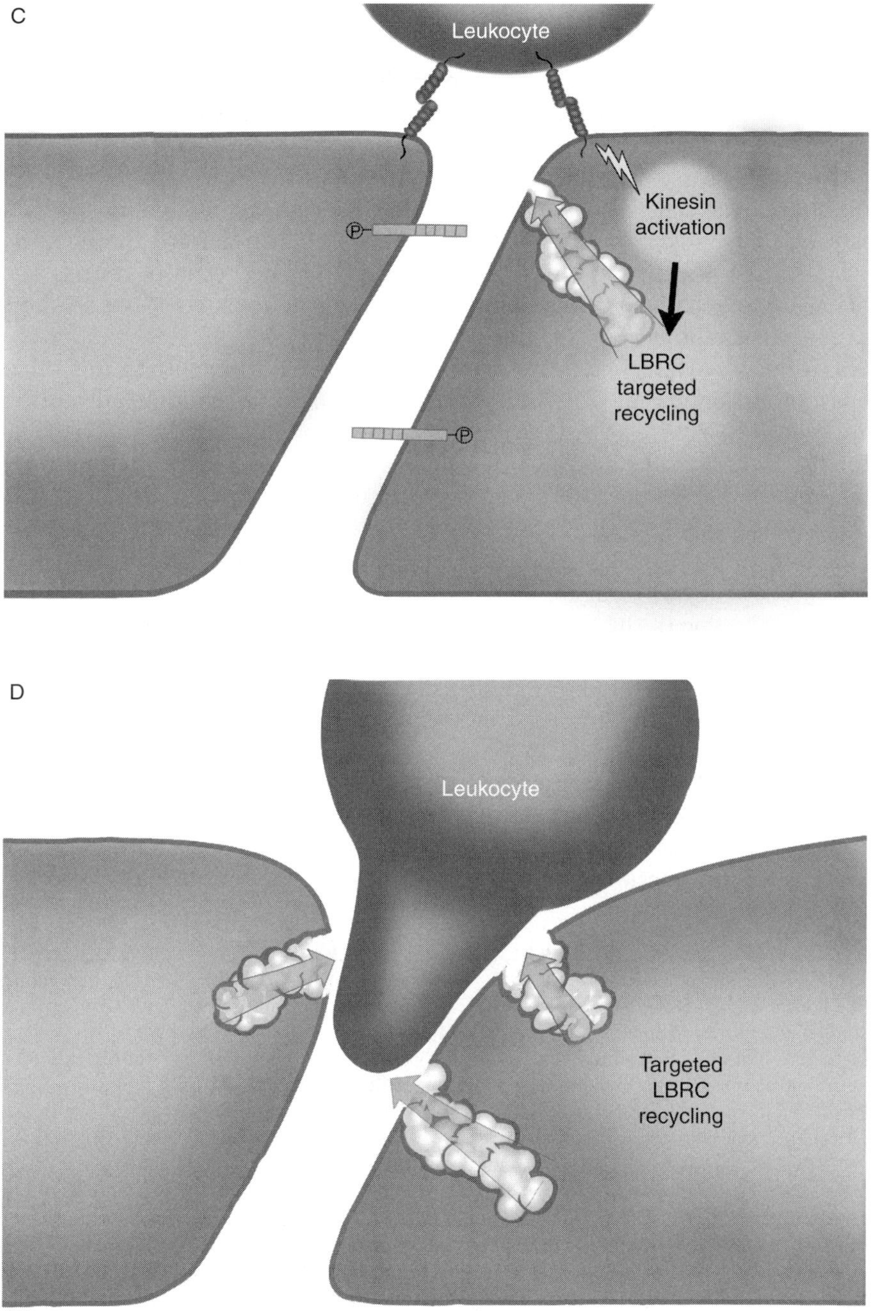

FIGURE 1 (continued)

ICAM-1 and VCAM-1 signaling simultaneously result in weakening of the endothelial junctions due to effects on phosphorylation of VE-cadherin. This dissociates VE-cadherin from its links to the actin cytoskeleton and it potentially (but not necessarily) becomes subject to endocytosis in a clathrin-dependent manner (Fig. 1B).

The increase in cytosolic-free calcium activates MLCK to induce tension in the endothelial cells. The activation of MLCK is augmented by the inactivation of PP1 phosphatase mediated by the RhoA activation stimulated by signals originated through ICAM-1 clustering. The net result of contraction of the endothelial cell body against weakened junctions would be to allow easier passage of leukocytes (Fig. 1B).

With leukocytes poised over weakened adherens junctions, the other homophilic junctional adhesion molecules still hold the endothelial borders apposed. PECAM–PECAM interactions between leukocyte and endothelial cell (Mamdouh et al., 2003) or other signals (Mamdouh et al., 2008) stimulate targeted trafficking of LBRC membrane to surround the leukocyte (Fig. 1C). Targeted recycling of the LBRC may displace components of the adherens junction laterally, providing increased surface area and unligated molecules that the leukocytes want to interact with (Fig. 1D). It is possible, and even likely, that some of the many junctional molecules discussed earlier are also part of the LBRC or function to recruit it. That is, the LBRC may be one of the hypothetical multimolecular complexes controlling transmigration while other multimolecular complexes may function to recruit it to the site of TEM and reinternalize it after TEM.

The signals that trigger targeted recycling are not known, nor is it clear how the membrane is directed to the site of transmigration. However, weakening of the endothelial cell adherens junctions by brief calcium chelation leads to diffuse exteriorization of the LBRC along the endothelial cell border

FIGURE 1 A unified schematic view of transendothelial migration (see text for details). (A) Clustering of ICAM-1 and VCAM-1 through engagement of their leukocyte integrin counter-receptors ($\alpha\beta$ in diagram) initiates activation of *src*, Rho A, and Rac-1, as well as increased cytosolic-free calcium ion. Phosphorylation of cortactin by *src* stimulates f-actin rearrangements in the cortical cytoplasm, which facilitates more ICAM-1 clustering. (B) These signals lead to activation of MLCK, inactivation of PP1c, and phosphorylation of VE-cadherin, inducing release of the associated catenins. (C) Leukocyte PECAM engagement of endothelial cell PECAM and/or other leukocyte/endothelial cell interactions at the apical surface of the endothelial border activate kinesin molecular motors in the endothelial cell and stimulate targeted trafficking of LBRC membrane to the vicinity of the leukocyte. (D) Targeted trafficking of LBRC membrane continues as the leukocyte moves into the border between endothelial cells, now enlarged by the contribution of membrane from the LBRC. This process continues until transmigration is complete. CaM, calmodulin; ROS, reactive oxygen species; circled P, phosphorylated state.

(unpublished data). It is possible that local weakening of the adherens junctions at the site of leukocyte engagement may allow for localized exteriorization of the LBRC.

V. EPILOGUE: UNANSWERED QUESTIONS

While much has been learned about the mechanisms that regulate TEM, many important unanswered questions remain: Why are so many endothelial cell molecules implicated in this process? Where is actin–myosin contraction tension exerted *in vivo*, since endothelial cells in postcapillary venules do not have stress fibers? What directs LBRC targeted recycling? Finally, how are all of the molecules and mechanisms that have been identified to participate in transendothelial migration coordinated to ensure efficient leukocyte emigration with minimal leakage of soluble vascular contents?

Acknowledgments

Supported by grants from the National Institutes of Health, R01HL046849 and R37HL064774.

References

Abdala-Valencia, H., & Cook-Mills, J. M. (2006). VCAM-1 signals activate endothelial cell protein kinase Calpha via oxidation. *Journal of Immunology, 177*, 6379–6387.

Alcaide, P., Newton, G., Auerbach, S., Sehrawat, S., Mayadas, T. N., Golan, D. E., et al. (2008). p120-Catenin regulates leukocyte transmigration through an effect on VE-cadherin phosphorylation. *Blood, 112*, 2770–2779.

Allingham, M. J., van Buul, J. D., & Burridge, K. (2007). ICAM-1-mediated, Src- and Pyk2-dependent vascular endothelial cadherin tyrosine phosphorylation is required for leukocyte transendothelial migration. *Journal of Immunology, 179*, 4053–4064.

Allport, J. R., Muller, W. A., & Luscinskas, F. W. (2000). Monocytes induce reversible focal changes in vascular endothelial cadherin complex during transendothelial migration under flow. *Journal of Cell Biology, 148*, 203–216.

Bardin, N., Blot-Chabaud, M., Despoix, N., Kebir, A., Harhouri, K., Arsanto, J. P., et al. (2009). CD146 and its soluble form regulate monocyte transendothelial migration. *Arteriosclerosis, Thrombosis, and Vascular Biology, 29*, 746–753.

Barreiro, O., Yanez-Mo, M., Serrador, J. M., Montoya, M. C., Vicente-Manzanares, M., Tejedor, R., et al. (2002). Dynamic interaction of VCAM-1 and ICAM-1 with moesin and ezrin in a novel endothelial docking structure for adherent leukocytes. *Journal of Cell Biology, 157*, 1233–1245.

Bird, I. N., Spragg, J. H., Ager, A. H., & Matthews, N. (1993). Studies of lymphocyte transendothelial migration: Analysis of migrated cell phenotypes with regard to CD31 (PECAM-1), CD45RA and CD45RO. *Immunology, 80*, 553–560.

Bixel, G., Kloep, S., Butz, S., Petri, B., Engelhardt, B., & Vestweber, D. (2004). Mouse CD99 participates in T cell recruitment into inflamed skin. *Blood, 104*, 3205–3213.

Bixel, M. G., Petri, B., Khandoga, A. G., Khandoga, A., Wolburg-Buchholz, K., Wolburg, H., et al. (2007). A CD99-related antigen on endothelial cells mediates neutrophil, but not lymphocyte extravasation *in vivo*. *Blood, 109*, 5327–5336.

Bogen, S., Pak, J., Garifallou, M., Deng, X., & Muller, W. A. (1994). Monoclonal antibody to murine PECAM-1 [CD31] blocks acute inflammation *in vivo*. *Journal of Experimental Medicine, 179*, 1059–1064.

Carman, C. V., Jun, C. D., Salas, A., & Springer, T. A. (2003). Endothelial cells proactively form microvilli-like membrane projections upon intercellular adhesion molecule 1 engagement of leukocyte LFA-1. *Journal of Immunology, 171*, 6135–6144.

Carman, C. V., Sage, P. T., Sciuto, T. E., de la Fuente, M. A., Geha, R. S., Ochs, H. D., et al. (2007). Transcellular diapedesis is initiated by invasive podosomes. *Immunity, 26*, 784–797.

Carman, C. V., & Springer, T. A. (2004). A transmigratory cup in leukocyte diapedesis both through individual vascular endothelial cells and between them. *Journal of Cell Biology, 167*, 377–388.

Carman, C. V., & Springer, T. A. (2008). Trans-cellular migration: Cell–cell contacts get intimate. *Current Opinion in Cell Biology, 20*, 533–540.

Cernuda-Morollon, E., & Ridley, A. J. (2006). Rho GTPases and leukocyte adhesion receptor expression and function in endothelial cells. *Circulation Research, 98*, 757–767.

Chavakis, T., Keiper, T., Matz-Westphal, R., Hersemeyer, K., Sachs, U. J., Nawroth, P. P., et al. (2004). The junctional adhesion molecule-C promotes neutrophil transendothelial migration *in vitro* and *in vivo*. *Journal of Biological Chemistry, 279*, 55602–55608.

Cook-Mills, J. M. (2002). VCAM-1 signals during lymphocyte migration: Role of reactive oxygen species. *Molecular Immunology, 39*, 499–508.

Cook-Mills, J. M., Johnson, J. D., Deem, T. L., Ochi, A., Wang, L., & Zheng, Y. (2004). Calcium mobilization and Rac1 activation are required for VCAM-1 (vascular cell adhesion molecule-1) stimulation of NADPH oxidase activity. *Biochemical Journal, 378*, 539–547.

Dasgupta, B., Dufour, E., Mamdouh, Z., & Muller, W. (2009). A novel and critical role for tyrosine 663 in PECAM trafficking and transendothelial migration. *Journal of Immunology, 182*, 5041–5051.

Dasgupta, B., & Muller, W. A. (2008). Endothelial Src kinase regulates membrane recycling from the lateral border recycling compartment during leukocyte transendothelial migration. *European Journal of Immunology, 38*, 3499–3507.

Dufour, E. M., Deroche, A., Bae, Y., & Muller, W. A. (2008). CD99 is essential for leukocyte diapedesis *in vivo*. *Cell Communication & Adhesion, 16*, 1–13.

Duncan, G. S., Andrew, D. P., Takimoto, H., Kaufman, S. A., Yoshida, H., Spellberg, J., et al. (1999). Genetic evidence for functional redundancy of platelet/endothelial cell adhesion molecule-1 (PECAM-1): CD31-deficient mice reveal PECAM-1-dependent and PECAM-1-independent functions. *Journal of Immunology, 162*, 3022–3030.

Durieu-Trautmann, O., Chaverot, N., Cazaubon, S., Strosberg, A. D., & Couraud, P. O. (1994). Intercellular adhesion molecule 1 activation induces tyrosine phosphorylation of the cytoskeleton-associated protein cortactin in brain microvessel endothelial cells. *Journal of Biological Chemistry, 269*, 12536–12540.

Feng, D., Nagy, J. A., Pyne, K., Dvorak, H. F., & Dvorak, A. M. (1998). Neutrophils emigrate from venules by a transendothelial cell pathway in response to fMLP. *Journal of Experimental Medicine, 187*, 903–915.

Gavard, J., & Gutkind, J. S. (2006). VEGF controls endothelial-cell permeability by promoting the beta-arrestin-dependent endocytosis of VE-cadherin. *Nature Cell Biology, 8*, 1223–1234.

Gotsch, U., Borges, E., Bosse, R., Boggemeyer, E., Simon, M., Mossmann, H., et al. (1997). VE-cadherin antibody accelerates neutrophil recruitment *in vivo*. *Journal of Cell Science, 110*, 583–588.

Hixenbaugh, E. A., Goeckeler, Z. M., Papaiya, N. N., Wysolmerski, R. B., Silverstein, S. C., & Huang, A. J. (1997). Chemoattractant-stimulated neutrophils induce regulatory myosin light chain phosphorylation and isometric tension development in endothelial cells. *American Journal of Physiology, 273*, H981–H988.

Huang, M. T., Larbi, K. Y., Scheiermann, C., Woodfin, A., Gerwin, N., Haskard, D. O., et al. (2006). ICAM-2 mediates neutrophil transmigration *in vivo*: Evidence for stimulus specificity and a role in PECAM-1-independent transmigration. *Blood, 107*, 4721–4727.

Huang, A. J., Manning, J. E., Bandak, T. M., Ratau, M. C., Hanser, K. R., & Silverstein, S. C. (1993). Endothelial cell cytosolic free calcium regulates neutrophil migration across monolayers of endothelial cells. *Journal of Cell Biology, 120*, 1371–1380.

Jin, S., Umemoto, E., Tanaka, T., Shimomura, Y., Tohya, K., Kunizawa, K., et al. (2008). Nepmucin/CLM-9, an Ig domain-containing sialomucin in vascular endothelial cells, promotes lymphocyte transendothelial migration *in vitro*. *FEBS Letters, 582*, 3018–3024.

Johnson-Leger, C., Aurrand-Lions, M., Beltraminelli, N., Fasel, N., & Imhof, B. A. (2002). Junctional adhesion molecule-2 (JAM-2) promotes lymphocyte transendothelial migration. *Blood, 100*, 2479–2486.

Ley, K., Laudanna, C., Cybulsky, M. I., & Nourshargh, S. (2007). Getting to the site of inflammation: The leukocyte adhesion cascade updated. *Nature Reviews Immunology, 7*, 678–689.

Liao, F., Huynh, H. K., Eiroa, A., Greene, T., Polizzi, E., & Muller, W. A. (1995). Migration of monocytes across endothelium and passage through extracellular matrix involve separate molecular domains of PECAM-1. *Journal of Experimental Medicine, 182*, 1337–1343.

Liu, Y., Nusrat, A., Schnell, F. J., Reaves, T. A., Walsh, S., Pochet, M., et al. (2000). Human junction adhesion molecule regulates tight junction resealing in epithelia. *Journal of Cell Science, 113*, 2363–2374.

Lorenzon, P., Vecile, E., Nardon, E., Ferrero, E., Harlan, J. M., Tedesco, F., et al. (1998). Endothelial cell E- and P-selectin and vascular cell adhesion molecule-1 function as signaling receptors. *Journal of Cell Biology, 142*, 1381–1391.

Lou, O., Alcaide, P., Luscinskas, F. W., & Muller, W. A. (2007). CD99 is a key mediator of the transendothelial migration of neutrophils. *Journal of Immunology, 178*, 1136–1143.

Majno, G., & Palade, G. E. (1961). Studies on inflammation. 1. The effect of histamine and serotonin on vascular permeability: An electron microscopic study. *Journal of Biophysical and Biochemical Cytology, 11*, 571–605.

Mamdouh, Z., Chen, X., Pierini, L. M., Maxfield, F. R., & Muller, W. A. (2003). Targeted recycling of PECAM from endothelial cell surface-connected compartments during diapedesis. *Nature, 421*, 748–753.

Mamdouh, Z., Kreitzer, G. E., & Muller, W. A. (2008). Leukocyte transmigration requires kinesin-mediated microtubule-dependent membrane trafficking from the lateral border recycling compartment. *Journal of Experimental Medicine, 205*, 951–966.

Martin-Padura, I., Lostaglio, S., Schneemann, M., Williams, L., Romano, M., Fruscella, P., et al. (1998). Junctional adhesion molecule, a novel member of the immunoglobulin superfamily that distributes at intercellular junctions and modulates monocyte transmigration. *Journal of Cell Biology, 142*, 117–127.

Masedunskas, A., King, J. A., Tan, F., Cochran, R., Stevens, T., Sviridov, D., et al. (2006). Activated leukocyte cell adhesion molecule is a component of the endothelial junction involved in transendothelial monocyte migration. *FEBS Letters, 580*, 2637–2645.

Millan, J., Hewlett, L., Glyn, M., Toomre, D., Clark, P., & Ridley, A. J. (2006). Lymphocyte transcellular migration occurs through recruitment of endothelial ICAM-1 to caveola- and F-actin-rich domains. *Nature Cell Biology, 8*, 113–123.

Miller, J., Knorr, R., Ferrone, M., Houdei, R., Carron, C. P., & Dustin, M. L. (1995). Intercellular adhesion molecule-1 dimerization and its consequences for adhesion mediated by lymphocyte function associated-1. *Journal of Experimental Medicine, 182*, 1231–1241.

Muller, W. A. (2001). Migration of leukocytes across endothelial junctions: Some concepts and controversies. *Microcirculation, 8*, 181–193.

Muller, W. A. (2003). Leukocyte–endothelial cell interactions in leukocyte transmigration and the inflammatory response. *Trends in Immunology, 24*, 326–333.

Muller, W. A. (2007). PECAM: Regulating the start of diapedesis. In K. Ley (Ed.), *Adhesion molecules: Function and inhibition* (pp. 201–220). Basel: Birkhauser Verlag AG.

Muller, W. A., Weigl, S. A., Deng, X., & Phillips, D. M. (1993). PECAM-1 is required for transendothelial migration of leukocytes. *Journal of Experimental Medicine, 178*, 449–460.

Nasdala, I., Wolburg-Buchholz, K., Wolburg, H., Kuhn, A., Ebnet, K., Brachtendorf, G., et al. (2002). A transmembrane tight junction protein selectively expressed on endothelial cells and platelets. *Journal of Biological Chemistry, 277*, 16294–16303.

Nottebaum, A. F., Cagna, G., Winderlich, M., Gamp, A. C., Linnepe, R., Polaschegg, C., et al. (2008). VE-PTP maintains the endothelial barrier via plakoglobin and becomes dissociated from VE-cadherin by leukocytes and by VEGF. *Journal of Experimental Medicine, 205*, 2929–2945.

Ostermann, G., Weber, K. S. C., Zernecke, A., Schroder, A., & Weber, C. (2002). JAM-1 is a ligand of the b_2 integrin LFA-1 involved in transendothelial migration of leukocytes. *Nature Immunology, 3*, 151–158.

Park, S. H., Shin, Y. K., Suh, Y. H., Park, W. S., Ban, Y. L., Choi, H. S., et al. (2005). Rapid divergency of rodent CD99 orthologs: Implications for the evolution of the pseudoautosomal region. *Gene, 353*, 177–188.

Reilly, P. L., Woska , J. R., Jr., Jeanfavre, D. D., McNally, E., Rothlein, R., & Bormann, B. J. (1995). The native structure of intercellular adhesion molecule-1 (ICAM-1) is a dimer. Correlation with binding to LFA-1. *Journal of Immunology, 155*, 529–532 . [published erratum appears in *Journal of Immunology*, 15 April 1996;*156*(8):following 3088].

Reiss, Y., Hoch, G., Deutsch, U., & Engelhardt, B. (1998). T cell interaction with ICAM-1-deficient endothelium *in vitro*: Essential role for ICAM-1 and ICAM-2 in transendothelial migration of T cells. *European Journal of Immunology, 28*, 3086–3099.

Reymond, N., Imbert, A. M., Devilard, E., Fabre, S., Chabannon, C., Xerri, L., et al. (2004). DNAM-1 and PVR regulate monocyte migration through endothelial junctions. *Journal of Experimental Medicine, 199*, 1331–1341.

Schenkel, A. R., Chew, T. W., Chlipala, E., Harbord, M. W., & Muller, W. A. (2006). Different susceptibilities of PECAM-deficient mouse strains to spontaneous idiopathic pneumonitis. *Experimental and Molecular Pathology, 81*, 23–30.

Schenkel, A. R., Chew, T. W., & Muller, W. A. (2004a). Platelet endothelial cell adhesion molecule deficiency or blockade significantly reduces leukocyte emigration in a majority of mouse strains. *Journal of Immunology, 173*, 6403–6408.

Schenkel, A. R., Dufour, E. M., Chew, T. W., Sorg, E., & Muller, W. A. (2007). The murine CD99-related molecule CD99-like 2 (CD99L2) is an adhesion molecule involved in the inflammatory response. *Cell Communication & Adhesion, 14*, 227–237.

Schenkel, A. R., Mamdouh, Z., Chen, X., Liebman, R. M., & Muller, W. A. (2002). CD99 plays a major role in the migration of monocytes through endothelial junctions. *Nature Immunology, 3*, 143–150.

Schenkel, A. R., Mamdouh, Z., & Muller, W. A. (2004b). Locomotion of monocytes on endothelium is a critical step during extravasation. *Nature Immunology, 5*, 393–400.

Seidman, M. A., Chew, T. W., Schenkel, A. R., & Muller, W. A. (2009). PECAM-independent thioglycollate peritonitis is associated with a locus on murine chromosome 2. *PLoS ONE, 4*, e4316.

Shaw, S. K., Bamba, P. S., Perkins, B. N., & Luscinskas, F. W. (2001). Real-time imaging of vascular endothelial-cadherin during leukocyte transmigration across endothelium. *Journal of Immunology, 167*, 2323–2330.

Shaw, S. K., Ma, S., Kim, M. B., Rao, R. M., Hartman, C. U., Froio, R. M., et al. (2004). Coordinated redistribution of leukocyte LFA-1 and endothelial cell ICAM-1 accompany neutrophil transmigration. *Journal of Experimental Medicine, 200*, 1571–1580.

Smith, M. J., Goodfellow, P. J., & Goodfellow, P. N. (1993). The genomic organisation of the human pseudoautosomal gene *MIC2* and the detection of a related locus. *Human Molecular Genetics, 2*, 417–422.

Stefanidakis, M., Newton, G., Lee, W. Y., Parkos, C. A., & Luscinskas, F. W. (2008). Endothelial CD47 interaction with SIRPgamma is required for human T-cell transendothelial migration under shear flow conditions *in vitro*. *Blood, 112*, 1280–1289.

Suh, Y. H., Shin, Y. K., Kook, M. C., Oh, K. I., Park, W. S., Kim, S. H., et al. (2003). Cloning, genomic organization, alternative transcripts and expression analysis of CD99L2, a novel paralog of human CD99, and identification of evolutionary conserved motifs. *Gene, 307*, 63–76.

Turowski, P., Martinelli, R., Crawford, R., Wateridge, D., Papageorgiou, A. P., Lampugnani, M. G., et al. (2008). Phosphorylation of vascular endothelial cadherin controls lymphocyte emigration. *Journal of Cell Science, 121*, 29–37.

van Buul, J. D., Allingham, M. J., Samson, T., Meller, J., Boulter, E., Garcia-Mata, R., et al. (2007a). RhoG regulates endothelial apical cup assembly downstream from ICAM1 engagement and is involved in leukocyte trans-endothelial migration. *Journal of Cell Biology, 178*, 1279–1293.

van Buul, J. D., Kanters, E., & Hordijk, P. L. (2007b). Endothelial signaling by Ig-like cell adhesion molecules. *Arteriosclerosis, Thrombosis, and Vascular Biology, 27*, 1870–1876.

van Wetering, S., van Buul, J. D., Quik, S., Mul, F. P., Anthony, E. C., ten Klooster, J. P., et al. (2002). Reactive oxygen species mediate Rac-induced loss of cell–cell adhesion in primary human endothelial cells. *Journal of Cell Science, 115*, 1837–1846.

van Wetering, S., van den Berk, N., van Buul, J. D., Mul, F. P., Lommerse, I., Mous, R., et al. (2003). VCAM-1-mediated Rac signaling controls endothelial cell–cell contacts and leukocyte transmigration. *American Journal of Physiology Cell Physiology, 285*, C343–C352.

Wakelin, M. W., Sanz, M.-J., Dewar, A., Albelda, S. M., Larkin, S. W., Boughton-Smith, N., et al. (1996). An anti-platelet/endothelial cell adhesion molecule-1 antibody inhibits leukocyte extravasation from mesenteric microvessels *in vivo* by blocking the passage through basement membrane. *Journal of Experimental Medicine, 184*, 229–239.

Weber, C., Fraemohs, L., & Dejana, E. (2007). The role of junctional adhesion molecules in vascular inflammation. *Nature Reviews Immunology, 7*, 467–477.

Wegmann, F., Petri, B., Khandoga, A. G., Moser, C., Khandoga, A., Volkery, S., et al. (2006). ESAM supports neutrophil extravasation, activation of Rho, and VEGF-induced vascular permeability. *Journal of Experimental Medicine, 203*, 1671–1677.

Woodfin, A., Voisin, M. B., Imhof, B. A., Dejana, E., Engelhardt, B., & Nourshargh, S. (2009). Endothelial cell activation leads to neutrophil transmigration as supported by the sequential roles of ICAM-2, JAM-A and PECAM-1. *Blood, 113*, 6246–6257.

Yang, L., Froio, R. M., Sciuto, T. E., Dvorak, A. M., Alon, R., & Luscinskas, F. W. (2005). ICAM-1 regulates neutrophil adhesion and transcellular migration of TNF-alpha-activated vascular endothelium under flow. *Blood, 106*, 584–592.

Yang, L., Kowalski, J. R., Zhan, X., Thomas, S. M., & Luscinskas, F. W. (2006). Endothelial cell cortactin phosphorylation by Src contributes to polymorphonuclear leukocyte transmigration *in vitro*. *Circulation Research, 98*, 394–402.

SECTION 4

METHODS

CHAPTER 12

Fluorescence Resonance Energy Transfer in the Studies of Integrin Activation

Craig T. Lefort and Minsoo Kim
Department of Microbiology and Immunology, David H. Smith Center for Vaccine Biology and Immunology, University of Rochester, Rochester, New York 14642, USA

ABBREVIATIONS

CFP, cyan fluorescent protein; FACS, fluorescence-activated sell sorter; FITC, fluorescein isothiocyanate; FLIM, fluorescence lifetime microscopy; FP, fluorescent protein; FRET, fluorescence resonance energy transfer; GFP, green fluorescent protein; ICAM-1, intercellular adhesion molecule-1; mAb, monoclonal antibody; ORB, octadecyl rhodamine B; PA, photoactivatable; PC, photoconvertible; PE, phycoerythrin; PLL, poly-L-lysine; ROI, region of interest; SBT, spectral bleed-through; VCAM-1, vascular cell adhesion molecule-1; YFP, yellow fluorescent protein.

Current Topics in Membranes, Volume 64

1063-5823/09 $35.00
DOI: 10.1016/S1063-5823(09)64012-6

I. OVERVIEW

The activation of integrin receptors is essential for the recruitment of leukocytes from the circulation to sites of inflammation. Like many other proteins that form the component parts of membrane structures, integrins are not fixed in a particular conformation. Instead, they reversibly equilibrate between the bent, low-affinity and extended, high-affinity conformation, with several apparent intermediate conformations. Recent advances in fluorescence microscope techniques provide various methods to define the structure and function of integrins on the living cell surface. One of the most widely used fluorescence techniques to measure dynamic integrin activation with nanometer resolution is Förster (or fluorescence) resonance energy transfer (FRET). Advances in optics, image processing, highly sensitive photodetectors, and the development of mutant forms of green fluorescent protein (GFP) have improved the application of FRET microscopy to the study of complex processes in living specimens.

In this chapter, we review the different techniques that utilize fluorescent biomolecules and discuss the technical aspects and limitations of each method. In particular, we focus on FRET microscopy and its applications to studying leukocyte integrin activation. Many of the techniques that we describe are accessible to the nonspecialized research laboratory.

II. FLUORESCENT BIOMOLECULES

Fluorescence is a photophysical phenomenon in which a molecule that absorbs light and becomes electrically excited subsequently reverts to its original state by emitting light at a wavelength longer than that at which it was absorbed (Fig. 1). The difference in absorbance (or excitation) and emission wavelength is called the Stokes shift, after the nineteenth-century scientist Sir George G. Stokes who first described fluorescence when he observed the emission of red light when the mineral fluorspar was impinged with ultraviolet light.

Fluorescent molecules have characteristic spectra that indicate their tendency to absorb and emit photons of light at specific wavelengths. The molar extinction coefficient, ε, is a direct measure of the ability of a molecule to absorb photons of light at the peak excitation wavelength. The overall efficiency of fluorescence, called the quantum yield, is expressed by the ratio of photons absorbed to photons emitted. The quality of a fluorophore is often measured as its "brightness," the product of its molar extinction coefficient and quantum yield.

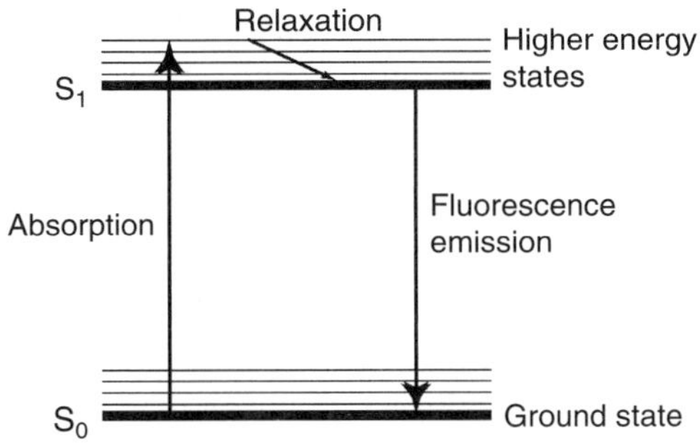

FIGURE 1 A simplified Perrin-Jablonski diagram depicts the energy state changes during photon absorption and fluorescence emission.

The investigation of cellular processes using imaging or other spectroscopic techniques requires the ability to discriminate between objects of interest and background. The direct or indirect conjugation of a fluorescent biomolecule to the object of interest greatly enhances the contrast between signal and noise, and increases the resolution at which biological events can be observed. There is a vast array of fluorescent biomolecules currently available, each with inherent properties that determine their suitability for specific biological research applications.

A. Fluorescent Dyes and Quantum Dots

The most common fluorophores used in early fluorescence applications were organic dyes. Fluorescein and rhodamine were two fluorescent dyes initially used for labeling proteins. Isothiocyanate is a common reactive group used to conjugate these dyes to a molecule of interest by reacting with free amino groups on proteins (Waggoner, 2006). Fluorescein isothiocyanate (FITC) remains a broadly used dye for the direct conjugation of antibodies for use in immunofluorescence applications because of its ease of use, despite its sensitivity to pH. Texas Red, a sulfonated form of rhodamine, was also an early redshifted fluorophore that was used in conjunction with FITC (Titus, Haugland, Sharrow, & Segal, 1982).

With the advancement of fluorescence techniques came the demand for probes with distinct spectral parameters for use in multicolor applications. Phycoerythrin (PE) proved to be an ideal dye because its excitation wavelength was similar to that of FITC, but its emission was shifted to 580 nm, significantly longer than that for FITC, allowing for the simultaneous detection of FITC and PE labels (Waggoner, 2006). Another class of fluorescent dyes, cyanine synthetic dyes, represented an improvement over other labels because of their low molecular weight, redshifted emission spectra, and the ability to tune their spectral properties (Ernst, Gupta, Mujumdar, & Waggoner, 1989; Waggoner). Cy3, Cy5, Cy7, and some intermediate wavelength dyes are common labels for use with multiparameter flow cytometry.

In the last decade, the palette of fluorescence dyes has greatly expanded. The Alexa fluors are a family of commercially produced dyes that are derived from the sulfonation of organic dyes, including those described above (Panchuk-Voloshina et al., 1999). Alexa dyes exhibit more favorable photostability and are less sensitive to pH than their parent fluorescent dyes.

Recently, quantum dots have joined the growing ranks of fluorescence probes. Quantum dots are fluorescent nanocrystals, on the order of several nanometers, with unique optical properties. The wavelength of fluorescence emission depends on the size of the quantum dot, a parameter that can be precisely controlled during their synthesis (Michalet et al., 2005). Quantum dots also exhibit a high quantum yield, are resistant to photobleaching, a broad excitation spectrum at short wavelengths, and a relatively narrow, symmetric emission peak (Michalet et al.; Resch-Genger, Grabolle, Cavaliere-Jaricot, Nitschke, & Nann, 2008). These properties make quantum dots a promising tool for advanced imaging applications, especially for applications where single-molecule sensitivity is needed. The primary shortcomings of quantum dots are their large size, several nanometers compared to $\sim$0.5 nm for organic dyes, and potential toxicity due to heavy metal release during photolysis (Michalet et al.; Resch-Genger et al.). The development of quantum dots as a research tool is in its infancy, but it is only a matter of time before quantum dots are applied to the many fluorescence techniques currently employed in the laboratory.

B. Aequorea*-Derived Fluorescent Proteins*

The cloning of GFP from the jellyfish *Aequorea victoria* (Shimomura, Johnson, & Saiga, 1962) and its subsequent mutation to a photostable protein with excitation and emission properties similar to that of FITC (Heim, Cubitt, & Tsien, 1995) marked the advent of genetically encoded fluorescence as a common tool in the laboratory. Subsequently, many fluorescent proteins (FPs) homologous to GFP that have been isolated from marine species, including the DsRed FPs from the red and orange spectral classes, have added to the growing list of fluorescent biomolecules (Matz et al., 1999). In addition, the repertoire of FPs has greatly expanded (Table I) through the introduction of mutations to the amino acid sequence of GFP and its homologues and derivatives (Pakhomov & Martynov, 2008; Shaner,

TABLE I
Spectral Properties of Fluorescent Proteins

FP	EX (nm)	EM (nm)	QY	ε	References
EBFP	383	445	0.25	31	Patterson, Day, and Piston (2001)
EBFP2	383	448	0.56	32	Ai, Shaner, Cheng, Tsien, and Campbell (2007)
Azurite	384	450	0.55	26.2	Mena, Treynor, Mayo, and Daugherty (2006)
mKalama1	385	456	0.45	36	Ai et al. (2007)
mCerulean	433	475	0.62	43	Rizzo et al. (2004)
ECFP	434	477	0.4	26	Heim, Prasher, and Tsien (1994) and Patterson et al. (2001)
CyPet	435	477	0.51	35	Nguyen and Daugherty (2005)
mTFP1	462	492	0.85	64	Ai, Henderson, Remington, and Campbell (2006)
MiCy	472	495	0.9	27.25	Karasawa, Araki, Nagai, Mizuno, and Miyawaki (2004)
EGFP	484	507	0.6	53	Heim et al. (1995) and Patterson, Knobel, Sharif, Kain, and Piston (1997)
GFP	475, 395	508	0.79	25	Ormo et al. (1996) and Patterson et al. (1997)
T-Sapphire	399	511	0.60	44	Zapata-Hommer and Griesbeck (2003)
Dronpa	503	518	0.85	95	Ando, Mizuno, and Miyawaki (2004) and Andresen et al. (2007)
EYFP	514	527	0.61	84	Ormo et al. (1996) and Wachter, Elsliger, Kallio, Hanson, and Remington (1998)
mVenus	515	528	0.57	92.2	Nagai et al. (2002)
mCitrine	516	529	0.76	77	Griesbeck, Baird, Campbell, Zacharias, and Tsien (2001)
YPet	517	530	0.77	104	Nguyen and Daugherty (2005)
mHoneydew	487/504	537/562	0.12	17	Shaner et al. (2004)
PhiYFP	525	537	0.6	115	Shagin et al. (2004)
mBanana	540	553	0.7	6	Shaner et al. (2004)
mKO	548	561	0.6	51.6	Karasawa et al. (2004)
mOrange	548	562	0.69	71	Shaner et al. (2004)
dTomato	554	581	0.69	69	Shaner et al. (2004)

(*Continued*)

TABLE I (*Continued*)

FP	EX (nm)	EM (nm)	QY	ε	References
DsRed	558	583	0.79	75	Matz et al. (1999)
TagRFP	555	584	0.48	100	Merzlyak et al. (2007)
mTangerine	568	585	0.3	38	Shaner et al. (2004)
mStrawberry	574	596	0.29	90	Shaner et al. (2004)
mRFP1	584	607	0.25	44	Campbell et al. (2002)
mCherry	587	610	0.22	72	Shaner et al. (2004)
mRaspberry	598	625	0.15	86	Wang, Jackson, Steinbach, and Tsien (2004)
mKate	588	635	0.34	65	Shcherbo et al. (2007)
mPlum	590	649	0.1	41	Wang et al. (2004)

EX, excitation peak; EM, emission peak; QY, quantum yield; ε, molar extinction coefficient ($\times 10^3\ M^{-1}\ cm^{-1}$).

Patterson, & Davidson, 2007; Zhang, Campbell, Ting, & Tsien, 2002). These FP variants exhibit more favorable properties, including varied excitation and emission spectra, enhanced fluorescence intensity, increased photostability, abrogated tendency to form oligomers, and a muted response to changes in pH. Some GFP derivatives exhibit enhanced responses to fluctuations in pH (Llopis, McCaffery, Miyawaki, Farquhar, & Tsien, 1998) or sensitivity to changes in membrane potential (Siegel & Isacoff, 1997), properties that can be advantageous for studying certain cellular processes.

The genetically encoded FPs are widely used in the study of dynamic cellular processes because of several factors. First, FPs induce minimal cellular toxicity, as exemplified by the plethora of transgenic mice that have been engineered to express FPs (Hadjantonakis & Nagy, 2001). Second, the FPs exhibit no intrinsic enzymatic activity or intracellular targeting. In most cases, the concatenation of a FP to a protein of interest does not alter that protein's localization or activity. The inert nature of FPs has made their utilization in studying cellular processes more accessible to the researcher that does not have the resources or specialization to perform extensive control experiments. In contrast to small molecule fluorescent dye conjugation, FP tagging results in a product with a known stoichiometry and site of labeling.

There are several considerations that should be made when choosing which fluorescent probe, or set of probes, to employ. The application for which the FP will be utilized will narrow the possible FPs that can be used.

For example, the excitation and emission spectra will be the principal factor in determining whether two FPs can be used as FRET pairs. Variants of cyan FP (CFP) and yellow FP (YFP) are the classical partners for FRET studies. In contrast to choosing fluorophores for FRET, if the goal is to distinctly visualize multiple FP-tagged proteins in the same sample, the excitation and emission spectra of the FPs should exhibit minimal overlap, or spectral bleed-through (SBT). SBT is discussed in further detail in the FRET section of this chapter. FPs exhibit a tendency to dimerize or form higher-order oligomers (Yang, Moss, & Phillips, 1996). However, mutation of hydrophobic residues within the dimerization interface to positively charged amino acids results in monomeric FPs (Zacharias, Violin, Newton, & Tsien, 2002). The monomeric form of the FPs should be used if FP-tagged proteins will be in close proximity or at high local concentrations ($K_d = 0.11$ mM for GFP dimerization; Zacharias et al.). It should be noted, however, that even low FP oligomerization tendencies can affect FRET measurements and the monomeric character of a FP should be closely scrutinized (Shaner et al., 2007).

C. Optical Highlighter FPs

Genetic manipulations performed in search of better FP properties have yielded a unique group of variant FPs called optical highlighters whose fluorescence emission changes in response to irradiation with specific wavelengths of light. Photoactivatable (PA) FPs experience a large enhancement in fluorescence emission, but no change in emission wavelength, while photoconvertible (PC) FPs undergo a change in emission wavelength (Lippincott-Schwartz, Altan-Bonnet, & Patterson, 2003; Lukyanov, Chudakov, Lukyanov, & Verkhusha, 2005; Mizuno et al., 2003).

PA GFP, the first photoactivatable FP to be described, is initially dark and exhibits a greater than 100-fold increase in fluorescence emission intensity when irradiated with 413-nm light (Patterson & Lippincott-Schwartz, 2002). Redshifted PA variants of mRFP (Verkhusha & Sorkin, 2005) and mCherry (Subach et al., 2009) have also been engineered to have greatly enhanced fluorescence when irradiated with ultraviolet light.

Most of the currently described PC FPs exhibit green-to-red photoconversion. These FPs include Kaede (Ando, Hama, Yamamoto-Hino, Mizuno, & Miyawaki, 2002), KikGR (Tsutsui, Karasawa, Shimizu, Nukina, & Miyawaki, 2005), tdEos (Nienhaus et al., 2006), and Dendra2 (Gurskaya et al., 2006), all of which contain the His–Tyr–Gly tripeptide chromophore (Shaner et al., 2007). In this case, photoconversion occurs with exposure to ultraviolet light or short wavelength visible light, and involves rearrangement of the chromophore His residue (Shaner et al.). In addition to the green-to-red PC FPs, there is a CFP variant that switches from a cyan emission at 468 nm to green emission at 511 nm (Chudakov et al., 2004).

Optical highlighter PA and PC FPs are useful tools for monitoring protein dynamics and mobility (Lippincott-Schwartz et al., 2003). Optical highlighter FPs can be photoactivated or photoconverted in distinct regions of the cell and subsequently tracked over time. Thus, the optical highlighter FPs can be used as an alternative to fluorescence recovery after photobleaching (FRAP) and fluorescence loss in photobleaching (FLIP) techniques.

III. FLUORESCENCE TECHNIQUES

A. *Immunofluorescence*

One of the first research techniques that achieved widespread use of fluorescent biomolecules was immunofluorescence. Developed by Coons and colleagues in the 1930s (Coons, 1961), immunofluorescence involves the conjugation of antigen-specific antibodies to a fluorescent dye. With this technique, the contrast between proteins of interest and background signal was greatly enhanced under the microscope. Immunofluorescence was subsequently applied in the development of the fluorescence-activated cell sorter (FACS) by Herzenberg and colleagues in the 1960s (Hulett, Bonner, Barrett, & Herzenberg, 1969). The ability to analyze the protein fingerprint of individual cells within a population using FACS is perhaps the most important technical development in the field of modern immunology.

Technical advances of the early 1970s introduced the first commercially available epifluorescence microscopes with sensitive dichroic filter systems. These instruments allowed biologists to take full advantage of immunofluorescence techniques and to image cell structures with relatively high resolution. In addition to these technical advances in microscopic systems, the development of more specific antibodies to a variety of cellular proteins permitted the investigation of the remarkably complex cellular architecture and of associations between molecules within the same subcellular compartment. However, the bulk of imaging studies were performed on samples that were fixed and stained, and thus static, providing only a brief snapshot of the organization and properties of these cellular structures.

Integrins are heterodimeric transmembrane receptors that mediate adhesive interactions that are critical for leukocyte recruitment from the vasculature (Ley, Laudanna, Cybulsky, & Nourshargh, 2007). The affinity of the extracellular ligand-binding domain is rapidly upregulated during a process termed "inside-out activation" in which intracellular signaling, often triggered by G protein-coupled receptors, results in the induction of global conformational changes (Luo, Carman, & Springer, 2007). To study the conformational changes in integrin structure that occur during activation, monoclonal antibodies (mAbs) were developed that recognize epitopes that are accessible only in the active integrin conformer (Humphries, 2000). While these mAbs are still widely used to determine integrin activation state, more advanced techniques for monitoring the conformational state of integrins have been developed using fluorescence imaging techniques, such as FRET. These new methods, described later in this chapter, have allowed the kinetics of integrin activation to be monitored in real time and in living specimens.

B. FRET

Although FRET was originally described and developed in the mid-twentieth century, its use to study biological processes was uncommon until the development of a wide range of fluorescent biomolecules that exhibited favorable properties for energy transfer and the widespread accessibility of advanced high resolution imaging instruments. In recent years, FRET microscopy studies have been used as a powerful tool to validate protein–protein interactions observed by biochemical methods, such as coimmunoprecipitation or yeast two hybrid, in a living cell or organism, as well as to elucidate the structural changes that occur in proteins and protein complexes.

FRET is a spectroscopic phenomenon that involves the migration of energy from one fluorophore to a second fluorophore or molecule through a nondestructive and nonradiative process. For this transfer of energy to occur, the two fluorescent species must have overlapping spectra and be in close proximity (Fig. 2). With conventional epifluorescence microscopy, the limit of resolution between fluorescent biomolecules is greater than 200 nm, while FRET occurs only at distances less than 10 nm (100 Å). This makes FRET a superior method for analyzing intermolecular or intramolecular interactions.

A set of fluorescent molecules, such as CFP and YFP, for which the emission spectrum of the "donor" (CFP) overlaps with the excitation spectrum of the "acceptor" (YFP) are called a FRET pair (Table II). The efficiency with which the acceptor quenches the donor emission is inversely related to the sixth power of the distance between the fluorophores, according to the following equation:

$$E = \frac{1}{1 + (r/R_0)^6},$$

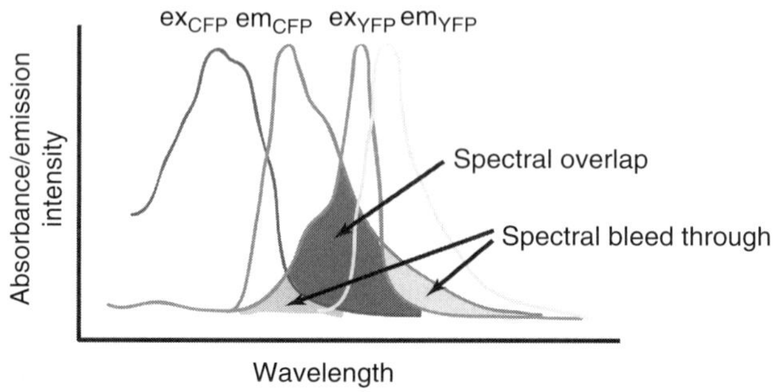

FIGURE 2 Spectral properties of the CFP–YFP FRET pair. The overlap of the CFP emission and YFP excitation spectra indicate that energy transfer is possible between the two fluorophores. The FRET signal, YFP fluorescence emission upon excitation of CFP, contains SBT signals that need to be accounted for.

TABLE II
Donor–Acceptor Pairs for FRET

Donor–acceptor	EX_D (nm)	EM_A (nm)	Method	References
CFP–YFP	434	527	ALL	
BFP–GFP	383	507	se, ff	Daelemans et al. (2004) and Heim and Tsien (1996)
GFP–mRFP	484	607	ff	Peter et al. (2005)
GFP–mCherry	484	610	ff	Tramier, Zahid, Mevel, Masse, and Coppey-Moisan (2006)
FITC–Rhodamine	488	610	r, ff	Chigaev et al. (2003) and Wallrabe and Periasamy (2005)
BFP–YFP	383	527	se, ff	Day et al. (2003) and Pollok and Heim (1999)
Cy3–Cy5, Cy5.5	550	670, 715	se	Clegg (1992) and Hohng, Joo, and Ha (2004)
MiCy–mKO	472	561	se	Karasawa et al. (2004)
CyPet–YPet	435	530	se	Nguyen and Daugherty (2005)
mTFP1–mVenus	462	528	ff	Day, Booker, and Periasamy (2008)

EX_D, donor excitation peak; EM_A, acceptor emission peak; ap, acceptor photobleaching FRET; se, sensitized emission FRET; r, ratiometric FRET; ff, FLIM–FRET; ALL, all FRET methods.

where R_0 is the Förster distance, a characteristic of the FRET pair that indicates the distance at which energy migration efficiency is 50%. R_0 is determined by the spectral overlap (J) and the donor quantum yield (Q_D), and is calculated according to (Lakowicz, 1999)

$$R_0 = \left(\frac{JQ_D\kappa^2}{n^4}\right)^{1/6} (9.79 \times 10^3\,\text{Å}),$$

where κ^2 is the dipole–dipole orientation factor, commonly taken as 2/3 (Lakowicz, 1999). For example, R_0 is 4.9 nm for the CFP–YFP FRET pair (Patterson, Piston, & Barisas, 2000). Therefore, in the presence of YFP at a distance of 4.9 nm, a CFP molecule in its excited state will transfer its energy to an YFP molecule with 50% probability. For a FRET pair comprised of fluorescent biomolecules derived from GFP, the closest possible proximity of the donor and acceptor is approximately 30 Å, as each fluorophore is buried within the FP molecule (Lippincott-Schwartz, Snapp, & Kenworthy, 2001). This limits the maximum FRET efficiency possible with many GFP-based FRET pairs to about 80–90%, whereas small organic fluorochromes can achieve higher FRET efficiencies.

C. *FRET Techniques*

When FRET occurs, the energy from the donor fluorophore is transferred to the acceptor. The most widely used methods for quantifying FRET efficiency measure either the decrease in the donor emission, the increase in acceptor emission, or both. Included within each of these measurement realms are multiple modalities for collecting and displaying FRET efficiency data, including sensitized FRET, acceptor photobleaching FRET, and ratiometric FRET. The best FRET measurement method for any given experiment depends upon the goal of the experiment and the limitations in the FRET system. In the following section, we will describe several FRET techniques, discuss their advantages and drawbacks, and briefly detail how each method has been used to study leukocyte integrin activation and function.

1. Sensitized Emission FRET

Sensitized emission FRET involves the measurement of FRET efficiency by monitoring the emission of the acceptor when the donor undergoes excitation. This is the most widely used and straightforward approach to measuring a FRET signal using fluorescence microscopy. Sensitized emission FRET is also called the three-cube FRET method, as the necessary images can be obtained using three different microscope filter sets (Gordon, Berry, Liang, Levine, & Herman, 1998). The first filter set measures the raw FRET signal by exciting the donor fluorophore and receiving light at the emission wavelength of the acceptor. The raw FRET signal then needs to be corrected by acquiring the emission of the donor and acceptor when each is excited.

Correction of the raw FRET signal is required due to two different types of SBT (Fig. 2). If the excitation spectrum of the acceptor is wide, the excitation of the donor may result in the direct excitation of the acceptor. Likewise, if the emission spectrum of the donor is wide, the acquisition of the FRET signal at the emission wavelength of the acceptor may be partially affected by the donor emission. These control experiments need to be performed on samples in which only the donor or acceptor fluorophores are present, at concentrations comparable to the sample from which the FRET signal is obtained. Validation of the accuracy of measuring FRET efficiency using the sensitized emission FRET technique has been performed by using it in combination with other FRET methods, such as acceptor photobleaching FRET (Zal & Gascoigne, 2004; Zal, Zal, & Gascoigne, 2002) or fluorescence lifetime imaging (Hoppe, Christensen, & Swanson, 2002).

The correction of the FRET signal is necessary to obtain quantitative information about the FRET efficiency between the donor and acceptor. Advanced methods and algorithms have been developed for correcting signal contamination by SBT (Berney & Danuser, 2003; Elangovan et al., 2003; Periasamy & Day, 2005), the scope of which is beyond this chapter. However, in many cases qualitative information about changes in FRET efficiency are sufficient for making conclusions about the spatial relationship of the donor and acceptor fluorophores. Under these circumstances, the FRET signal should be normalized to the fluorescence intensity of the acceptor.

The sensitized emission FRET method has distinct advantages over other techniques for measuring FRET that make it the most popular measurement modality. As described above, the direct acquisition of the FRET signal can be performed on a simple wide-field fluorescence microscope with the appropriate filter set. Image acquisition is fast relative to other FRET methods, and multiple time-lapse images can be obtained to observe the relationship of specific protein–protein interactions with distinct cellular processes, such as membrane extension. For intramolecular FRET sensors in which the donor and acceptor fluorophores are expressed within a single construct, sensitized emission FRET may be an ideal method for measuring FRET, as a constant 1:1 donor–acceptor ratio greatly simplifies signal correction for SBT. Finally, the sensitized emission FRET technique, despite its need for SBT corrections, is the best method for performing more advanced FRET experiments, such as three-chromophore FRET microscopy for analyzing interactions between more than two proteins (Galperin, Verkhusha, & Sorkin, 2004).

The main drawback to using sensitized emission FRET is the difficulty in obtaining accurate absolute values for FRET efficiency, even if extensive controls and FRET signal corrections are performed. Experimental conditions in which the relative abundance of donor and acceptor significantly varies throughout the cell population require more extensive and time-consuming SBT corrections. Furthermore, the nonspecific collision of donor and acceptor molecules must be accounted for when there is heterologous expression of donor and acceptor. Finally, for experiments in which FRET is monitored over time and multiple images are acquired, controls should also be performed to determine the extent of unintended photobleaching of both the donor and acceptor.

2. Sensitized Emission FRET: VLA-4 Activation

We have utilized the CFP–YFP FRET pair in developing a system to analyze integrin activation by detecting changes in the spacing of the integrin cytoplasmic tails (Fig. 3A) (Kim, Carman, & Springer, 2003). The rapid upregulation of integrin affinity during activation involves chemokine signaling-induced global conformational changes, including the spatial separation of the α- and β-subunit cytoplasmic tails (Kim et al.) and extension of the extracellular domain (Chigaev, Buranda, Dwyer, Prossnitz, & Sklar, 2003; Takagi, Petre, Walz, & Springer, 2002). It should be noted that changes in the distance between integrin α- and β-subunit tails might not always reflect activation, and *vice versa*. The structural changes that occur during integrin activation are still a topic of active investigation. VLA-4 (integrin $\alpha_4\beta_1$) plays an important role in the trafficking of T-lymphocytes as well as in the pathogenesis of several autoimmune and inflammatory disorders (Hyun, Lefort, & Kim, 2009). Our lab has employed the intracellular integrin activation FRET sensor to measure the kinetics of VLA-4 activation using the sensitized emission FRET technique (Hyun et al., 2009).

To investigate the spatial dynamics of VLA-4 activation, the β_1-integrin-deficient GD25 cell line was transfected with DNA plasmids encoding the α_4-integrin subunit with mCFP fused to its cytoplasmic carboxy-terminus and the β_1-subunit similarly coupled to mYFP. Cells expressing the VLA-4-fused FRET pair were allowed to adhere and spread on a substrate composed of vascular cell adhesion molecule-1 (VCAM-1), a VLA-4 ligand, while the activation state of VLA-4 was monitored by sensitized emission FRET. The loss of FRET between the α_4-mCFP donor and β_1-mYFP acceptor indicates a separation of the VLA-4 cytoplasmic tails, a readout of the active state of the integrin. A low FRET signal was detected over a broad region of the leading edge of cells spreading on VCAM-1. In contrast, cells spreading on poly-L-lysine (PLL), an adhesive substrate that does not engage integrins, exhibit a narrow region of VLA-4 activation restricted to the tip of the lamellipodium. The data suggested that VLA-4 engagement by VCAM-1 stabilized the active form of the integrin, as the accumulation of VLA-4 with a low FRET signal in cells on VCAM-1 was more than twice as great as that of cells on PLL. The use of sensitized emission

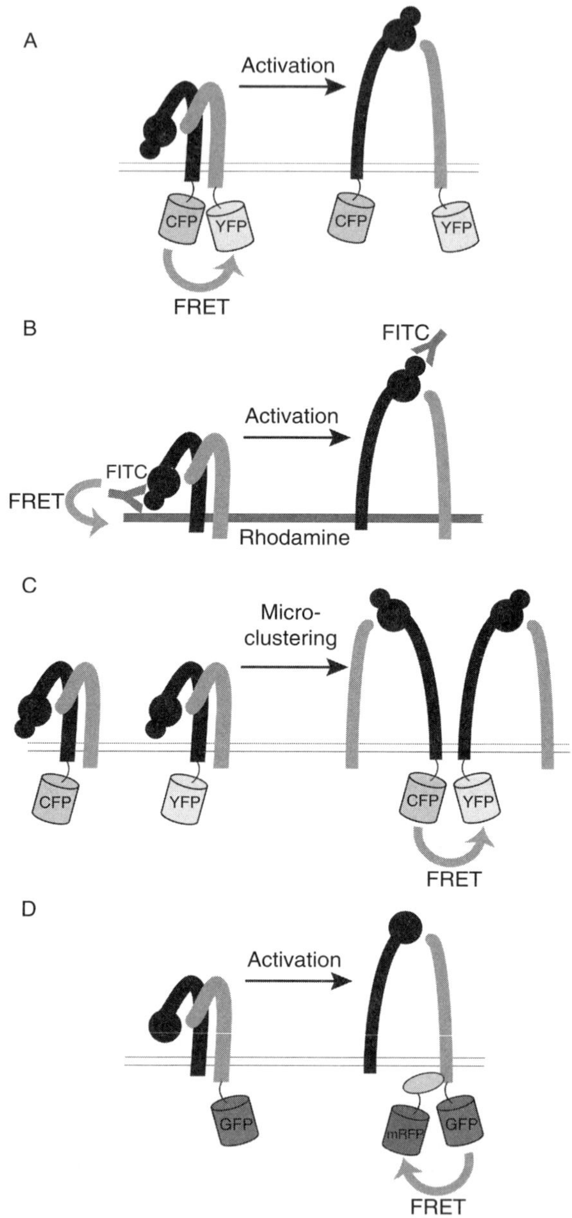

FIGURE 3 Experimental systems used for FRET analysis of integrin activation, clustering, and intracellular protein recruitment. (A) Loss of FRET between CFP- and YFP-tagged integrin cytosolic tails indicates spatial separation during integrin activation. (B) Loss of FRET between

FRET allowed us to observe the dynamics of VLA-4 activation in a single cell as the lamellipodium developed, advanced, and retracted over time, providing useful information about the role of VLA-4 in cell motility.

As discussed above, corrections in the YFP emission intensity need to be performed to obtain an accurate FRET efficiency metric. SBT coefficients were reached using cells expressing the α_4-mCFP donor alone or the β_1-mYFP acceptor cotransfected with their untagged wild-type α/β-subunit counterpart. In addition, control studies analyzing the same region of interest (ROI) using multiple FRET methods indicated that the FRET efficiency calculated by measuring the sensitized emission of YFP fluorescence closely correlated with the FRET efficiency obtained using the acceptor photobleaching method.

3. Acceptor Photobleaching FRET

The simplest endpoint FRET measurement technique is the acceptor photobleaching FRET method. The acceptor photobleaching FRET modality provides a more accurate measurement of FRET efficiency than other techniques, as it directly quantifies the donor emission in the presence and absence of acceptor molecules. Like sensitized emission FRET, acceptor photobleaching FRET can be performed on the wide-field fluorescence microscope with the appropriate filter sets. However, for acceptor photobleaching FRET the necessary filter sets include the standard excitation/emission filter for both the donor and acceptor, as well as a specialized filter for the destruction, or photobleaching, of the acceptor fluorophore. In acceptor photobleaching FRET, the emission of the donor and acceptor are acquired before and after photobleaching of the acceptor. After correction for the background signal, the digital subtraction of the donor image after photobleaching from the donor image before photobleaching results in an image of the donor signal that was quenched by the acceptor due to FRET. The calculation of FRET efficiency is simple:

$$E = 1 - \frac{F_{\mathrm{D,Pre}}}{F_{\mathrm{D,Post}}},$$

where $F_{\mathrm{D,Pre}}$ and $F_{\mathrm{D,Post}}$ are the donor emission intensity before and after acceptor photobleaching, respectively.

FITC-conjugated anti-integrin I-domain antibodies and the membrane dye ORB indicates extracellular domain extension during integrin activation. (C) FRET between CFP- and YFP-tagged integrin α-subunits indicates integrin microclustering. (D) FRET between GFP-tagged integrins and mRFP-tagged proteins, such as talin and paxillin, indicates adhesion complex formation.

There are several advantages to employing the acceptor photobleaching FRET method for measuring FRET efficiency. The emission from the donor fluorophore is the only signal used to calculate FRET efficiency. Therefore, this method does not suffer from many of the SBT issues that interfere with the accurate measurement of the acceptor emission in other FRET techniques. However, controls in which only the donor or acceptor is present should be performed, as some investigators have observed changes in the donor FP spectra as a result of acceptor photobleaching (Valentin et al., 2005).

Despite the fact that the acceptor photobleaching FRET method produces relatively accurate measurements of FRET efficiency, there are some negative characteristics of this method. The main drawback to acceptor photobleaching FRET is that it is an endpoint assay. Only a single measurement can be made on any given ROI, as the photobleaching process irreversibly destroys the acceptor fluorophore. The accuracy of this technique correlates with the completeness of acceptor photobleaching. In fact, a technical study found that bleaching the acceptor by 70% introduces up to 50% error in FRET efficiency measurement (Berney & Danuser, 2003). Complete photobleaching of the acceptor requires exposure to high intensity light for several minutes. Depending on the stability of the intermolecular or intramolecular interaction that is being studied, or the mobility of donor fluorophore within the system, changes not due to the dequenching of FRET may occur during the time required to completely photobleach the acceptor. For example, acceptor photobleaching FRET would not be the ideal method for measuring VLA-4 activation as described in the previous section, as dynamic changes in the lamellipodium and movement of VLA-4 within the membrane would introduce significant error in the measurement of FRET.

4. Acceptor Photobleaching FRET: LFA-1 and Mac-1 Activation

As described in the previous section for VLA-4, FRET has been used to report the active conformation of the integrin (Fig. 3A) (Kim et al., 2003). While the studies of VLA-4 activation kinetics in adherent cells required the use of sensitized emission FRET to resolve integrin conformational changes over time and space, the investigation of LFA-1 (integrin $\alpha_L\beta_2$) and Mac-1 (integrin $\alpha_M\beta_2$) activation was performed using acceptor photobleaching FRET on cells in suspension. Using the nonadherent myeloid K562 cell line to probe the conformational mechanisms of integrin activation is appropriate, as LFA-1 and Mac-1 activation precedes the transition of leukocytes from selectin-mediated rolling on the endothelium to integrin-mediated adhesion, migration, and diapedesis (Ley et al., 2007). The use of nonadherent cells allowed us to employ acceptor photobleaching FRET since the ROI would not change in location or morphology during the course of a measurement.

Using the CFP–YFP FRET pair genetically fused to the α_L- and β_2-integrin subunits (Fig. 3A), Kim et al. (2003) were the first to demonstrate that integrins undergo a spatial separation of their cytoplasmic tails during activation. These first studies were performed using the acceptor photobleaching FRET technique and indicated a FRET efficiency of 20% between mCFP and mYFP in the resting state of the integrin (Kim et al.). Similarly, α_M-mCFP and β_2-mYFP exhibited a FRET efficiency of 20% in the inactive Mac-1 integrin (Lefort et al., in preparation). To photobleach the YFP acceptor fluorophore, the sample was exposed to a maximum intensity of light at an excitation wavelength of 535 nm for 3 min. These settings resulted in a consistent reduction of YFP emission by greater than 90% (C.T. Lefort and M. Kim, unpublished observations). To prevent the drifting or movement of cells in suspension, glass microscope slides were coated with PLL and cells were allowed to settle on the substrate before measurements were taken. Utilizing this experimental system and the acceptor photobleaching FRET method, the conformational separation of both LFA-1 and Mac-1 cytoplasmic tails was observed in response to binding of the physiologic ligand intracellular adhesion molecule-1 (ICAM-1) or an anti-β_2-integrin activating monoclonal antibody (Kim et al.; Lefort et al.). CFP–YFP-tagged integrin subunits have been used by others to monitor LFA-1 activation by acceptor photobleaching FRET (Vararattanavech et al., 2008) and Mac-1 heterodimerization by sensitized emission FRET (Fu et al., 2006).

5. Ratiometric FRET

The donor–acceptor ratio in an intramolecular FRET construct, by definition, is equal to 1. In contrast, intermolecular FRET systems where a donor and acceptor reside on distinct proteins or molecules introduce variability in the relative abundance of each. As technical studies have shown, donor–acceptor ratios that significantly vary from 1 lead to large errors in measuring FRET efficiency using either the sensitized emission or acceptor photobleaching FRET methods (Berney & Danuser, 2003). For genetically encoded FP-based FRET constructs, introduction of equal amounts of plasmid DNA can result in relatively equal expression of donor and acceptor for the entire cell population. However, the donor–acceptor ratio may be highly variable for any given cell within the population. This is an especially cumbersome issue when employing sensitized emission FRET, as the SBT signals that need to be corrected for will also vary from cell to cell. To address this issue, many investigators will make FRET measurements only on cells that fall within a given range of donor and acceptor fluorescence intensity (Kim et al., 2003).

Alternatively, one could use the ratiometric FRET method for quantifying FRET efficiency. The ratiometric FRET technique takes advantage of the dependence of FRET on the donor–acceptor ratio. While the abundance of the donor fluorophore is held relatively constant, the donor emission, or ratio of FRET signal to donor emission, is measured for increasing concentrations of the acceptor. As the abundance of the acceptor relative to the donor is increased, the probability that an acceptor is available for transfer of energy from the donor also increases. Obtaining a quantitative measure of FRET efficiency using the ratiometric FRET method is somewhat more elusive than with the previously described techniques. At first glance, it would seem that calculation of the FRET efficiency would be similar to that of the acceptor photobleaching FRET method, as donor emission intensities are acquired in both the absence and presence of acceptor molecules. However, models of this method indicate that, in certain circumstances, measurements of FRET efficiency become inaccurate for high acceptor densities (Wolber & Hudson, 1979), as increasing acceptor density enhances FRET even between noninteracting proteins (Zacharias et al., 2002). Therefore, FRET data are often collected at appropriate acceptor levels and presented as a relative measure that compares the change in distance between the donor and acceptor under different experimental conditions.

The ratiometric FRET method is not as widely used as the sensitized emission and acceptor photobleaching FRET measurement techniques. Often, ratiometric FRET is used when the experimental system limits the use of other methods. For example, it may be the ideal option when the acceptor-labeled protein competes with the endogenous protein (Periasamy & Day, 2005). A caveat when varying the acceptor concentration to perform ratiometric FRET measurements is that, depending on the experimental system, one may also be varying the expression of the protein being studied. Protein overexpression may affect cell function as well as the specific protein–protein interaction being analyzed. Similar to sensitized emission FRET, measurements taken using the ratiometric FRET method need to be corrected for SBT.

6. Ratiometric FRET: Integrin Extension

During inside-out activation, integrins undergo conformational rearrangement of their cytoplasmic tails, as described in the previous sections, and their extracellular stalk domains (Luo et al., 2007). Electron micrograph and X-ray crystal studies have demonstrated that isolated integrins exist in a compact bent structure (Takagi et al., 2002; Xiong et al., 2001), suggesting that the ligand-binding integrin head is in close proximity with the plasma membrane in the inactive molecule. These studies also suggested that integrin activation induced a switchblade-like extension of the integrin head away from the plasma membrane (Takagi et al.).

Studies utilizing the ratiometric FRET method have verified the conformational extension of the extracellular domain during activation of the leukocyte integrins VLA-4 (Chigaev et al., 2003, 2004, 2009; Chigaev, Waller, Amit, &

Sklar, 2008; Chigaev, Waller, Zwartz, Buranda, & Sklar, 2007), LFA-1 (Larson et al., 2005), and Mac-1 (Lefort et al., in preparation). In each of these studies, a fluorescent mAb or ligand bound to the ligand-binding integrin head domain acted as a FRET donor and octadecyl rhodamine B (ORB), a lipid dye that incorporates into the plasma membrane, was used as an acceptor (Fig. 3B). In addition, these experiments used flow cytometry to enable measurement of donor fluorescence over the entire cell surface and for a large number of cells. FRET between FITC-conjugated donors and ORB dye was observed on resting cells. Cell stimulation with various integrin agonists induced a decrease in the quenching of donor fluorescence by the membrane-integrated acceptor molecules, indicating that the integrin head domain undergoes a conformational rearrangement during activation resulting in its extension away from the plasma membrane. These data validated studies on isolated molecules suggesting that integrins "stand up" upon activation, demonstrating that conformational extension of the extracellular domain occurs in living cells. The experiments were also the first to show the relationship between integrin conformation and affinity by combining kinetic analysis of ligand binding with FRET (Chigaev et al., 2003).

Ratiometric FRET was the appropriate technique for measuring FRET by flow cytometry in these studies for several reasons. First, the use of the lipid dye as an acceptor allowed for the monitoring of FRET in real time, but would have introduced a high background fluorescence in the solution containing the cells if FRET was analyzed under the microscope (Chigaev et al., 2003). The flow cytometer is able to discern between fluorescent biomolecules that are cell-bound versus in solution (Nolan & Sklar, 1998). While the chemical properties of ORB made it an ideal FRET acceptor for these experiments, the spectral properties of rhodamine made acceptor photobleaching FRET an inferior method because of difficulties in photobleaching the ORB signal. By using ratiometric FRET, the donor fluorescence in the absence of acceptors was acquired before addition of the fluorescent dye acceptor.

7. Fluorescence Lifetime Microscopy (FLIM)

Similar to acceptor photobleaching FRET, FLIM–FRET relies on the measurement of donor emission to determine the efficiency of energy transfer to the acceptor. However, in contrast to the acceptor photobleaching FRET method, FLIM–FRET is not an endpoint assay and does not require manipulation of the acceptor fluorophore. FLIM–FRET takes advantage of the effect of FRET on the exponential decay rate of fluorescence emission (Bastiaens & Squire, 1999; Oida, Sako, & Kusumi, 1993; Wallrabe & Periasamy, 2005). The fluorescence spontaneously emitted by a fluorophore in its excited state decays according to the equation (Lakowicz, 1999)

$$F(t) = F_0 e^{-t/\tau},$$

where the initial fluorescence is F_0 and the fluorescence lifetime, τ, is the average time that a molecule remains in its excited state. The exponential decay of the donor fluorophore emission is enhanced when FRET occurs with an acceptor molecule. Therefore, donor–acceptor FRET efficiency can be obtained by monitoring the decay in donor emission and comparing the fluorescence lifetime of the donor in the presence and absence of acceptors. Calculation of the FRET efficiency using FLIM is straightforward (Wallrabe & Periasamy, 2005):

$$E = 1 - \frac{\tau_{DA}}{\tau_D},$$

where τ_{DA} and τ_D are the fluorescence lifetime of the donor in the presence and absence of acceptors, respectively.

Typically, FLIM measurements are taken by impinging the sample with a short pulse of light at the donor excitation wavelength and then detecting the emission fluorescence intensity at multiple nanosecond time points after excitation (Wallrabe & Periasamy, 2005). This modality of collecting fluorescence lifetime data is called the time-domain or pulse method (Wallrabe & Periasamy). FLIM data can also be obtained using the frequency-domain method, which requires a modulated light source as well as a modulated detector (Wallrabe & Periasamy).

FLIM–FRET has several advantages over other methods for measuring FRET efficiency. The fluorescence lifetime of a fluorophore is independent of its intensity and relative abundance. The time resolution of FLIM–FRET is on the order of nanoseconds and FRET can be monitored over a time course. Like acceptor photobleaching FRET, FLIM–FRET does not require correction for SBT since only the donor fluorescence is monitored. There are some caveats to be aware of when utilizing FLIM–FRET analyses. The fluorescence lifetime of a fluorophore is impacted not only by FRET, but also by other factors, such as pH. In addition, some FP species have been shown to exhibit two-component fluorescence lifetimes (Periasamy & Day, 2005; Rizzo, Springer, Granada, & Piston, 2004), a potential source for measurement error. Finally, a significant drawback to the FLIM–FRET method is the specialized equipment that is required.

8. FLIM–FRET: Integrin–Effector Binding

The association of intracellular proteins with the integrin cytoplasmic tail occurs during both inside-out activation of integrin affinity and outside-in triggering of signal transduction pathways. Talin binds and activates β_1-, β_2-, and β_3-integrins (Calderwood et al., 1999; Kim et al., 2003). Paxillin is an adaptor protein that can bind directly to the β_1-integrin cytoplasmic tail (Schaller, Otey, Hildebrand, & Parsons, 1995). VLA-4 function and focal adhesion formation are also regulated by the association of paxillin with the α_4-integrin subunit (Han et al., 2001; Han, Rose, Woodside, Goldfinger, & Ginsberg, 2003; Liu et al., 1999).

FLIM–FRET studies have recently been performed using β_1- and α_4-integrins fused to a GFP FRET donor and mRFP-tagged effector proteins talin and paxillin as FRET acceptors (Fig. 3D) (Parsons, Messent, Humphries, Deakin, & Humphries, 2008). The investigators showed that the fluorescence lifetime of the β_1-GFP donor is decreased in the presence of the mRFP-tagged talin rod domain under conditions where the integrin is activated, demonstrating a correlation between talin recruitment and integrin activation. Likewise, α_4-GFP and paxillin–mRFP exhibited FRET in cells adherent to fibronectin, but not PLL. Further, Parsons and colleagues demonstrated that small molecule inhibitors of VLA-4 could act as integrin agonists and induce the recruitment of focal adhesion proteins, an event potentially detrimental to their therapeutic activity.

FRET was also recently used to demonstrate interaction of the actin-binding protein α-actinin with the β_2-integrin subunit of LFA-1 (Stanley et al., 2008). In contrast to the FLIM–FRET study described above, the authors employed acceptor photobleaching FRET to measure FRET between a GFP-tagged β_2-integrin subunit donor and α-actinin labeled with an Alexa546-conjugated mAb. Using this experimental setup, LFA-1 and α-actinin were found to exhibit FRET, suggesting a direct interaction, at the leading edge of migrating T-lymphocytes (Stanley et al., 2008), but not at the midcell region where talin has been shown to bind LFA-1 (Smith et al., 2005).

9. Photoquenching FRET

Another recently developed variation of the FRET methods that monitor the donor emission in the absence and presence of acceptor molecules is called photoquenching FRET (Demarco, Periasamy, Booker, & Day, 2006). Photoquenching FRET is a technique for the manipulation of the FRET pair to obtain energy transfer data, rather than a distinct method for measuring FRET signals. Thus, it is used in combination with one of the FRET methods described in this chapter. Briefly, the technique involves the activation of acceptor fluorescence during the course of monitoring donor emission. Photoactivated GFP acts as an acceptor whose ability to quench the fluorescence of a CFP donor fluorophore is greatly enhanced when it is activated with a pulse of 400-nm light (Demarco et al.). A great advantage to using photoquenching FRET is the added ability to track the mobility of a population of photoactivated GFP-tagged proteins while simultaneously measuring their interaction with donor-tagged proteins by FRET.

10. Studying Integrin Clustering with FRET

In addition to the conformational regulation of integrin activity, the lateral organization of integrins in the plasma membrane affects their adhesive and signaling functions (van Kooyk & Figdor, 2000). Lateral clustering of integrins on the scale of >200 nm, called "macroclustering," can be resolved by standard light microscopy (Kim, Carman, Yang, Salas, & Springer, 2004). However, more advanced methods are required to analyze integrin "microclustering," or oligomerization on the nanometer scale.

Two different FRET methods have been used to investigate integrin microclustering. Measurement of plasma membrane protein clustering by FRET was first achieved while investigating lipid-enriched microdomains called lipid rafts (Zacharias et al., 2002). These studies were performed with lipid-modified FPs and utilized acceptor photobleaching FRET to measure energy transfer efficiency. To quantify microclustering, the FRET donor and acceptor densities were varied and the FRET efficiency versus acceptor intensity relationship was analyzed using a saturable one-site binding model (Zacharias et al.). In this manner, the shape of the FRET efficiency versus acceptor intensity curve indicates the extent of density-independent FRET, or microclustering, between donors and acceptors. Similar studies were undertaken with the integrin LFA-1, using the CFP–YFP FRET pair fused to α_L-integrin subunits (α_L-mCFP and α_L-mYFP) and coexpressed with the wild-type β_2-integrin subunit (Fig. 3C) (Kim et al., 2004). Using this FRET system, the investigators demonstrated that microclustering of LFA-1 occurs as a consequence of multivalent ligand binding, and not in response to conformational activation or cytoskeletal disruption alone (Kim et al.) as had been suggested by previous studies (Li et al., 2003).

The most recent study to apply FRET methods for analyzing LFA-1 integrin microclustering used a similar experimental setup to obtain FRET efficiencies between α_L-mCFP and α_L-mYFP (Fig. 3C) (Vararattanavech, Lin, Torres, & Tan, 2009). In this study, disruption of the transmembrane interface between the α_L- and β_2-integrin subunits by mutation was shown to significantly enhance α_L-mCFP donor fluorescence in response to YFP acceptor photobleaching (Vararattanavech et al.). In contrast to the results from the study described above (Kim et al., 2004), these data suggested that separation of the LFA-1 transmembrane domains induces microclustering (Vararattanavech et al.). However, it is possible that LFA-1 clustering was driven in part by ligand binding or homotypic cell aggregation, since the mutant integrin also exhibited enhanced ligand affinity (Vararattanavech et al.).

Finally, a third study to utilize FRET in analyzing integrin receptor microclustering employed sensitized emission FRET with the donor–acceptor pair of mVenus and mCherry (Smith, Bunch, & Brower, 2007).

The authors describe a novel system of FRET reporters consisting of integrin transmembrane and cytosolic domains alone, with the FP fluorophores as an extracellular component of the constructs. Thus, the reporters may be used to test the effects of a series of integrin mutants on microclustering while requiring only a single cloning step. FRET efficiencies were calculated by acquiring the FRET signal, the emission of the acceptor mCherry upon excitation of the donor mVenus, and SBT corrections were applied. The assay system was used to analyze the effects of several *Drosophila* integrin mutations on lateral oligomerization.

11. Other Considerations for FRET

FRET is a powerful technique that can provide useful information about protein interactions and molecular structure. However, in addition to the caveats described in the previous sections, there are several general considerations that should be recognized when interpreting FRET measurements. First, a positive FRET signal alone does not indicate a direct protein interaction. FRET is a complementary method in a living specimen that usually supports biochemical observations. Unless resolved on the single-molecule level (Roy, Hohng, & Ha, 2008), FRET efficiency does not necessarily indicate the absolute distance between donor and acceptor since most methods are unable to distinguish between an increase in FRET efficiency and an increase in the concentration of FRET species (Wallrabe & Periasamy, 2005). Moreover, FRET efficiency is a metric for all donor molecules, while under most conditions only a fraction of donors are undergoing FRET. The issue of SBT and correcting for artificial signals is a topic of intense research. Some studies indicate that SBT corrections for FRET measurements on confocal microscopes are unstable, and even further caution should be used when employing certain instrumentation (Tadross, Park, Veeramani, & Yue, 2009; van Rheenen, Langeslag, & Jalink, 2004).

IV. SUMMARY

The development of genetically encoded FPs has triggered a new era in biomedical research. Dynamic processes on the cellular and molecular scale can be observed and quantified using today's fluorescence microscopy instrumentation. Photophysical phenomena, such as FRET and fluorescence lifetime, can be exploited to obtain spatiotemporal information about intermolecular and intramolecular interactions. FRET has been used to study the global conformational changes associated with integrin activation and lateral oligomerization, cellular processes important for regulating leukocyte adhesion and migration.

As novel fluorescence probes and techniques are developed and advanced imaging equipment becomes more accessible, the complexity of cellular events that can be analyzed and the resolution at which they can be observed increases. In the future, improved variants of GFP and FPs derived from other marine species, as well as the further development of quantum dots, will provide a palette of fluorophores that exhibit more favorable characteristics for FRET. These advances will enable multichromophore FRET, for measuring interactions within protein complexes or multidimensional intramolecular conformational changes. Additionally, FPs with more favorable spectral properties will reduce errors associated with SBT and increase the accuracy of FRET efficiency measurements. Developments in imaging technology such as multiphoton excitation will also allow for accurate FRET imaging *in vivo* in real time, providing valuable information about integrin activation kinetics in a living specimen.

Acknowledgments

Due to limitations in space, the authors would like to apologize for not being able to cite all of the important research that has contributed to this field. We are indebted to funding provided by the National Institutes of Health to M.K. (HL087088 and HL18208).

References

Ai, H. W., Henderson, J. N., Remington, S. J., & Campbell, R. E. (2006). Directed evolution of a monomeric, bright and photostable version of Clavularia cyan fluorescent protein: Structural characterization and applications in fluorescence imaging. *Biochemical Journal, 400*(3), 531–540.

Ai, H. W., Shaner, N. C., Cheng, Z., Tsien, R. Y., & Campbell, R. E. (2007). Exploration of new chromophore structures leads to the identification of improved blue fluorescent proteins. *Biochemistry, 46*(20), 5904–5910.

Ando, R., Hama, H., Yamamoto-Hino, M., Mizuno, H., & Miyawaki, A. (2002). An optical marker based on the UV-induced green-to-red photoconversion of a fluorescent protein. *Proceedings of the National Academy of Sciences of the United States of America, 99*(20), 12651–12656.

Ando, R., Mizuno, H., & Miyawaki, A. (2004). Regulated fast nucleocytoplasmic shuttling observed by reversible protein highlighting. *Science, 306*(5700), 1370–1373.

Andresen, M., Stiel, A. C., Trowitzsch, S., Weber, G., Eggeling, C., Wahl, M. C., et al. (2007). Structural basis for reversible photoswitching in Dronpa. *Proceedings of the National Academy of Sciences of the United States of America, 104*(32), 13005–13009.

Bastiaens, P. I., & Squire, A. (1999). Fluorescence lifetime imaging microscopy: Spatial resolution of biochemical processes in the cell. *Trends in Cell Biology, 9*(2), 48–52.

Berney, C., & Danuser, G. (2003). FRET or no FRET: A quantitative comparison. *Biophysical Journal, 84*(6), 3992–4010.

Calderwood, D. A., Zent, R., Grant, R., Rees, D. J., Hynes, R. O., & Ginsberg, M. H. (1999). The talin head domain binds to integrin beta subunit cytoplasmic tails and regulates integrin activation. *Journal of Biological Chemistry, 274*(40), 28071–28074.

Campbell, R. E., Tour, O., Palmer, A. E., Steinbach, P. A., Baird, G. S., Zacharias, D. A., et al. (2002). A monomeric red fluorescent protein. *Proceedings of the National Academy of Sciences of the United States of America, 99*(12), 7877–7882.

Chigaev, A., Buranda, T., Dwyer, D. C., Prossnitz, E. R., & Sklar, L. A. (2003). FRET detection of cellular alpha4-integrin conformational activation. *Biophysical Journal, 85*(6), 3951–3962.

Chigaev, A., Waller, A., Amit, O., Halip, L., Bologa, C. G., & Sklar, L. A. (2009). Real-time analysis of conformation-sensitive antibody binding provides new insights into integrin conformational regulation. *Journal of Biological Chemistry, 284*(21), 14337–14346.

Chigaev, A., Waller, A., Amit, O., & Sklar, L. A. (2008). Galphas-coupled receptor signaling actively down-regulates alpha4beta1-integrin affinity: A possible mechanism for cell de-adhesion. *BMC Immunology, 9*, 26.

Chigaev, A., Waller, A., Zwartz, G. J., Buranda, T., & Sklar, L. A. (2007). Regulation of cell adhesion by affinity and conformational unbending of alpha4beta1 integrin. *Journal of Immunology, 178*(11), 6828–6839.

Chigaev, A., Zwartz, G. J., Buranda, T., Edwards, B. S., Prossnitz, E. R., & Sklar, L. A. (2004). Conformational regulation of alpha 4 beta 1-integrin affinity by reducing agents. "Inside-out" signaling is independent of and additive to reduction-regulated integrin activation. *Journal of Biological Chemistry, 279*(31), 32435–32443.

Chudakov, D. M., Verkhusha, V. V., Staroverov, D. B., Souslova, E. A., Lukyanov, S., & Lukyanov, K. A. (2004). Photoswitchable cyan fluorescent protein for protein tracking. *Nature Biotechnology, 22*(11), 1435–1439.

Clegg, R. M. (1992). Fluorescence resonance energy transfer and nucleic acids. *Methods in Enzymology, 211*, 353–388.

Coons, A. H. (1961). The beginnings of immunofluorescence. *Journal of Immunology, 87*, 499–503.

Daelemans, D., Costes, S. V., Cho, E. H., Erwin-Cohen, R. A., Lockett, S., & Pavlakis, G. N. (2004). In vivo HIV-1 Rev multimerization in the nucleolus and cytoplasm identified by fluorescence resonance energy transfer. *Journal of Biological Chemistry, 279*(48), 50167–50175.

Day, R. N., Booker, C. F., & Periasamy, A. (2008). Characterization of an improved donor fluorescent protein for Forster resonance energy transfer microscopy. *Journal of Biomedical Optics, 13*(3), 031203.

Day, R. N., Voss, T. C., Enwright III, J. F., Booker, C. F., Periasamy, A., & Schaufele, F. (2003). Imaging the localized protein interactions between Pit-1 and the CCAAT/enhancer binding protein alpha in the living pituitary cell nucleus. *Molecular Endocrinology, 17*(3), 333–345.

Demarco, I. A., Periasamy, A., Booker, C. F., & Day, R. N. (2006). Monitoring dynamic protein interactions with photoquenching FRET. *Nature Methods, 3*(7), 519–524.

Elangovan, M., Wallrabe, H., Chen, Y., Day, R. N., Barroso, M., & Periasamy, A. (2003). Characterization of one- and two-photon excitation fluorescence resonance energy transfer microscopy. *Methods, 29*(1), 58–73.

Ernst, L. A., Gupta, R. K., Mujumdar, R. B., & Waggoner, A. S. (1989). Cyanine dye labeling reagents for sulfhydryl groups. *Cytometry, 10*(1), 3–10.

Fu, G., Yang, H. Y., Wang, C., Zhang, F., You, Z. D., Wang, G. Y., et al. (2006). Detection of constitutive heterodimerization of the integrin Mac-1 subunits by fluorescence resonance energy transfer in living cells. *Biochemical and Biophysical Research Communication, 346*(3), 986–991.

Galperin, E., Verkhusha, V. V., & Sorkin, A. (2004). Three-chromophore FRET microscopy to analyze multiprotein interactions in living cells. *Nature Methods, 1*(3), 209–217.

Gordon, G. W., Berry, G., Liang, X. H., Levine, B., & Herman, B. (1998). Quantitative fluorescence resonance energy transfer measurements using fluorescence microscopy. *Biophysical Journal, 74*(5), 2702–2713.

Griesbeck, O., Baird, G. S., Campbell, R. E., Zacharias, D. A., & Tsien, R. Y. (2001). Reducing the environmental sensitivity of yellow fluorescent protein. Mechanism and applications. *Journal of Biological Chemistry, 276*(31), 29188–29194.

Gurskaya, N. G., Verkhusha, V. V., Shcheglov, A. S., Staroverov, D. B., Chepurnykh, T. V., Fradkov, A. F., et al. (2006). Engineering of a monomeric green-to-red photoactivatable fluorescent protein induced by blue light. *Nature Biotechnology, 24*(4), 461–465.

Hadjantonakis, A. K., & Nagy, A. (2001). The color of mice: In the light of GFP-variant reporters. *Histochemistry and Cell Biology, 115*(1), 49–58.

Han, J., Liu, S., Rose, D. M., Schlaepfer, D. D., McDonald, H., & Ginsberg, M. H. (2001). Phosphorylation of the integrin alpha 4 cytoplasmic domain regulates paxillin binding. *Journal of Biological Chemistry, 276*(44), 40903–40909.

Han, J., Rose, D. M., Woodside, D. G., Goldfinger, L. E., & Ginsberg, M. H. (2003). Integrin alpha 4 beta 1-dependent T cell migration requires both phosphorylation and dephosphorylation of the alpha 4 cytoplasmic domain to regulate the reversible binding of paxillin. *Journal of Biological Chemistry, 278*(37), 34845–34853.

Heim, R., Cubitt, A. B., & Tsien, R. Y. (1995). Improved green fluorescence. *Nature, 373*(6516), 663–664.

Heim, R., Prasher, D. C., & Tsien, R. Y. (1994). Wavelength mutations and posttranslational autoxidation of green fluorescent protein. *Proceedings of the National Academy of Sciences of the United States of America, 91*(26), 12501–12504.

Heim, R., & Tsien, R. Y. (1996). Engineering green fluorescent protein for improved brightness, longer wavelengths and fluorescence resonance energy transfer. *Current Biology, 6*(2), 178–182.

Hohng, S., Joo, C., & Ha, T. (2004). Single-molecule three-color FRET. *Biophysical Journal, 87*(2), 1328–1337.

Hoppe, A., Christensen, K., & Swanson, J. A. (2002). Fluorescence resonance energy transfer-based stoichiometry in living cells. *Biophysical Journal, 83*(6), 3652–3664.

Hulett, H. R., Bonner, W. A., Barrett, J., & Herzenberg, L. A. (1969). Cell sorting: Automated separation of mammalian cells as a function of intracellular fluorescence. *Science, 166*(906), 747–749.

Humphries, M. J. (2000). Integrin structure. *Biochemical Society Transactions, 28*(4), 311–339.

Hyun, Y. M., Lefort, C. T., & Kim, M. (2009). Leukocyte integrins and their ligand interactions. *Immunologic Research,* [Epub ahead of print].

Hyun, Y. M., Chung, H. L., McGrath, J. L., Waugh, R. E., & Kim, M. (2009). Activated integrin VLA-4 localizes to the lamellipodia and mediates T cell migration on VCAM-1. *Journal of Immunology, 183*(1), 359–369.

Karasawa, S., Araki, T., Nagai, T., Mizuno, H., & Miyawaki, A. (2004). Cyan-emitting and orange-emitting fluorescent proteins as a donor/acceptor pair for fluorescence resonance energy transfer. *Biochemical Journal, 381*(Pt. 1), 307–312.

Kim, M., Carman, C. V., & Springer, T. A. (2003). Bidirectional transmembrane signaling by cytoplasmic domain separation in integrins. *Science, 301*(5640), 1720–1725.

Kim, M., Carman, C. V., Yang, W., Salas, A., & Springer, T. A. (2004). The primacy of affinity over clustering in regulation of adhesiveness of the integrin {α}L{β}2. *Journal of Cell Biology, 167*(6), 1241–1253.

Lakowicz, J. R. (1999). *Principles of fluorescence spectroscopy.* pp. xxiii, 698. New York/London: Kluwer Academic/Plenum.

Larson, R. S., Davis, T., Bologa, C., Semenuk, G., Vijayan, S., Li, Y., et al. (2005). Dissociation of I domain and global conformational changes in LFA-1: Refinement of small molecule-I domain structure-activity relationships. *Biochemistry, 44*(11), 4322–4331.

Lefort, C. T., Hyun, Y. M., Schultz, J. B., Law, F. Y., Waugh, R. E., Knauf, P. A., et al. (In preparation). Outside-in signal transmission by conformational changes in integrin Mac-1. *Journal of Immunology*.

Ley, K., Laudanna, C., Cybulsky, M. I., & Nourshargh, S. (2007). Getting to the site of inflammation: The leukocyte adhesion cascade updated. *Nature Reviews Immunology, 7*(9), 678–689.

Li, R., Mitra, N., Gratkowski, H., Vilaire, G., Litvinov, R., Nagasami, C., et al. (2003). Activation of integrin alphaIIbbeta3 by modulation of transmembrane helix associations. *Science, 300*(5620), 795–798.

Lippincott-Schwartz, J., Altan-Bonnet, N., & Patterson, G. H. (2003). Photobleaching and photoactivation: Following protein dynamics in living cells. *Nature Cell Biology, 5*(Suppl.), S7–S14.

Lippincott-Schwartz, J., Snapp, E., & Kenworthy, A. (2001). Studying protein dynamics in living cells. *Nature Reviews Molecular Cell Biology, 2*(6), 444–456.

Liu, S., Thomas, S. M., Woodside, D. G., Rose, D. M., Kiosses, W. B., Pfaff, M., et al. (1999). Binding of paxillin to alpha4 integrins modifies integrin-dependent biological responses. *Nature, 402*(6762), 676–681.

Llopis, J., McCaffery, J. M., Miyawaki, A., Farquhar, M. G., & Tsien, R. Y. (1998). Measurement of cytosolic, mitochondrial, and Golgi pH in single living cells with green fluorescent proteins. *Proceedings of the National Academy of Sciences of the United States of America, 95*(12), 6803–6808.

Lukyanov, K. A., Chudakov, D. M., Lukyanov, S., & Verkhusha, V. V. (2005). Innovation: Photoactivatable fluorescent proteins. *Nature Reviews Molecular Cell Biology, 6*(11), 885–891.

Luo, B. H., Carman, C. V., & Springer, T. A. (2007). Structural basis of integrin regulation and signaling. *Annual Review of Immunology, 25*, 619–647.

Matz, M. V., Fradkov, A. F., Labas, Y. A., Savitsky, A. P., Zaraisky, A. G., Markelov, M. L., et al. (1999). Fluorescent proteins from nonbioluminescent Anthozoa species. *Nature Biotechnology, 17*(10), 969–973.

Mena, M. A., Treynor, T. P., Mayo, S. L., & Daugherty, P. S. (2006). Blue fluorescent proteins with enhanced brightness and photostability from a structurally targeted library. *Nature Biotechnology, 24*(12), 1569–1571.

Merzlyak, E. M., Goedhart, J., Shcherbo, D., Bulina, M. E., Shcheglov, A. S., Fradkov, A. F., et al. (2007). Bright monomeric red fluorescent protein with an extended fluorescence lifetime. *Nature Methods, 4*(7), 555–557.

Michalet, X., Pinaud, F. F., Bentolila, L. A., Tsay, J. M., Doose, S., Li, J. J., et al. (2005). Quantum dots for live cells, in vivo imaging, and diagnostics. *Science, 307*(5709), 538–544.

Mizuno, H., Mal, T. K., Tong, K. I., Ando, R., Furuta, T., Ikura, M., et al. (2003). Photo-induced peptide cleavage in the green-to-red conversion of a fluorescent protein. *Molecular Cell, 12*(4), 1051–1058.

Nagai, T., Ibata, K., Park, E. S., Kubota, M., Mikoshiba, K., & Miyawaki, A. (2002). A variant of yellow fluorescent protein with fast and efficient maturation for cell-biological applications. *Nature Biotechnology, 20*(1), 87–90.

Nguyen, A. W., & Daugherty, P. S. (2005). Evolutionary optimization of fluorescent proteins for intracellular FRET. *Nature Biotechnology, 23*(3), 355–360.

Nienhaus, G. U., Nienhaus, K., Holzle, A., Ivanchenko, S., Renzi, F., Oswald, F., et al. (2006). Photoconvertible fluorescent protein EosFP: Biophysical properties and cell biology applications. *Photochemistry and Photobiology, 82*(2), 351–358.

Nolan, J. P., & Sklar, L. A. (1998). The emergence of flow cytometry for sensitive, real-time measurements of molecular interactions. *Nature Biotechnology, 16*(7), 633–638.

Oida, T., Sako, Y., & Kusumi, A. (1993). Fluorescence lifetime imaging microscopy (flimscopy). Methodology development and application to studies of endosome fusion in single cells. *Biophysical Journal, 64*(3), 676–685.

Ormo, M., Cubitt, A. B., Kallio, K., Gross, L. A., Tsien, R. Y., & Remington, S. J. (1996). Crystal structure of the *Aequorea victoria* green fluorescent protein. *Science, 273*(5280), 1392–1395.

Pakhomov, A. A., & Martynov, V. I. (2008). GFP family: Structural insights into spectral tuning. *Chemistry & Biology, 15*(8), 755–764.

Panchuk-Voloshina, N., Haugland, R. P., Bishop-Stewart, J., Bhalgat, M. K., Millard, P. J., Mao, F., et al. (1999). Alexa dyes, a series of new fluorescent dyes that yield exceptionally bright, photostable conjugates. *Journal of Histochemistry and Cytochemistry, 47*(9), 1179–1188.

Parsons, M., Messent, A. J., Humphries, J. D., Deakin, N. O., & Humphries, M. J. (2008). Quantification of integrin receptor agonism by fluorescence lifetime imaging. *Journal of Cell Science, 121*(Pt. 3), 265–271.

Patterson, G., Day, R. N., & Piston, D. (2001). Fluorescent protein spectra. *Journal of Cell Science, 114*(Pt. 5), 837–838.

Patterson, G. H., Knobel, S. M., Sharif, W. D., Kain, S. R., & Piston, D. W. (1997). Use of the green fluorescent protein and its mutants in quantitative fluorescence microscopy. *Biophysical Journal, 73*(5), 2782–2790.

Patterson, G. H., & Lippincott-Schwartz, J. (2002). A photoactivatable GFP for selective photolabeling of proteins and cells. *Science, 297*(5588), 1873–1877.

Patterson, G. H., Piston, D. W., & Barisas, B. G. (2000). Forster distances between green fluorescent protein pairs. *Analytical Biochemistry, 284*(2), 438–440.

Periasamy, A., & Day, R. N. (2005). *Molecular imaging: FRET microscopy and spectroscopy. Methods in physiology series.* pp. xv, 312. New York: Oxford University Press.

Peter, M., Ameer-Beg, S. M., Hughes, M. K., Keppler, M. D., Prag, S., Marsh, M., et al. (2005). Multiphoton-FLIM quantification of the EGFP–mRFP1 FRET pair for localization of membrane receptor–kinase interactions. *Biophysical Journal, 88*(2), 1224–1237.

Pollok, B. A., & Heim, R. (1999). Using GFP in FRET-based applications. *Trends in Cell Biology, 9*(2), 57–60.

Resch-Genger, U., Grabolle, M., Cavaliere-Jaricot, S., Nitschke, R., & Nann, T. (2008). Quantum dots versus organic dyes as fluorescent labels. *Nature Methods, 5*(9), 763–775.

Rizzo, M. A., Springer, G. H., Granada, B., & Piston, D. W. (2004). An improved cyan fluorescent protein variant useful for FRET. *Nature Biotechnology, 22*(4), 445–449.

Roy, R., Hohng, S., & Ha, T. (2008). A practical guide to single-molecule FRET. *Nature Methods, 5*(6), 507–516.

Schaller, M. D., Otey, C. A., Hildebrand, J. D., & Parsons, J. T. (1995). Focal adhesion kinase and paxillin bind to peptides mimicking beta integrin cytoplasmic domains. *Journal of Cell Biology, 130*(5), 1181–1187.

Shagin, D. A., Barsova, E. V., Yanushevich, Y. G., Fradkov, A. F., Lukyanov, K. A., Labas, Y. A., et al. (2004). GFP-like proteins as ubiquitous metazoan superfamily: Evolution of functional features and structural complexity. *Molecular Biology and Evolution, 21*(5), 841–850.

Shaner, N. C., Campbell, R. E., Steinbach, P. A., Giepmans, B. N., Palmer, A. E., & Tsien, R. Y. (2004). Improved monomeric red, orange and yellow fluorescent proteins derived from *Discosoma* sp. red fluorescent protein. *Nature Biotechnology, 22*(12), 1567–1572.

Shaner, N. C., Patterson, G. H., & Davidson, M. W. (2007). Advances in fluorescent protein technology. *Journal of Cell Science, 120*(Pt. 24), 4247–4260.

Shcherbo, D., Merzlyak, E. M., Chepurnykh, T. V., Fradkov, A. F., Ermakova, G. V., Solovieva, E. A., et al. (2007). Bright far-red fluorescent protein for whole-body imaging. *Nature Methods, 4*(9), 741–746.

Shimomura, O., Johnson, F. H., & Saiga, Y. (1962). Extraction, purification and properties of aequorin, a bioluminescent protein from the luminous hydromedusan, *Aequorea*. *Journal of Cellular and Comparative Physiology, 59*, 223–239.

Siegel, M. S., & Isacoff, E. Y. (1997). A genetically encoded optical probe of membrane voltage. *Neuron, 19*(4), 735–741.

Smith, E. A., Bunch, T. A., & Brower, D. L. (2007). General *in vivo* assay for the study of integrin cell membrane receptor microclustering. *Analytical Chemistry, 79*(8), 3142–3147.

Smith, A., Carrasco, Y. R., Stanley, P., Kieffer, N., Batista, F. D., & Hogg, N. (2005). A talin-dependent LFA-1 focal zone is formed by rapidly migrating T lymphocytes. *Journal of Cell Biology, 170*(1), 141–151.

Stanley, P., Smith, A., McDowall, A., Nicol, A., Zicha, D., & Hogg, N. (2008). Intermediate-affinity LFA-1 binds alpha-actinin-1 to control migration at the leading edge of the T cell. *EMBO Journal, 27*(1), 62–75.

Subach, F. V., Patterson, G. H., Manley, S., Gillette, J. M., Lippincott-Schwartz, J., & Verkhusha, V. V. (2009). Photoactivatable mCherry for high-resolution two-color fluorescence microscopy. *Nature Methods, 6*(2), 153–159.

Tadross, M. R., Park, S. A., Veeramani, B., & Yue, D. T. (2009). Robust approaches to quantitative ratiometric FRET imaging of CFP/YFP fluorophores under confocal microscopy. *Journal of Microscopy, 233*(1), 192–204.

Takagi, J., Petre, B. M., Walz, T., & Springer, T. A. (2002). Global conformational rearrangements in integrin extracellular domains in outside-in and inside-out signaling. *Cell, 110*(5), 599–611.

Titus, J. A., Haugland, R., Sharrow, S. O., & Segal, D. M. (1982). Texas Red, a hydrophilic, red-emitting fluorophore for use with fluorescein in dual parameter flow microfluorometric and fluorescence microscopic studies. *Journal of Immunological Methods, 50*(2), 193–204.

Tramier, M., Zahid, M., Mevel, J. C., Masse, M. J., & Coppey-Moisan, M. (2006). Sensitivity of CFP/YFP and GFP/mCherry pairs to donor photobleaching on FRET determination by fluorescence lifetime imaging microscopy in living cells. *Microscopy Research and Technique, 69*(11), 933–939.

Tsutsui, H., Karasawa, S., Shimizu, H., Nukina, N., & Miyawaki, A. (2005). Semi-rational engineering of a coral fluorescent protein into an efficient highlighter. *EMBO Reports, 6*(3), 233–238.

Valentin, G., Verheggen, C., Piolot, T., Neel, H., Coppey-Moisan, M., & Bertrand, E. (2005). Photoconversion of YFP into a CFP-like species during acceptor photobleaching FRET experiments. *Nature Methods, 2*(11), 801.

van Kooyk, Y., & Figdor, C. G. (2000). Avidity regulation of integrins: The driving force in leukocyte adhesion. *Current Opinion in Cell Biology, 12*(5), 542–547.

van Rheenen, J., Langeslag, M., & Jalink, K. (2004). Correcting confocal acquisition to optimize imaging of fluorescence resonance energy transfer by sensitized emission. *Biophysical Journal, 86*(4), 2517–2529.

Vararattanavech, A., Lin, X., Torres, J., & Tan, S. M. (2009). Disruption of the integrin alphaLbeta2 transmembrane domain interface by beta2 Thr-686 mutation activates alphaL-beta2 and promotes micro-clustering of the alphaL subunits. *Journal of Biological Chemistry, 284*(5), 3239–3249.

Vararattanavech, A., Tang, M. L., Li, H. Y., Wong, C. H., Law, S. K., Torres, J., et al. (2008). Permissive transmembrane helix heterodimerization is required for the expression of a functional integrin. *Biochemical Journal, 410*(3), 495–502.

Verkhusha, V. V., & Sorkin, A. (2005). Conversion of the monomeric red fluorescent protein into a photoactivatable probe. *Chemistry & Biology, 12*(3), 279–285.

Wachter, R. M., Elsliger, M. A., Kallio, K., Hanson, G. T., & Remington, S. J. (1998). Structural basis of spectral shifts in the yellow-emission variants of green fluorescent protein. *Structure, 6*(10), 1267–1277.

Waggoner, A. (2006). Fluorescent labels for proteomics and genomics. *Current Opinion in Chemical Biology, 10*(1), 62–66.

Wallrabe, H., & Periasamy, A. (2005). Imaging protein molecules using FRET and FLIM microscopy. *Current Opinion in Biotechnology, 16*(1), 19–27.

Wang, L., Jackson, W. C., Steinbach, P. A., & Tsien, R. Y. (2004). Evolution of new nonantibody proteins via iterative somatic hypermutation. *Proceedings of the National Academy of Sciences of the United States of America, 101*(48), 16745–16749.

Wolber, P. K., & Hudson, B. S. (1979). An analytic solution to the Forster energy transfer problem in two dimensions. *Biophysical Journal, 28*(2), 197–210.

Xiong, J. P., Stehle, T., Diefenbach, B., Zhang, R., Dunker, R., Scott, D. L., et al. (2001). Crystal structure of the extracellular segment of integrin alpha Vbeta3. *Science, 294*(5541), 339–345.

Yang, F., Moss, L. G., & Phillips, G. N. ?Jr., (1996). The molecular structure of green fluorescent protein. *Nature Biotechnology, 14*(10), 1246–1251.

Zacharias, D. A., Violin, J. D., Newton, A. C., & Tsien, R. Y. (2002). Partitioning of lipid-modified monomeric GFPs into membrane microdomains of live cells. *Science, 296*(5569), 913–916.

Zal, T., & Gascoigne, N. R. (2004). Photobleaching-corrected FRET efficiency imaging of live cells. *Biophysical Journal, 86*(6), 3923–3939.

Zal, T., Zal, M. A., & Gascoigne, N. R. (2002). Inhibition of T cell receptor–coreceptor interactions by antagonist ligands visualized by live FRET imaging of the T-hybridoma immunological synapse. *Immunity, 16*(4), 521–534.

Zapata-Hommer, O., & Griesbeck, O. (2003). Efficiently folding and circularly permuted variants of the Sapphire mutant of GFP. *BMC Biotechnology, 3*, 5.

Zhang, J., Campbell, R. E., Ting, A. Y., & Tsien, R. Y. (2002). Creating new fluorescent probes for cell biology. *Nature Reviews Molecular Cell Biology, 3*(12), 906–918.

Index

D

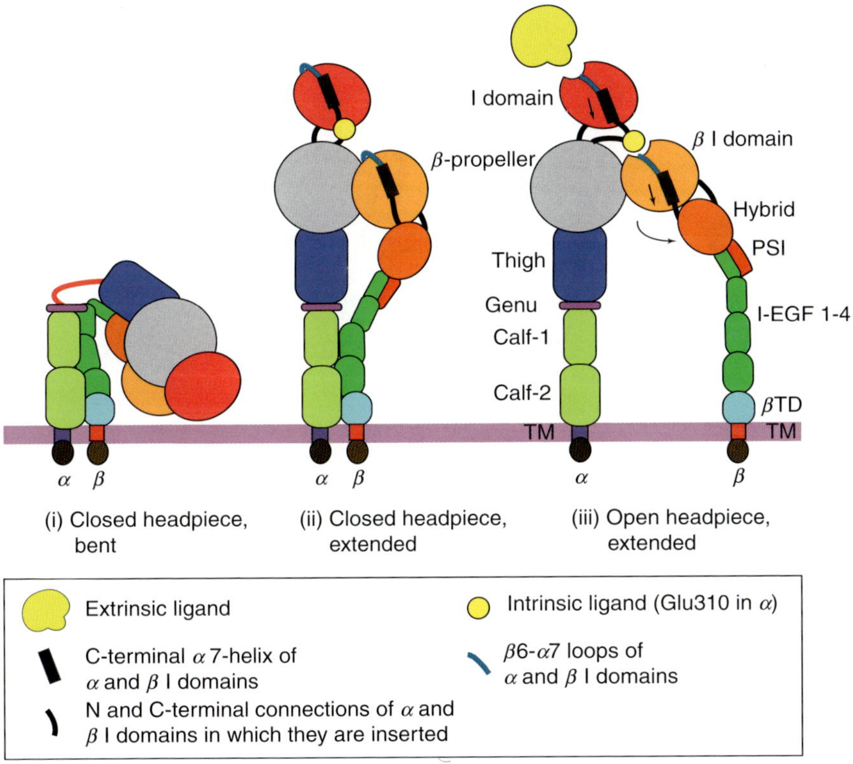

Please refer to Fig. 4.1 in text for figure legend.

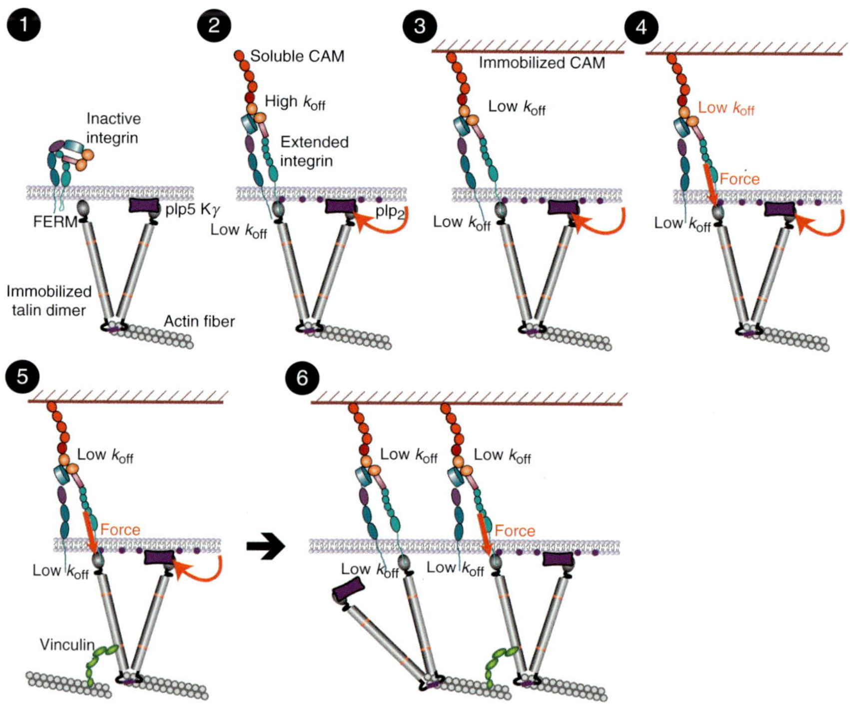

Please refer to Fig. 6.3 in text for figure legend.

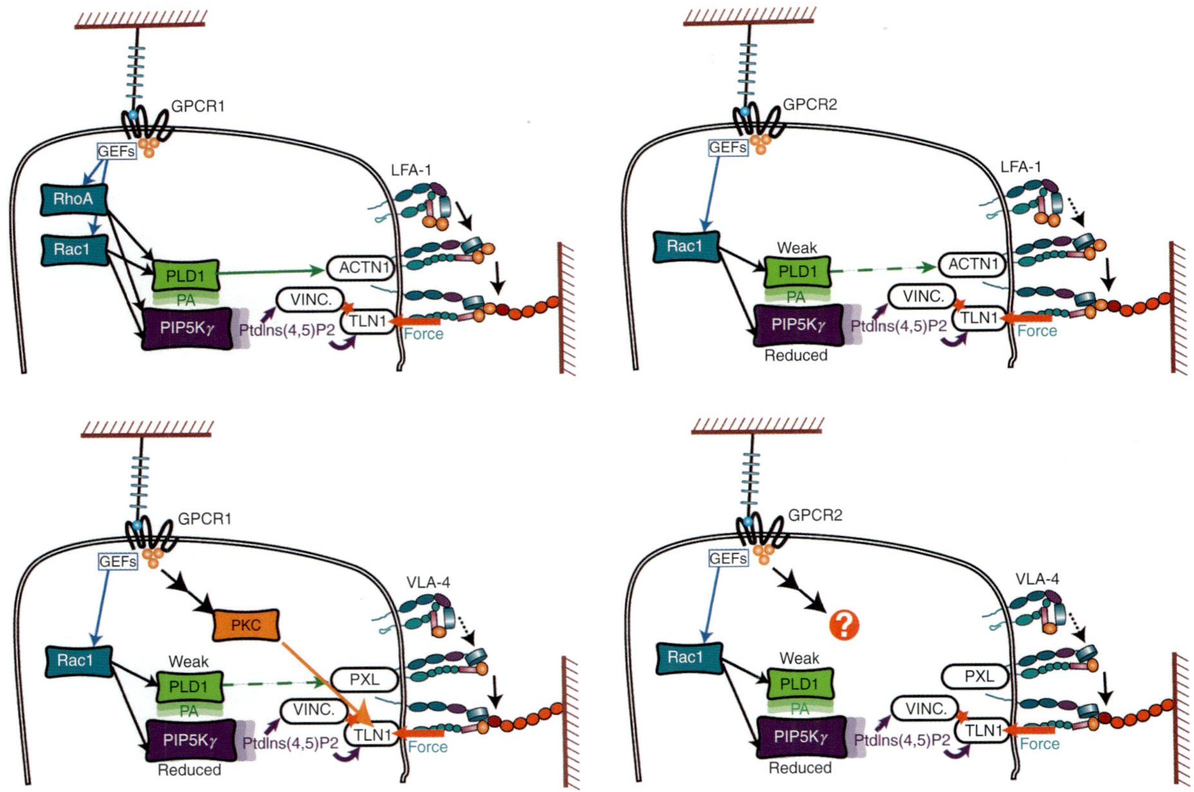

Please refer to Fig. 6.4 in text for figure legend.

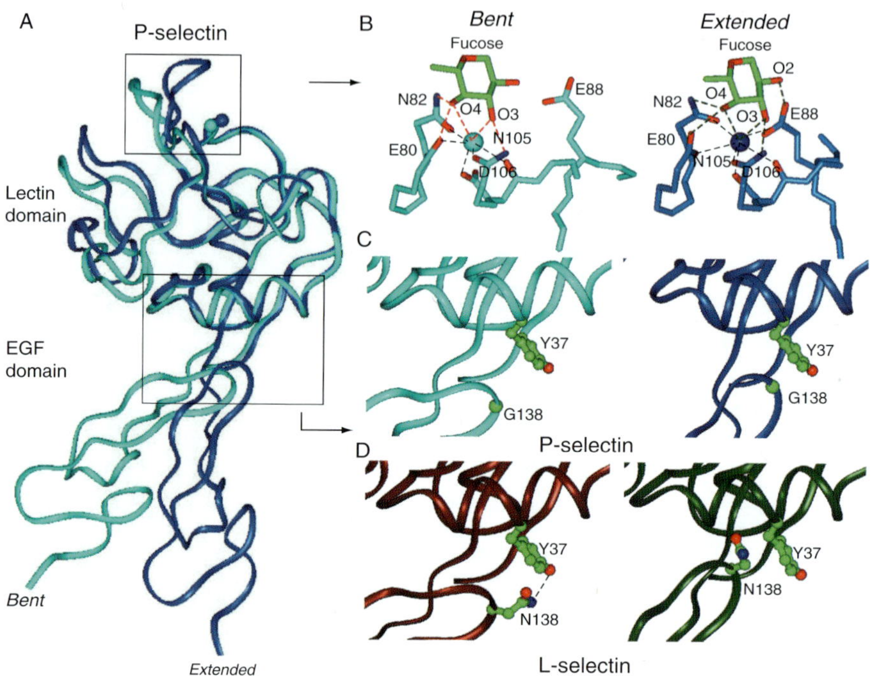

Please refer to Fig. 7.6 in text for figure legend.

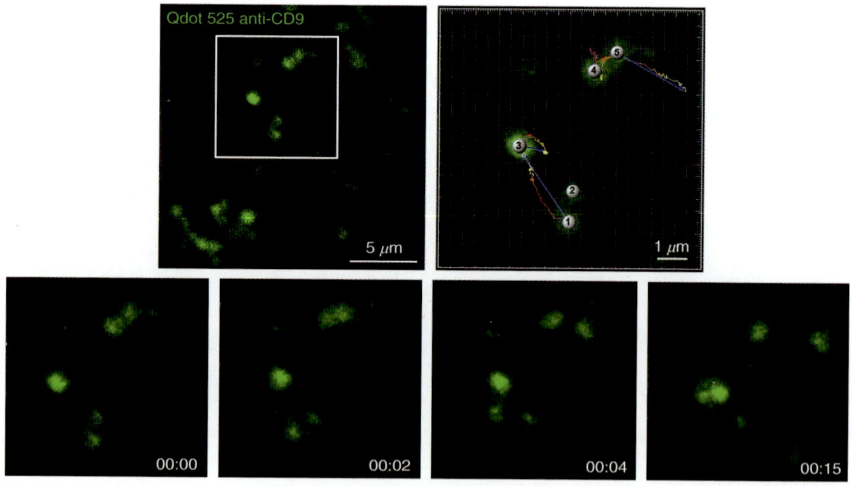

Please refer to Fig. 9.1 in text for figure legend.

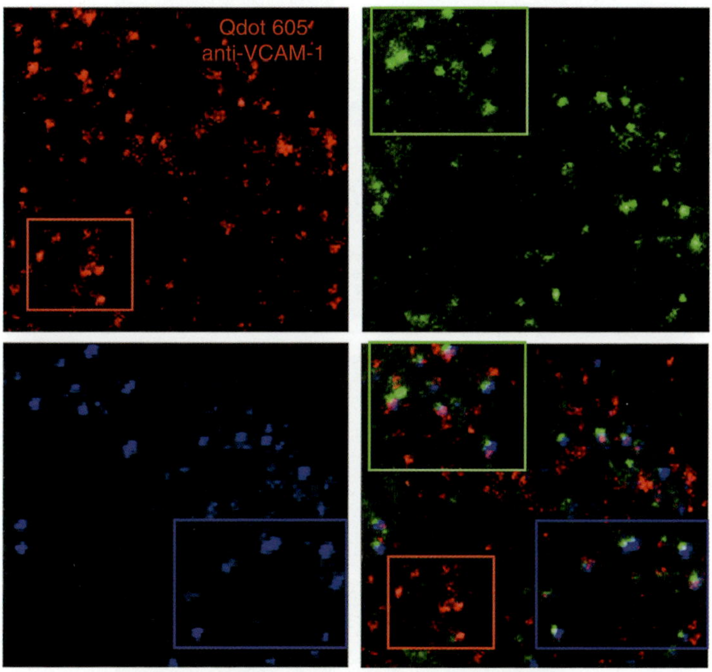

Please refer to Fig. 9.2 in text for figure legend.

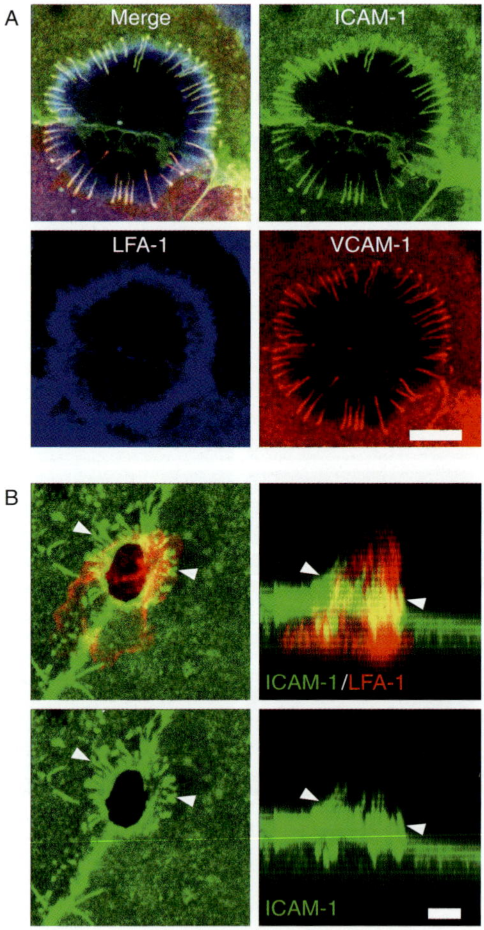

Please refer to Fig. 10.2 in text for figure legend.

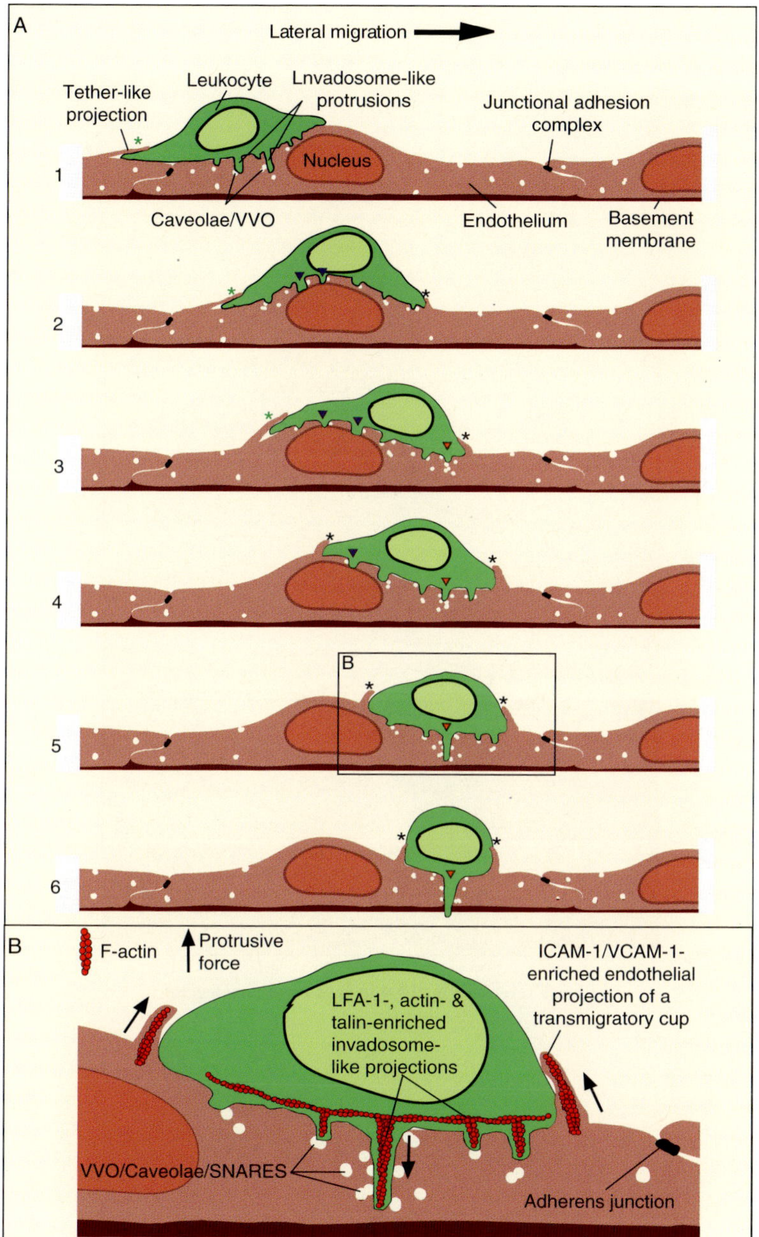

Please refer to Fig. 10.3 in text for figure legend.

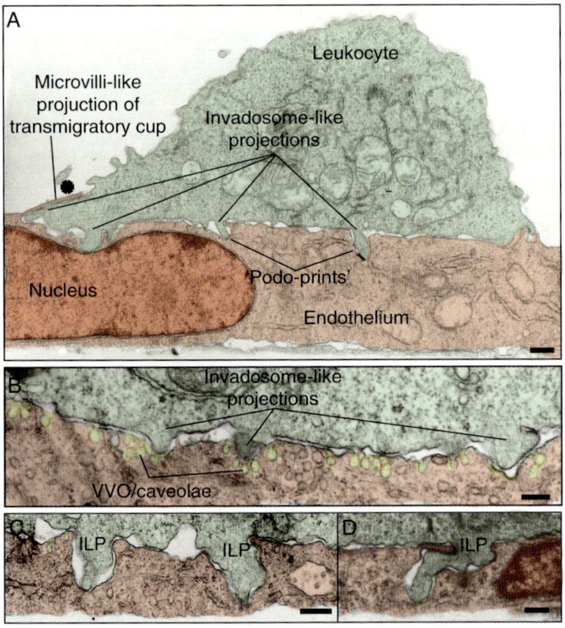

Please refer to Fig. 10.4 in text for figure legend.